D1783109

ENCYCLOPEDIA
OF
ELECTROCHEMISTRY
OF THE ELEMENTS

VOLUME III

ORGANIZATIONAL CHART

	Symbol	Element No.	Volume	Chapter
Actinium	Ac	24		
Aluminum	Al	21		
Americium	Am	24		
Antimony	Sb	11		
Argon	Ar	1	I	8
Arsenic	As	9	II	2
Astatine	At	6	I	3
Barium	Ba	18	I	7
Berkelium	Bk	24		
Beryllium	Be	16		
Bismuth	Bi	12		
Boron	B	14	II	1
Bromine	Br	5	I	2
Cadmium	Cd	20	I	4
Calcium	Ca	18	I	7
Californium	Cf	24		
Carbon	C			
Cerium	Ce	22		
Cesium	Cs	15		
Chlorine	Cl	5	I	1
Chromium	Cr	31		
Cobalt	Co	36	III	2
Copper	Cu	38	II	6
Curium	Cm	24		
Dysprosium	Dy	22		
Einsteinium	Es	24		
Erbium	Er	22		
Europium	Eu	22		
Fermium	Fm	24		
Fluorine	F	4		
Francium	Fr	15		
Gadolinium	Gd	22		
Gallium	Ga	14		
Germanium	Ge	13		
Gold	Au	40		
Hafnium	Hf	26		
Helium	He	1	I	8
Holmium	Ho	22		
Hydrogen	H	2		
Indium	In	14		
Iodine	I	6	I	3
Iridium	Ir	41		
Iron	Fe	37		
Krypton	Kr	1	I	8
Lanthanum	La	22		
Lawrencium	Lw	24		
Lead	Pb	28	I	5
Lithium	Li	15		
Lutetium	Lu	22		
Magnesium	Mg	17		
Manganese	Mn	34	I	6
Mendelevium	Md	24		

ORGANIZATIONAL CHART

	Symbol	Element No.	Volume	Chapter
Mercury	Hg	10		
Molybdenum	Mo	32		
Neodymium	Nd	22		
Neon	Ne	1	I	8
Neptunium	Np	24		
Nickel	Ni	35	III	3
Niobium	Nb	30	II	3
Nitrogen	N	8		
Nobelium	No	24		
Osmium	Os	41		
Oxygen	O	3	II	5
Palladium	Pd	41		
Phosphorus	P	9	III	1
Platinum	Pt	41		
Plutonium	Pu	24		
Polonium	Po	7		
Potassium	K	15		
Praseodymium	Pr	22		
Promethium	Pm	22		
Protactinium	Pa	24		
Radium	Ra	18	I	7
Radon	Rn	1	I	8
Rhenium	Re	42	II	4
Rhodium	Rh	41		
Rubidium	Rb	15		
Ruthenium	Ru	41		
Samarium	Sm	22		
Scandium	Sc	22		
Selenium	Se	7		
Silicon	Si	13		
Silver	Ag	39		
Sodium	Na	15		
Strontium	Sr	18	I	7
Sulfur	S	7		
Tantalum	Ta	30	II	3
Technetium	Tc	42	II	4
Tellurium	Te	7		
Terbium	Tb	22		
Thallium	Tl	14		
Thorium	Th	24		
Thulium	Tm	22		
Tin	Sn	27		
Titanium	Ti	25		
Tungsten	W	33		
Uranium	U	24		
Vanadium	V	29		
Xenon	Xe	1	I	8
Ytterbium	Yb	22		
Yttrium	Y	22		
Zinc	Zn	19		
Zirconium	Zr	26		

ADVISORY BOARD

RALPH N. ADAMS
Department of Chemistry, The University of Kansas, Lawrence, Kansas

N. A. BALASHOVA
Institute of Electrochemistry, Academy of Sciences (USSR), Moscow

ANDRE DEBETHUNE
Department of Chemistry, Boston College, Chestnut Hill, Massachusetts

K. S. G. DOSS
Department of Chemical Engineering, Indian Institute of Technology, Madras

NORMAN HACKERMAN
Rice University, Houston, Texas

HENNING LUND
Department of Organic Chemistry, University of Aarhus, Denmark

G. MILAZZO
Laboratorio di Chimica, Istituto Superiore di Sanità, Rome, Italy

ROGER PARSONS
School of Chemistry, University of Bristol, England

M. POURBAIX
Centre Belge d'Etude de la Corrosion, Brussels, Belgium

K. VETTER
Freie Universität Berlin, Institüt für Physikalische Chemie, Berlin, West Germany

PETR ZUMAN
Department of Chemistry, Clarkson College of Technology, Potsdam, New York

NOBUYUKI TANAKA
Department of Chemistry, Tohoku University, Sendai, Japan

ENCYCLOPEDIA OF ELECTROCHEMISTRY OF THE ELEMENTS

EDITOR

Allen J. Bard

Department of Chemistry
University of Texas
Austin, Texas

VOLUME III

Co
Ni
P

MARCEL DEKKER, INC. New York

COPYRIGHT © 1975 BY MARCEL DEKKER, INC.

ALL RIGHTS RESERVED

Neither this book nor any part may be reproduced or transmitted in any form or by any means, electronic or mechanical, including photocopying, microfilming, and recording, or by any information storage and retrieval system, without permission in writing from the publisher.

MARCEL DEKKER, INC.
270 Madison Avenue, New York, New York 10016

LIBRARY OF CONGRESS CATALOG CARD NUMBER 73-88796

ISBN 0-8247-6137-5

PRINTED IN THE UNITED STATES OF AMERICA

QD
551
.B37
v.3

CONTENTS

WILLIAM F. MAAG LIBRARY
YOUNGSTOWN STATE UNIVERSITY

363045

LIST OF CONTRIBUTORS

Alejandro Jorge Arvia, Instituto de Investigaciones Fiscoquimicas Teóricas y Aplicadas, División Electroquímica, Universidad Nacional de La Plata, La Plata, Argentina

N. E. Chomutov, Department of Physical Chemistry, D. I. Mendeleev Institute of Chemistry and Technology, Moscow, USSR

Nobufumi Maki, Department of Chemistry, Shizuoka University, Hamamatsu City, Japan

Dionisio Posadas, Instituto de Investigaciones Fisicoquimicas Teóricas y Aplicadas, División Electroquímica, Universidad Nacional de La Plata, La Plata, Argentina

Nobuyuki Tanaka, Department of Chemistry, Tohoku University, Sendai, Japan

A. P. Tomilov, Department of Physical Chemistry, D. I. Mendeleev Institute of Chemistry and Technology, Moscow, USSR

INTRODUCTION

The aim of this series is to provide a critical, systematic, and comprehensive review of the electrochemical behavior of the elements and their compounds. Part I deals with inorganic electrochemistry and Part II with organic electrochemistry. The field of electrochemistry has undergone extensive growth in recent years and electrochemical techniques and concepts have been applied to many areas of basic and applied research and technology. While the fundamentals of electrochemistry and the recent advances in this field have been described in a number of textbooks, monographs, and review series, the vast literature concerning the descriptive electrochemistry of inorganic and organic compounds has been relatively neglected. This series is designed to provide the natural starting point for new electrochemical investigations and to suggest areas where further research is needed. Each chapter, written by experts on that subject, contains the best available information on the electrochemical behavior and applications of the element.

The classification of elements generally follows the scheme used in Gmelin, with some grouping and rearrangement. Each element is assigned a number (n). This number determines in what chapter a compound or alloy will be considered; a chapter on element n will discuss compounds of element n with elements of number lower than n. For example, zinc-amalgam is treated in the zinc chapter (element number 19), rather than in the mercury chapter (element number 10). A listing of the elements and the element numbers is given in the organizational chart at the front of the book. To minimize delays in publication of the chapters, these are published in the order received. The volume and chapter number for each element are also located in the chart.

The chapters are generally organized into five sections:

1. Introduction and Standard Potentials
    1.1. Aqueous Solutions
    1.2. Nonaqueous Solvents
    1.3. Fused Salts
    1.4. Other Data
2. Voltammetric Characteristics
    Presentation of Polarographic and Other Voltammetric Results

3. Kinetic Parameters and Double-Layer Properties
   Presentation of available rate constants, exchange current densities, transfer coefficients, potential of zero charge, double-layer capacities, etc.
4. Electrochemical Studies
   A survey and critical review of the electrochemical reactions of the element and its compounds. Description of known mechanisms of the electrode reactions, oxidation and reduction products, current efficiencies, reaction orders, etc. Discussion of passivation phenomena, oxide films, and anodization.
5. Applied Electrochemistry
   The use of electrochemistry in the isolation or purification of the element, the electrochemical production of compounds, and the use and behavior of the element and its compounds in electrochemical devices.
   5.1. Electrowinning and Electrorefining
   5.2. Electrodeposition, Electroplating, and Electropolishing
   5.3. Electrosynthesis
   5.4. Corrosion
   5.5. Batteries and Cells
   5.6. Other

Conventions generally follow the recommendations of IUPAC [1]. Half reactions corresponding to a standard electrode potential are uniformly written as reductions and the sign of the electrode potential follows the Gibbs-Stockholm convention [2]. Notation generally follows that in the "Manual of Symbols and Terminology for Physicochemical Quantities and Units" [3]. Modified and frequently-used symbols are given below.

1. POTENTIALS

E: potential vs NHE, SCE, or other reference electrode. When corresponding cell is to be indicated, it may be written

$E(Zn^{2+}/Zn)$ or $E(1/2)Zn^{2+} + e = (1/2)Zn$

E°: standard electrode potential. It is the EMF of the cell in which the reaction is the reduction of the oxidized species by hydrogen, all reactants being in their standard state, e.g.,

$(1/2)Zn^{2+} + (1/2)H_2 = (1/2)Zn + H^+$ $E^\circ (Zn^{2+}/Zn)$

$E^{\circ\prime}$: formal electrode potential or conditional electrode potential corresponding to unit concentration quantities.

η: overpotential, the deviation of the potential of an electrode from its equilibrium value. Negative values are associated with reductions and positive values are associated with oxidations.

Polarographic and Voltammetric Potentials

$E_{1/2}$: half-wave potential in dc polarography, i.e., potential at which $i = (1/2)i_d$.

$E_{1/4}$, $E_{3/4}$: 1/4 and 3/4 wave potentials in dc polarography.

E_p, $E_{p/2}$: peak and half-peak potential, e.g., in linear scan or cyclic voltammetry. Cathodic and anodic waves are indicated by E_{pc} and E_{pa}.

$E_{\tau/4}$: quarter-wave potential in chronopotentiometry.

E_Z: point of zero charge.

E_{mix}: mixed potential.

2. CURRENT

i: current, instantaneous current, total current.

i_a: anodic current.

i_c: cathodic current.

i_o: exchange current.

i_d: diffusion current.

i_l: limiting current.

i_p: peak current.

Anodic or cathodic waves are indicated by i_{pa} and i_{pc}, i_{da} and i_{dc}, etc. Average value of current is indicated by a bar over the appropriate symbol, e.g., $\bar{i}_d$: average diffusion current.

j: current density, subscripted as above to indicate j_a, anodic; j_c, cathodic; j_o, exchange, etc.

I, $\bar{I}$: diffusion current constant in dc polarography

$$I = i_d/m^{2/3}t_d^{1/6} \qquad \bar{I} = \bar{i}_d/m^{2/3}t_d^{1/6}$$

with i_d in μA, m in mg/s and t in s.

3. CONCENTRATION

C, C_O, C_R: concentration of O, R, ..., at electrode surface. Unless stated otherwise, [O], [R], ..., may be used in discussing equilibria in solution.

C°, C_R°: bulk concentrations, bulk concentration of species R.

$C_R(x, t)$: concentration of species R at distance x from electrode at time t.

4. TIME

t: time.

t_d: drop time of DME.

τ: transition time in chronopotentiometry.

5. RATE CONSTANTS

For the reaction O + ne = R:

$$j = i/A = nF(k_aC_R - k_cC_O)$$

n: charge number of elementary step.

k_a: formal or conditional rate constant of anodic (oxidation) reaction at potential E vs reference electrode, in cm/sec (k_a° refers to value of k_a at E = 0 vs reference electrode).

k_c: formal or conditional rate constant of cathodic (reduction) reaction at potential E vs reference electrode, in cm/sec (k_c° refers to value of k_c at E = 0 vs reference electrode).

The current convention corresponds to an anodic current positive.

$$k_c = k^\circ \exp\{-\alpha zF(E - E^{\circ\prime})/RT\}$$

$$k_a = k^\circ \exp\{(1 - \alpha)zF(E - E^{\circ\prime})/RT\}$$

k°: formal or conditional standard rate constant for the electrode reaction.

α: the electrochemical charge transfer coefficient for the cathodic reaction.

$1 - \alpha$: the electrochemical charge transfer coefficient for the anodic reaction.

$$j_0 = i_0/A = nFk^\circ C_O^{1-\alpha}C_R^\alpha$$

i_0: formal or conditional exchange current (A).

j_0: formal or conditional exchange current density (A/cm^2).

If the rate constants are determined by extrapolation to zero ionic strength so that they are written in terms of activities (a_O, a_R) in place of concentrations (C_O, C_R), the symbols are primed: k_c', k_a', $k^{\circ\prime}$, j_0', i_0'.

For rate constants corrected for double layer effects, the subscript t is used; k_t°, j_t° where $k_t^\circ = k^\circ \exp -\{(\alpha n - z)F\phi_2/RT\}$.

6. OTHER

A = electrode area (cm^2).

F = Faraday constant (C/mol).

u_i: mobility of species i (cm^2/V/sec).

z_i: charge number of species i, positive for cations and negative for anions.

t_i: transport number of species i.

D_O, D_R: diffusion coefficient of species O, R.

δ: thickness of diffusion layer (cm).

m_O, m_R: mass transport constant of species O, R (cm/sec)

$$m_R = D_R/\delta_R$$

Q = quantity of electricity, charge.

C = capacitance (μF/cm^2).

C_{dl} = double-layer capacitance.

m = rate of mercury flow in DME (mg/sec).

v = potential scan rate (V/sec).

ω = angular frequency of rotation (sec^{-1}).

Abbreviations for words and journals follow the American Chemical Society Handbook for Authors and the Chemical Abstracts Service Guide for Abbreviating Periodical Titles. Some special abbreviations used in the tables in this compilation are given below.

1. TECHNIQUES

E swp: potential sweep.
i swp: current sweep.
CV: cyclic voltammetry.
E stp: potentiostatic or potential step.
V stp: voltage step.
i stp: chronopotentiometry, galvanostatic, or current step.
ac pol: ac polarography.
ac harm: ac harmonic method.
farad. imp: faradaic impedance (ac bridge) method.
chramp: chronoamperometry.
chrcoul: chronocoulometry.
coul: coulostatic method.
farad. rect: faradaic rectification method.
dc pol: dc polarography.

2. ELECTRODES

rot: rotating.
hng: hanging.
vib: vibrating.
stat: stationary.
stir: stirred.
bub: bubbling.
soln: solution.
drp: drop.
dsk: disk.
wr: wire.
pl: pool.
pwd: powder.
pst: paste.
fl: foil.
rd: rod.
DME: dropping mercury electrode.
SCE: saturated calomel electrode.
NHE: normal hydrogen electrode.

Formula of element or compound is used to complete description of electrode, e.g., rot Pt wr (rotating platinum wire electrode), C pst - stir soln (carbon paste electrode in stirred solution), etc.

3. DESCRIPTIONS OF REACTIONS AND WAVES

rev: reversible.
irr: irreversible.

sl: slightly.
q: quasi.
w: well-defined.
i: ill-defined.
fw: fairly well-defined.
fi: fairly ill-defined.
do: drawn out.
mb: merges with background.

The general philosophy of this compilation was to let each author choose the extent and scope of his chapter, subject only to limitations of format and notation, since it would clearly be impossible to prescribe too closely the type of treatment appropriate for the different elements. Supplementary volumes and, eventually, new editions are planned which will contain new results and corrections. The authors and the editor would welcome comments, suggestions, and corrections for inclusion in the supplements. The editor is indebted to the members of the advisory board, his colleagues, and his students for suggestions and assistance in preparing this volume. Special thanks are due to Mrs. Gaynel Klingemann for her secretarial and administrative efforts.

REFERENCES

1. J. Electroanal. Chem., 7, 417 (1964).
2. A. J. deBethune, J. Electrochem. Soc., 102, 288C (1955); T. S. Licht and A. J. deBethune, J. Chem. Educ., 34, 433 (1957).
3. Pure Appl. Chem., 21, 3 (1970).

Allen J. Bard
Austin, Texas

Chapter III-1

PHOSPHORUS

A. P. TOMILOV and N. E. CHOMUTOV
Department of Physical Chemistry
D. I. Mendeleev Institute of Chemistry and Technology
Moscow, USSR

with the assistance of
I. M. OSADCHENKO

1. STANDARD POTENTIALS

1.1. AQUEOUS SOLUTIONS

The element phosphorus has several modifications: α-white, red (triclinic) crystalline, red amorphous, and black crystalline. Red crystalline phosphorus is the most stable modification, but white phosphorus is taken to be the standard state of this element.

The most common degrees of oxidation of phosphorus are (-3) in gaseous hydrogen phosphide PH_3 (phosphine), (-2) in liquid hydrogen phosphide P_2H_4, (+3) in phosphorus anhydride P_2O_3 and in orthophosphorous acid H_3PO_3, (+4) in phosphorous tetraoxide P_2O_4 and in hypophosphoric acid $H_4P_2O_6$, (+5) in phosphoric anhydride P_2O_5 and in H_3PO_4, HPO_3, and $H_4P_2O_7$ (ortho-, meta-, and pyrophosphoric acids, respectively).

Other degrees of oxidation of this element, (-0.5), (+1), (+6), and (+7), are also known to occur.

No published data are available on methods for preparing a reversible galvanic cell with the participation of phosphorus and its compounds, or on methods for directly measuring the electrode potentials in experiments with electrode equilibrium. Equilibrium in the systems involving compounds of phosphorus is established very slowly.

The standard electrode potentials E° pertaining to P(-3)-P(0)-P(+3) and P(+3)-P(+5) systems have been published. The first systematic values of E° were calculated from thermodynamic data by Latimer [1] for a series of electrode reactions involving compounds of phosphorus. Pourbaix et al. [2] calculated the values of E° for many electrode reactions.

A large list of E° values and their temperature coefficients is included by de Bethune and Loud in the Encyclopedia of Electrochemistry [3]. Wagman et al. [4] have compiled a table of thermodynamic data containing the improved values of the Gibbs energy of formation for some compounds.

A complete list of the thermodynamic parameters for different substances, including the compounds of phosphorus, has been compiled by Karapetyans and Karapetyans [5]. The temperature coefficients of standard electrode potentials are known only for a few electrode reactions involving phosphorus and its compounds. They have been calculated by de Bethune, Licht, and Swendeman [6]. Chomutov [7] classified the E° values with respect to their accuracy, and it has been found that the accuracy of E° in electrode reactions involving the compounds of phosphorus does not exceed ±0.01 V. Milazzo [8] has compiled a fuller table of E° values and their temperature coefficients.

The E° values [1, 2] listed in Tables 1.1.1 and 1.1.2 were calculated with the use of the following values of Gibbs energy of formation ΔG_f° (kcal/mole): PH_3(g) (4.36), P_2H_4(l) (9.0), P_4H_2(s) (16.0), H_3PO_2(aq) (-125.1), $H_2PO_2^-$(aq) (-122.4), H_3PO_3(aq) (-204.8), $H_2PO_3^-$(aq) (-202.35), HPO_3^{2-}(aq) (-194.0), $H_4P_2O_6$(aq) (-392.0), $H_3P_2O_6^-$(aq) (-389.0), $H_2P_2O_6^{2-}$(aq) (-385.2), $HP_2O_6^{3-}$(aq) (-375.3), $P_2O_6^{4-}$(aq) (-361.7), H_3PO_4(aq) (-274.2), $H_2PO_4^-$(aq) (-271.3), HPO_4^{2-}(aq) (-261.5), and PO_4^{3-}(aq) (-245.1).

Improved values of ΔG_f° of P(red, cryst) (-2.9) and PH_3(gas) (3.2) are given in Ref. 4. By using the data given in Ref. 4 it is possible to obtain the exact values of E° for the reactions involving these compounds in the range from a few millivolts up to 50 mV.

Table 1.1.1 does not give the E° values for reactions involving PH_4^+ and $P_2H_5^+$ ions. In Ref. 1 the Gibbs formation energies of PH_3(g) and PH_4^+(aq) are assumed to be the same, and therefore the E° values are identical for the reactions involving PH_3(g) and PH_4^+(aq) ions. The value of Gibbs formation energy is given as equal to 16.2 kcal/mole for the PH_4^+

TABLE 1.1.1. Standard Potentials and Their Temperature Coefficients in Aqueous Solutions at 25°C[a]

Half-reaction	Standard potential (V)	Temperature coefficient dE°/dT Thermal mV/°C	Isothermal mV/°C	Refs.
1. $4P(r) + 2H^+ + 2e = P_4H_2$	-0.633			[2], [3], [8]
2. $4P(w) + 2H^+ + 2e = P_4H_2$	-0.347			[1], [2], [8]
3. $P_4H_2 + 10H^+ + 10e = 4PH_3$	-0.006			[2], [8]
4. $P(r) + 3H^+ + 3e = PH_3$	-0.111			[2], [8]
5. $P(w) + 3H^+ + 3e = PH_3$	-0.063	+0.767	-0.104	[1-3], [6-8]
6. $P_4H_2 + 6H^+ + 6e = 2P_2H_4$	-0.014			[2], [8]
7. $2P(r) + 4H^+ + 4e = P_2H_4$	-0.169			[2], [8]
8. $2P(w) + 4H^+ + 4e = P_2H_4$	-0.100			[2], [8]
9. $P_2H_4 + 2H^+ + 2e = 2PH_3$	+0.006			[2], [8]
10. $4H_3PO_2 + 6H^+ + 6e = P_4H_2 + 8H_2O$	-0.455			[2]
11. $4H_2PO_2^- + 10H^+ + 6e = P_4H_2 + 8H_2O$	-0.376			[2], [8]
12. $4H_3PO_3 + 14H^+ + 14e = P_4H_2 + 12H_2O$	-0.480			[2], [8]
13. $4H_2PO_3^- + 18H^+ + 14e = P_4H_2 + 12H_2O$	-0.450			[2], [8]
14. $4HPO_3^{2-} + 22H^+ + 14e = P_4H_2 + 12H_2O$	-0.346			[2], [8]
15. $4H_3PO_4 + 22H^+ + 22e = P_4H_2 + 16H_2O$	-0.406			[2], [8]
16. $4H_2PO_4^- + 26H^+ + 22e = P_4H_2 + 16H_2O$	-0.383			[2], [8]
17. $4HPO_4^{2-} + 30H^+ + 22e = P_4H_2 + 16H_2O$	-0.305			[2], [8]
18. $4PO_4^{3-} + 34H^+ + 22e = P_4H_2 + 16H_2O$	-0.176			[2], [8]
19. $H_3PO_2 + H^+ + e = 2H_2O + P(r)$	-0.365			[2], [8]
20. $H_3PO_2 + H^+ + e = 2H_2O + P(w)$	-0.508	+0.45	-0.42	[1-3], [6-8]
21. $H_2PO_2^- + 2H^+ + e = 2H_2O + P(r)$	-0.248			[2], [8]
22. $H_2PO_2^- + 2H^+ + e = 2H_2O + P(w)$	-0.391			[2], [8]
23. $H_3PO_3 + 3H^+ + 3e = 3H_2O + P(r)$	-0.454			[2], [8]
24. $H_3PO_3 + 3H^+ + 3e = 3H_2O + P(w)$	-0.502			[2], [8]
25. $H_2PO_3^- + 4H^+ + 3e = 3H_2O + P(r)$	-0.419			[2], [8]

(continued)

TABLE 1.1.1 (continued)

Half-reaction	Standard potential (V)	Temperature coefficient dE°/dT Thermal mV/°C	Isothermal mV/°C	Refs.
26. $H_2PO_3^- + 4H^+ + 3e = 3H_2O + P(w)$	-0.467			[2], [8]
27. $HPO_3^{2-} + 5H^+ + 3e = 3H_2O + P(r)$	-0.298			[2], [8]
28. $HPO_3^{2-} + 5H^+ + 3e = 3H_2O + P(w)$	-0.346			[2], [8]
29. $H_3PO_4 + 5H^+ + 5e = 4H_2O + P(r)$	-0.383			[2], [8]
30. $H_3PO_4 + 5H^+ + 5e = 4H_2O + P(w)$	-0.411			[2], [8]
31. $H_2PO_4^- + 6H^+ + 5e = 4H_2O + P(r)$	-0.358			[2], [8]
32. $H_2PO_4^- + 6H^+ + 5e = 4H_2O + P(w)$	-0.386			[2], [8]
33. $HPO_4^{2-} + 7H^+ + 5e = 4H_2O + P(r)$	-0.288			[2], [8]
34. $HPO_4^{2-} + 7H^+ + 5e = 4H_2O + P(w)$	-0.316			[2], [8]
35. $PO_4^{3-} + 8H^+ + 5e = 4H_2O + P(r)$	-0.128			[2], [8]
36. $PO_4^{3-} + 8H^+ + 5e = 4H_2O + P(w)$	-0.156			[2], [8]
37. $H_3PO_2 + 4H^+ + 4e = PH_3 + 2H_2O$	-0.174			[2], [8]
38. $H_2PO_2^- + 5H^+ + 4e = PH_3 + 2H_2O$	-0.145			[2], [8]
39. $H_3PO_3 + 6H^+ + 6e = PH_3 + 3H_2O$	-0.282			[2], [8]
40. $H_2PO_3^- + 7H^+ + 6e = PH_3 + 3H_2O$	-0.265			[2], [8]
41. $HPO_3^{2-} + 8H^+ + 6e = PH_3 + 3H_2O$	-0.205			[2], [8]
42. $H_3PO_4 + 8H^+ + 8e = PH_3 + 4H_2O$	-0.281			[2], [8]
43. $H_2PO_4^- + 9H^+ + 8e = PH_3 + 4H_2O$	-0.265			[2], [8]
44. $HPO_4^{2-} + 10H^+ + 8e = PH_3 + 4H_2O$	-0.212			[2], [8]
45. $PO_4^{3-} + 11H^+ + 8e = PH_3 + 4H_2O$	-0.123			[2], [8]
46. $H_3PO_3 + 2H^+ + 2e = H_3PO_2 + H_2O$	-0.499	+0.51	-0.36	[1-3], [6-8]
47. $H_2PO_3^- + 3H^+ + 2e = H_3PO_2 + H_2O$	-0.446			[2], [8]
48. $HPO_3^{2-} + 3H^+ + 2e = H_2PO_2^- + H_2O$	-0.504			[2], [8]

(continued)

TABLE 1.1.1 (continued)

Half-reaction	Standard potential (V)	Temperature coefficient dE°/dT Thermal mV/°C	Isothermal mV/°C	Refs.
49. $HPO_3^{2-} + 3H^+ + 2e = H_2PO_2^- + H_2O$	-0.323			[2], [8]
50. $H_4P_2O_6 + 2H^+ + 2e = 2H_3PO_3$	+0.380			[2], [8]
51. $H_4P_2O_6 + 2e = 2H_2PO_3^-$	+0.275			[2], [8]
52. $H_3P_2O_6^- + H^+ + 2e = 2H_2PO_3^-$	+0.340			[2], [8]
53. $H_2P_2O_6^{2-} + 2H^+ + 2e = 2H_2PO_3^-$	+0.423			[2], [8]
54. $H_2P_2O_6^{2-} + 2e = 2HPO_3^{2-}$	-0.061			[2], [8]
55. $HP_2O_6^{3-} + H^+ + 2e = 2HPO_3^{2-}$	+0.275			[2], [8]
56. $P_2O_6^{4-} + 2H^+ + 2e = 2HPO_3^{2-}$	+0.570			[2], [8]
57. $H_3PO_4 + 2H^+ + 2e = H_3PO_3 + H_2O$	-0.276	+0.51	-0.36	[2], [3], [6-8]
58. $H_3PO_4 + H^+ + 2e = H_2PO_3^- + H_2O$	-0.329			[2], [8]
59. $H_2PO_4^- + 2H^+ + 2e = H_2PO_3^- + H_2O$	-0.260			[2], [8]
60. $H_2PO_4^- + H^+ + 2e = HPO_3^{2-} + H_2O$	-0.447			[2], [8]
61. $HPO_4^{2-} + 2H^+ + 2e = HPO_3^{2-} + H_2O$	-0.234			[2], [8]
62. $PO_4^{3-} + 3H^+ + 2e = HPO_3^{2-} + H_2O$	+0.121			[2], [8]
63. $2H_3PO_4 + 2H^+ + 2e = H_4P_2O_6 + 2H_2O$	-0.933			[1], [2], [8]
64. $2H_2PO_4^- + 4H^+ + 2e = H_4P_2O_6 + 2H_2O$	-0.807			[2], [8]
65. $2H_2PO_4^- + 3H^+ + 2e = H_3P_2O_6^- + 2H_2O$	-0.872			[2], [8]
66. $2H_2PO_4^- + 2H^+ + 2e = H_2P_2O_6^{2-} + 2H_2O$	-0.955			[2], [8]
67. $2HPO_4^{2-} + 4H^+ + 2e = H_2P_2O_6^{2-} + 2H_2O$	-0.551			[2], [8]
68. $2HPO_4^{2-} + 3H^+ + 2e = HP_2O_6^{3-} + 2H_2O$	-0.744			[2], [8]
69. $2HPO_4^{2-} + 2H^+ + 2e = P_2O_6^{4-} + 2H_2O$	-1.039			[2], [8]
70. $2PO_4^{3-} + 4H^+ + 2e = P_2O_6^{4-} + 2H_2O$	-0.328			[2], [8]

[a] P(r) = red phosphorus. P(w) = white phosphorus. E° on molality and molarity scale.

TABLE 1.1.2. Standard Potentials in Basic Aqueous Solutions at 25°C[a] E° on molality and molarity scale.

Half-reaction	Standard potential E° (V)	Temperature coefficient dE°/dT Thermal mV/°C	Isothermal mV/°C	Refs.
$H_2PO_2^- + e = P(w) + 2OH^-$	-2.05	-	-	[3]
$HPO_3^{2-} + 2H_2O + 2e = H_2PO_2^- + 3OH^-$	-1.57	-	-	[3]
$PO_4^{3-} + 2H_2O + 2e = HPO_3^{2-} + 3OH^-$	-1.12	+0.38	-0.49	[3], [6]
$P(w) + 3H_2O + 3e = PH_3 + 3OH^-$	-0.89	-0.067	-0.938	[3], [6]

[a] P(w) = white phosphorus. E° on molality and molarity scale.

ion in Ref. 4. On comparing the values of ΔG_f° for PH_3 and PH_4^+, it is obvious that the PH_4^+ ion is unstable in aqueous solutions and therefore should not exist.

In Ref. 1 the ΔG_f° value for the $H_2PO_2^-$ ion has been calculated with the help of the dissociation constant of H_3PO_3 acid, $K_1 = 0.01$, found by Kolthoff [9]. The values of ΔG_f° for the $H_2PO_3^-$ ion were calculated in Ref. 4 by making use of the first and second degree dissociation constants of orthophosphorous acid, $K_1 = 1.6 \times 10^{-2}$ and $K_2 = 7 \times 10^{-7}$, given in Ref. 9. The values of ΔG_f° given in Refs. 1 and 2 for $H_2PO_4^-$ and HPO_4^{3-} ions agree with the third degree dissociation constants of orthophosphoric acid, $K_1 = 7.5 \times 10^{-3}$, $K_2 = 6.2 \times 10^{-8}$, and $K_3 = 10^{-12}$, which were determined by Pitzer [10]. Grzybowsky [11] found that $pK_2 = 7.2$.

The value $pK_3 = 12$ is close to the value $pK_3 = 11.82$ found later by Konopic and Leberl [12], as well as the value $pK_3 = 12.38$ given by Vanderzee and Quist [13]. The values of ΔG_f° given in Ref. 4 for H_3PO_4 (-273.1), PO_4^{3-} (-243.5), and HPO_4^{2-} (-260.34) are significantly different from those given in Refs. 1 and 2. The use of these values may change the E° value of some reactions by an amount approximately up to 60 mV. The value of ΔG_f° given in Ref. 4 for $H_2PO_4^-$ ion (-260.17) is, perhaps, erroneous. Reference 4 does not give these data for hypophosphoric acid and its ions.

Phosphorus forms compounds with halogens of the type PX_3 with a (+3) degree of oxidation. By making use of the ΔG_f° values published in Ref. 4 for the pure phase compounds PX_3, we can calculate E° of some electrode reactions involving the participation of PX_3. However, the E° values so obtained cannot be used as a behavior characteristic of real systems. The PX_3 solutions are totally hydrolyzed, and therefore the redox properties of aqueous solutions of halogenides of phosphorus are identical with the analogous properties of their hydrolysis products—orthophosphorous acid, monoorthophosphite ions, and diorthophosphite ions.

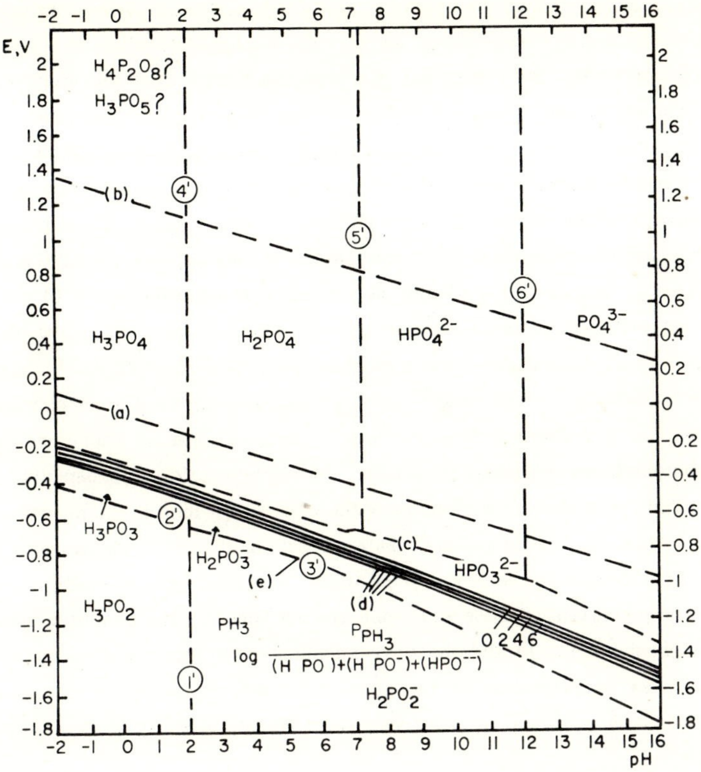

FIG. 1.1.1. Potential vs pH equilibrium for the phosphorus-water system at 25°C.

In Fig. 1.1.1 the potentials of the water-phosphorus system and its compounds are given in the form of pH functions. The lines (a) and (b) show the reversible potentials of hydrogen and oxygen electrodes. The set of lines (c) gives the potentials in the H_3PO_4/H_3PO_3 system described by Reactions (57)-(62) of Table 1.1.1. The family of lines (d) shows the potentials in PH_3-H_3PO_3, $H_2PO_3^-$, and HPO_3^{2-} systems for different ratios of the oxidizing and reducing forms. These lines (d) are described by Reactions (39)-(41) of Table 1.1.1. The lines (e) show the potentials of the system H_3PO_3, $H_2PO_3^-$, HPO_3^{2-}, PO_3^{3-}-H_3PO_2, and $H_2PO_2^-$ described by the Eqs. (46)-(49). The lines (1^1), (2^1), (3^1), (4^1), (5^1), and (6^1) pass through the limits of pH regions related to the predominance of one of the forms corresponding to the following equilibria: $H_3PO_2/H_2PO_2^-$, $H_3PO_3/H_2PO_3^-$, $H_2PO_3^-/HPO_3^{2-}$, $H_3PO_4/H_2PO_4^-$, $H_2PO_4^-/HPO_4^{2-}$, and HPO_4^{2-}/PO_4^{3-}. The pH dependence of E° shown in Fig. 1.1.1 can be used in the absence of substances with which phosphorus yields complexes or insoluble salts. According to Charlot [14], phosphorus compounds give rise to a large number of complexes with cations. In general, the hypophosphates, phosphates, metaphosphates, and pyrophosphates exhibit a strong tendency for complexing.

WILLIAM F. MAAG LIBRARY
YOUNGSTOWN STATE UNIVERSITY
362045

From Table 1.1.1 it is evident that phosphorus is thermodynamically unstable in the presence of water and is a strong reducing agent. The region of stability of PH_3 lies below (a), and it characterizes this compound as a reducing agent. It is thermodynamically very unstable. The compounds P_2H_4 and P_4H_2 are thermodynamically unstable as compared to PH_3, therefore there are no regions where P_2H_4 and P_4H_2 exist on Fig. 1.1.1, and accordingly hypophosphorous acid and the hypophosphites are powerful reducers. They should evolve hydrogen from water. Because of some kinetic reasons, these processes do not take place in practice in the absence of catalysts, but they become very intense in the presence of catalysts. Figure 1.1.1 clearly shows that orthophosphorous acid and orthophosphites are thermodynamically unstable in aqueous solutions over the entire pH range. They are reducers to a lesser degree than hypophosphoric acid ($H_4P_2O_6$) and hypophosphates. $H_3P_2O_6^-$, $H_2P_2O_6^{2-}$, and $P_2O_6^{4-}$ are unstable as compared to phosphorous and phosphites, as well as in relation to phosphoric acid and phosphates. Therefore there are no regions for the existence of these compounds. Phosphoric acid and phosphates are thermodynamically stable in aqueous solutions over the entire pH range in the presence or absence of oxygen. Phosphoric acid and phosphates can be oxidized to monoperphosphoric acid (H_3PO_5) and perphosphoric acid ($H_4P_2O_8$) either by electrolytic or chemical oxidation. The standard potentials of these reactions are not known. The conditions for the formation of perphosphoric acid and the corresponding references are given by Gmelin [15]. The dependence of potentials on the current density in the anode processes of formation of perphosphoric acid and its salts have been investigated by Khutchaturyan and Kruwtchinsky [16].

1.2. NONAQUEOUS SOLUTIONS

No data have been published on E° values for the electrode reactions involving phosphorus and its compounds in nonaqueous solutions; similarly, very scanty information is available on the values of ΔG_f° and other thermodynamical parameters for the phosphorus compounds in nonaqueous solutions. Nonetheless, useful data are currently being published. Moltchanowa and Dulowa [17] measured the pK values for the dissociation reaction: $H_3PO_4 = H_2PO_4^- + H^+$ at 22°C in n-butanol (6.66), isobutanol (8.53), tert-butanol (7.38), cyclohexanol (8.27 at 27°), acetone (6.72), and cyclohexanone (5.72).

1.3. FUSED SALTS

No published data are available on the electrode potentials, thermodynamical properties of phosphorus, and its compounds in fused media. Nevertheless, electroreduction of fused phosphates was investigated in detail by Franks and Inman [18], Caton and Frend [19], and Andreeva [20]. Franks and Inman [18] studied the electroreduction of phosphates ($NaPO_3$,

Na_3PO_4, $Na_4P_2O_4$, $Na_5P_3O_{10}$) by means of voltammetric and chronopotentiometric methods with a fused mixture of sodium phosphate, KCl, and NaCl in the temperature interval from 400 to 800°C. The chronopotentiometric studies revealed three electroactive particles being reduced at higher anode potentials as the ortho-, tri-, and pyrophosphate ions. During electroreduction the pentavalent phosphate is reduced to trivalent phosphate. All the products, including the phosphorus so formed in this process, are probably the result of secondary chemical reactions.

2. VOLTAMMETRIC CHARACTERISTICS

2.1. POLAROGRAPHIC CHARACTERISTICS

Methanol and ethanol solutions of white phosphorus are polarographically active [21, 22]. The wave height is proportional to the concentration of phosphorus in the interval from 1×10^{-4} to 2×10^{-3} g-atom/l. The wave is diffusion controlled and corresponds to the transfer of three electrons to the phosphorus atom (12 electrons to the P_4 molecule).

In aprotic solvents — acetonitrile and dimethylformamide — the number of electrons participating in the process decreases; thus n = 2 (for P_4) at a phosphorus concentration from 1×10^{-4} to 6×10^{-4} g-atom/l, while it reduces to 1.5 in the interval from 6×10^{-4} to 2×10^{-3} g-atom/l. The number of electrons participating in the process is not dependent on temperature. At a phosphorus concentration above 2×10^{-3}, the value of n is approximately unity, and the reduction process proceeds according to

$$P_4 + e \rightarrow P_4^-$$

In more dilute solutions the double-charge dianion P_4^{2-} may be formed, and it is further reduced with prior protonization [23].

On a mercury drop cathode, PCl_3 and $POCl_3$ undergo one- and two-stage reductions, respectively [24]. The wave heights are proportional to the concentration of the respective compounds in the interval from 1×10^{-4} to 1.8×10^{-3} mole/l. The polarographic method is a very suitable technique for detecting the compounds when they occur together.

Polarographic activity is exhibited by solutions of sodium salts of oxygen-containing acids of phosphorus in nonaqueous glycerine [25] containing phenol as the proton donor. The wave heights are proportional to the salt concentration in the interval from 5×10^{-3} to 2.5×10^{-1} M. The waves are generated as a result of the reduction of anions and largely depend on the concentration of H^+ ions. Distorted waves are observed in binary mixtures of different oxygen-containing acids [26].

The compounds of tri- and pentavalent phosphorus — phosphines, phosphine oxides and esters of oxygen, and sulfur-containing acids of phosphorus — exhibit polarographic activity in aprotic solvents only when the tri- or tetra-coordinated phosphorus atom has phenyl or styryl radicals [27-29]. The compounds containing aliphatic groups are polarographically inactive [30]. The communication reported by Saikiņa [31] on the polarographic activity of such compounds is, in all probability, erroneous. The observed reduction waves are of a diffusional nature and are proportional to the concentration of the compound in the interval from 1 X 10^{-4} to 2 X 10^{-3} mole/l [23]. The number of observed polarographic waves depends on the structure of the compound.

Oxialkyl- [32] and carboxyalkylphosphines [33], $(C_6H_5)_3P$ [34], and PH_3 in the presence of Pt^{4+} [35], as well as the primary and secondary phosphines in the presence of Co^+ [36], are capable of generating waves due to catalytically evolved hydrogen.

The polarographic parameters of the compounds of phosphorus studied are listed in Table 2.1.1 which does not contain those phosphorus compounds whose polarographic activity is caused by the reduction of other functional groups, for instance, phosphorylated aldehydes [37, 50], ketones [38, 39], and nitro compounds [40-42].

2.2. VOLTAMMETRIC CHARACTERISTICS

In oscillopolarographical experiments, phosphorus is reduced on a mercury electrode at E_p = -1.88 to 1.9 V vs mercury pool [21]. A similar picture is observed at the hanging mercury drop; moreover, if the drop is washed with methanol, the peak height is reduced by an insignificant amount but is not eliminated altogether. This fact can be regarded as a proof for the formation of a chemisorbed layer on the mercury electrode surface [22].

Yellow phosphorus in alkaline and acidic solutions gives rise to distinct oxidation waves at anodes made of carbon paste or enamelled silver [55] in the form of phosphorus impregnated matrices. Coulometric measurements show that phosphorus is oxidized to H_3PO_4 and H_3PO_3 to a qualitative amount in acidic and neutral media. In alkaline solution the oxidation process is complicated due to the oxidation of PH_3.

The secondary and tertiary phosphines, as well as compounds with the P—P bond, are oxidized at the carbon anode in acetonitrile solutions [56]. The oxidation currents are hump-shaped. In the pH range from 5 to 9 the value of $E_{1/2}$ remains constant, but in higher acidic solutions the oxidation current drops in value and $E_{1/2}$ shifts toward greater positive potentials. On adding 50% water to the solution, $E_{1/2}$ shifts by about 100 to 70 mV toward less positive potentials. The mercury anode begins to disintegrate, even at cathodic potentials, owing to the formation of mercury complexes [32, 53]. Aqueous solutions of peroxomonophosphoric and peroxodiphosphoric acids give rise to one irreversible diffusion wave at a platinized electrode [57]. The plot of $E_{1/2}$ vs pH of the media is shown in Fig. 2.2.1.

Adsorption waves are formed on ac polarograms owing to the high adsorption capacity of polyphosphate [58], and esters of phosphorous and thiophosphorous acids [59-62]. Two peaks, corresponding to the commencement and the end of adsorption of the compound are generally observed. These adsorption waves can be of use in analytical investigations. The high adsorbability of organophosphorus compounds is responsible for the suppression of the waves of hydrated Cu^{2+} and Cd^{2+} ions or their complexes with EDTA. The waves of Cu^{2+} inhibited by the triethylphosphate complex are shown in Fig. 2.2.2. The magnitude of drop in the limiting current peak is proportional to the concentration of the compound, and it can be used as a tool in analytical investigations [63-65].

The voltammetric characteristics of phosphorus and its compounds are listed in Table 2.2.1.

TABLE 2.1.1. Polarographic Characteristics

Compound	Product (n)	Conditions[a]	$E_{1/2}$ (V vs SCE)	I	$E_{3/4} - E_{1/4}$	Remarks	Refs.
			Cathode process				
P_4 (yellow)	PH_3	EtOH(95%), 0.1 M LiCl	-1.7w (vs HgPl)	-	irr		[22]
P_4 (yellow)	2.7-1.5	DMF, Bu_4NClO_4	-1.55 (vs HgPl)	-	irr		[23]
P_4 (yellow)	2.0-1.2	CH_3CN, Bu_4NClO_4	-1.73 (vs HgPl)	-	irr		[23]
$NaH_2PO_2.H_2O$	0.7	Glycerol, 5×10^{-3} Et_4NClO_4	-0.64w	-	rev	20.0°C	[25]
$NaH_2PO_3.2H_2O$	0.4	Glycerol, 5×10^{-3} Et_4NClO_4	-0.605w	-	rev	20.0°C	[25]
$Na_2HPO_3.5H_2O$	0.7	Glycerol, 5×10^{-3} Et_4NClO_4	-0.42w	-	rev	20.0°C	[25]
$Na_2HPO_4.2H_2O$	0.5	Glycerol, 5×10^{-3} Et_4NClO_4	-0.225w	-	rev	20.0°C	[25]
$Na_3PO_4.6H_2O$	0.3	Glycerol, 5×10^{-3} Et_4NClO_4	-0.310w	-	rev	20.0°C	[25]
$Na_2H_2P_2O_5$	0.2	Glycerol, 5×10^{-3} Et_4NClO_4	-0.605w	-	rev	20.0°C	[25]
$Na_3HP_2O_5.6H_2O$	0.8	Glycerol, 5×10^{-3} Et_4NClO_4	-0.340w	-	rev	20.0°C	[25]
$Na_2H_2P_2O_6.6H_2O$	0.9	Glycerol, 5×10^{-3} Et_4NClO_4	-0.370w	-	rev	20.0°C	[25]
$Na_3HP_2O_6.8H_2O$	0.2	Glycerol, 5×10^{-3} Et_4NClO_4	-0.370w	-	rev	20.0°C	[25]
$Na_2H_2P_2O_7.6H_2O$	1.0	Glycerol, 5×10^{-3} Et_4NClO_4	-0.330w	-	rev	20.0°C	[25]
PCl_3	3	CH_3CN, Me_4NCl	-0.93 (vs HgPl)	-	-	-	[24]
$POCl_3$	2	CH_3CN, Me_4NCl	-0.89 (vs HgPl)	-	-	-	[24]
$POCl_3$	1	CH_3CN, Me_4NCl	-2.74 (vs HgPl)	-	-	-	[24]

(continued)

TABLE 2.1.1 (continued)

Compound	Product (n)	Conditions[a]	$E_{1/2}$ (V vs SCE)	I	$E_{3/4} - E_{1/4}$	Remarks	Refs.
$(C_6H_5)_2PCl$	1	Monoglyme, Bu_4NClO_4	-3.3 vs $Ag/AgClO_4$	-	-	-	[43]
$(C_6H_5)PCl_2$	2	Monoglyme, Bu_4NClO_4	-2.5 vs $Ag/AgClO_4$	-	-	-	[43]
$(C_6H_5)_3P$	1	DMF, 0.1 $\underline{M}$ Bu_4NI	-2.70w	0.60	62	$i_a/i_k = 0.69$	[28], [29] [43], [44]
$(C_{10}H_7)_3P$	-	DMF, 0.03 $\underline{M}$ Et_4NI	-2.240	0.58	-	-	[29]
$(C_{10}H_7)_3P$	-	DMF, 0.09 $\underline{M}$ Et_4NI	-2.497	0.64	-	-	[29]
$(C_6H_{11})(CH_2CH(OH)CH_2)_2P$	-	CH_3OH, 0.2 $\underline{M}$ LiCl	-0.270	-	-	-	[32]
$(C_6H_{11})(CH_2CH_2OH)_2P$	-	CH_3OH, 0.2 $\underline{M}$ LiCl	-0.200	-	-	-	[32]
$(C_6H_5)(o\text{-}C_6H_{10}OH)_2P$	-	CH_3OH, 0.2 $\underline{M}$ LiCl	-0.170	-	-	-	[32]
$(C_6H_5)(CH_2CH(OH)CH_3)_2P$	-	CH_3OH, 0.2 $\underline{M}$ LiCl	-0.140	-	-	-	[32]
$(C_6H_5)(CH_2CH_2OH)_2P$	-	CH_3OH, 0.2 $\underline{M}$ LiCl	-0.110	-	-	-	[32]
$(C_6H_{11})_2(o\text{-}C_6H_{10}OH)P$	-	CH_3OH, 0.2 $\underline{M}$ LiCl	-0.440 vs Hg pool	5.4	-	-	[32]

$(C_6H_{11})_2(CH_2CH_2OH)P$	-	CH_3OH, 0.2 M LiCl	-0.430 vs Hg pool	-	-	-	[32]
$(C_6H_{11})_2(CH_2CH_2CH_2OH)P$	-	CH_3OH, 0.2 M LiCl	-0.420	-	-	-	[32]
$(C_6H_{11})_2(o\text{-}C_6H_{10}OH)P$	-	CH_3OH, 0.2 M LiCl	-0.200	-	-	-	[32]
$(C_6H_5)_2(CH_2CH_2OH)P$	-	CH_3OH, 0.2 M LiCl	-0.130	-	-	-	[32]
$(C_6H_{11})(o\text{-}C_6H_{10}OH)_2P$	-	CH_3OH, 0.2 M LiCl	-0.430	-	-	-	[32]
$(C_6H_5)_2PCH_2C_6H_5$	1	DMF, 0.091 M Et_4NI	-2.565	1.0	-	-	[32]
$(C_6H_5)_2P\text{-}CH_2\text{-}P(C_5H_5)_2$	1	DMF, 0.091 M Et_4NI	-2.655	0.88	-	-	[32]
$(C_6H_5)_2P(\alpha C_{10}H_7)$	1	DMF, 0.091 M Et_4NI	-2.27	0.60	-	-	[32]
$(C_6H_5)_2P(\alpha C_{10}H_7)_2$	1	DMF, 0.091 M Et_4NI	-2.492	0.67	-	-	[32]
$(C_6H_5)_3PO$		DMF, 0.1 M Bu_4NI	-2.51; -2.84	0.56	60-62	$i_a/i_k = 0.68$, $A_n = 4.1$ k/m	[28]
$(C_6H_5)(C_6H_5)(C_5H_5CH_2)P{=}O$	2	95% DMF, 0.04 M Et_4NI	-2.53w	-	-	-	[45]

(continued)

TABLE 2.1.1 (continued)

Compound	Product (n)	Conditions[a]	$E_{1/2}$ (V vs SCE)	I	$E_{3/4} - E_{1/4}$	Remarks	Refs.
$(C_6H_5CH_2)_2P(H)=O$	2	95% DMF, 0.04 M Et_4NI	-2.93	-	-	-	[45]
$(C_6H_5)_2P(O)CH_3$	1	DMF, 0.1 M Et_4NI	-2.575w	0.6	-	-	[29]
$(C_6H_5)_2P(O)C_2H_5$	1	DMF, 0.1 M Et_4NI	-2.570w	0.54	-	-	[29]
$(C_6H_5)_2P(O)CH_2C_6H_5$	1	DMF, 0.1 M Et_4NI	-2.570w	0.54	-	-	[29]
$(C_6H_5)_2P(O)(C_6H_{10}OH)$	1	DMF, 0.1 M Et_4NI	-2.51w	0.68	-	-	[29]
$(C_6H_5CH{=}CH)_3PO$	1	DMF, 0.08 M Et_4NI	-1.44w vs HgPl	-	-	-	[46]
$(C_6H_5CH{=}CH)_3PO$	1	DMF, 0.08 M Et_4NI	-1.62 vs HgPl	-	-	-	[46]
$(C_6H_5)_3PS$	1	DMF	-1.87 vs HgPl	-	-	$i_a/i_k = 0.28$	[46]
$(C_6H_5CH{=}CH)_3PS$	1	DMF	-1.41 vs HgPl	-	-	$i_a/i_k = 0.42$	[46]
$(C_6H_5CH{=}CH)_3PS$	1	DMF	-1.60 vs HgPl	-	-	-	[46]
$(C_6H_5CH{=}CH)_3PS$	1	DMF, 0.08 M Et_4NI	-1.95 vs HgPl	-	-	-	[46]
$(C_6H_5)_2P(S)C_6H_{11}$	1	DMF, 0.1 M Et_4NI	-2.485	0.60	-	-	[29]
$(C_6H_5)_2P(S)C_6H_{11}$	1	DMF, 0.1 M Et_4NI	-2.700	0.58	-	-	[29]
$(C_6H_5)_2P(S)(\alpha C_6H_{10}OH)$	2	DMF, 0.1 M Et_4NI	-2.470	1.14	-	-	[29]
$(C_6H_5)_2P(S)C_2H_5$	1	DMF, 0.1 M Et_4NI	-2.500	0.58	-	-	[29]
$(C_6H_5)_2P(S)C_2H_5$	1	DMF, 0.1 M Et_4NI	-2.700	0.44	-	-	[29]
$(C_6H_5)_2P(S)CH_2C_6H_5$	2	DMF, 0.1 M Et_4NI	-2.370	0.95	-	-	[29]
$(C_6H_5)_2P(S)C_6H_5$	1	DMF, 0.1 M Et_4NI	-2.415	0.61	-	-	[29]

$(C_6H_5)_2P(S)C_6H_5$	1	DMF, 0.1 M Et_4NI	-2.530	0.56	-	-	[29]
$(C_6H_5)_2P(S)C_6H_5$	1	DMF, 0.1 M Et_4NI	-2.700	0.15	-	-	[29]
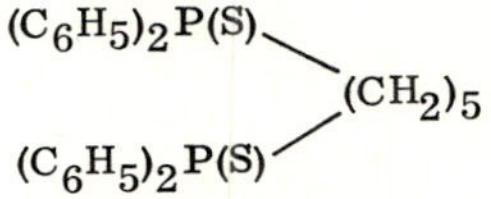	-	DMF, 0.1 M Et_4NI	-2.517	-	-	-	[29]
$(C_2H_5)_2\underset{\overset{\Vert}{S}}{P}-\underset{\overset{\Vert}{S}}{P}(C_2H_5)$	-	DMF, 0.1 M Et_4NI	-2.450	-	-	-	[47]
$(nC_3H_7)\underset{\overset{\Vert}{S}}{P}-\underset{\overset{\Vert}{S}}{P}(nC_3H_7)_2$	-	DMF, 0.1 M Et_4NI	-2.482	-	-	-	[47]
$(nC_4H_9)_2\underset{\overset{\Vert}{S}}{P}-\underset{\overset{\Vert}{S}}{P}(nC_4H_9)$	-	DMF, 0.1 M Et_4NI	-2.517	-	-	-	[47]
$(C_6H_5)_2\underset{\overset{\Vert}{S}}{P}-P(C_6H_5)_2$	1	DMF, 0.1 M Et_4NI	-1.795	-	-	-	[47]
$(C_6H_5)_2\underset{\overset{\Vert}{S}}{P}-\underset{\overset{\Vert}{S}}{P}(C_6H_5)_2$	-	DMF, 0.1 M Et_4NI	-1.535	2.7	-	-	[29], [47]
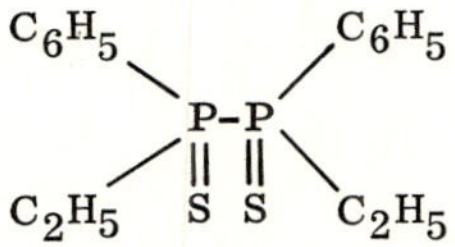	-	DMF, 0.1 M Et_4NI	-1.777	-	-	-	[45]
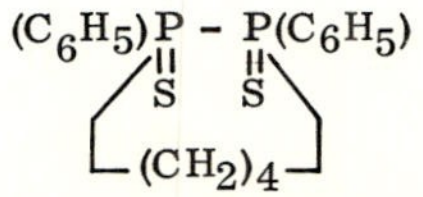	-	DMF, 0.1 M Et_4NI	-1.880	-	-	-	[45]

(continued)

TABLE 2.1.1 (continued)

Compound	Product (n)	Conditions[a]	$E_{1/2}$ (V vs SCE)	I	$E_{3/4} - E_{1/4}$	Remarks	Refs.
$(C_6H_5)(C_4H_9)P(S)-P(S)(C_6H_5)(C_4H_9)$	-	DMF, 0.1 <u>M</u> Et_4NI	-1.910	-	-	-	[45]
$(C_6H_5CH_2)_2P(S)-P(S)(CH_2C_6H_5)_2$	-	DMF, 0.1 <u>M</u> Et_4NI	-2.110	-	-	-	[47]
$(iC_3H_7)_2P-P(S)(iC_3H_7)_2$	-	DMF, 0.1 <u>M</u> Et_4NI	-2.315	-	-	-	[47]
$(C_6H_{11})_2P(S)-P(S)(C_6H_{11})_2$	-	DMF, 0.1 <u>M</u> Et_4NI	-2.335	-	-	-	[47]
$(C_6H_5)_2P(Se)C_2H_5$	2	DMF, 0.1 <u>M</u> Et_4NI	-2.280	1.14	-	-	[29]
$(C_6H_5)_2P(Se)CH_2C_6H_5$	2	DMF, 0.1 <u>M</u> Et_4NI	-2.22	0.97	-	-	[29]
$(C_6H_5)_2P(Se)CH_2C_6H_5$	2	DMF, 0.1 <u>M</u> Et_4NI	-2.57	0.91	-	-	[29]
$(C_6H_5)_2P(Se)C_6H_5$	2	DMF, 0.1 <u>M</u> Et_4NI	-2.145	1.02	-	-	[29]
$(C_6H_5)_2P(Se)C_6H_5$	1	DMF, 0.1 <u>M</u> Et_4NI	-2.690	0.54	-	-	[29]
$C_6H_5P(O)(OC_2H_5)_2$	1	DMF, 0.08 <u>M</u> Et_4NI	-2.06 vs HgPl	-	-	25 ± 1°C	[27]
$C_6H_5P(O)(OC_2H_5)(C_2H_5)$	1	DMF, 0.08 <u>M</u> Et_4NI	-2.08 vs HgPl	-	-	25 ± 1°C	[27]
$C_6H_5PH(O)OC_2H_5$	1	DMF, 0.08 <u>M</u> Et_4NI	-2.93 vs HgPl	-	-	25 ± 1°C	[27]

$C_6H_5P(S)(OC_2H_5)_2$	1	DMF, 0.08 M Et_4NI	-1.53 vs HgPl	-	-	25 ± 1°C	[27]
$C_6H_5P(S)(OC_2H_5)_2$	1	DMF, 0.08 M Et_4NI	-1.95 vs HgPl	-	-	25 ± 1°C	[27]
$C_6H_5CH{=}CHP(O)$ C_2H_5O OC_2H_5	1	DMF, 0.08 M Et_4NI	-1.45 vs HgPl	-	-	25 ± 1°C	[27]
$C_6H_5CH{=}CHP(O)$ C_2H_5O OC_2H_5	1	DMF, 0.08 M Et_4NI	-1.81 vs HgPl	-	-	25 ± 1°C	[27]
$C_6H_5CH{=}CHP(O)$ C_3H_7O OC_3H_7	1	DMF, 0.08 M Et_4NI	-1.47 vs HgPl	-	-	25 ± 1°C	[27]
$C_6H_5CH{=}CHP(O)$ C_3H_7O OC_3H_7	1	DMF, 0.08 M Et_4NI	-1.83 vs HgPl	-	-	25 ± 1°C	[27]
$C_6H_5CH{=}CHP(O)$ C_4H_9O OC_4H_9	1	DMF, 0.08 M Et_4NI	-1.48 vs HgPl	-	-	25 ± 1°C	[27]
$C_6H_5CH{=}CHP(O)$ C_4H_9O OC_4H_9	1	DMF, 0.08 M Et_4NI	-1.84 vs HgPl	-	-	25 ± 1°	[27]
$C_6H_5CH{=}CHP(O)$ C_2H_5O C_2H_5	1	DMF, 0.08 M Et_4NI	-1.52 vs HgPl	-	-	25 ± 1°C	[27]
$C_6H_5CH{=}CHP(O)$ C_2H_5O C_2H_5	1	DMF, 0.08 M Et_4NI	-1.90 vs HgPl	-	-	25 ± 1°C	[27]
$C_6H_5CH{=}CHP(O)$ C_4H_9O C_4H_9	1	DMF, 0.08 M Et_4NI	-1.53 vs HgPl	-	-	25 ± 1°C	[27]

(continued)

TABLE 2.1.1 (continued)

Compound	Product (n)	Conditions[a]	$E_{1/2}$ (V vs SCE)	I	$E_{3/4} - E_{1/4}$	Remarks	Refs.
$C_6H_5CH{=}CHP(O)(OC_4H_9)(C_4H_9)$	1	DMF, 0.08 M Et_4NI	-1.87 vs HgPl	-	-	25 ± 1°C	[27]
$C_6H_5CH{=}CH{-}P(O)(OC_2H_5)(H)$	1	DMF, 0.08 M Et_4NI	-1.40 vs HgPl	-	-	25 ± 1°C	[27]
$C_6H_5CH{=}CH{-}P(O)(OC_2H_5)(H)$	1	DMF, 0.08 M Et_4NI	-2.04 vs HgPl	-	-	25 ± 1°C	[27]
$C_6H_5CH{=}CHP(S)(OC_2H_5)(OC_2H_5)$	1	DMF, 0.08 M Et_4NI	-1.39 vs HgPl	-	-	25 ± 1°C	[27]
$C_6H_5CH{=}CHP(S)(OC_2H_5)(OC_2H_5)$	1	DMF, 0.08 M Et_4NI	-1.78 vs HgPl	-	-	25 ± 1°C	[27]
$C_6H_5CH{=}CHP(OC_2H_5)_2$	1	DMF, 0.08 M Et_4NI	-1.45 vs HgPl	-	-	25 ± 1°C	[27]
$C_6H_5CH{=}CHP(OC_2H_5)_2$	1	DMF, 0.08 M Et_4NI	-1.85 vs HgPl	-	-	25 ± 1°C	[27]
$C_6H_5CH{=}CH{-}P(O)(OC_2H_5)_2$ [$(C_2H_5O)_2P{=}O$ on CH]	1	DMF, 0.08 M Et_4NI	-1.48 vs HgPl	-	-	$i_a/i_k = 0.75$, $A_n = 3.1$	[46]
$C_6H_5P(O)(OC_2H_5)_2$ [C_6H_5, $(C_2H_5O)_2$ on P=O]	1	DMF, 0.08 M Et_4NI	-2.02 vs HgPl	-	rev	$i_a/i_k = 0.64$	[46]

$CH_2=CH$ $P=O$ $(C_2H_5O)_2$	2	DMF, 0.08 M Et_4NI	-2.02 vs HgPl	-	-	-	[46]
$(CH_2=CH)_2$ $P=O$ C_2H_5O	2	DMF, 0.08 M Et_4NI	-1.87 vs HgPl	-	-	-	[46]
$(p\text{-}NO_2C_6H_4O)_3PO$	2	DMF	-0.8 vs aq SCE	3.5	0.08	diff	[48]
$(p\text{-}NO_2C_6H_4O)_3PO$	2	DMF	-1.07i vs aq SCE	-	-	-	[48]
$(p\text{-}NO_2C_6H_4O)_3PO$	2	DMF	-1.90i vs aq SCE	-	-	max	[48]
$(p\text{-}NO_2C_6H_4O)_3PO$	2	DMF	-2.20i vs aq SCE	-	-	max	[48]
$(p\text{-}NO_2C_6H_4)_2PO(OH)$	2	DMF, 0.1 M Bu_4NI	-0.94 vs aq SCE	0.9	-	ads	[49]
$(p\text{-}NO_2C_6H_4)_2PO(OH)$	4	DMF, 0.1 M Bu_4NI	-1.23 vs aq SCE	-	-	diff	[49]
$(p\text{-}NO_2C_6H_4)_2PO(OH)$	4	DMF, 0.1 M Bu_4NI	-2.60 vs aq SCE	-	-	max	[49]
$(C_6H_5O)_3P=O$	2	95% DMF, 0.04 M Et_4NI	-2.72w	-	-	-	[45]
$[o\text{-}(CH_3)_2C_6H_3O]P=O$	2	95% DMF, 0.04 M Et_4NI	-2.80w	-	-	-	[45]
$[C_6H_5CH_2O]_3P=O$	2	95% DMF, 0.04 M Et_4NI	-2.60w	-	-	-	[45]
$[C_6H_5CH_2O]_3P=O$	2	95% DMF, 0.04 M Et_4NI	-2.93i	-	-	-	[45]

(continued)

TABLE 2.1.1 (continued)

Compound	Product (n)	Conditions[a]	$E_{1/2}$ (V vs SCE)	I	$E_{3/4} - E_{1/4}$	Remarks	Refs.
$\underset{\underset{C_6H_5}{\overset{\|}{(CH_3O)_2PSCHC}}}{\overset{S}{\|}} \overset{O}{\underset{OC_2H_5}{\|}}$	2	DMF, 0.04 $\underline{M}$ Bu$_4$NClO$_4$	-2.16	-	irr	-	[51]
(C$_6$H$_5$)$_3$P=NC$_6$H$_5$	1	DMF, 0.2 $\underline{M}$ Bu$_4$NI	-1.86 vs HgCl	-	0.06	-	[52]
(C$_6$H$_5$)$_3$P=NC$_6$H$_5$	1	DMF, 0.2 $\underline{M}$ Bu$_4$NI	-2.25 vs HgPl	-	-	-	[52]
CH$_3$NHCSP(C$_6$H$_5$)$_2$	-	DMF, 0.1 $\underline{M}$ Et$_4$NI	-2.150	1.5	-	-	[29]
C$_6$H$_5$NHCSP(C$_6$H$_5$)$_2$	-	DMF, 0.1 $\underline{M}$ Et$_4$NI	-1.785	1.7	-	-	[29]
C$_6$H$_5$NHCSP(C$_6$H$_5$)$_2$	-	DMF, 0.1 $\underline{M}$ Et$_4$NI	-2.670	-	-	-	[29]
C$_6$H$_7$NHCSP(C$_6$H$_5$)$_2$	-	DMF, 0.1 $\underline{M}$ Et$_4$NI	-1.755	1.2	-	-	[29]
C$_6$H$_7$NHCSP (C$_6$H$_5$)$_2$	-	DMF, 0.1 $\underline{M}$ Et$_4$NI	-2.525	1.4	-	-	[29]
CH$_3$NHCSP (C$_6$H$_{11}$)$_2$	-	DMF, 0.1 $\underline{M}$ Et$_4$NI	-2.355	1.5	-	-	[29]
C$_2$H$_5$NHCSP (C$_6$H$_{11}$)$_2$	-	DMF, 0.1 $\underline{M}$ Et$_4$NI	-2.355	1.5	-	-	[29]
C$_6$H$_5$NHCSP (C$_6$H$_{11}$)$_2$	-	DMF, 0.1 $\underline{M}$ Et$_4$NI	-1.925	1.8	-	-	[29]
CH$_2$=CHCH$_2$NHCS (C$_6$H$_{11}$)$_2$P	-	DMF, 0.1 $\underline{M}$ Et$_4$NI	-2.270	1.9	-	-	[29]
CH$_3$NHCSP(O)(C$_6$H$_{11}$)$_2$	-	DMF, 0.1 $\underline{M}$ Et$_4$NI	-1.890	1.76	-	-	[29]
CH$_3$NHCSP(O)(C$_6$H$_{11}$)$_2$	-	DMF, 0.1 $\underline{M}$ Et$_4$NI	-2.665	1.68	-	-	[29]
CH$_2$=CHCH$_2$NHCS (C$_6$H$_{11}$)$_2$P(O)	-	DMF, 0.1 $\underline{M}$ Et$_4$NI	-2.015	2.0	-	-	[29]

$C_6H_5NHCSP(S)(C_6H_5)_2$	-	DMF, 0.1 M Et_4NI	-1.640	1.3	-	-	[29]
$C_6H_5NHCSP(S)(C_6H_5)_2$	-	DMF, 0.1 M Et_4NI	-2.505	1.2	-	-	[29]
			Anodic process				
$(C_6H_5)_3P$	1	CH_3CN, 0.1 M $LiClO_4$	+0.12	3.3	-	n = 96%	[53]
	$[(C_6H_5)_3P]Hg(ClO_4)_2$						
$(p\text{-}CH_3C_6H_4)_3P$	-	CH_3CN, 0.1 M $LiClO_4$	+0.10	2.4	-	n = 68%	[53]
	$[(CH_3C_6H_4)_3P]_2.Hg(ClO_4)_2$						
$(p\text{-}ClC_6H_4)_3P$	1	CH_3CN, 0.1 M $LiClO_4$	+0.24	3.0	-	-	[53]
$(p\text{-}CH_3OC_6H_4)_3P$	1	CH_3CN, 0.1 M $LiClO_4$	+0.29	2.0	-	n = 70	[53]
	$[(CH_3OC_6H_4)P]_2.Hg(ClO_4)_2$						
$(p\text{-}CH_3OC_6H_4)_2P(C_6H_5)$	1	CH_3CN, 0.1 M $LiClO_4$	+0.08	1.2	-	-	[53]
$(C_6H_5)_2PH$	1	CH_3CN, 0.1 M $LiClO_4$	+0.22	-	-	-	[53]
$(C_2H_5O)_3P$	1	CH_3CN, 0.1 M $LiClO_4$	+0.19	-	-	-	[53]
$[(CH_3)_2N]_3P$	1	CH_3CN, 0.1 M $LiClO_4$	+0.28	2.0	-	-	[53]
$(C_6H_5)_3P$	-	CH_3OH, 0.1 M $LiClO_4$	+0.080	-	-	-	[54]
$(C_2H_5)_3P$	-	CH_3OH, 0.1 M $LiClO_4$	-0.3	-	-	-	[54]

[a] DMF = N, N-dimethylformamide, Et_4N^+ = tetraethylammonium. Bu_4N = tetra-n-butylammonium.

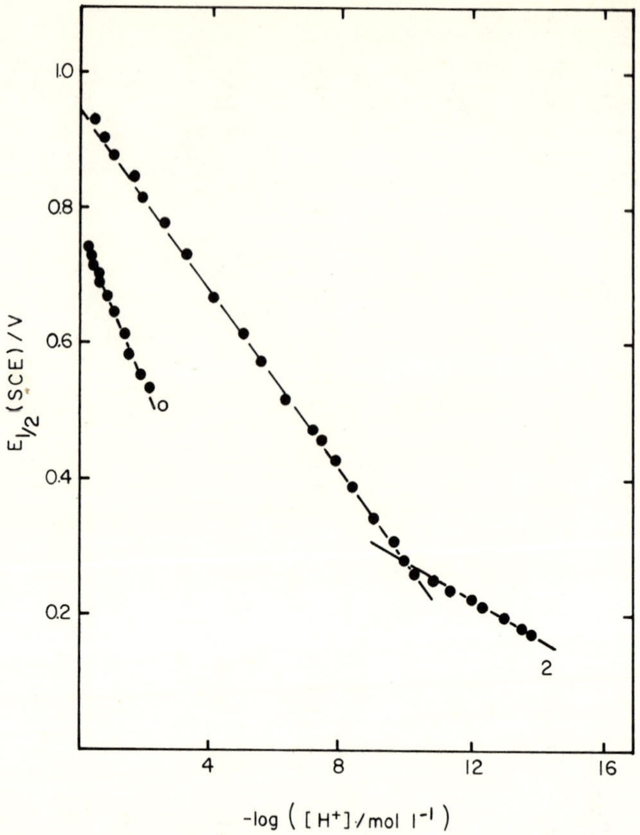

FIG. 2.2.1. Change in the half-wave potential $E_{1/2}$ referred to SCE as a function of H^+ for reduction wave. (1) 10^{-3} M H_3PO_5. (2) 10^{-3} M $H_4P_2O_8$.

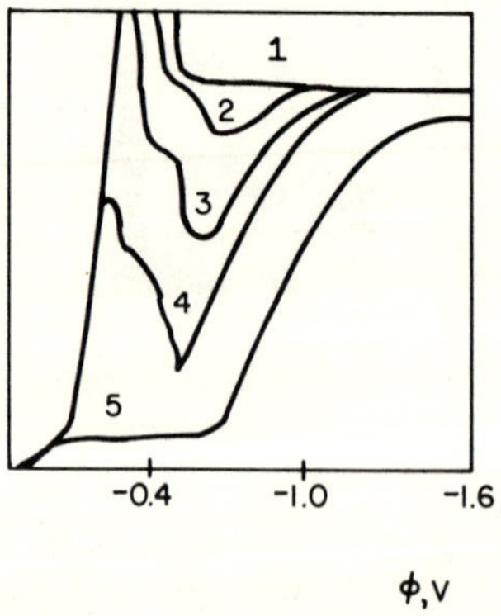

FIG. 2.2.2. Suppression of Cu^{2+} wave by triethylphosphite complex 0.1 N Na_2SO_4, 3 X 10^{-3} $CuSO_4$. Concentration of triethylphosphite: (1) 0; (2) 2 X 10^{-3}; (3) 4 X 10^{-3} (4) 8 X 10^{-3}; (5) 4 X 10^{-3}.

TABLE 2.2.1. Voltammetric Characteristics

Substance	Product	Conditions	Electrode	Technique	Potential (V vs SCE)	Rev	Remarks	Refs.
P_4 (yellow)	-	CH_3OH, 0.1 M LiCl	Hg	dc pol	-1.5 vs HgPl	-	-	[22]
PCl_3	-	DMF, 1 M $LiNO_3$	Pt rot	i stp	-0.96 vs Au el	irr	$D = 1.0 \times 10^{-3}$, i/ac = 1254	[66]
$POCl_3$	-	DMF, 1 M $LiNO_3$	Pt rot	i stp	-0.925 vs Au el	irr	$D = 0.3 \times 10^{-5}$, i/ac = 1377	[66]
PCl_5	-	DMF, 1 M $LiNO_3$	Pt rot	i stp	-0.73 vs Au el	irr	$D = 2.1 \times 10^{-5}$, i/ac = 3620	[66]
$(C_6H_5)_3P$	-	DMF, 0.1 M Bu_4NI	Hg	ac pol	$E_{pc} = -2.75$ $E_{pa} = -2.68$	irr	-	[28]
$(C_6H_5)_3P$	-	CH_3CN, 0.1 M Bu_4NI	Hg	ac pol	$E_{pc} = -1.60$	-	-	[28]
$(C_6H_5)_3PO$	-	DMF, 0.1 M Bu_4NI	Hg	ac pol	$E_{pc} = -2.60$ $E_{pa} = -2.44$	-	-	[28]
$(C_6H_5)_3PO$	-	$(CH_3)_2SO_2$, 0.1 M Bu_4NI	Hg	ac pol	$E_{pc} = -2.44$	-	-	[28]
$(n\text{-}NO_2C_6H_4)_3PO$	-	DMF, 0.1 M Et_4NI	Pt dsk	ac pol	$E_{pc} = -1.13$	irr	Scan rate potential 67 mV/sec	[48]
$(n\text{-}NO_2C_6H_4)_3PO$	-	DMF, 0.1 M Et_4NI	Pt dsk	ac pol	$E_{pa} = -1.04$	-	Scan rate potential 67 mV/sec	[48]
$(n\text{-}NO_2C_6H_4)_3PO$	-	DMF, 0.1 M Et_4NI	Pt dsk	ac pol	$E_{pc} = -1.96$	-	Scan rate potential 67 mV/sec	[48]
$(n\text{-}NO_2C_6H_4)_3PO$	-	DMF, 0.1 M Et_4NI	Pt dsk	ac pol	$E_{pa} = -1.86$	3.12	Scan rate potential 67 mV/sec	[48]
$(n\text{-}NO_2C_6H_4)_3PO$	-	DMF, 0.1 M Et_4NI	Pt dsk	ac pol	$E_{pc} = -2.24$	0.86	Scan rate potential 67 mV/sec	[48]

(continued)

TABLE 2.2.1 (continued)

Substance	Product	Conditions	Electrode	Technique	Potential (V vs SCE)	Rev	Remarks	Refs.
$(n\text{-}NO_2C_6H_4)_3PO$	-	DMF, 0.1 M Et_4NI	Pt dsk	ac pol	$E_{pc} = -2.51$	6.1	Scan rate potential 67 mV/sec	[48]
$(p\text{-}NO_2C_6H_4)_2P(O)OH$	-	DMF, 0.1 M Bu_4NI	Hg	ac pol	$E_{pc} = -1.02$	2.2	Scan rate potential 91.4 mV/sec	[49]
$(p\text{-}NO_2C_6H_4)_2P(O)OH$	-	DMF, 0.1 M Bu_4NI	Hg	ac pol	$E_{pc} = -1.35$	0.73	Scan rate potential 91.4 mV/sec	[49]
$(p\text{-}NO_2C_6H_4)_2P(O)OH$	-	DMF, 0.1 M Bu_4NI	Hg	ac pol	$E_{pc} = -2.51$	2.00	Scan rate potential 91.4 mV/sec	[49]
P_4 (yellow)	-	10 M NaOH	Ag-Hg	dc pol	1.08	-	60°C	[55]
P_4 (yellow)	-	1.0 M NaOH	Ag-Hg	dc pol	0.90	-	60°C	[55]
P_4 (yellow)	-	10 M NaOH	Ag-Hg	dc pol	1.07	-	40°C	[55]
P_4 (yellow)	-	1.0 M NaOH	Ag-Hg	dc pol	0.89	-	40°C	[55]
P_4 (yellow)	-	0.1 M NaOH	Ag-Hg	dc pol	0.83	-	40°C	[55]
P_4 (yellow)	-	0.5 M Na_2SO_4	Ag-Hg	dc pol	0.30	-	40°C	[55]
P_4 (yellow)	-	0.5 M Na_2SO_4, 0.005 M H_2SO_4	Ag-Hg	dc pol	0.18	-	40°C, pH = 5.5	[55]
P_4 (yellow)	-	0.05 M H_2SO_4	Ag-Hg	dc pol	0.13	-	40°C	[55]
P_4 (yellow)	-	0.49 M H_2SO_4	Ag-Hg	dc pol	0.08	-	40°C	[55]
P_4 (yellow)	-	10 M H_2SO_4	Ag-Hg	dc pol	0.95	-	40°C	[55]
P_4 (yellow)	-	1.02 M NaOH	C pst	dc pol	0.84	-	40°C	[55]

P_4 (yellow)	-	0.49 $\underline{M}$ H_2SO_4	Ag-Hg	dc pol	0.04	-	40°C	[55]
PH_3	-	9.99 $\underline{M}$ NaOH	Ag-Hg	dc pol	1.12	-	60°C	[55]
PH_3	-	1 $\underline{M}$ NaOH	Ag-Hg	dc pol	0.94	-	40°C	[55]
PH_3	-	0.49 $\underline{M}$ NaOH	Ag-Hg	dc pol	0.11	-	40°C	[55]
$H_2PO_2^-$	-	10.0 $\underline{M}$ NaOH	Ag-Hg	dc pol	0.78	-	60°C	[55]
H_3PO_2	-	0.5 $\underline{M}$ H_2SO_4	Pd on Au	dc pol	0.47	-	60°C	[67], [68]
H_3PO_3	-	0.5 $\underline{M}$ H_2SO_4	Pd on Au	dc pol	0.52	-	60°C	[67], [68]
$(C_6H_5)_3P$	-	CH_3CN, 0.1 $\underline{M}$ $NaClO_4$	C pst	dc pol	1.000	-	60°C	[56]
$(C_6H_5)_2PCH_2C_6H_5$	-	CH_3CN, 0.1 $\underline{M}$ $NaClO_4$	C pst	dc pol	0.990	-	-	[56]
$(C_6H_5)_2P(\alpha\text{-}C_{10}H_7)$	-	CH_3CN, 0.1 $\underline{M}$ $NaClO_4$	C pst	dc pol	0.980	-	-	[56]
$(C_6H_5)_2PCH_3$	-	CH_3CN, 0.1 $\underline{M}$ $NaClO_4$	C pst	dc pol	0.955	-	-	[56]
$(C_6H_5)_2PC_2H_5$	-	CH_3CN, 0.1 $\underline{M}$ $NaClO_4$	C pst	dc pol	0.953	-	-	[56]
$(C_6H_5)P(C_5H_5)_2$	-	CH_3CN, 0.1 $\underline{M}$ $NaClO_4$	C pst	dc pol	0.890	-	-	[56]
$(C_2H_5)_3P$	-	CH_3CN, 0.1 $\underline{M}$ $NaClO_4$	C pst	dc pol	0.885	-	-	[56]
$(n\text{-}C_3H_7)_3P$	-	CH_3CN, 0.1 $\underline{M}$ $NaClO_4$	C pst	dc pol	0.880	-	-	[56]
$(n\text{-}C_4H_9)_3P$	-	CH_3CN, 0.1 $\underline{M}$ $NaClO_4$	C pst	dc pol	0.880	-	-	[56]
$(C_6H_5)_2P(CH_2)_3OH$	-	CH_3CN, 0.1 $\underline{M}$ $NaClO_4$	C pst	dc pol	0.850	-	-	[56]
$(C_6H_5)P(CH_2CH_2CH_2OH)_2$	-	CH_3CN, 0.1 $\underline{M}$ $NaClO_4$	C pst	dc pol	0.800	-	-	[56]
$C_6H_{11}P(CH_2)_3OH$	-	CH_3CN, 0.1 $\underline{M}$ $NaClO_4$	C pst	dc pol	0.870	-	-	[56]

(continued)

Table 2.2.1 (continued)

Substance	Product	Conditions	Elec-trode	Tech-nique	Potential (V vs SCE)	Rev	Remarks	Refs.
Cyclo-$C_8H_{17}P(CH_2)_3OH$	-	CH_3CN, 0.1 M $NaClO_4$	C pst	dc pol	0.850	-	-	[56]
$C_6H_5CHCH_2COOCH_3$ $P(C_6H_5)_2$	-	CH_3CN, 0.1 M $NaClO_4$	C pst	dc pol	0.950	-	-	[56]
$C_6H_5CHCH_2COC_6H_5$ $P(C_6H_5)_2$	-	CH_3CN, 0.1 M $NaClO_4$	C pst	dc pol	0.950	-	-	[56]
$(C_6H_5)_2PH$	-	CH_3CN, 0.1 M $NaClO_4$	C pst	dc pol	0.990	-	-	[56]
$C_4H_9P<^{C_6H_5}_{H}$	-	CH_3CN, 0.1 M $NaClO_4$	C pst	dc pol	0.915	-	-	[56]
$(C_6H_5)_2PH$	-	CH_3CN, 0.1 M $NaClO_4$	C pst	dc pol	0.845	-	-	[56]
$(C_6H_5)_2P-P(C_6H_5)_2$	-	CH_3CN, 0.1 M $NaClO_4$	C pst	dc pol	0.915	-	-	[56]
$C_6H_5PCH=CHC_6H_5$ $C_6H_5PCH=CHC_6H_5$	-	CH_3CN, 0.1 M $NaClO_4$	C pst	dc pol	0.860	-	-	[56]
$C_6H_5PC_2H_5$ $C_6H_5PC_2H_5$	-	CH_3CN, 0.1 M $NaClO_4$	C pst	dc pol	0.760	-	-	[56]
$(C_2H_5)_2P-P(C_2H_5)_2$	-	CH_3CN, 0.1 M $NaClO_4$	C pst	dc pol	0.765	-	-	[56]
$(n-C_3H_7)_2P-P(n-C_3H_7)_2$	-	CH_3CN, 0.1 M $NaClO_4$	C pst	dc pol	0.760	-	-	[56]
$(n-C_4H_9)_2P-P(n-C_4H_9)_2$	-	CH_3CN, 0.1 M $NaClO_4$	C pst	dc pol	0.735	-	-	[56]
$(i-C_3H_7)_2P-P(i-C_3H_7)_2$	-	CH_3CN, 0.1 M $NaClO_4$	C pst	dc pol	0.770	-	-	[56]
$(t-C_4H_9)_2P-P(t-C_4H_9)_2$	-	CH_3CN, 0.1 M $NaClO_4$	C pst	dc pol	0.685	-	-	[56]
$[C_6H_5P]_5$	-	CH_3CN, 0.1 M $NaClO_4$	C pst	dc pol	0.900	-	-	[56]

3. KINETIC PARAMETERS

3.1. KINETIC PARAMETERS

Mayer and Brown carried out detailed investigations on the kinetics of electroreduction of chlorides of phosphorus in acetonitrile on the platinum disk cathode [66]. The basic results obtained in chronopotentiometric measurements are listed in Table 3.1.1.

TABLE 3.1.1. Kinetic Parameters for the Reduction of Halogenides of Phosphorus in Acetonitrile in 1 M $LiNO_3$

Compound	Concentration (moles/ml) X 10^6	i(A/cm^2) X 10^3	Transition time (sec)	E_0	Approx slope	αn_a	D_{O_2} (cm/sec) X 10^5	Log rate const (cm/sec)	Bond energy (kcal)
PCl_5	3.85	0.893	11.36	-0.730	0.076	0.536	2.10	- 9.53	40
PCl_3	3.95	0.327	10.80	-0.960	0.053	0.776	1.01	-16.68	86
$POCl_3$	3.87	0.466	6.11	-0.925	0.084	0.434	0.30	-10.02	--

According to thermodynamic calculations, phosphorous acid (H_3PO_3) and hypophosphorous acid (H_3PO_2) must readily oxidize in aqueous solutions. However, practical measurements carried out by Hickling and Johnson [69] show that these acids do not undergo anode oxidation on almost all of the electrodes that were used, except on palladium electrodes (see Table 3.1.2).

TABLE 3.1.2. Oxidation of Phosphorous and Hypophosphorous Acids (c = 0.1 M; i = 0.05 A/cm^2; t = 25°C; pH = 7)

	Anode current efficiency							
Electrolyte	Pt	PtPt	PdPt	Pd	C	Pb	Ni	PbO_2
Na_2HPO_3	0	0	0	1	2	2	5	6
$NaHPO_2$	2	4	45	28	15	7	4	10

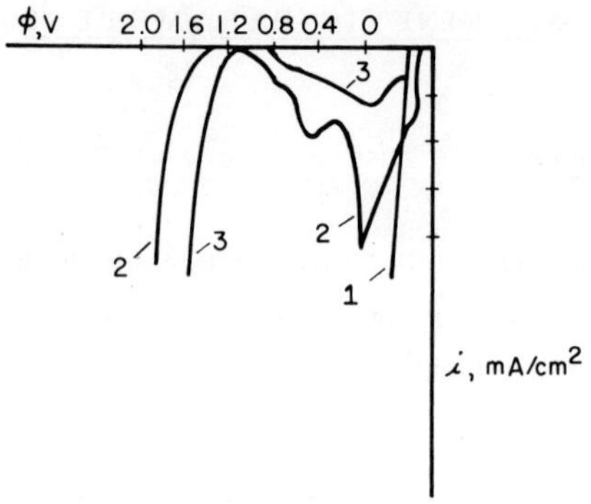

FIG. 3.1.1. Potentiostatic curves of Pd electrode in NaH_2PO_4 solutions. Concentration of hypophosphite: (1) 1.0 M; (2) 0.1 M; (3) 0.01 M.

Such an anomalous behavior is explained by the fact that these acids exist in two forms, the active (containing P^{3+}) and the inactive (P^{5+}) forms:

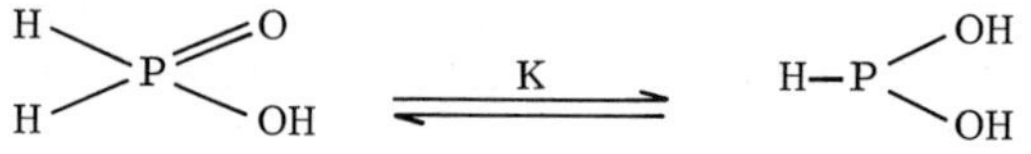

$$\begin{array}{c} H \\ H \end{array} P \begin{array}{c} =O \\ OH \end{array} \quad \underset{}{\overset{K}{\rightleftharpoons}} \quad H-P \begin{array}{c} OH \\ OH \end{array} \quad K = 3.66 \text{ mole}^{-1}$$

Since the equilibrium is shifted toward the inactive form, the concentration of molecules capable of undergoing oxidation is very low. But on the palladium electrode the inactive form is subjected to catalytic dehydration, and this fact is corroborated by potentiostatic measurements. The potentiostatic curve (Fig. 3.1.1) has a distinctive maximum at an anode potential close to zero. At higher positive potentials the oxidation process is suppressed by the formation of oxides of palladium. Preparative electrolysis showed that in the potential range from -0.3 to -0.15 V, the oxidation of hypophosphite requires 1.16 to 1.25 F/mole, and that of phosphite 1.57 to 1.80 F/mole.

Trasatti and Alberti [67, 68] studied the polarization of a gold-plated palladium electrode. Typical polarization curves obtained for the solutions of phosphorous and hypophosphorous acids are shown in Fig. 3.1.2. The slope of Tafel line for a solution of hypophosphorous acid is greater than 0.12 V/decade, while that for phosphorous acid is close to 0.12 V/decade.

With the help of the plot of log i vs reciprocal of absolute temperature, it was found that the activation energy of the processes responsible for the limiting current at a potential of 550 mV is 17.0 kcal/mole for phosphorous acid and 13.0 kcal/mole for hypophosphorous acid.

The linear relation between log i and log c clearly shows that the oxidation process is first order for H_3PO_3 and 1/3 order for H_3PO_2.

The curve of oxidation reaction overpotential vs current for a 0.5-M solution of H_3PO_2 on a Pd electrode is shown in Fig. 3.1.3.

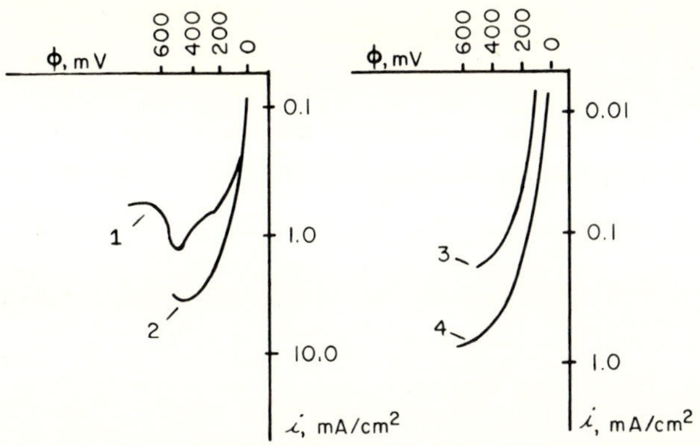

FIG. 3.1.2. Anode polarization curves on Pd anode in 0.5 M sulfuric acid. Concentration: (1) 0.02 M H_3PO_2; (2) 0.5 M H_3PO_2; (3) 0.1 M H_3PO_3; (4) 0.5 M H_3PO_3.

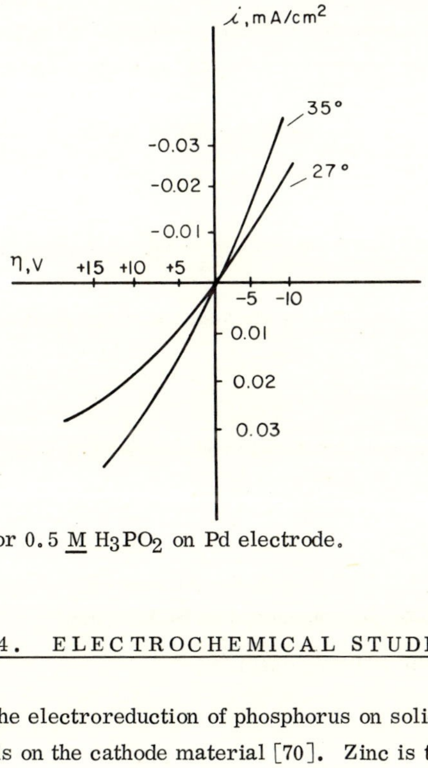

FIG. 3.1.3. η vs i for 0.5 M H_3PO_2 on Pd electrode.

4. ELECTROCHEMICAL STUDIES

The final product of the electroreduction of phosphorus on solid metals is PH_3, and its yield significantly depends on the cathode material [70]. Zinc is the most active electrode material. The reduction process is not so active on Cd and Pb electrodes. One specific feature of the reduction process on a lead cathode is that PH_3 is not evolved during the initial moments of electrolysis, but thereafter the PH_3 yield is stabilized at the 75% level, and finally at the end of the electrolysis the yield sharply increases and exceeds 100% with respect to the current. These results are interpreted as proof of the formation of low hydrides of phosphorus (e.g., H_4P_4) which are subsequently reduced to PH_3 [71].

4.1. COMPOUNDS OF TRIVALENT PHOSPHORUS

The first stage in the reduction of aromatic phosphines is the addition of one electron with the formation of an unstable anion-radical [28, 29]:

$$(C_6H_5)_3P + e \rightarrow (C_6H_5)_3P^-$$

Thereafter, the P—C bond is broken down with the cleavage of phenyl radicals which dimerize to diphenyl:

$$(C_6H_5)_3P^-. + H^+ \rightarrow (C_6H_5)_2PH + C_6H_5.$$

$$2C_6H_5. \rightarrow C_6H_5.C_6H_5$$

Diphenyl is subjected to further reduction during electrolysis. Phosphine oxides are reduced in a similar manner [29].

The electroreduction of phosphine sulfides is complicated by the fact that the anion radical formed at the first stage is partially subjected to further reduction:

$$(C_6H_5)_3PS \xrightarrow{e} \begin{cases} (C_6H_5)_2PS^- + C_6H_5^- \\ (C_6H_5)_3P + S^{2-} \end{cases}$$

Because of this, a short postwave, in addition to the main wave, is observed on the polarogram.

Phosphine selenides are reduced by the addition of two electrons and form phosphines [29]:

$$R_3PS_e + 2e \rightarrow R_3P + Se^{2-}$$

Phosphorous acid in aqueous solution is reduced to hypophosphorous acid [72]:

$$H_2PO_3^- + 2H^+ + 2e \rightarrow H_2PO_2^- + H_2O$$

and hydrogen phosphide [73].

On a platinum anode the phosphorus of phosphorous acid is oxidized to the pentavalent state [73], probably via dimerization according to the following scheme [68]:

$$H_3PO_3 \rightarrow H_2PO_3 + H(Pd) \text{ (limiting stage)}$$

$$H(Pd) \rightleftharpoons H^+ + e$$

$$2H_2PO_3 + H_2O \rightleftharpoons H_3PO_3 + H_3PO_4$$

It is generally believed that the deactivation of palladium observed in prolonged polarization is caused by the formation of palladium phosphide as a by-product.

Similarly, hypophosphorous acid is also oxidized via catalytic dehydration stage, the intermediate products of oxidation being metaphosphoric acid and the alkyl form of phosphorous acid [68]:

$$H_3PO_2 \rightleftharpoons H_2PO_2 + H(Pd)$$

$$H_2PO_2 \rightleftharpoons HPO_2 + H(Pd)$$

$$HPO_2 + H_2O \rightleftharpoons H_2PO_3 + H(Pd)$$

$$H(Pd) \longrightarrow H^+ + e \text{ (limiting stage)}$$

Phosphorous acid undergoes further oxidation according to the mechanism considered above.

The discharged cations of tertiary salts of phosphonium disintegrate at the mercury cathode via the intermediate formation of onium amalgam, and the cathode process can be represented by [74]:

$$(Hg)_n[PR_4]^+ \rightleftharpoons (Hg)_n[PR_4] \overset{e}{\rightleftharpoons} (Hg)_n[PR_4]^- \overset{H}{\longrightarrow} (Hg)_nHPR_4$$

$$\downarrow$$

$$R{-}R \longleftarrow R\cdot + (Hg)_nPR_3 \overset{e}{\longrightarrow} (Hg)_nPR_3^+ + R \overset{H}{\longrightarrow} (Hg)_n + PR_3 + RH$$

In some cases ylides are formed, and their formation mechanism is obscure. For the formation of ylide from $(C_6H_5)_3PCH_2CN$, it is supposed that the formation of $\dot{C}H_2CN$, which reacts with the ammonium cation, is the first stage of the process [75]:

$$(C_6H_5)_3\overset{+}{P}CH_2CN + \dot{C}H_2CN \longrightarrow (C_6H_5)_3P{=}CHCN + CH_3CN$$

According to another hypothesis, ylide is formed as a result of charge transfer [76]

$$(C_6H_5)_3\overset{+}{P}CH_2R \overset{+e}{\longrightarrow} [(C_6H_5)_3PCH_2R] \longrightarrow [(C_6H_5)_3P{=}CHR] + H$$

The formation of a radical which splits the proton from the solvent is the first stage in the electroreduction of phosphine halides $(Ph_2)PHal$ and $PhPHal_2$ [43]:

$$(Ph_2)PHal + e \longrightarrow Ph_2P^\cdot + Hal^-$$

$$\downarrow \text{(solvent)}$$

$$Ph_2PH$$

Cyclic voltammetric measurements show that the lifetime of the radical $Ph_2P^\cdot$ is less than 10 msec.

4.2. COMPOUNDS OF PENTAVALENT PHOSPHORUS

Orthophosphoric acid can be electrolytically reduced but the yield is very insignificant [73]; however, potentiodynamic measurements show that H_3PO_3 is deposited on the platinum electrode surface polarized to $\phi = 0.05$ V in 5 M H_3PO_4 [77]. Unlike aqueous solutions, H_3PO_4 readily undergoes reduction in nonaqueous media [73, 78]. In order to explain this difference, a hypothesis is advanced to the effect that in solution phosphoric acid is protonized and binds the hydroxyl ions with the formation of molecules carrying penta-coordinated phosphorus:

$$\text{(HO)}_3\text{P=O} \xrightarrow{H^+} \text{(HO)}_3\overset{+}{\text{P}}-\text{OH} \xrightarrow{OH^-} \text{(HO)}_3\text{P(OH)}_2$$
$$\qquad\qquad\qquad\qquad\qquad A \qquad\qquad\qquad\qquad B$$

$$+e \swarrow$$

$$\text{(HO)}_3\dot{\text{P}}-\text{OH} \xrightarrow{H^+} \text{(HO)}_3\text{P}\genfrac{}{}{0pt}{}{H}{OH} \xrightarrow{H_2O} \text{(HO)}_2\text{P}\genfrac{}{}{0pt}{}{H}{O}$$
$$\qquad C \qquad\qquad\qquad\qquad D \qquad\qquad\qquad\qquad E$$

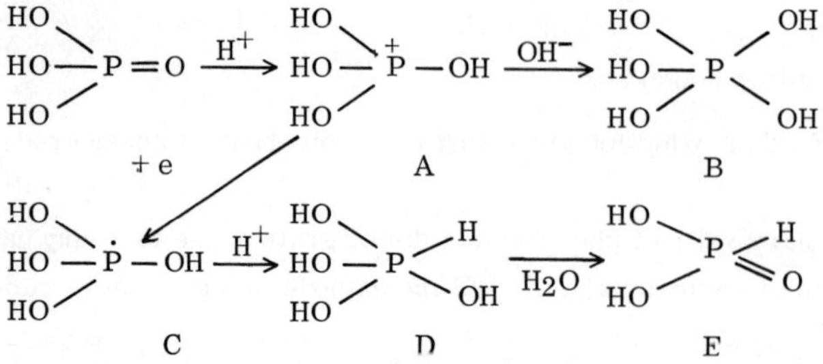

In aqueous solution the hydrated form B binds the electron and disintegrates with the evolution of hydrogen and the initial acid molecules. In nonaqueous medium the electron is transferred to the protonated form A which binds the atomic hydrogen and splits the water molecule with the formation of phosphorous acid.

The peroxomonophosphoric and peroxodiphosphoric acids are reduced according to the following equations [57]:

$$H_3PO_5 + 2H^+ + 2e \rightarrow H_3PO_4 + H_2O$$

$$H_4P_2O_8 + 2H^+ + 2e \rightarrow 2H_3PO_4$$

The ESR method showed that the formation of anion radical is the first stage in the electroreduction of phosphine oxides [46, 79]:

$$(Ar)_3PO + e \rightarrow (Ar)_3PO^{-\bullet}$$

The formation of anion radicals was also detected during the electroreduction of phosphazine [52, 80]. The esters of phosphoric acid are reduced with the cleavage of the carbon-oxygen bond [81, 82]. The reduction of tri(p-nitrophenyl) phosphate is a many-stage process. The first pre-wave and the second main wave correspond to the reduction of the adsorbed and dissolved compound [48]:

$$(NO_2C_6H_4O)_3PO + 2e \rightarrow (NO_2C_6H_4O)_3PO^{2-}$$

The anion so formed is protonated with the splitting of the C−P bond:

$$(NO_2C_6H_4O)_3PO^{2-} + 2H^+ \rightarrow (NO_2C_6H_4O)\underset{OH}{\overset{OH}{\text{PO}}} + NO_2C_6H_4\cdot C_6H_4NO_2$$

Then the products formed undergo further reduction with the formation of the following waves:

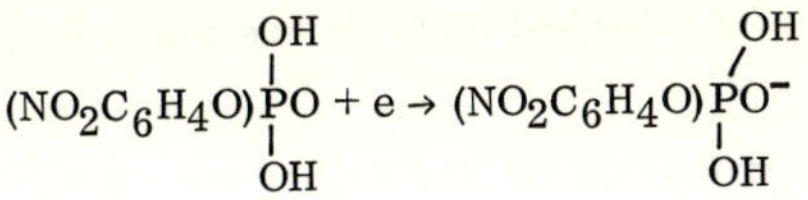

$$(NO_2C_6H_4O)\underset{OH}{\overset{OH}{P}}O + e \rightarrow (NO_2C_6H_4O)\underset{OH}{\overset{OH}{P}}O^-$$

$$NO_2C_6H_4C_6H_4NO_2 + e \rightarrow NO_2C_6H_4C_6H_4NO_2^-$$

Phosphoric acid is the final product of the electroreduction of di(p-nitrophenyl)phosphate [49]:

$$(NO_2C_6H_4O)_2\underset{OH}{P}{=}O + 2e \rightarrow (NO_2C_6H_4O)_2\underset{OH}{P}{=}O^{2-}$$

$$(NO_2C_6H_4O)_2\underset{OH}{P}{=}O^{2-} + H^+ \rightarrow NO_2C_6H_4C_6H_4NO_2 + HO\underset{OH}{\overset{OH}{P}}{=}O$$

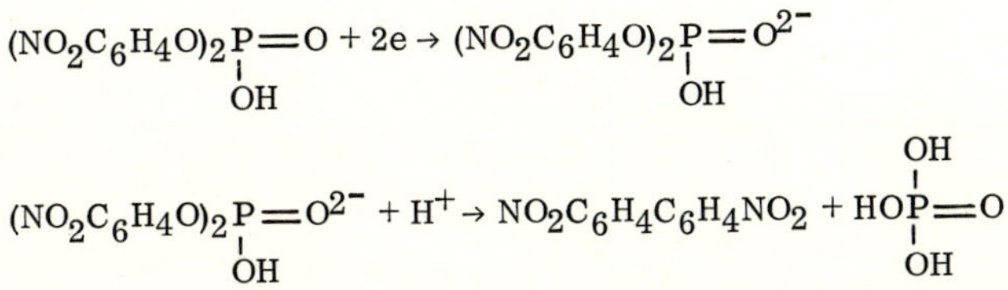

Dialkylaroylphosphonates in acetonitrile are reduced with the splitting of the C—P bond and the formation of hydrobenzoins (Path A):

$$\underset{}{Ar\overset{O}{\overset{\|}{C}}{-}\overset{O}{\overset{\|}{P}}(OR)_2} \xrightarrow{e} Ar\overset{O^-}{\overset{|}{C}}{-}\overset{P}{\overset{\|}{P}}(OR)_2 \rightarrow ArCO + H\overset{O}{\overset{\|}{P}}(OR)_2$$

$$\downarrow 2H^+ \qquad\qquad\qquad\qquad\qquad Ar\overset{O}{\overset{\|}{C}}{-}\overset{O}{\overset{\|}{C}}Ar \xrightarrow{2e} Ar\overset{O}{\overset{\|}{C}}{-}\overset{OH}{\overset{|}{C}}HAr$$

$$Ar\underset{OH}{\overset{}{C}}H{-}P(OR)_2$$

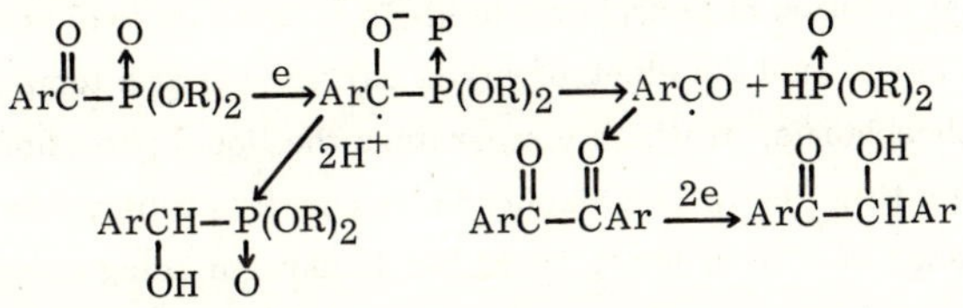

B A

In the presence of a proton donor, two electrons participate in the cathode process, and the reduction product is α-oxyarylmethylphosphonate (Path B) [82].

5. APPLIED ELECTROCHEMISTRY

5.1. ELECTROWINNING AND PURIFICATION

Metal-phosphorus alloys may be formed during the electrodeposition of metals from solutions containing sodium hypophosphates. The most important alloys from the practical viewpoint are Ni—P, Co—P, and Ni—Co—P, which possess valuable magnetic properties. The alloy is formed either as a result of cathode reduction of hypophosphites to phosphines with the subsequent disintegration of phosphonium salts [83], or by the reduction of hypophosphites to free phosphorus [84]:

$$H_2PO_2^- + 2H^+ + e \rightarrow P + 2H_2O$$

The alloy Ni—Co—P containing about 3% P exhibits a maximum coercive force. Such an alloy is generally deposited from a solution containing (g/l) 120 to 140 $NiCl_2 \cdot 6H_2O$, 120 to 140 $CoCl_2 \cdot 6H_2O$, 80 to 100 NH_4Cl, 8 to 10 NaH_2PO_2, at i = 10 to 15 A/cm^2, pH = 3.0 to 6.0, and temperature 40 to 60°C [85]. A large number of chemical methods are available for depositing the magnetic alloys [86].

The electrochemical method for purifying phosphorus consists in charging red or yellow phosphorus and chemically pure nitric acid in a 1:10 ratio into the middle chamber of a three-section electrolyzer in which all electrical energy is consumed in separating the impurities from phosphorus which does not take part in the current transfer during the electrolysis [87]. Phosphorus up to 99.9999% purity can be obtained by this method.

5.2. ELECTROSYNTHESIS

5.2.1. Cathode Processes

Electroreduction of yellow phosphorus is the most promising method for preparing hydrogen phosphide [88]. Maximum yield of hydrogen phosphide is obtained from the aqueous solutions of acids, mostly from a 0.25-M solution of phosphoric acid [70]. Zinc or lead is the metal that is generally used for the cathode, and the electrolysis is carried out at a temperature higher than the melting point of phosphorus, at which temperature the liquid phosphorus melts and covers the cathode with a thin film [88]. The diaphragm is usually prepared from fiber glass, polypropylene, polyurethane, etc. in order to avoid the diaphragm being clogged with phosphorus [89, 90]. It is very necessary to add some dopant say, for instance, hydrogen sulfide, carbon dioxide gas, or prussic acid to the electrolyte for the purpose of reducing the viscosity of phosphorus [91]. Under optimal conditions the yield of hydrogen phosphide may reach 87% with respect to the current, and the content of PH_3 up to 90% in the cathode gas. A fraction of phosphorus is spent on secondary processes, of which the most important is oxidation due to the oxygen entering from the anode chamber, and leading to the formation of phosphoric, phosphorous, and hypophosphorous acids. About 12% of phosphorus is decomposed by oxidation if the diaphragm is made of alundum [92].

The electroreduction of yellow phosphorus can be used for directly synthesizing organophosphorus compounds. Primary and secondary phosphines, as well as phosphine oxides, are produced during the electrolysis of a water suspension of yellow phosphorus in the presence of styrol [93], aldehydes [94], and ketones [95], as well as in the electrolysis of alcoholic suspension of phosphorus in the presence of alkyl halides [96]. However, if the electrolysis is carried out in aprotic media, compounds containing the P—P bond are also produced side by side with phosphines [97] (Table 5.2.1).

Orthophosphoric acid is reduced in nonaqueous media on a lead cathode to phosphorous acid with 50% yield [25]. This yield increases on adding P_2O_5 to the solution because this dopant binds the water formed during the process. The following method is suggested for preparing compounds of trivalent phosphorus from phosphorous acid [98]: a mixture containing $2H_3PO_4$ (2 g), C_2H_5OH (20 ml), and $(C_2H_5O)_3PO$ (2 ml), when electrolyzed on lead electrodes at 70°C, yields 38% of the desired product.

TABLE 5.2.1. Electrosynthesis of Organophosphorus Compounds from Elementary Phosphorus

Initial compound (g)	Composition of catholyte (g), cathode	Reaction products	Yield % per P	Refs.
$C_6H_5CH{=}CH_2$ (40)	CH_3COOH (50); KOH (5); P_4 (5); H_2O (300); Pb cathode	$(C_6H_5CH_2CH_2)PH_2$	26.7	[93]
		$(C_6H_5CH_2CH_2)_3P$	1.7	
		$[(C_6H_5CH_2CH_2)_2P]_2$	4.0	
$C_9H_{19}I$ (81)	CH_3OH (160); P_4 (2.5); graphite cathode	$(C_9H_{19})_3P$	5.1	[96]
C_4H_9I (390)	DMF (1170); P_4 (39.1); graphite cathode	$C_4H_9PH_2$	6.3	[97]
		$(C_4H_9)_2PH$	1.9	
		$(C_4H_9)_3P$	6.35	
		$[(C_6H_9)_2P]_2$, $(C_4H_9P)_4$	9	
HC(=O)H (20)	P_4((10.2) (aq); H_3PO_4	$OP(CH_2OH)_3$	64	[94]
CH_3C(=O)H (27)	P_4 (10.2) (aq); H_3PO_4	$OP(CH_3CHOH)_3$	42	[94]
Cyclo-$C_6H_{10}O$ (38.1)	CH_3COOH (650 ml 50%); hydrochloric acid (60 ml); $(CH_3COO)_2$ (1); Zn cathode	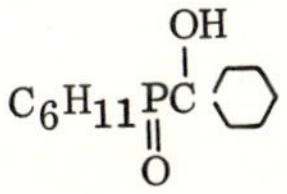	21.3	[95]

The electrolysis of quarternary salts of phosphonium is a convenient method for preparing tertiary phosphines. Different radicals can be arranged in the following increasing sequence with respect to the ease with which the P—C bond in the phosphonium ion can be split up [99-101]: $CH_3 > C_6H_5 > C_2H_5 >$ n-$C_4H_9 >$ i-$C_3H_7 >$ i-$C_4H_9 >$ tert-$C_4H_9 > CH_2CH_2OH > C_6H_5CH_2$. This series is valid for Hg, Ni, Pt, and Cd electrodes.

If the optically active quarternary salts of phosphonium are electrolyzed, then optically active phosphines are formed with the same configuration [102, 103]. If no diaphragm is used, then the phosphines formed during the electrolysis are oxidized to phosphine oxides.

In some cases when the quarternary salts of phosphonium are electrolyzed, the following ylides are formed [75, 76]:

$$(C_6H_5)_3PCH_2R \xrightarrow{+e} [(C_6H_5)_3PCH_2R] \longrightarrow (C_6H_5)_3P{=}CHR + H$$

It is more convenient to use aldehydes or ketones (benzaldehyde, butyraldehyde, cyclohexanol) as a solvent in the preparation of ylides. Table 5.2.2 lists the ylides that can be synthesized by the electrochemical method.

TABLE 5.2.2. Electrochemically Synthesized Ylides

Phosphonium salt	Solvent	Ylide	Yield (%)	Refs.
$(C_6H_5)_3PCH_2CN$		$(C_6H_5)_3P{=}CHCN$		[75]
$(C_6H_5)_3PCH_3I$	Benzaldehyde	$(C_6H_5)_3P{=}CH_2$	84	[76]
$(C_6H_5)_3PC_2H_5I$	Benzaldehyde	$(C_6H_5)_3P{=}CHCH_3$	95	[76]
$(C_6H_5)_3PCH_2C_6H_5Br$	Cyclohexanol	$(C_6H_5)_3P{=}CHC_6H_5$	95	[76]
$(C_6H_5)_3PCH_2COOCH_3Cl$	Benzaldehyde	$(C_6H_5)_3P{=}CHCOOCH_3$	75	[76]

Styrene was dimerized by the radical formed as an intermediate during the electrolysis of quarternery salts of phosphonium of the type $[(C_6H_5)_3P(CH_2)_nCN]Br$ [75]:

$$2(CH_2)_nCN + 2C_6H_5CH{=}CH_2 \longrightarrow \begin{array}{l} CN(CH_2)_{n+1}CHC_6H_5 \\ CN(CH_2)_{n+1}CHC_6H_5 \end{array}$$

A polymer of the oxide of phosphorus [104, 105] was obtained on electrolyzing a solution of triethylamine hydrochloride in nonaqueous phosphorus oxychloride at 0°C on a platinum cathode by the following reactions:

$$POCl_3 \rightarrow POCl_2^+ + Cl^-$$

$$nPOCl_2^+ + 3ne \rightarrow (PO)_n + 2nCl^-$$

Diphenylphosphine was obtained in qualitative amounts by the electroreduction of diphenylchlorophosphine on a mercury electrode at a potential of -3.4 V. Phosphazine $(C_6H_5)_3P{=}NC_6H_5$ gives rise to $(C_6H_5)_3P$ with 96% yield on a mercury cathode in hydrochloric solution [106]. In methanol solution on a mercury cathode, phenyldibenzilphosphine oxide is reduced to cyclohexadienyldibenzilphosphine oxide [107].

5.2.3. Anode Processes

All acids containing phosphorus in low oxidation degrees (less than P^{5-}) undergo anodic oxidation [73]. The phosphites are oxidized to phosphates in alkaline media on Pt and graphite anodes with a yield of about 30 to 40%, the optimal conditions being 1.0 to 1.5 M NaOH, i = 0.04 to 0.06 A/cm^2 [108, 109]. The rate of oxidation of phosphites significantly increases on adding KI or HCl to the solution [78, 109a].

Phosphoric acid is oxidized to the pyrophosphoric form on a PbO_2 anode. The design of the electrolytic chamber for carrying out this process is given in Ref. 110.

Phosphates give rise to perphosphates up to 53% yield [111, 112] on electrolyzing a solution containing (g/l) NaH_2PO_4 (545), KOH (390), KF (53 to 116), and K_2CrO_4 (0.0035) on a Pt anode at a temperature of 10 to 25°C, the current being in the range 0.65 to 1.3 A/cm^2. An essential condition is to maintain the pH value within the range from 13.0 to 13.5.

Chlorides of phosphorus — PCl_3, PCl_5, and $POCl_3$ — are formed at the anode chamber on electrolyzing fused calcium chloride in an electrolyzer and by introducing calcium phosphate and carbon in the anode chamber [113].

On oxidizing the aromatic phosphines in a solution of lithium perchlorate in acetonitrile on mercury cathode, the corresponding mercury salts are formed [53]:

$$2(C_6H_5)_3P \xrightarrow{-2e} 2(C_6H_5)_3P\cdot \xrightarrow{Hg} [(C_6H_5)_3P]_2\ Hg^{2+}$$

High yields of trialkylphosphates are obtained by electrolyzing a suspension of red phosphorus in aliphatic alcohols saturated with HCl [114].

Tertiary phosphines with saturated aliphatic groups are prepared by electrolyzing ester solutions of Grignard reagent with black phosphorus at the anode [115].

Anodic oxidation of methylphosphonic acid is suggested for decontaminating waste waters [116].

5.3. INHIBITION OF CORROSION

Phosphoric acid, when it comes in contact with metals, forms a chemically stable layer of phosphates; therefore it is used either in pure form or in a mixture as a corrosion inhibitor to protect the carbon- and chrome-nickel steels [117].

5.4. USE IN FUEL CELLS

When PCl_3 reacts with Cl_2 to form PCl_5 at 25°C, an emf of 0.27 V is developed in the open circuit; PCl_3 and Cl_2 can be regenerated thermally by decomposing PCl_5 at 300°C or by a photochemical method, and thus returned to the fuel cycle [118]. An active catalyst composed of a fine Ni-P alloy powder has been patented for the fuel elements operating on hydrazine [119].

REFERENCES

1. W. M. Latimer, The Oxidation States of the Elements and Their Potentials in Aqueous Solutions, Prentice-Hall, New York, 1952.
2. M. Pourbaix, Atlas of Electrochemical Equilibria in Aqueous Solutions, Oxford Univ. Press, Oxford, 1966, p. 516.
3. A. J. de Bethune and N. S. Loud, in Encyclopedia of Electrochemistry, Vol. 5 (C. A. Hampel, ed.), Reinhold, New York, 1964, p. 415.

4. D. D. Wagman, W. H. Evans, V. B. Parker, I. Halow, S. M. Baily, and R. H. Schumm, Selected Values of Chemical Thermodynamic Properties, Technical Note 270-3, Issued January 1968, Washington, D.C.

5. M. H. Karapetyans and M. L. Karapetyans, Fundamental Thermodynamic Constants of Inorganic and Organic Substances, Moscow, Khimiya, 1968.

6. A. J. de Bethune, T. S. Licht, and N. Swendeman, J. Electrochem. Soc., 108, 616 (1961).

7. N. E. Chomutov, Itogi Nauki: Elektrochemia, Moscow, 1, 1 (1965).

8. G. Milazzo, In Preparation, To Be Published as an Addendum to the Encyclopedia of the Electrochemistry of the Elements.

9. I. M. Kolthoff, Rec. Trav. Chim. Pays-Bas, 46, 350 (1927).

10. K. S. Pitzer, J. Amer. Chem. Soc., 59, 2365 (1937).

11. A. K. Grzybowsky, J. Phys. Chem., 62, 555 (1958).

12. N. Konopik and O. Leberl, Monatsh. Chem., 80, 655 (1949).

13. C. E. Vanderzee and A. S. Quist, J. Phys. Chem., 65, 118 (1961).

14. G. Charlot, Theory et methode nouvelles d'analyse qualitative, 3rd ed., Masson, Paris, 1949.

15. Gmelin Handbuch der anorganischen Chemie Phosphor, 8th ed., Vol. 16, Verlag Chemie, GMBH, Weinheim.

16. O. B. Khutchaturyan and A. P. Kruwtchinsky, Chimija Perekisnih Soedinenii, Moscow, 1963, p. 162.

17. P. R. Moltchanowa and W. I. Dulowa, J. Phys. Chem. (Russia), 64, 1542 (1970).

18. E. Franks and S. Inman, J. Electroanal. Chem., 26, 13, 1970.

19. R. D. Caton and H. Frend, Anal. Chem., 35, 2103 (1963).

20. W. N. Andreeva, Ukr. Chim. J., 21, 569 (1955).

21. A. P. Tomilov and I. M. Osadchenko, Zh. Anal. Khim., 21, 1498 (1966).

22. A. P. Tomilov, I. N. Brago, and I. M. Osadchenko, Elektrokhimiya, 4, 1153 (1968).

23. I. N. Brago and A. P. Tomilov, ibid., 4, 697 (1968).

24. L. F. Filimonova and A. P. Tomilov, Zh. Vses. Khim. Obshchest., 15, 352 (1970).

25. M. Baudler and A. Bougardt, Z. Anorg. Allg. Chem., 350, 186 (1967).

26. M. Baudler and A. Bougardt, ibid., 350, 202 (1967).

27. Ja. A. Levin, Iu. M. Kargin, V. S. Galeev, and V. V. Sannikov, Izv. Akad. Nauk SSSR, Ser. Khim., 1968, 411.

28. K. S. V. Santhanann and A. Bard, J. Amer. Chem. Soc., 90, 1118 (1968).

29. H. Matschiner, A. Tschach, and A. Steinert, Z. Anorg. Allg. Chem., 373, 237 (1970).

30. P. Giang and R. Coswell, Agr. Food. Chem., 5, 753 (1957).

31. M. K. Saikina, Uch. Zap. Kazansk. Gos. Univ., 116, 121 (1956).

32. K. Issleib, H. Matschiner, and M. Hoppe, Z. Anorg. Allg. Chem., 351, 251 (1967).

33. K. Issleib and H. Metscher, ibid., 340, 34 (1965).

34. D. Shamsel and T. Sabar, Electrochim. Acta, 15, 711 (1970).

35. V. Vojiř, Collect. Czech. Chem. Commun., 26, 289 (1961).

36. K. Issleib, H. Matshiner, and S. Naumann, J. Electroanal. Chem., 16, 563 (1968).

37. A. N. Razumov, G. A. Savicheva, and G. K. Budnikov, Zh. Obshch. Khim., 35, 1454 (1965).

38. G. A. Savicheva, M. B. Gazizov, A. V. Ilyasov, and A. N. Razumov, ibid., 37, 2785 (1962).

39. J. Saveant and H. Veilard-Royer, Bull. Soc. Chim. Fr., 1967, 2415.

40. G. Kosolapoff and Ch. Jenking, J. Chem. Soc., 1957, 3430.

41. C. Bowan and F. Edwards, Anal. Chem., 22, 706 (1950).

42. Hon Thon Guk and Kin Thun Fu, ibid., I, 7 (1962).

43. R. Dessy and W. Kitching, J. Amer. Chem. Soc., 88, 467 (1966).

44. L. Horner, I. Ertel, H-D. Reprecht, and O. Berlovsky, Chem. Ber., 103, 1582 (1970).

45. V. G. Mairanovskii, L. P. Fokina, L. A. Vakulova, and G. P. Samokhvalov, Zh. Obshch. Khim., 36, 1345 (1966).

46. A. V. Ilysaov, Ju. M. Kargin, Ja. A. Levin, I. D. Morozova, B. V. Melnikov, A. A. Vafina, H. H. Sotnikova, and V. S. Galeev, Izv. Akad. Nauk SSSR, Ser. Khim., 1971, 770.

47. H. Matschiner, F. Krech, and A. Steinert, Z. Anorg. Allg. Chem., 371, 256 (1969).

48. K. S. V. Santhanam, L. Wheeber, and A. J. Bard, J. Amer. Chem. Soc., 89, 3386 (1967).

49. K. S. V. Santhanam and A. J. Bard, J. Electroanal. Chem., 25, App. 9 (1970).

50. A. I. Razumov, G. A. Savicheva, and G. K. Budnikov, Dokl. Akad. Nauk SSSR, 158, 423 (1964).

51. E. S. Kosmatii and M. T. Tretjak, Elektrokhimiya, 5, 965 (1969).

52. V. V. Penkovskii, Iu. P. Egorov, and G. S. Chapoval, Zh. Obshch. Khim., 41, 738 (1971).

53. L. Horner and J. Haufe, Chem. Ber., 101, 2921 (1968).

54. L. Horner and H. Nickel, ibid., 89, 1681 (1956).

55. M. Barry and Ch. Tobias, Electrochem. Technol., 4, 502 (1966).

56. H. Matschiner, R. Krause, and F. Krech, Z. Anorg. Allg. Chem., 373, 1 (1970).

57. M. Venturini, A. Indelli, and G. Caspi, J. Electroanal. Chem., 33, N 1, 99 (1971).

58. V. Vetterl and J. Bahaček, ibid., 16, 313 (1968).

59. H. Sohr, Chem. Zvesti, 16, 316 (1962).

60. H. Gehrig, Z. Phys. Chem., 225, 116 (1964).

61. B. I. Ribakov, Zavod. Lab., 33, 946 (1967).

62. P. Naugniot, Anal. Chim. Acta, 31, 166 (1964).

63. H. Sohr, J. Electroanal. Chem., 11, 188 (1966).

64. H. Sohr and Kh. Lohs, ibid., 13, 107, 114 (1967).

65. H. Sohr and Kh. Lohs, ibid., 14, 227 (1967).

66. S. Mayer and W. Brown, J. Electrochem. Soc., 110, 306 (1963).

67. S. Trasatti and A. Alberti, Ric. Sci., Parte Z, Ser. A, 8, 1012 (1965).

68. S. Trasatti and A. Alberti, J. Electroanal. Chem., 12, 236 (1966).

69. A. Hickling and D. Johnson, ibid., 13, 101 (1967).

70. I. M. Osadchenko, Zh. Prikl. Khim., 43, 1255 (1970).

71. N. Ja. Chandrinov and A. P. Tomilov, Elektrokhimiya, 4, 237 (1968).

72. French Patent 1,130,548 (1957).

73. M. Baudler and D. Shellenberg, Z. Anorg. Allg. Chem., 340, 113 (1965).
74. L. Horner, F. Rottger, and H. Fuchs, Chem. Ber., 96, 3141 (1963).
75. J. Wagenecht and M. Baizer, J. Org. Chem., 31, 3885 (1966).
76. T. Shono and M. Mitani, J. Amer. Chem. Soc., 90, 2728 (1968).
77. J. Bravacos, M. Bonnemay, E. Zevart, and A. Pilla, C. R. Acad. Sci., Paris, Ser. C, 265, 313 (1968).
78. M. Baudler and D. Sihellnberg, Z. Anorg. Allg. Chem., 356, 140 (1968).
79. A. V. Ilyasov, Ju. M. Kargin, Ja. A. Levin, B. V. Melnikov, and V. S. Galeev, Izv. Akad. Nauk SSSR, Ser. Khim., 1968, 2841.
80. V. V. Penkovskii, B. N. Kuzminskii, Ju. P. Egorov, I. N. Gmureva, and A. P. Martiniyk, Teor. Eksp. Khim., 6, 94 (1970).
81. E. Meter and J. Chambers, J. Electroanal. Chem., 25, 435 (1970).
82. K. Berlin, D. Pulison, and P. Arthur, Anal. Chem., 41, 1554 (1969).
83. N. P. Fedotiev and P. M. Viactheslavov, Tr. Leningr. Tekhnol. Inst. im. Lensoveta, 53, 30 (1959).
84. K. M. Gorbunova and A. A. Nikiforova, Zashchita Metall., 5, 195 (1969).
85. B. Ja. Kaznacthei and V. M. Gogina, Tr. Inst. Zvukozapisi, 1, 91 (1957); 6, 119 (1959).
86. V. V. Bondar, V. V. Melnikova, and Ju. M. Polukarov, Itogi Nauki: Elektrochimya 1964. VINITI, 1966, 114.
87. N. D. Talanov and S. L. Livchitz, USSR Patent 172,273 (1965).
88. I. M. Osadchenko and A. P. Tomilov, Usp. Khim., 38, 1089 (1969).
89. U.S. Patent 3,312,610 (1967).
90. U.S. Patent 3,404,076 (1968).
91. U.S. Patent 3,109,785-791; 793-795 (1963).
92. I. M. Osadchenko and A. P. Tomilov, Zh. Prikl. Khim., 42, 1404 (1969).
93. L. V. Kaabak, N. Ja. Chandrinov, and A. P. Tomilov, Zh. Obshch. Khim., 40, 548 (1970).
94. I. M. Osadchenko and A. P. Tomilov, ibid., 40, 698 (1970).
95. I. M. Osadchenko and A. P. Tomilov, ibid., 39, 469 (1969).
96. L. F. Filimonova, L. V. Kaabak, and A. P. Tomilov, ibid., 39, 2174 (1969).
97. L. V. Kaabak, M. I. Kabacthnik, A. P. Tomilov, and S. L. Varchavskii, ibid., 36, 2060 (1966).
98. British Patent 1,115,548 (1968); Chem. Abstr., 69, 40750a (1968).
99. L. Horner and A. Mentrup, Justus Liebigs Ann. Chem., 646, 65 (1961).
100. L. Horner and A. Mentrup, German Patent 1,114,190 (1962).
101. M. Finkelstein, J. Org. Chem., 27, 4076 (1962).
102. L. Horner, H. Fuchs, H. Winkler, and A. Rapp, Tetrahedron Lett., 1963, 965.
103. L. Horner, ibid., 1961, 161.
104. H. Spandau and A. Beyer, Naturwissenschaften, 46, 400 (1959).
105. H. Spandau, A. Beyer, and F. Prengschaft, Z. Anorg. Allg. Chem., 306, 14 (1960).
106. H. Bestmann and F. Seng, Tetrahedron, 21, 1373 (1965).
107. L. Horner and H. Neuman, Chem. Ber., 102, 3953 (1969).
108. A. M. Lunetckas, USSR Patent 181,063 (1965).

109. A. M. Lunetckas, Tr. Akad. Nauk. Lit. SSR, Ser. B, 1965, 3.

109a. I. M. Osadchenko, A. P. Tomilov, and V. V. Rublev, Electrochimiya, 9, 1492 (1973).

110. M. Kuva and T. Sadzi, Denki Kagaku, 35, 343 (1967).

111. N. E. Khomutov, O. B. Khatchjaturyan and T. P. Kotova, USSR Patent 247,258 (1967).

112. A. V. Ianuch and A. K. Gorbatchev, USSR Patent 231,533 (1966).

113. Czechoslovakian Patent 131,293 (1963).

114. A. P. Tomilov, Iu. D. Smirnov, and S. L. Varchavskii, Zh. Vses. Khim. Obshchest., 7, 598 (1962).

115. U. S. Patent 3,079,311 (1963).

116. I. M. Osadchenko, A. P. Tomilov, N. Ch. Fuks, and E. G. Bondarenko, Zh. Obshch. Khim., 39, 932 (1969).

117. O. J. Bergman, Inhibitor of Corrosion, Moscow, 1966.

118. J. N. Pitts, J. D. Margerum, and W. E. McKee, J. Amer. Rocket Soc., 31, 890 (1961).

119. U. S. Patent 3,411,953 (1969).

Chapter III-2

COBALT

NOBUFUMI MAKI

Department of Chemistry
Shizuoka University
Hamamatsu City, Japan

and

NOBUYUKI TANAKA

Department of Chemistry
Tohoku University
Sendai, Japan

1. STANDARD AND FORMAL POTENTIALS

1.1. AQUEOUS SOLUTION

On the basis of potentiometric titration data obtained from cells, Bjerrum [1] computed a standard potential of -0.259 V at 30°C in 1 M KCl for the $[Co\ en_3]^{2+}/[Co\ en_3]^{3+}$ couple. Buckingham and Sargeson [2], however, suggested that the standard potential of the $[Co\ en_3]^{2+}/$

$[Co\ en_3]^{3+}$ couple may be in error on the basis of their observation that silver wool in the presence of chloride ion reduces the $[Co\ en_3]^{3+}$ to the $[Co\ en_3]^{2+}$ ion. That Bjerrum's value (-0.259 V at 30°C) may not be reliable was also suggested by Rock [3] based on an estimate of the standard potential obtained from a semitheoretical procedure involving the use of ligand field theory parameters. After then, Kim and Rock [4] reported that the standard oxidation potential of the $[Co\ en_3]^{2+}/[Co\ en_3]^{3+}$ couple is -0.180 ± 0.002 V at 25°C, using cells without liquid junction. The following two cells were used:

$$Au(s)\,|\,[Co\ en_3]Cl_3(l),\ [Co\ en_3]Cl_2(l),\ NaCl(l),\ en(l)\,|\,glass \qquad (1.1.1)$$

$$Hg(l)\,|\,Hg_2Cl_2(s)\,|\,NaCl(l)\,|\,glass \qquad (1.1.2)$$

In the net cell reaction of the hypothetical double cell combined with the above two cells, all of the properties of the particular glass electrode used cancel out, and in addition the double cell has no liquid junctions. They determined dE°/dT = +1.39 ± 0.09 mV/deg. From this observed value of dE°/dT together with other available thermodynamic data for the half-reaction, $[Co\ en_3]^{3+} + e \rightarrow [Co\ en_3]^{2+}$ [5], the values of ΔS°, ΔG°, and ΔH° for the $[Co\ en_3]^{2+}/[Co\ en_3]^{3+}$ (aq) couple were estimated to be +40 eu, +4.1 kcal, and +11.7 kcal, respectively.

On the other hand, it has been well known that the hexaquocobalt(III) ion, $[Co^{III}(OH_2)_6]^{3+}$, is one of the most oxidizing agents in aqueous media, and that the standard potential of the $[Co(OH_2)_6]^{2+}/[Co(OH_2)_6]^{3+}$ couple is about +1.8 V [6]. Rotinyan, Kheifets, and Nikolaeva [7], however, have criticized the measurements on which this value was based, and Kheifets et al. [8] have asserted, based on studies of the equilibrium between cobalt(III) hydroxide and a solution containing aquocobalt(III) and -cobalt(II) sulfates at pH 2.2 to 2.6 in the presence of air that the true standard potential is as low as +1.3 V, assuming that $[Co(OH_2)_6]^{3+}(aq)$ is the only cobalt(III) species present in the solution. Johnson and Sharpe [9] also cast suspicion on the validity of this assumption; it is worth noting there is strong evidence for the formation of $[Co(OH)(OH_2)_5]^{2+}$ and binuclear species in dilute solutions of cobalt(III) in weakly acidic perchlorate media [10, 11], and complexing by sulfate ion also occurs [12]. In order to determine the formal oxidation potential of the aquo couple, Johnson and Sharpe [9] have chosen the following redox system with 4 M perchloric acid at 15°C as a working medium:

$$[Co(OH_2)_6]^{3+} + [Fe(OH_2)_6]^{2+} \rightarrow [Co(OH_2)_6]^{2+} + [Fe(OH_2)_6]^{3+}$$

In this medium, the heat of the reaction was found to be ΔH = -26.3 kcal. The free energy change derived from the changes in heat content and magnetic entropy for this system leads to the difference between the standard potentials of the $[Fe(OH_2)_6]^{2+}/[Fe(OH_2)_6]^{3+}$ and $[Co(OH_2)_6]^{2+}/[Co(OH_2)_6]^{3+}$ systems (+1.18 V at 15°C). From this value, combining with that of +0.745 V for the formal $[Fe(OH_2)_6]^{2+}/[Fe(OH_2)_6]^{3+}$ potential in 0.5 M perchloric acid at 15°C, the formal $[Co(OH_2)_6]^{2+}/[Co(OH_2)_6]^{3+}$ oxidation potential is obtained as +1.93 V at 15°C. Comparison with the Fe^{2+}/Fe^{3+} system suggests that the true standard potential at 25°C would be about +0.02 V lower. That is, the standard potential of the $[Co(OH_2)_6]^{2+}/[Co(OH_2)_6]^{3+}$ couple is estimated as E° = +1.95 ± 0.1 V at 25°C. Independent of this work, Huchital, Sutin, and Warnqvist [13] have determined directly the formal potential of the $[Co(OH_2)_6]^{2+}/[Co(OH_2)_6]^{3+}$ couple to be +1.92 ± 0.02 V in 4 M $HClO_4$ at 25°C. The equilibrium constant for the reaction cited below can also be calculated from the formal potentials of Ag(I)/Ag(II) and Co(II)/Co(III) couples:

$$Co(III) + Ag(I) \rightleftharpoons Co(II) + Ag(II)$$

The formal potentials of these two couples are +2.00 V in 4 M $HClO_4$ [14] and +1.84 V in 4 M HNO_3 [15], respectively, at 25°C. From these values, they calculated that K = (4 ± 2) x 10^{-2}, in satisfactory agreement with the equilibrium constant obtained from the kinetic measurements.

As mentioned previously, there is still some uncertainty concerning the problem of whether or not the predominant aquo-cobalt(III) species in perchloric acid solutions is the dimeric and/or polymeric complex [16]. Recently, however, Warnqvist [17] concluded from the potentiometric study of the aquo couple that the cobalt(III) is mainly monomeric, viz., $[Co(OH_2)_6]^{3+}$, under the experimental conditions where a 3-M $NaClO_4$ solution contains 0.50 M $HClO_4$ and 0.5 x 10^{-3} to 5 x 10^{-3} M Co(III) ion at 23 and 3°C, respectively. That is, if Co(III) exists in solution in the form of dimeric ions, the "Nernst" slope would be half of that expected for monomeric cobalt(III) at a constant cobalt(II) concentration. As a result, the slope obtained agreed with that predicted from the monomeric cobalt(III) species. The cell used for the emf measurements was

ref. electrode || 0.50 M $HClO_4$, 3 M $NaClO_4$(aq) | 0.50 M $HClO_4$, 3 M $NaClO_4$, $[Co(OH_2)_6]^{3+}$ $[Co(OH_2)_6]^{2+}$ (Ag^+) | Au(s)

The reference electrode was a calomel electrode in which the KCl had been replaced with 4.0 M NaCl. As a potential mediator, a few drops of $AgClO_4$ solution were used in order to obtain the electrode equilibrium quickly. Table 1.1.1 shows the upper limit of the cobalt(III) dimerization quotient, K_D, and the formal potentials of the $[Co(OH_2)_6]^{2+}/[Co(OH_2)_6]^{3+}$ couple in the 3 M $NaClO_4$ solution containing h M $HClO_4$ on the hydrogen scale.

TABLE 1.1.1. Upper Limit Estimates of Cobalt(III) Dimerization Quotient, K_D, and Standard Potentials for $[Co(OH_2)_6]^{2+}/[Co(OH_2)_6]^{3+}$ in 3 M $NaClO_4$[a]

Temp (°C)	h (M)	K_D (approx upper limit), (M^{-1})	E° (V)[b]
23	3	20	+1.86
23	0.5	500	+1.85
3	3	10	+1.83
3	0.5	20	+1.83
3	0.1	20	+1.82
3	0.05	10	+1.83

[a] $[Co(III)\text{-dimer}]/[Co(III)\text{-monomer}]^2 = K_D$; brackets are used to denote concentrations.

[b] Error estimated at ±(0.01 - 0.02) V.

In general, the observed potential E° of the Co(II)/Co(III) couple varies considerably with the kind of ligands. Rock [3] has attempted to account for this observed shift in E° values with ligands by using ligand field theory. Rock [3] pointed out that two effects of entropy and ligand field stabilization energies operate mainly on the E° values of the Co(II)/Co(III) couple. The entropy effect is most important for complexes of different valence types and/or widely different between the E° values of $[Co(NH_3)_6]^{2+}/[Co(NH_3)_6]^{3+}$ and $[Co\ en_3]^{2+}/[Co\ en_3]^{3+}$ couples and the $[Co(OH_2)_6]^{2+}/[Co(OH_2)_6]^{3+}$ couple. This can be ascribed primarily to a difference of ligand field stabilization energies (see Table 1.1.2). Above all, when π-bonding is important between the cobalt and ligands, the effect of the latter is maximum for a given ligand when the cobalt ion has the configuration t_{2g}^6. Since for cobalt this configuration occurs in the Co(III) state whereas for iron it occurs in the Fe(II) state, a π-bonding ligand like phen (1,10-phenanthroline) shifts the Co(II)/Co(III) potential to a lower value, and conversely the Fe(II)/Fe(III) potential to a higher value relative to the aquo couple. Data for the Co(II)/Co(III) couples as collected by Rock [3] are given in Table 1.1.2. The ΔH°_{LF} enthalpies were calculated from the data using the expressions given in the table for LFSE (ligand field stabilization energies).

TABLE 1.1.2. Thermodynamic and Optical Data for Co(II)-Co(III) Systems at 25°C[a]

Couple	Δ (kK)[b]	E° (V)	Spin type	ΔH°_{LF} (kcal)[c]	ΔS° (gibbs)[d]
$[Co\ en_3]^{2+}/[Co\ en_3]^{3+}$	11.0, 23.2	-0.26	h, l	+41.2	(+39)
$[Co(NH_3)_6]^{2+}/[Co(NH_3)_6]^{3+}$	10.1, 22.9	+0.10	h, l	+41.0	+39
$[Co\ phen_3]^{2+}/[Co\ phen_3]^{3+}$	-	+0.42	h, l	-	-5
$[Co\ ox_3]^{4-}/[Co\ ox_3]^{3-}$	9.1, 18.0	+0.57	h, l	+9.2	(-39)
$[Co\ edta]^{2-}/[Co\ edta]^{-}$	10.2, 19.7	+0.60	h, l	+18.9	-9
$[Co(OH_2)_6]^{2+}/[Co(OH_2)_6]^{3+}$	9.3, 18.2	+1.81	h, l	+10.3	+44

[a]Co(II): $t_{2g}^5e_g^2$(h); $t_{2g}^6e_g^1$(l). LFSE: -0.70Δ; ~-1.80Δ + P. Co(III): $t_{2g}^4e_g^2$(h); t_{2g}^6(l). LFSE: -0.40Δ; -2.40Δ + 2P.

[b]Optical data taken from C. K. Jørgensen, Absorption Spectra and Chemical Bonding in Complexes, Pergamon, London, 1962, Chap. 15. Several Δ values given were estimated from data therein.

[c]Pairing energies for low-spin Co(II) and Co(III) complexes were taken as 51.5 and 48.0 kcal, respectively. Crystal field stabilization for Co(II) high-spin complexes was taken as -0.70Δ.

[d]Values in parentheses were estimated by Rock.

The potential of the mercury-mercury(I) hexacyanocobaltate(III) electrode was also determined by Rock [18] in the cell

$$Hg(l)\,|Hg_2Cl_2(s)\,|KCl(aq)\,|\underset{0.01\%}{K(Hg)}\,|K_3[Co(CN)_6](aq)\,|(Hg_2)_3[Co(CN)_6]_2(s)\,|Hg(l)$$

From the observed cell voltages and the mean molal activity coefficients of $K_3[Co(CN)_6]$ and KCl, the standard potential of the $Hg(l)/Hg_6[Co(CN)_6]_2(s)$ was calculated to be

$$(Hg_2)_3[Co(CN)_6]_2(s) + 6e = 6Hg(l) + 2[Co(CN)_6]^{3-}(aq,\ a = 1) \qquad E^\circ = +0.427 \pm 0.001\ V$$

In the calculation of this E°, the solid phase was assumed to be the anhydrous salt, although it was probably the tetrahydrate.

Similarly, the standard potential of the $Ag(s)/Ag_3[Co(CN)_6](s)$ electrode was computed to be

$$Ag_3[Co(CN)_6](s) + 3e = 3Ag(s) + [Co(CN)_6]^{3-}(aq,\ a = 1) \qquad E^\circ = +0.298 \pm 0.001\ V$$

Furthermore, from the G° values cited below, Rock [18] computed the solubility products of $Ag_3[Co(CN)_6](s)$ and $(Hg_2)_3[Co(CN)_6]_2(s)$ to be 3.9×10^{-26} and 1.9×10^{-37}, respectively:

$$3Ag^+ + [Co(CN)_6]^{3-} = Ag_3[Co(CN)_6](s) \qquad \Delta G^\circ = -34.66\ kcal$$

$$3Hg_2^{2+} + 2[Co(CN)_6]^{3-} = (Hg_2)_3[Co(CN)_6]_2(s) \qquad \Delta G^\circ = -50.09\ kcal$$

These E° values of Ag/Ag(I) and Hg/Hg(I) couples were added here only in connection with the hexacyanocobaltate(III) ion.

The standard potential of the $[Co(NH_3)_6]^{2+}/[Co(NH_3)_6]^{3+}$ couple was calculated to be $\sim$+0.06 V from the free energies of formation (-55.2 kcal) [5]. It is worthwhile mentioning here that the hexamminecobalt(II) ion is not present in solution to an appreciable extent. Calculations using Bjerrum's stability constant data for the $[Co(NH_3)_6]^{2+}$ ion reveal that even at $NH_3(aq)$ concentrations as high as 6 <u>M</u>, $[Co(NH_3)_6]^{2+}(aq)$ comprises less than 50% of the cobalt(II)-aquoammine species present in an equilibrium state [4]. The change of the electronic configuration from the low-spin to the high-spin state accompanies the redox electrode process, and a slow and irreversible process could always be expected for the $[Co(NH_3)_6]^{3+}$-$[Co(NH_3)_6]^{2+}$ system in aqueous media.

By contrast, since the outer-sphere redox reactions of Co(II)/Co(III) systems with π-bonding ligands do not involve bond cleavage during electron transfer, it is possible to study the kinetics of oxidation of a common reductant, $[Co\ terpy_2]^{2+}$, by a series of cobalt(III) oxidants, $[Co\ dip_3]^{3+}$, $[Co\ phen_3]^{3+}$, and $[Co\ tmp_3]^{3+}$, or of reduction of a common oxidant, $[Co\ py_4Cl_2]^+$, by a series of cobalt(II) reductants, $[Co\ dip_3]^{2+}$, $[Co\ phen_3]^{2+}$, and $[Co\ terpy_2]^{2+}$ (tmp = 3, 5, 6, 8-tetramethyl-1, 10-phenanthroline; dip = 2, 2'-dipyridyl; phen = 1, 10-phenanthroline; terpy = 2, 2'2"-terpyridyl). The effect of free energy drive, the relative reactivity of aquo and hydroxo species, and electronic configuration, especially low- and high-spin cobalt(II), on the rates of these outer-sphere redox reactions has been systematically examined by Farina and Wilkins [19]. The standard potentials of several Co(II)/Co(III) couples have been determined and compared with each other (see Tables 1.1.3 and 1.1.4).

TABLE 1.1.3. The Standard Potentials of the Co(II)/Co(III) Systems with π-Bonding Ligands at 0°C

Half-reaction	E° (V)	Refs.
$[Co\ terpy_2]^{3+} + e = [Co\ terpy_2]^{2+}$	+0.31	[19]
$[Co\ dip_3]^{3+} + e = [Co\ dip_3]^{2+}$	+0.34	[19]
$[Co\ phen_3]^{3+} + e = [Co\ phen_3]^{2+}$	+0.40	[19]
$[Co(OH_2)_2phen_2]^{3+} + e = [Co(OH_2)_2phen_2]^{2+}$	+0.68	[19]
$[Co(OH_2)_4dip]^{3+} + e = [Co(OH_2)_4dip]^{2+}$	+0.84	[19]
$[Co(OH_2)_4phen]^{3+} + e = [Co(OH_2)_4phen]^{2+}$	+0.84	[19]
$[Co(OH_2)_6]^{3+} + e = [Co(OH_2)_6]^{2+}$	+1.92 ± 0.02 (25°C)	[13]

TABLE 1.1.4. The Standard Potentials of the Co(II)/Co(III) Systems with σ-Bonding Ligands at 25°C[a]

Half-reaction	E° (V)	Refs.
$[Co\ ox_3]^{3-} + e = [Co\ ox_3]^{4-}$	+0.57 ± 0.02 (μ = 1.0 M)	[20]
$[Co\ ox_2(OH_2)_2]^{-} + e = [Co\ ox_2(OH_2)_2]^{2-}$	+0.78 ± 0.04 (μ = 1.0 M)	[20]
α-$[Co\ gly_3] + e = \alpha$-$[Co\ gly_3]^{-}$	+0.20 (μ = 1.0 M)	[20]
$[Co\ en_3]^{3+} + e = [Co\ en_3]^{2+}$	-0.18	[4]
$[Co\ edta]^{-} + e = [Co\ edta]^{2-}$	+0.38 ± 0.01	[20]
$[Co(NH_3)_6]^{3+} + e = [Co(NH_3)_6]^{2+}$	+0.06	[5]
$[Co(OH_2)(NH_3)_5]^{3+} + e = [Co(OH_2)(NH_3)_5]^{2+}$	+0.37	[21]
$[Co(OH_2)_6]^{3+} + e = [Co(OH_2)_6]^{2+}$	+1.92	[13]

[a] ox = oxalate; gly = glycine.

TABLE 1.1.5. Standard or Formal Oxidation Potentials of Co(II)/Co(III), Co(c)/Co(II), and Co(0)/Co(I) Couples[a]

Half-reaction	Standard or formal potential (V)	Conditions	Refs.
$[Co(NH_3)_6]^{3+} + e = [Co(NH_3)_6]^{2+}$	+0.06	25°C	[1]
	+0.108	25°C	[27]
	+0.10	25°C	[3], [5]
$[Co(OH_2)(NH_3)_4]^{3+} + e = [Co(OH_2)(NH_3)_4]^{2+}$	+0.33	25°C, 1 M NH_4ClO_4	[21]
	+0.37	25°C, 1 M NH_4NO_3	[21]
$[Co(OH_2)_6]^{3+} + e = [Co(OH_2)_6]^{2+}$	+1.817	25°C, 2 M H_2SO_4	[28]
	+1.774	0°C, 2 M H_2SO_4	[28]
	+1.795	16°C, 2 M H_2SO_4	[28]
	+1.808		[27], [29]
	+1.816	0°C, 4 M HNO_3	[15]
	+1.850	25°C, 4 M HNO_3	[15]
	+1.808	0°C, 3 M HNO_3	[15]
	+1.842	25°C, 3 M HNO_3	[15]
	+1.800	0°C, 1 M HNO_3	[15]
	+1.80	0°C, 1 M HNO_3	[30]
	+1.92 ± 0.02	25°C, 4 M $HClO_4$	[13], [31]
	+1.95 ± 0.10	25°C, 4 M $HClO_4$	[9]
	+1.81	25°C	[3]
	+1.829 ± 0.001	25°C, 0.50 M $HClO_4$ $NaClO_4$ (μ = 3 M)	[17]
$[Co\ ox_3]^{3-} + e = [Co\ ox_3]^{4-}$	+0.57 ± 0.02	25°C, 0.100 M KHC_2O_4, KCl (pH = 3.65) (μ = 1.0 M)	[20]

(continued)

TABLE 1.1.5 (continued)

Half-reaction	Standard or formal potential (V)	Conditions	Refs.
$[Co(OH_2)_2ox_2]^- + e = [Co(OH_2)_2ox_2]^{2-}$	+0.78 ± 0.04	25°C, 0.060 M $K_2C_2O_4$ 0.060 M KHC_2O_4 (pH = 3.63)	[20]
$[Co\ en_3]^{3+} + e = [Co\ en_3]^{2+}$	-0.180 ± 0.002		[4]
$[Co\ dip_3]^{3+} + e = [Co\ dip_3]^{2+}$	+0.370	25°C	[32]
	+0.34	0°C (μ = 0.05 M)	[19]
$[Co(OH_2)_4dip]^{3+} + e = [Co(OH_2)_4dip]^{2+}$	+0.84 ± 0.02	0°C (μ = 0.05 M)	[19]
$[Co\ phen_3]^{3+} + e = [Co\ phen_3]^{2+}$	+0.42	25°C	[32]
	+0.40 ± 0.02	0°C (μ = 0.05 M)	[19]
	+0.42	25°C	[3]
$[Co(OH_2)_2phen_2]^{3+} + e = [Co(OH_2)_2phen_2]^{2+}$	+0.68 ± 0.02	0°C (μ = 0.05 M)	[19]
$[Co(OH_2)_4phen]^{3+} + e = [Co(OH_2)_4phen]^{2+}$	+0.84 ± 0.02	0°C (μ = 0.05 M)	[19]
$[Co\ terpy_2]^{3+} + e = [Co\ terpy_2]^{2+}$	-0.31 ± 0.02	0°C (μ = 0.05 M)	[19]
$[Co\ edta]^- + e = [Co\ edta]^{2-}$	+0.13 (vs SCE)	25°C, 0.05 M acetate buffer (pH = 4.6 to 5.4) (μ = 0.2 M)	[33]
	+0.135 (vs SCE)	25°C, 0.1 M acetate buffer (pH = 5.0)	[34],[35]
	+0.60 (vs NHE)	25°C	[3]
$[Co\ pdta]^- + e = [Co\ pdta]^{2-}$	+0.12 (vs SCE)	25°C, 0.05 M acetate buffer (pH = 4.6 to 5.4) (μ = 0.2 M)	[33]
$[Co\ trdta]^- + e = [Co\ trdta]^{2-}$	+0.05 (vs SCE)	25°C, 0.05 M acetate buffer (pH = 4.6 to 5.4) (μ = 0.2 M)	[33]

$[Co\ cydta]^- + e = [Co\ cydta]^{2-}$	+0.120 (vs SCE)	25°C, 0.1 M acetate buffer (pH = 4.0 to 6.0)	[35]
$1/2\ [Co(CO)_4]_2 + e = [Co(CO)_4]^-$	-0.40 (vs NHE)	20°C	[36]
$[Co(C_5H_5)_2]^+ + e = [Co(C_5H_5)_2]$	-0.918 ± 0.010	25°C, 0.1 M KNO_3	[37]
$[Co\ trendisal]^+ + e = [Co\ trendisal]$	-0.463 ± 0.010	25°C, 0.1 M KNO_3	[37]
$[Co(OH_2)_6]^{2+} + 2e = Co(s)$	-0.277	25°C	[27]
	-0.287 ± 0.01	25°C	[22], [23]
$CoCO_3(aq) + 2e = Co(s) + CO_3^{2-}$	-0.64	25°C, 4 M $HClO_4$	[13], [31]
$Co(OH)_2(s) + 2e = Co(s) + 2OH^-$	-0.73	25°C	[27], [5]
$Co(OH)_3(s) + e = Co(OH)_2(s) + OH^-$	+0.17	25°C	[27], [5]
$[Co(OH_2)_6]^{3+} + e = HCoO_2^- + 3H^+$	-0.0065	25°C	[29]
$Co(OH)_2(s) + 2H^+ + 2e = Co(s) + 2H_2O$	+0.095	25°C	[29]
$CoO(s) + 2H^+ + 2e = Co(s) + H_2O$	+0.166	25°C	[29]
$Co_3O_4(s) + 2H_2O + 2H^+ + 2e = 3Co(OH)_2(s)$	+0.993	25°C	[29]
$Co_3O_4(s) + 2H^+ + 2e = 3CoO(s) + H_2O$	+0.777	25°C	[29]
$3Co_2O_3(s) + 2H^+ + 2e = 2Co_3O_4(s) + H_2O$	+1.018	25°C	[29]
$2CoO_2(s) + 2H^+ + 2e = Co_2O_3(s) + H_2O$	+1.477	25°C	[29]
$HCoO_2^- + 3H^+ + 2e = Co(s) + 2H_2O$	+0.659	25°C	[29]
$Co_3O_4(s) + 8H^+ + 2e = 3Co^{2+}(aq) + 4H_2O$	+2.112	25°C	[29]
$Co_3O_4(s) + 2H_2O + 2e = 3HCoO_2^- + H^+$	-0.700	25°C	[29]
$Co_3O_4(s) + 6H^+ + 9H_2O + 2e = 2[Co(OH_2)_6]^{2+}$	+1.746	25°C	[29]

(continued)

TABLE 1.1.5 (continued)

Half-reaction	Standard or formal potential (V)	Conditions	Refs.
$Co_2O_3(s) + H_2O + 2e = 2HCoO_2^-$	-0.128	25° C	[29]
$CoO_2(s) + 4H^+ + 4H_2O + 2e = [Co(OH_2)_6]^{2+}$	+1.612	25° C	[29]
$CoO_2(s) + 4H^+ + 4H_2O + e = [Co(OH_2)_6]^{3+}$	+1.416	25° C	[29]
$CoO_2(s) + H_2O + e = CoOOH(s) + OH^-$	+0.7	90° C	[5], [38]
$CoOOH(s) + H_2O + e = Co(OH)_2(s) + OH^-$	+0.17	25° C	[5]
$[Co(NH_3)_6]^{2+} + 2e = Co(s) + 6NH_3(aq)$	-0.42		[5]
$CoS(\alpha) + 2e = Co(s) + S^{2-}$	-0.90		[5]
$CoS(\beta) + 2e = Co(s) + S^{2-}$	-1.07		[5]
$CoCO_3 + 2e = Co(s) + CO_3^{2-}$	-0.64		[5]

[a]Abbreviations used: edta = ethylenediaminetetraacetate ion; pdta = propylenediaminetetraacetate ion; trdta = trimethylenediaminetetraacetate ion; cydta = 1,2-cyclohexanediaminetetraacetate ion; trendisal = triethylenetetraminedisalicylidine,

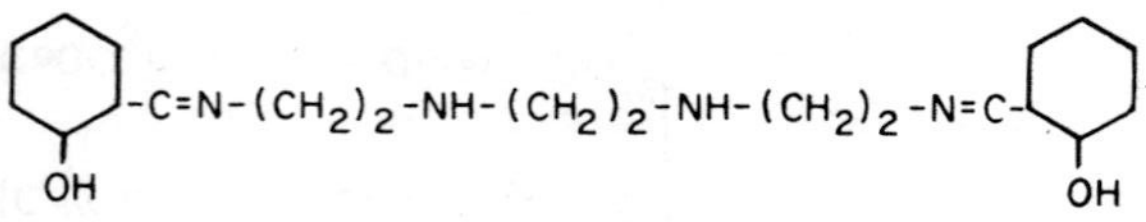

; these abbreviations denote the ligands from which all hydrogen ions capable of dissociating are detached.

From Tables 1.1.3 and 1.1.4 it can be seen that the replacement of chelated ligand by water in the cobalt(III) complexes leads to an increase in the value of the standard potential (an increase in oxidizing power). This would naturally be expected since the final member of the series, $[Co(OH_2)_6]^{3+}$, is an extremely powerful oxidizing agent [3].

On the other hand, Larson, Cerutti, Garber, and Hepler [22] calculated the standard oxidation potential of the $Co(s)/[Co(OH_2)_6]^{2+}$ couple to be $E° = -0.287 \pm 0.01$ V, which is in satisfactory agreement with the three higher potentials cited by Latimer [5] but is not in good agreement with the lower potentials. Previously, Goldberg et al. [23] reported $\Delta S° = -1.9$ gibbs/mole for solution of $[Co(OH_2)_6]SO_4.H_2O$ (c). Using the most recent entropies of SO_4^{2-}(aq) [5] and H_2O(l), Larson et al. [22] repeated the earlier calculation leading to the entropy of $[Co(OH_2)_6]^{2+}$(aq) to obtain $\Delta S = -26.6$ gibbs/mole for $[Co(OH_2)_6]^{2+}$(aq). The recalculation of the result by Adami and King [24] gave the heat of formation $\Delta H_f^° = -212.55$ kcal/mole for the $[Co(OH_2)_6]SO_4.H_2O$(c). Combination of this value with the previously determined heat of solution of $[Co(OH_2)_6]SO_4.H_2O$(c) [23] and $\Delta H_f^°$ for SO_4^{2-}(aq) [5] leads to $\Delta H_f^° = -14.0$ kcal/mole for $[Co(OH_2)_6]^{2+}$(aq). Combination of $\Delta H_f^°$ and ΔS values for $[Co(OH_2)_6]^{2+}$(aq) with entropies for Co(c) [25] and H_2(g) [26] leads to $\Delta G_f^° = -13.2$ kcal/mole for $[Co(OH_2)_6]^{2+}$(aq). The $\Delta G_f^°$ gives $E° = -0.287$ V as the standard potential for $Co^{2+}(aq) + 2e = Co(c)$ using $\Delta G_f^° = -nFE°$.

In Table 1.1.5 the standard or formal oxidation potentials of the Co(II)/Co(III), Co(c)/Co(II), and Co(0)/Co(I) couples which have so far been obtained in aqueous solutions are briefly summarized.

1.2. NONAQUEOUS SOLUTION

The half-wave potential of the reversible tris(2,2'-dipyridyl) cobalt(III)-cobalt(II) system at the dropping mercury electrode (DME) was determined in the acetonitrile (AN) containing 0.05 M tetraethylammonium perchlorate at 25°C, as shown in Table 1.2.1 [39].

From these observed values the standard potential of the $[Co\ dip_3]^{2+}/[Co\ dip_3]^{3+}$ couple was calculated roughly to be +0.320 V (vs NHE) in AN at 25°C [40].

TABLE 1.2.1. Anodic and Cathodic Half-Wave Potentials of $[Co^{II}\ dip_3]^{2+}$ and $[Co^{III}\ dip_3]^{3+}$ Couple in AN[a]

Reaction	$E_{1/2}$ (V vs SCE)	Slope (mV)
Oxidation process	+0.241	62
Reduction process	+0.235	65

[a]Concentration of the complex: 5×10^{-3} M (25°C).

Sameč and Němec [40] have examined the voltammetric behavior of the [Co dip$_3$]$(ClO_4)_3 \cdot 3H_2O$ complex at a platinum rotating disk electrode (RDE) in AN and showed that the standard oxidation potential of the [Co dip$_3$]$^{2+}$/[Co dip$_3$]$^{3+}$ couple can be calculated from the half-wave potential, $E_{1/2}$, which can be expressed by

$$E_{1/2} = E^\circ + (RT/nF)\ln(D_{red}/D_{ox})^{2/3} + (RT/nF)\ln(f_{ox}/f_{red}) + E_L \qquad (1.2.1)$$

where E° is the standard potential, D_i is the diffusion coefficients of both the oxidized and the reduced forms, f_i is the activity coefficients of both oxidation forms, and E_L is the liquid-junction potential. The liquid-junction potential between the aqueous and the AN solutions was estimated to be about 190 mV at a $NaClO_4$ concentration of 0.1 M. The half-wave potentials of the voltammetric waves on which the calculations were based are given as

$$E_{1/2} = +0.222 \text{ V (A)}; \qquad E_{1/2} = +0.214 \text{ V (B)}$$

vs SCE with a saturated solution of NaCl. The direction of polarization of (A) is from negative to positive potentials; the direction of (B) is the opposite.

A comparison of the results which have thus far been obtained in aqueous and in AN solutions is given in Table 1.2.2. The difference between the E° values obtained from the $E_{1/2}$ of polarographic (+0.320 V) and voltammetric (+0.339 V) waves is probably due to complex formation of 2,2'-dipyridyl, which is liberated by the dissociation of the tris(2,2'-dipyridyl)-cobalt(II) ion with mercury [37].

TABLE 1.2.2. The Standard Potentials of the [Co dip$_3$]$^{2+}$/[Co dip$_3$]$^{3+}$ Couple in Aqueous and in Nonaqueous Solutions

Half-reaction	E° (V vs NHE)	Conditions	Refs.
[Co dip$_3$]$^{3+}$ + e = [Co dip$_3$]$^{2+}$	+0.339	In AN	[40]
	+0.320	In AN	[39]
	+0.370	In H_2O	[32]
	+0.34	0°C in H_2O	[19]

Table 1.2.3 summarizes the standard or formal electrode potentials of the Co(0)/Co(II), Co(0)/Co(I), and Co(II)/Co(III) couples which have thus far been measured in nonaqueous media.

TABLE 1.2.3. Standard or Formal Electrode Potentials of the Co(0)/Co(II), Co(0)/Co(I), and Co(II)/Co(III) Couples[a]

Half-reaction	Standard or formal potential (V)	Conditions	Refs.
$Co^{2+} + 2e = Co(s)$	-0.233	25°C, in methanol	[41],[42]
$[Co\ dip_3]^{3+} + e = [Co\ dip_3]^{2+}$	+0.320	25°C, in AN, 0.05 <u>M</u> $[(C_2H_5)_4N]ClO_4$	[39],[40]
$[Co\ dip_3]^{3+} + e = [Co\ dip_3]^{2+}$	+0.339	20°C, in AN, 0.1 <u>M</u> $NaClO_4$	[40]
$[Co(C_5H_5)_2]^{+} + e = [Co^0(C_5H_5)_2](1)$	-0.790 ± 0.012	25°C, in formamide (FA), 0.1 <u>M</u> $LiNO_3$	[37]
	-1.146 ± 0.012	25°C, in AN, 0.03 <u>M</u> $LiNO_3$	[37]
	-0.930 ± 0.012	25°C, in methanol, 0.1 <u>M</u> $LiNO_3$	[37]
$[Co^{I}trendisal]^{+} + e = [Co^{0}trendisal](1)$	-0.455 ± 0.012	25°C, in FA, 0.1 <u>M</u> $LiNO_3$	[37]
	-0.880 ± 0.012	25°C, in AN, 0.03 <u>M</u> $LiNO_3$	[37]
	-0.550 ± 0.012	25°C, in methanol, 0.1 <u>M</u> $LiNO_3$	[37]

[a] C_5H_5 = cyclopentadienyl; trendisal = triethylenetetraminedisalicylidine.

1.3. MOLTEN SALTS

The standard electrode potential of the Co(s)/Co(II) couple has been determined to be -0.127 V [vs Ag/Ag(I)] in molten alkali thiocyanates by Cescon, Marassi, Bartocci, and Fiorani [43]. That is, the emf measurements of the following cells were carried out with a cobalt-metal stick at 415 to 470°K:

$$Pt|Co(s)|Co^{2+}, (K, Na)NCS|glass|AgNO_3(Li, K, Na)NO_3|Ag \quad (1.3.1)$$

$$|AgNO_3, (Li, K, Na)NO_3|glass|Ag^{+}, (K, Na)NCS|Ag|Pt \quad (1.3.2)$$

The potential of a cobalt-metal stick was measured vs the glass reference electrode, standardized by means of Cell (1.3.2). The activity of the cobalt was calculated by extrapolating the thermodynamic selected data [44] at every temperature assuming the pure cobalt metal as standard state, 1 mole/kg. The emf values (E°_{Co}, $E^{\circ\prime}_{Ag}$) of Cells (1.3.1) and (1.3.2) were used to obtain the standard electrode potential (E°_{Co}) of the Co(s)/Co(II) couple from the difference between the quantities $E^{\circ\prime}$ and E°_{Ag} where

$$E^{\circ\prime}_{Co} = E_V - E^{\circ}_{Co}$$

$$E^{\circ\prime}_{Ag} = E_V - E^{\circ}_{Ag}$$

E_V is the potential of the glass reference electrode and E°_{Ag} has been assumed, conventionally, as equal to zero. All measured potentials were expressed as being referred to the standard Ag/Ag(I) (1m or mole/kg) electrode. Here the cobalt ion (Co^{2+} was introduced into the cell as pellets of a (K, Na)NCS mixture of a cobalt salt.

Gaur and Jindal [45] have examined the polarographic behavior of cobalt(II) chloride at a platinum microelectrode (PME) and a platinum wire microelectrode (PWE) in molten $MgCl_2$-KCl (475 ± 1°C). The reduction wave of the process Co(II) → Co(metal) with $E_{1/2}$ = -1.028 V [vs Pt(II)/Pt electrode] was observed at a concentration of 14.52 mM of cobalt(II) chloride. The ~ 0.01 M Pt(II)/Pt in the fused salt was employed as a reference. The nature of the microelectrode material as well as the physical state of the cobalt metal deposited on it was found to determine whether the Heyrovský-Ilkovič or the Kolthoff-Lingane equation [46] expresses the current-voltage functions of this wave in molten salt solutions. At low concentrations the current-voltage curves show a limiting current plateau although a tendency for the current to increase at more negative potentials is noticeable. On increasing the cobalt concentration, the limiting current plateau becomes less defined and the increase in wave height is not proportional to the cobalt concentration in the molten salt. The cobalt metal deposited on the PME or WME could be anodically stripped, the current decreasing to the residual value at that potential in 3 to 5 min. On a stripped microelectrode surface the current is reproducible within 3 to 5% in the pre-"increasing current" region. Furthermore, Gaur and Jindal [47] have determined the apparent electrode potential of the Co(0)/Co(II) couple to be -1.046 V (vs the standard chlorine reference) in the fused ternary eutectic $MagCl_2$-NaCl-KCl (475°C). The standard state for the cobalt metal is chosen as the physical state, and that for cobalt(II) chloride in solution as the unit mole fraction of the solute at the working temperature and pressure. The standard state for chlorine is the ideal gas at unit atmospheric pressure, and the standard potential of the Cl_2(graphite)/Cl^- in this melt is arbitrarily given a value of 0.000 V. The following galvanic cell was set up:

Co|$CoCl_2$, $MgCl_2$-NaCl-KCl|glass frit|$MgCl_2$-NaCl-KCl|Cl_2 (graphite)

Indicator half-cell   Reference half-cell

The cobalt(II) chloride was either generated in situ by anodic dissolution of the cobalt metal using a constant current source or the anhydrous chloride was externally added. The cell emf was measured as a function of the cobalt ion concentration in the indicator half-cell. The apparent standard electrode potentials of a number of metal/metal ion couples in molten $MgCl_2$-NaCl-KCl (50:30:20 mole %) vs unit molal Pt(II)/Pt(0) in the melt as reference

TABLE 1.3.1. Apparent Standard Electrode Potentials of the Co(0)/Co(II) Couple in Various Molten Salts

Half-reaction	Apparent standard potential (V)	Conditions	Refs.
$Co^{2+} + 2e = Co(0)$	-1.046	$MgCl_2$-NaCl-KCl	[48-52]
	-1.207	LiCl-KCl (450°C)	[53]
	-1.277	NaCl-KCl (450°C)	[54]
	-1.156	Calcd as standard potential	[55]
$Pt^{2+} + 2e = Pt(0)(s)$	-0.076	$MgCl_2$-NaCl-KCl	
	-0.216	LiCl-KCl (450°C)	[53]
	-0.284	Calcd	[55]
$Cl_2(g) + 2e = 2Cl^-$	0.000	$MgCl_2$-NaCl-KCl	
		LiCl-KCl	
		NaCl-KCl	

[a]V vs Cl_2(graphite)/Cl^- electrode.

have already been compiled by Gaur and Behl [48, 49]. After the electrode potentials earlier reported vs Pt(II)/Pt(0) had been converted to the standard chlorine/chloride as reference, the comparison of the apparent standard electrode potentials of the Co(0)/Co(II) couple in molten $MgCl_2$-NaCl-KCl, in molten NaCl-KCl, and in molten LiCl-KCl was made with each other, as shown in Table 1.3.1.

From Table 1.3.1 it can be said that the magnitude of the potential increases in the order of NaCl-KCl > LiCl-KCl > $MgCl_2$-NaCl-KCl.

2. VOLTAMMETRIC CHARACTERISTICS

2.1. POLAROGRAPHIC CHARACTERISTICS

The cobalt(III) and cobalt(II) complexes are the most familiar ones encountered in the polarographic redox reactions in aqueous and in nonaqueous media. More than 600 kinds of mixed cobalt(III) complexes with various ligands have already been prepared, but only a few of them have been examined polarographically. Discussions of compounds known in earlier years may be found in the standard reference works of Friend [56] and Gmelin [57-59] or in the monographs of Young [60, 61] and Pyatiniskii [62]. Among the cobalt compounds, cobalt (III) is characterized by the formation of six-coordinate, octahedral d^6-complexes ($d\epsilon^6$) which exhibit the structural integrity usually attributed to the formation of covalent bonds, whereas cobalt(II) is practically confined to the aquated d^7-complexes ($d\epsilon^4 d\gamma^3$) of the high-spin type except for a few organocobalt(II) complexes with ligands of the π-bonding character. This is in marked contrast to the stable low-spin type cobalt(III) complexes which are of the "substitution-inert" type toward solvolysis. In general, the dissociation of the inert cobalt(III) complexes can substantially be neglected from the standpoint of polarography, while that of the "labile" cobalt(II) complexes occurs so rapidly, compared with the scanning rate of applied potential, that even the oscillopolarographic techniques cannot follow it. Consequently, the cobalt(II) complex results in the formation of an aquated cobalt(II) mixture in an equilibrium state with water molecules (as ligands) or with free ligands present in aqueous solutions. Such a disruption of the resulting cobalt(II) complex can be discerned and is detectable by observing the effect of free ligands on the $E_{1/2}$ of the second step Co(II) → Co(metal); that is, the $E_{1/2}$ of the second step is always near to that (-1.24 V vs SCE in 1 M KCl) of the $[Co^{II}(OH_2)_6]^{2+}$ ion in noncomplexing aqueous solutions when the bond rupture in the cobalt(II) complex occurs completely. Conversely, the difference of $E_{1/2}$ from the value (-1.24 V) for the $[Co(OH_2)_6]^{2+}$ ion indicates no bond-breaking of the particular ligand at a cobalt(II) state. For example, the $E_{1/2}$ of the second reduction step of the $[Co(S_2O_3)(NH_3)_5]^+$ ion is -0.965 V (vs SCE), fairly different from that of the $[Co(OH_2)_6]^{2+}$ ion, suggesting that a $S_2O_3^{2-}$ ion is still bound to the cobalt(II) ion as a ligand during reduction [63]. In aqueous solution containing free ligands capable of coordinating (complexing agents), the $E_{1/2}$ may shift to more negative potentials. This is probably because the vacant $d\epsilon$ (3d) orbitals of the cobalt(II) ion are protected against the attack of the electrode (DME) through screening of them by the donor electrons of coordinated ligands. In complexing media, the process Co(III) → Co(II) → Co(metal) is most frequently observed for ammine, amine, and their analogous mixed cobalt(III) complexes with ligands of σ-bonding character. Upon the reduction of these complexes, the presence of a large excess of free ligands may certainly play the role of preventing the aquation of the primary cobalt(II) complex formed midway on the process Co(III) → Co(metal) to some extent, and of retaining the original structure as far as possible at the cobalt(II) state. Nevertheless, a complete retention of the former configuration throughout the electrode reaction is not possible even in the presence of an overwhelmingly large excess of free ligands, and as a matter of fact the parent cobalt complex would release some of its ligands in solution as it is reduced to the lower oxidation states of cobalt, especially at higher temperatures. Therefore the interpretation of polarographic behavior becomes more difficult for the second wave appearing at negative potentials than for the first wave, and little can be said of the fate of the primary cobalt(II) product since the labile cobalt(II) complex undergoes some further chemical reactions prior to the electron transfer in the second step.

In contrast to the reduction of the "inert-labile" type for the complexes with σ-bonds, the cobalt(III) complexes with ligands of π-bonding character, i.e., with the ligands involving a double or triple bond adjacent to the donor atom, show polarographically the step-by-step reduction of Co(III) → Co(II) → Co(I) → Co(0), suggesting the inertness of the cobalt complexes with π-bonds at lower oxidation states in aqueous solutions. For example, the molecules or ions cited below act as the π-bonding ligands which react with cobalt(III) ion to form such "inert" cobalt(III) complexes: $Co \equiv N^-$ (cyanide), $C = N - R$ (isocyanide) [64], dip (2,2'-dipyridyl), phen (1,10-phenanthroline), oximes (dimethylglyoxime, salicylaldoxime, etc.), $P \equiv \phi_3$ (triphenylphosphine), $As \equiv \phi_3$ (triphenylarsine), $C \equiv O$ (carbonyl), $C_5H_5^-$ (cyclopentadienyl), and other similar ligands.

Their coordinate bonds have a strong π-bonding character due to the "back donation" from the ligands and exhibit a property of conducting electrons readily. The most significant feature of cobalt(III) complexes of this type is that the lower oxidation states of cobalt [e.g., Co(I) and Co(0)] are stabilized and detectable during reduction by means of conventional polarography; the configuration of the cobalt(III) complexes remains intact throughout the reduction of the formal oxidation state of cobalt even in aqueous media, suggesting the reduction of the "inert-inert" type in stepwise fashion. This tendency is seen more strongly in aprotic nonaqueous solvents; i.e., there is an indication that not only the cobalt(0) complex but also the cobalt(-I) complex in which π-bonding ligands are still tightly bound to the cobalt ion could exist and be stable during reduction in AN and in DMSO, although the valence electrons in orbitals are actually delocalized over the entire complex ion and, needless to say, the cobalt(-I) state should be interpreted as implying only a formal oxidation state of the central cobalt. The possibility of the existence of the $[Co\ dip_3]^-$ anion in the AN solution with an excess of free 2,2'-dipyridyl was suggested on the course of polarographic reduction by Tanaka and Sato [39] and of the formation of the $[Co\ phen_3]^-$ anion was found during reduction in the DMSO (100%) solution without any free ligands in excess by Maki [65].

In general, the stable intermediates, i.e., cobalt complexes in lower oxidation states, give rise to well-defined anodic waves, corresponding to the counterpart of the step-by-step reduction, which can be observed either by means of the technique of Kalousek commutator [66], by use of cyclic voltammetry, or by Heyrovsky-Forejt-type oscillopolarography, since no liberation of ligands occurs even at cobalt(I) and cobalt(0) complexes with a few exceptions. For this reason, much more information on the kinetic data of the processes and on the structures of the intermediate products is provided and is available for complexes involving π-bonding ligands rather than for those involving σ-bonding ones in interpretation of the electrode reaction in nonaqueous media. The electrode pathways for these complexes in aprotic nonaqueous solvents are thought of as being the most simple, in the sense that no subsequent chemical reactions other than the electron-transfer process proper occur, except under particular conditions, such as at extremely high temperatures, and that inevitably the divergence of the electrode pathways due to the secondary products need not be taken into account in interpretation of the main process of the electrode reaction. For example, the trans-$[Co^{III}(dgH)_2(NH_3)_2]^+$ ion gives rise to three step waves of the reduction, Co(III) → Co(II) → Co(I) → Co(0), in aqueous media, but in DMSO only two step waves of the reduction, Co(III) → Co(II) → Co(I), at the DME, where dg^{2-} denotes the dimethylglyoxime ligand from which all hydrogen ions capable of dissociating are detached, obeying the custom of coordination chemistry. The discrepancy of polarographic behavior between them is attributed to the difference in the "lability" of the $[Co^{I}(dgH)_2(NH_3)_2]^-$ anion in water and in DMSO. Cobalt(III) and its degraded cobalt(II) complexes are thought of as inert types both in DMSO and in water with the exception of the Co(II)—NH_3 bonds (σ-bonds) in water, whereas the cobalt(I) complex seems most likely to be inert as a whole only in DMSO, but labile in water. Consequently, further reduction from the cobalt(I) to cobalt(0) in aqueous media is due to the aquation of the ligand, dgH^-, at the cobalt(I) state, since no further reduction to the cobalt(0) state occurs in DMSO for the complexes of the trans-$[Co^{III}(dgH)_2X_2]$ type, irrespective of the nature of the 5th and 6th axial ligands, X_2 [67]. Thus the change in the lability of the complex from an inert to a labile type depending on the nature of the medium suggests that the definition of the "lability" used polarographically is different in some respects from that defined by Taube [68], which is based on the rate of ligand exchange reactions with the same kind of ligands in solution. In this regard, the term "substitution-inert" or "substitution-labile" may be considered as a measure of representing the movability of the ligands in the coordination sites and reflects the extent of solvolysis (or aquation), which does not depend on the rate of substitution reactions with the same kind of ligands present in solution, but on the rate of the substitution reactions with solvent molecules through breaking (S_N1) or loosening (S_N2) of the Co—L bond.

The most striking feature and merit of adopting aprotic nonaqueous solvents is the possibility that the lability of cobalt complexes, especially at the reduced lower oxidation states, can be converted from a labile one to one inert to solvolysis. The reduced mixed ligand cobalt(II) and cobalt(I) complexes with σ-bonds, which would decompose immediately in water, may be surveyed by nonaqueous polarography without isolating the reduced species from solution, with the attendant advantages for interpretation of more complicated electrode processes in water, because the simplest process of an inert-inert type in nonaqueous media may be regarded as an ideal model of the electrode pathway which, if the solvolysis (aquation) of the intermediates was completely inhibited, would take place in a similar way in aqueous media. For example, upon the reduction of the $[Co^{III}(NH_3)_6]^{3+}$ ion (inert) in noncomplexing aqueous media, the primary product of the first reduction step, $[Co^{II}(NH_3)_6]^{2+}$ ion (labile), changes immediately into the $[Co^{II}(OH)(NH_3)_4]^+$ ion (the secondary product of five-coordinate) which could not exist as a stable species for a sufficient length of time to permit the direction reduction of the $[Co^{II}(OH)(NH_3)_4]^+$ ion (labile) to the metal, compared with the scanning rate of applied potential [69]. Accordingly, the cobalt(II) species finally responsible for the second step, Co(II) → Co(metal), is the aquated cobalt(II) ion, $[Co^{II}(OH_2)_6]^{2+}$, predominantly present in aqueous solutions. In DMSO (100%), however, it was shown that the primary product of the $[Co^{II}(NH_3)_6]^{2+}$ ion (inert) remains structurally intact upon the reduction to the metal, although the cobalt(0) complex with ammonia ligands could no longer exist, leading to disruption to the metal in DMSO. By analogy with the formation of the $[Co^{II}(OH)(NH_3)_4]^+$ ion in aqueous media [69], the monocyanotetramminecobalt(III) complexes of the trans-$[Co^{III}(CN)(NH_3)_4X]$ type were thought to have the tendency of converting to the penta-coordinate configuration, $[Co^{II}(CN)(NH_3)_4]^+$, in their reduced cobalt(II) state with liberation of the 6th ligand, X, in DMSO (100%) probably due to the trans-effect of cyanide on the counter ligand, except for the cases of X = NH_3, NH_2CH_3, or $NH_2C_2H_5$ (i.e., for X = NO_2^-, SO_3^{2-}, NO^-, NCS^-, pyridine, I^-, or N_3^-). Here it is worthwhile to add that upon reduction in aqueous media, these monocyanotetramminecobalt(III) complexes may be presumed most probably to take a course similar to that in DMSO, though there is neither any direct evidence to verify the existence of such a five-coordinate cobalt(II) complex nor direct evidence to illustrate a possible route of dissociation in aqueous solutions, due to the high lability of the primary cobalt(II) product, different from that in DMSO [70].

In any event, electrode pathways either completely of the inert-inert type or completely of the inert-labile type is consistent with the complex remaining structurally intact throughout the different oxidation states of cobalt or undergoing rapid dissociation (solvolysis), which in both extreme cases can most readily be understood and interpreted polarographically.

As a matter of fact, however, the complexity in elucidating the electrode process frequently arises from the necessity of taking into account the partly-labile type of mixed ligand cobalt complexes in their reduced states which possess both σ- and π-bonding ligands simultaneously in the same coordination sphere. Such difficulties must be overcome by a proper choice of experimental conditions, e.g., the presence of an excess of free ligands.

2.2. HALF-WAVE POTENTIALS

The correlation between the electrode process of the cobalt(III) and cobalt(II) complexes and their structures has been worked out from two points of view, one being expressed in terms of the half-wave potential, and the other in terms of the kinetic data of the electrode reaction. In connection with this, several attempts have been made to correlate the half-wave potential to the lowest unoccupied π-orbital [71], or to the ligand field strength (D_q) expressed as the absorption maximum or as the spectrochemical series of ligands for a series of very similar complexes from the point of view that the $E_{1/2}$ for a completely irreversible system (no occurrence of backward reaction) is a measure of the rate constant for the reduction process and has a meaning analogous to the rate constant of an electron exchange reaction, the only difference being that the $E_{1/2}$ for the reduction process relates only to the structure of the parent cobalt complex (oxidized form) and not to the nature of the reduced form [63, 72-75]. It should be noted, however, that the rate constant is not a unilateral function of the half-wave potential. Moreover, the half-wave potential for an irreversible system is a function not only of the kinetic parameters but also of the experimental conditions such as the duration of electrolysis, and accordingly the $E_{1/2}$ values are very important for characterizing the various electrode processes of cobalt complexes from a practical point of view, together with the number of electrons involved.

On the other hand, the half-wave potential for a reversible system is approximately equal to the normal redox potential. The $E_{1/2}$ values have a thermodynamic significance as a measure of the free-energy change of the electrode process: $nFE_{1/2} = -\Delta G^\circ$ [76]. In most cases, however, the low-spin $d\epsilon^6$ complexes of cobalt(III), in which the stabilized 3d subshell is fully occupied, are generally reduced irreversibly at fairly negative potentials, suggesting the occurrence of slow electron-transfer reactions. The situation is almost the same for the high-spin $d\epsilon^5 d\gamma^2$ complexes of cobalt(II). Though rare, there are a few reversible Co(III)/Co(II) redox systems including the following: $[Co\ dip_3]^{3+}/[Co\ dip_3]^{2+}$, $[Co\ terpy_2]^{3+}/[Co\ terpy_2]^{2+}$, $[Co\ phen_3]^{3+}/[Co\ phen_3]^{2+}$, $[Co\ bathophen_3]^{3+}/[Co\ bathophen_3]^{2+}$, $[Co\ tmp_3]^{3+}/[Co\ tmp_3]^{2+}$, $[Co\ edta]^{-}/[Co\ edta]^{2-}$, $[Co\ cydta]^{-}/[Co\ cydta]^{2-}$, vitamin B_{12a}[Co(III)]/vitamin B_{12a}[Co(II)], $[Co\ en_3]^{3+}/[Co\ en_3]^{2+}$, cis-$[Co(CN)_2tren]^{+}$/cis-$[Co(CN)_2tren]$, cis-$[Co(CN)_2tn_2]^{+}$/cis-$[Co(CN)_2tn_2]$ in DMSO and similar systems [77-84, 103, 106-108] (where the following abbreviations are used: terpy = 2,2',2''-terpyridyl; bathophen = 4,7-diphenyl-1,10-phenanthroline(bathophenanthroline); tmp = 3,5,6,8-tetramethyl-1,10-phenanthroline; cydta = 1,2-cyclohexanediaminetetraacetate; tren = triethylenetetramine; tn = trimethylenediamine). The feature of these systems is that the cobalt(II) complexes of the reduced form (a substitution-inert type) can exist stably in aqueous and nonaqueous solutions and also in the crystalline state in open air, the crystals of which are readily obtained by the usual syntheses (e.g., Ref. 85) with a few exceptions, such as the cobalt(II) complexes of the $[Co\ en_3]^{2+}$, $[Co(CN)_2tren]$, and $[Co(CN)_2tn_2]$ types. The other feature is the aromatic π-bond ring systems or the polydentate ligand, the complexes of which usually exhibit the outer-sphere redox reaction [86]. The $[Co\ en_3]^{3+}/[Co\ en_3]^{2+}$ system is certainly a reversible electrode process in the presence of excess ethylenediamines, but this system is not reversible polarographically [88, 89, 97, 284] (see Section 4.1.3). Tables 2.2.1-2.2.23 present the half-wave potentials of the irreversible process for the cobalt(II) and cobalt(II) complexes as gathered from the large, scattered mass of polarographic data in the literature. For the sake of simplicity, the half-wave potentials for the complexes which were synthesized simply by mixing hexaquocobalt(II) ion and complexing agents (free ligands in excess) at room temperatures are omitted because of the uncertainty of the structure of the electroactive species. The half-wave potentials appearing in earlier years are found in the work of Vlček [90].

TABLE 2.2.1. Effects of Supporting Electrolyte on the First Polarographic Reduction Wave, Co(III) → Co(II) for the Hexamminecobalt(III) Chloride [91]

0.001 M $[Co(NH_3)_6]Cl_3$, m = 1.67 mg/sec, t = 4.48 sec/drop (DME) in 0.1 M KCl soln (open circuit)

Supporting salt	Concn of supporting salt (M)	$E_{1/2}$ (V vs SCE)	Slope	i_d (μA)	t (sec/drop)	$i_d/Cm^{2/3}t^{1/6}$
KNO_3	0.1	-0.244	0.078	3.22	5.16	1.74
	0.9	-0.268	0.088	3.34	4.92	1.82
	0.1[a]	-0.284	0.060	3.07	5.00	1.67
KCl	0.1	-0.255	0.142	3.28	4.96	1.78
	0.2	-0.237	0.114	3.16	4.99	1.72
	0.3	-0.204	0.117	3.28	4.91	1.79
	0.4	-0.184	0.100	3.25	5.03	1.76
	0.5	-0.185	0.090	3.23	4.96	1.76
	0.8	-0.192	0.068	3.23	4.95	1.76
	0.9	-0.134	0.170	3.53	4.89	1.93
	0.1[a]	-0.279	0.060	3.24	4.95	1.76
	0.5[a]	-0.291	0.090	3.19	4.92	1.74
	0.8[a]	-0.312	0.112	3.30	4.90	1.80
K_2SO_4	0.1	-0.456	0.082	2.92	4.83	1.60
	0.2	-0.455	0.070	2.95	4.83	1.61
	0.3	-0.465	0.070	2.90	4.85	1.58
	0.4	-0.470	0.080	2.88	4.86	1.57
$K_2C_4H_4O_6$ (potassium tartrate)	0.1	-0.305	0.089	2.75	4.83	1.50
	0.5	-0.371	0.082	2.45	4.94	1.33
	0.9	-0.378	0.080	2.21	5.09	1.20
$K_3C_6H_5O_7$ (potassium citrate)	0.1	-0.270	0.066	2.59	4.89	1.41
	0.5	-0.356	0.065	2.25	4.79	1.23
	0.9	-0.382	0.060	1.92	4.95	1.04

[a]In the presence of 0.002% methyl red; all other experiments are in the absence of maximum suppressors.

TABLE 2.2.2. Effects of Supporting Electrolyte on the Second Polarographic Reduction Wave, Co(II) → Co(metal) for the Hexamminecobalt(III) Chloride [91]

0.001 M $[Co(NH_3)_6]Cl_3$, m = 1.67 mg/sec, t = 4.48 sec/drop in 0.1 M KCl soln (open circuit)

Supporting salt	Concn of supporting salt (M)	$E_{1/2}$ (V vs SCE)	Slope	i_d (μA)	t (sec/drop)	$i_d/Cm^{2/3}t^{1/6}$
KNO_3	0.1	-1.207	0.042	9.52	4.03	5.36
	0.9	-1.224	0.054	10.46	3.74	5.96
	0.1[a]	-1.200	0.062	8.89	3.80	5.06
KCl	0.1	-1.203	0.046	9.45	3.78	5.38
	0.2	-1.203	0.046	9.16	3.82	5.21
	0.3	-1.202	0.052	9.43	3.76	5.37
	0.4	-1.212	0.044	9.20	3.87	5.22
	0.5	-1.213	0.060	9.01	3.71	5.15
	0.8	-1.218	0.057	8.90	3.64	5.10
	0.9	-1.210	0.040	9.42	3.68	5.39
	0.1[a]	-1.203	0.062	9.17	3.84	5.21
	0.5[a]	-1.215	0.060	8.75	3.71	5.00
	0.8[a]	-1.225	0.060	8.88	3.63	5.09
K_2SO_4	0.1	-1.232	0.052	8.36	3.73	4.77
	0.2	-1.242	0.056	8.16	3.61	4.68
	0.3	-1.248	0.048	7.90	3.60	4.53
	0.4	-1.253	0.043	7.98	3.63	4.57
$K_2C_4H_4O_6$ (potassium tartrate)	0.1	-1.318	0.083	7.49	3.33	4.35
	0.5	-1.435	0.100	6.82	3.28	3.98
	0.9	-1.463	0.104	6.11	3.34	3.55
$K_3C_6H_5O_7$ (potassium citrate)	0.1	-1.422	0.067	6.78	3.39	3.89
	0.5	-1.519	0.089	6.18	2.97	3.66
	0.9	-1.529	0.122	5.11	3.35	2.97

[a] In the presence of 0.002% methyl red; all other experiments in the absence of maximum suppressors.

TABLE 2.2.3. Half-Wave Potentials of the First Step, Co(III) → Co(II), for the Polarography of the $[Co(NH_3)_6]Cl_3$ Complex and Reciprocal Slopes of the Plots of E vs $\log(i_d - i)/i$ Obtained with Noncomplexing Supporting Electrolytes (25°C) [92]

Supporting salt	Concn of supporting salt (M)	$E_{1/2}$ (V vs SCE)	Reciprocal slope of log plot
$NaNO_3$	1.0	-0.245	0.090
HNO_3	1.0	-0.280	0.076
$Ba(NO_3)_2$	0.2	-0.246	0.086
$NaClO_4$	0.1	-0.245	0.105
NaCl	1.0	-0.200	0.105
HCl	1.0	-0.216	0.116

TABLE 2.2.4. Half-Wave Potentials of the First Step, Co(III) → Co(II), for the Polarography of the $[Co(NH_3)_6]Cl_3$ Complex and Reciprocal Slopes of the Plots of E vs $\log(i_d - i)/i$ Obtained in the Presence of Various Concentrations of Ammonia (25°C) [92]

$[NH_3]$	$pa[NH_3]^a$	α_6	$E_{1/2}$ (V vs SCE)	$0.059 \log \alpha_6$	$E_{1/2}$ corr (V)	Reciprocal slope of log plot
0.71	+0.148	0.081	-0.2702	-0.0645	-0.3347	0.073
1.42	-0.175	0.212	-0.2754	-0.0397	-0.3151	0.072
2.13	-0.385	0.330	-0.2785	-0.0284	-0.3069	0.074
3.55	-0.655	0.503	-0.2812	-0.0175	-0.2987	0.073
5.69	-0.95	0.684	-0.2842	-0.0098	-0.2940	0.075
7.11	-1.11	0.760	-0.2912	-0.0071	-0.2982	0.075
9.25	-1.34	0.847	-0.2985	-0.0043	-0.3028	0.077
11.10	-1.53	0.90	-0.3040	-0.0027	-0.3067	0.078

[a]The activity of ammonia calculated on the basis of Bjerrum's data [1] as well as the value of α_6, which is the fraction of total cobalt(II) in the hexamminecobalt(II) form.

TABLE 2.2.5. Polarographic Reduction of 0.001 M $[Co(NH_3)_6]^{3+}$ Ion in Various Supporting Electrolytes (25°C)

m = 1.746 mg/sec, t = 4.68 sec (open circuit) [93]

Supporting salt	Concn of supporting salt (M)	$\eta^{1/2}$ [a]	i_d (μA)	$i_d\eta^{1/2}$	$E_{1/2}$ (V vs SCE)
		Univalent anions			
NaCl	1.0	1.047	3.59	3.75	-0.193
$CaCl_2$	1.0	1.073	3.31	3.55	-0.220
Na benzoate	1.0	1.285	2.68	3.41	-0.300
$NaClO_4$	1.0	1.023	3.56	3.65	-0.240
NH_4NO_3	1.0	0.985	3.63	3.57	-0.250
Na acetate	1.0	1.180	3.00	3.54	-0.345
NaOH	0.5	1.051	3.53	3.71	-0.355
$[Co(OH_2)_6]Cl_2$	1.0	1.097	3.13	3.43	-0.195
		Bivalent anions			
Na tartrate	1.0	1.155	2.22	2.57	-0.380
Na_2SO_4	0.7	1.074	2.95	3.17	-0.450
Na_2SO_4	1.0	1.110	2.64	2.94	-0.465
$[Co(OH_2)_6]SO_4 \cdot H_2O$	1.0	1.163	2.43	2.82	-0.440
$(NH_4)_2SO_4$	1.0	1.055	2.96	3.12	-0.465
K_2SO_4	0.7	1.036	3.09	3.20	-0.465
H_2SO_4	1.0	1.043	3.19	3.33	-0.385
Na_2SO_4	0.1	1.01	3.20	3.24	-0.430
Na_2CO_3	0.5	1.135	2.82	3.20	-0.455
$MgSO_4$	1.0	1.170	2.24	2.63	-0.452
		Trivalent anions			
Na_3 citrate	1.0	1.175	1.75	2.06	-0.350

[a] η = specific viscosity.

TABLE 2.2.6. Half-Wave Potentials for Mixed Amminecobalt(III) Complexes in Aqueous Media[a]

Compound	$E_{1/2}$ (V vs SCE)		Supporting electrolyte	Refs.
	Co(III) → Co(II)	Co(II) → Co(metal)		
$[Co(NO_2)(NH_3)_5]^{2+}$	−0.264	−1.288	1 M K_2SO_4	[94]
	-0.22_5		1 M $NaNO_3$	[95]
$[Co(NO_2)(NH_3)_5](ClO_4)_2$	−0.12	Maximum	0.1 M KCl	[63]
	−0.26	-1.28_5	0.5 M K_2SO_4	[63]
$[Co(ONO)(NH_3)_5]^{2+}$	−0.263	−1.279	1 M K_2SO_4	[94]
$[CoCl(NH_3)_5]^{2+}$	+[b]	−1.275	1 M K_2SO_4	[94]
$[CoCl(NH_3)_5]Cl_2$	+[b]	−1.24	0.1 M KCl	[63]
$[CoBr(NH_3)_5]Br_2$	+[b]	-1.24_5	0.1 M KCl	[63]
$[CoI(NH_3)_5]Cl_2$	+[b]	-1.24_5	0.1 M KCl	[63]
$[Co(OH_2)(NH_3)_5]^{3+}$	−0.474	−1.274	1 M K_2SO_4	[94]
$[Co(OH_2)(NH_3)_5]_2(SO_4)_3 \cdot 3H_2O$	−0.38	-1.24_5	0.1 M KCl	[63]
	−0.47	-1.27_5	0.5 M K_2SO_4	[63]
$[Co(OH)(NH_3)_5]^{2+}$	−0.473	−1.279	1 M K_2SO_4	[94]
$[Co(OH)(NH_3)_5]SO_4 \cdot 3/2\ H_2O$	−0.39	-1.25_5	0.1 M KCl	[63]
	-0.47_5	-1.27_5	0.5 M K_2SO_4	[63]
$[Co(NO_3)(NH_3)_5](NO_3)_2$	+[b]	-1.25_5	0.1 M KCl	[63]
	-0.05_5	−1.28	0.5 M K_2SO_4	[63]

(continued)

TABLE 2.2.6 (continued)

Compound	$E_{1/2}$ (V vs SCE) Co(III) → Co(II)	Co(II) → Co(metal)	Supporting electrolyte	Refs.
$[Co(N_3)(NH_3)_5]Cl_2$	$+^b$	-1.23	0.1 M KCl	[63]
	-0.12_5	-1.28_5	0.5 M K_2SO_4	[63]
$[Co(SO_4)(NH_3)_5]ClO_4$	-0.10_5	-1.23_5	0.1 M KCl	[63]
	-0.24_5	-1.27_5	0.5 M K_2SO_4	[63]
$[Co(S_2O_3)(NH_3)_5]Cl$	-0.13	-0.96_5	0.1 M KCl	[63]
	~ -0.12	Maximum	0.5 M K_2SO_4	[63]
$[Co\ ox\ (NH_3)_5]Br.3/2\ H_2O$	-0.21	-1.23_5	0.1 M KCl	[63]
	-0.33	-1.29	0.5 M K_2SO_4	[63]
$[Co(OCOCH_3)(NH_3)_5](ClO_4)_2$	-0.22_5	-1.24	0.1 M KCl	[63]
	-0.37	-1.27	0.5 M K_2SO_4	[63]
$[Co(OCOCH_3)(NH_3)_5](NO_3)_2$	-0.22_5	-1.23	0.1 M KCl	[96]
	-0.32	-1.26	0.25 M K_2SO_4	[96]
$[Co(CO_3)(NH_3)_5]NO_3$	-0.38_2	-1.25	0.1 M KCl	[63]
	-0.50	Maximum	0.5 M K_2SO_4	[63]
$[Co(HCO_3)(NH_3)_5](NO_3)_2$	-0.50_2	-1.28	0.5 M K_2SO_4	[63]
$[Co(CrO_4)(NH_3)_5]Cl$	$+^b$	Maximum	0.1 M KCl	[63]
	$+^b$	Maximum	0.5 M K_2SO_4	[63]
$[Co(NCS)(NH_3)_5]SO_4.2H_2O$	$+^b$	Maximum	0.5 M K_2SO_4	[63]

	$+^{b}$	Maximum	0.1 $\underline{M}$ KCl	[63]
$[Co(OCOCH_2Cl)(NH_3)_5](NO_3)_2$	-0.13	-1.23	0.1 $\underline{M}$ KCl	[96]
	-0.23	-1.26_5	0.25 $\underline{M}$ K_2SO_4	[96]
$[Co(OCOCHCl_2)(NH_3)_5](NO_3)_2$	-0.08_5	-1.22_5	0.1 $\underline{M}$ KCl	[96]
	-0.17	-1.27	0.25 $\underline{M}$ K_2SO_4	[96]
$[Co(OCOCCl_3)(NH_3)_5](NO_3)_2$	-0.05	-1.23	0.1 $\underline{M}$ KCl	[96]
	-0.09_5	-1.27_5	0.25 $\underline{M}$ K_2SO_4	[96]
$[Co(OCOH)(NH_3)_5](NO_3)_2$	-0.18_5	-1.22	0.1 $\underline{M}$ KCl	[96]
	-0.27_5	-1.26_5	0.25 $\underline{M}$ K_2SO_4	[96]
$[Co(OCOCH_2CH_3)(NH_3)_5](NO_3)_2$	-0.22_5	-1.23	0.1 $\underline{M}$ KCl	[96]
	-0.30	-1.27	0.25 $\underline{M}$ K_2SO_4	[96]
$[Co(OCOCH_2CH_2CH_3)(NH_3)_5](NO_3)_2$	-0.37_5	-1.23	0.1 $\underline{M}$ KCl	[96]
	-0.47_5	-1.26	0.25 $\underline{M}$ K_2SO_4	[96]
cis-$[Co(NO_2)_2(NH_3)_4]^+$	-0.043	-1.283	1 $\underline{M}$ K_2SO_4	[94]
	-0.075		1 $\underline{M}$ $NaNO_3$	[95]
cis-$[Co(NO_2)_2(NH_3)_4]ClO_4$	-0.048		0.1 $\underline{M}$ KCl	[74]
trans-$[Co(NO_2)_2(NH_3)_4]^+$	-0.207	-1.281	1 $\underline{M}$ K_2SO_4	[94]
	-0.219		1 $\underline{M}$ $NaNO_3$	[95]
trans-$[Co(NO_2)_2(NH_3)_4]ClO_4$	-0.21		0.1 $\underline{M}$ KCl	[74]

(continued)

TABLE 2.2.6 (continued)

Compound	$E_{1/2}$ (V vs SCE): Co(III) → Co(II)	$E_{1/2}$ (V vs SCE): Co(II) → Co(metal)	Supporting electrolyte	Refs.
cis-$[CoCl_2(NH_3)_4]^+$	+[b] -0.415 (aq species)	Maximum	1 M K_2SO_4	[94]
	+[b] -0.42 (aq species)	Maximum	0.5 M K_2SO_4	[74]
	+[b] -0.40 (aq species)	Maximum	0.1 M KCl	[74]
trans-$[CoCl_2(NH_3)_4]^+$	+[b] -0.415 (aq species)	Maximum	1 M K_2SO_4	[94]
	+[b] -0.42 (aq species)	Maximum	0.5 M K_2SO_4	[74]
	+[b] -0.40 (aq species)	Maximum	0.1 M KCl	[74]
cis-$NH_4[Co(SO_3)_2(NH_3)_4].3H_2O$	-0.63	Maximum	0.5 M K_2SO_4	[74]
	-0.57	-1.25	0.1 M KCl	[74]
cis-$[Co(OH_2)_2(NH_3)_4]_2(SO_4)_3.3H_2O$	-0.48_5	Maximum	0.5 M K_2SO_4	[74]
	-0.40_2	-1.25	0.1 M KCl	[74]
$[Co(IO_3)_2(NH_3)_4]IO_3.HIO_3$	+[b]	Hydrogen wave	0.5 M K_2SO_4	[74]
	+[b]	Hydrogen wave	0.1 M KCl	[74]
$[Co\,gly(NH_3)_4]SO_4.H_2O$	-0.40_5	Maximum	0.5 M K_2SO_4	[74]
	-0.23_4	-1.26	0.1 M KCl	[74], [72]
	-0.36	Maximum	0.1 M $(NH_4)_2$ ox	[72]

$[Co(CO_3)(NH_3)_4]_2SO_4.3H_2O$	-0.37	-1.28	0.5 M K_2SO_4	[74]
	-0.30_5	-1.24	0.1 M KCl	[74]
$[Co\ ala(NH_3)_4]SO_4$	-0.32	Maximum	0.5 M K_2SO_4	[74]
	-0.20_5	Maximum	0.1 M KCl	[74]
$[Co\ ox(NH_3)_4]Cl$	-0.29_5	Maximum	0.5 M K_2SO_4	[74]
	-0.21	-1.25	0.1 M KCl	[72]
	-0.21	-1.46	0.1 M $(NH_4)_2ox$	[72]
$[Co(leuc)(NH_3)_4](ClO_4)_2$	-0.25_5	Maximum	0.5 M K_2SO_4	[74]
	Maximum	Maximum	0.1 M KCl	[74]
$[Co(NO_2)_3(NH_3)_3]$	-0.026	-1.277	1 M K_2SO_4	[94]
	-0.10_1		1 M $NaNO_3$	[95]
$[Co(OH_2)_3(NH_3)_3]^{3+}$	-0.249	-1.274	1 M K_2SO_4	[94]
$[Co(NO_2)_4(NH_3)_2]^-$	-0.070	-1.276	1 M K_2SO_4	[94]
	-0.07_5		1 M $NaNO_2$	[95]
$NH4[Co\ ox_2(NH_3)_2].H_2O$	-0.03	-1.58	0.1 M $(NH_4)_2ox$	[72]
	-0.03	-1.23	0.1 M KCl	[72]
$[Co\ gly_2(NH_3)_2]Cl$	-0.23_0	-1.24	0.1 M KCl	[72]
	-0.28	-1.31	0.1 M K_2SO_4	[72]

[a] gly = glycinate; ala = alaninate; leuc = 1-leucinate; ox = oxalate; the abbreviations denote the ligands from which all hydrogen ions capable of dissociating are detached.

[b] The $E_{1/2}$ value is positive but is impossible to determine exactly due to the dissolution of mercury.

TABLE 2.2.7. Half-wave Potentials and Reversibility Values of $[Co\ en_3]^{3+}/[Co\ en_3]^{2+}$ Ion [97]

	Cathodic wave		Anodic wave	
Concn en (M)	$E_{1/2}$ (V vs SCE)	Reciprocal slope (V)	$E_{1/2}$ (V vs SCE)	Reciprocal slope (V)
0.5	-0.459	0.0604	-0.457	0.0594
0.4	-0.461	0.0587	-0.459	0.0567
0.2	-0.456	0.0564	-0.457	0.0578
0.1	-0.457	0.0570	-0.457	0.0579
0.1	-0.458	0.0602	-0.456	0.0586
0.07	-0.457	0.0580	-0.455	0.0585
0.05	-0.457	0.0570	-0.455	0.0583
0.03	-0.457	0.0585	-0.455	0.0582
0.02	-0.456	0.0572	-0.453	0.0589
0.01	-0.454	0.0581	-0.450	0.0584
0.007	-0.452	0.0584	-0.446	0.0574
0.005	-0.450	0.0578	-0.434	0.0595
0.004	-	-	-	-
0.003	-0.444	0.0575	-	-
0.001	-0.430	0.0626	-	-

TABLE 2.2.8. Effect of Na_2SO_4 Concentration on Half-Wave Potentials for the Solution of 1.00 mM $[Co\ en_3]Cl_3$ and 0.1 M en at 25°C [97]

Na_2SO_4 (M)	Activity of sulfate ion	$E_{1/2}$ (V vs SCE)
0.01	0.0095	-0.477
0.025	0.0230	-0.493
0.05	0.0442	-0.505
0.1	0.0844	-0.518
0.25	0.182	-0.537
0.5		-0.557
1.0		-0.580
2.5		-0.630
0.0		-0.420

TABLE 2.2.9. Effect of KCl Concentration on Half-Wave Potentials for the Solution Containing 1.00 mM [Co en_3]Cl_3 and 0.1 M en (25°C) [97]

KCl (M)	$Activity_{KCl}$	$E_{1/2}$ (V vs SCE)[a]	E_j (V)	$E_{1/2}$ corr (V vs SCE)
0.0	0.0			-0.420
0.01	0.009	-0.436	0.0031	-0.433
0.025	0.021	-0.442	0.0027	-0.439
0.05	0.041	-0.448	0.0024	-0.446
0.10	0.077	-0.456	0.0020	-0.454
0.20	0.144	-0.4652	0.0017	-0.4635
0.40	0.269	-0.4765	0.0013	-0.4752
0.70	0.447	-0.4880	0.0010	-0.4870
1.0	0.623	-0.4956	0.0008	-0.4948
2.0	1.225	-0.5125	0.0004	-0.5121
3.0	1.899	-0.5266	0.0002	-0.5264
4.0	2.686	-0.5390	0.0000	-0.5390

[a]The $E_{1/2}$ values were corrected for liquid junction potential (E_j) and for the activity coefficient ratio of the Co(III) and Co(II) complex ion.

TABLE 2.2.10. Variation of Half-Wave Potential and Slope of the First Step, Co(III) → Co(II), for 1 mM [Co en3]$(NO_3)_3$ with Concentration of Added Ethylenediamine (25°C) [97]

pH	Concn of en (mM)	Free en (mM)	$E_{1/2}$ (V vs SCE)	Slope
7.23	0.00	0.00	-0.38	0.086
12.0	0.00	0.00	-0.38	0.089
8.25	1.00	0.011	-0.39	0.082
8.71	1.00	0.034	-0.39	0.077
9.17	1.00	0.095	-0.40	0.070
9.73	1.00	0.28	-0.42	0.063
10.18	1.00	0.52	-0.43	0.063
9.44	10.0	1.60	-0.44	0.058
10.29	10.0	5.80	-0.45	0.059
10.70	10.0	7.80	-0.46	0.057
11.50	200	190	-0.46	0.061
10.39	500	320	-0.46	0.060
11.60	500	480	-0.46	0.061

TABLE 2.2.11. Half-Wave Potentials for Mixed Ethylenediaminecobalt(III) Complexes in Aqueous Media[a]

Compound	$E_{1/2}$ (V vs SCE) Co(III) → Co(II)	$E_{1/2}$ (V vs SCE) Co(II) → Co(metal)	Supporting electrolyte	Refs.
cis-$[Co(NO_2)_2en_2]NO_3$	-0.24 -0.41 (aq species)	max	0.1 M KCl	[99], [100]
cis-$[Co(NO_2)_2en_2]ClO_4$	-0.25 -0.5 (aq species)	max	0.5 M K_2SO_4	[73]
trans-$[Co(NO_2)_2en_2]NO_3$	-0.27 -0.40 (aq species) Co(III) → Co(II)	max	0.1 M KCl	[99], [100]
trans-$[Co(NO_2)_2en_2]Cl$	-0.26 -0.5 (aq species)	max	0.5 M K_2SO_4	[73]
cis-$[Co(NCS)(NO_2)en_2]Cl.H_2O$	-0.04 -0.38 (aq species)	max	0.1 M KNO_3	[99], [100]
trans-$[Co(NCS)(NO_2)en_2]Cl.H_2O$	-0.12 -0.36 (aq species)	max	0.1 M KNO_3	[99], [100]
cis-$[Co(NO_2)(NH_3)en_2]Br_2$	-0.12 -0.40 (aq species)	max	0.1 M KCl	[99], [100]
trans-$[Co(NO_2)(NH_3)en_2](NO_3)_2$	-0.20 -0.40 (aq species)	max	0.1 M KCl	[99], [100]
cis-$[Co(NCS)(NH_3)en_2](NCS)_2$	-0.13 -0.39 (aq species)	max	0.1 M KCl	[99], [100]
trans-$[Co(NCS)(NH_3)en_2](NCS)_2$	-0.10 -0.39 (aq species)	max	0.1 M KCl	[99], [100]
cis-$[Co(NH_3)_2en_2]I_3$	-0.31	max	0.1 M KCl	[99], [100]
cis-$[Co(NH_3)_2en_2](ClO_4)_3$	-0.45	max	0.5 M K_2SO_4	[73]
	-0.31	max	0.1 M KCl	[73]
trans-$[Co(NH_3)_2en_2]Cl_3.H_2O$	-0.31	max	0.1 M KCl	[99], [100]
	-0.45	max	0.5 M K_2SO_4	[73]
	-0.31	max	0.1 M KCl	[73]

cis-$[Co(NCS)_2en_2]Cl.H_2O$	$+^b$ -0.51 (aq species)	max	0.5 M K_2SO_4, 0.0016% Tween-80	[73]
	$+^b$	max	0.1 M KCl	[73]
trans-$[Co(NCS)_2en_2]Cl.H_2O$	$+^b$	max	0.5 M K_2SO_4	[73]
	$+^b$	max	0.1 M KCl	[73]
trans-$[Co(NCS)_2en_2]NO_3$	$+^b$ -0.37 (aq species) (pH = 7.92)	max	0.1 M KNO_3	[98]
	-0.39 (aq species) (pH = 9.20)	max	0.1 M KNO_3	[98]
	-0.40 (aq species) (pH = 10.28)	max	0.1 M KNO_3	[98]
trans-$[Co(NCS)(OH_2)en_2](NO_3)_2$	+0.08	max	0.1 M KNO_3	[98]
trans-$[Co(OH)(OH_2)en_2](NO_3)_2$	+0.02	max	0.1 M KNO_3	[98]
cis-$[Co(OH)(OH_2)en_2](NO_3)_2$	+0.02	max	0.1 M KNO_3	[98]
trans-$[Co(OH_2)(NH_3)en_2](NO_3)_3$	-0.08	max	0.1 M KNO_3	[98]
cis-$[CoCl_2en_2]Cl$	$+^b$ -0.5 (aq species)	max	0.5 M K_2SO_4	[73]
	$+^b$ -0.4 (aq species)	max	0.1 M KCl	[73]
cis-$[CoBr_2en_2]Br$	$+^b$ -0.5 (aq species)	max	0.5 M K_2SO_4	[73]
	$+^b$ -0.4 (aq species)	max	0.1 M KCl	[73]
trans-$[CoCl_2en_2]Cl$	$+^b$ -0.5 (aq species)	max	0.5 M K_2SO_4	[73]
	$+^b$ -0.4 (aq species)	max	0.1 M KCl	[73]
trans-$[CoBr_2en_2]Br$	$+^b$ -0.5 (aq species)	max	0.5 M K_2SO_4	[73]
	$+^b$ -0.4 (aq species)	max	0.1 M KCl	[73]

(continued)

TABLE 2.2.11 (continued)

Compound	$E_{1/2}$ (V vs SCE)		Supporting electrolyte	Refs.
	Co(III) → Co(II)	Co(II) → Co(metal)		
cis-$[Co(OH_2)_2en_2](NO_3)_3$	-0.50	max	0.5 M K_2SO_4	[73]
	-0.42	max	0.1 M KCl	[73]
	-0.01	max	1 M $HClO_4$	[73]
$[Co(CO_3)en_2]Cl.H_2O$	-0.45$_5$	max	0.5 M K_2SO_4	[73]
	-0.40	max	0.1 M KCl	
$[Co(SO_3)en_2]Cl.3H_2O$	-0.43 -0.5 (aq species)	max	0.5 M K_2SO_4	[73]
$[Co(S_2O_3)en_2]Br.3H_2O$	-0.49$_5$	max	0.5 M K_2SO_4	[73]
	max	max	0.1 M KCl	[73]
$[Co\ gly\ en_2]Cl_2.H_2O$	-0.41 -0.5 (aq species)	max	0.5 M K_2SO_4	[73]
	-0.36 -0.4 (aq species)	max	0.1 M KCl	[73]
$[Co\ ox\ en_2]Cl$	-0.31 -0.5 (aq species)	max	0.5 M K_2SO_4	[73]
	-0.32 -0.4 (aq species)	max	0.1 M KCl	[73]
$[Co\ leuc\ en_2](ClO_4)_2$	-0.30$_5$ -0.5 (aq species)	max	0.5 M K_2SO_4	[73]
	-0.33$_5$	max	0.1 M KCl	[73]
cis-$[Co(CN)_2en_2]NO_3$	-0.95 -1.38, Co(II) → Co(I)	max	0.5 M Na_2SO_4	[101]
	-0.86 -1.18, Co(II) → Co(I)	max	0.5 M Na_2SO_4 + 1 M en	[101]

trans-[$Co(CN)_2en_2$]NO_3	-0.71 -1.16, Co(II) → Co(I)	max	0.5 M Na_2SO_4	[102]
	-0.90			
	-0.78 -1.31, Co(II) → Co(I)	max	0.5 M Na_2SO_4 + 1 M en	[101]
trans-[$Co(CN)_2en(NH_3)_2$]$Cl.H_2O$	-0.67 -1.12	max	0.5 M Na_2SO_4	[102]
	-0.85			
	-0.79 -1.09, Co(II) → Co(I)	max	0.5 M Na_2SO_4 + 1 M en	[102]
cis-[$Co(CN)_2pn_2$]ClO_4	-0.90 -1.26, Co(II) → Co(I)	max	0.5 M Na_2SO_4	[103]
cis-[$Co(CN)_2tn_2$]ClO_4	-0.67 -1.14, Co(II) → Co(I)	max	0.5 M Na_2SO_4	[103]
[Co en $(NH_3)_4$]$_2(SO_4)_3.4H_2O$	-0.44	max	0.5 M K_2SO_4	[74]
	-0.27	max	0.1 M KCl	[74]
cis-[$Co(SO_3)_2en_2$]$^-$	-0.19 -0.35 (aq species)	max	0.1 M KNO_3	[111]
cis-[$Co(SO_3)(OH_2)en2$]$^+$	-0.20 -0.35	max	0.1 M KNO_3	[111]
cis-[$Co(SO_3)(NH_3)en_2$]$^+$	-0.52		0.1 M KNO_3 0.1 M phosphate buffer, pH = 1.10	[111]

[a]Cobalt(III) complexes with one or more ethylenediamines show always a maximum wave (denoted as "max") at the second step, Co(II) → Co(metal), at the DME in aqueous media. The dicyano cobalt(III) complexes exhibit a new wave at around -1.2 ~ -1.3 V due to the reduction of the tetracyanocobaltate(II) to cobaltate(I) complex, which would be formed through rearrangement of cyanides during reduction in aqueous solutions; these complexes were conveniently categorized in the cyanide complexes.

[b]The $E_{1/2}$ value is in the positive region of potential but it cannot be determined exactly due to dissolution of mercury.

TABLE 2.2.12. The Effect of Changes of pH on the Reduction of 0.001 M trans-$[Co(NO_2)_2en_2]NO_3$ in 0.2 M Phosphate Buffer (25°C) [104][a]

pH	First wave [Co(III) → Co(II)]			Wave for the aquated species [Co(III) → Co(II)]		
	$E_{1/2}$ (V)	i_d (μA)	Slope	$E_{1/2}$ (V)	i_d (μA)	Slope
5.78	-0.265	4.29	0.0625	-		
6.69	-0.265	4.34	0.0620			
8.20	-0.267	4.19	0.0630			
9.25	-0.263	3.60	0.0620	-0.466	4.13	0.0600
19.60	-0.262	3.04	0.0595	-0.462	4.06	0.0595
10.14	-0.260	3.08	0.0610	-0.472	4.30	0.0585
10.52	-0.267	2.93	0.0595	-0.489	4.20	0.0610
10.69	-0.264	2.96	0.0630	-0.497	4.26	0.0605
11.27	-0.269	2.84	0.0575	-0.516	4.20	0.0600

[a]V vs SCE.

TABLE 2.2.13. Half-Wave Potentials of Mixed Cobalt(III) and Cobalt(II) Complexes in Aqueous Media[a]

Compound	$E_{1/2}$ (V)	n	Change of valence	$E_{1/2}$ (V)	n	Change of valence	Supporting electrolyte	Refs.
$[Co(pim)_3]^{2+}$	-0.2	1	II → III				0.5 M Na_2SO_4 + borate buffer (pH = 10)	[109]
$[Co(pic)_3]^{3+}$	-0.30	1	III → II				1 M Na_2SO_4 + 0.1 M pic	[109]
$[Co\ pn_3]^{3+}$	-0.58	1	III → II	max			0.75 M Na_2SO_4 + 0.25 M pn H_2SO_4 + 0.75 M pn	[109]
$[Co(OH_2)_2tren]^{3+}$	-0.41	1	III → II				0.5 M phosphate buffer (pH = 8.1) + 3.99 M tren + Triton X 100	[110]
$[Co(dgH)_3].5/2\ H_2O$	-1.04	1	III → II				50% H_2O + ethanol 0.5 M LiCl + dgH_2	[112]
α-$[Co\ gly_3].2H_2O$	-0.04	1	III → II				0.1 M KCl	[99], [100]
	+0.04	1	III → II				70.72% $HClO_4$	[99], [100]
	-0.14	1	III → II				0.1 M KCl	[72]
	-0.07	1	III → II				0.1 M K_2SO_4	[72]
β-$[Co\ gly_3].H_2O$	-0.03	1	III → II				0.1 M KCl	[99], [100]
	+0.04	1	III → II				70.72% $HClO_4$	[99], [100]
α-$[Co\ ala_3]$	-0.08	1	III → II				70.72% $HClO_4$	[99], [100]
β-$[Co\ ala_3]$	-0.08	1	III → II				70.72% $HClO_4$	[99], [100]
$[Co\ gly_2en]Cl$	-0.237	1	III → II				0.1 M KCl	[103]
$K_3[Co\ ox_3]$	+	1	III → II	No reduction to the metal			0.1 M $(NH_4)_2ox$	[72]
	+	1	III → II	No reduction to the metal			0.1 M KCl	[72]

(continued)

TABLE 2.2.13 (continued)

Compound	$E_{1/2}$ (V)	n	Change of valence	$E_{1/2}$ (V)	n	Change of valence	Supporting electrolyte	Refs.
$NH_4[Co\ ox_2(NH_3)_2.H_2O$	-0.03	1	III → II	-1.58	1	II → 0	0.1 M $(NH_4)_2ox$	[72]
	-0.03	1	III → II	-1.23	1	II → 0	0.1 M KCl	[72]
$Na_2[Co^{II}edta]$	+0.140	1	II → III				pH = 4.5 to 10.5	[113]
$Na_2[Co(CN)edta].3H_2O$*	-0.34	1	III → II	No reduction to the metal			0.5 M Na_2SO_4	[103]
$Na_2[Co(NO_2)edta].3H_2O$	+0.13	1	III → II	No reduction to the metal			0.5 M Na_2SO_4	[103]
$K_2[CoCl\ edta].3H_2O$	+0.13	1	III → II	No reduction to the metal			0.5 M Na_2SO_4	[103]
$Na_2[CoBr\ edta].4H_2O$	-0.09	1	III → II	No reduction to the metal			0.5 M Na_2SO_4	[103]
$Na_2[Co(OH)edta].3H_2O$	+[b]	1	III → II	No reduction to the metal			0.5 M Na_2SO_4 (0°C)	[103]
$K[Co\ edta].2H_2O$	+0.14	1	III → II	No reduction to the metal			0.5 M Na_2SO_4	[103]
$[Co\ edta]^-$	+0.13	1	III → II				0.05 M acetate buffer (pH = 4.6 to 5.4) = 0.2 (KNO_3)	[33]
$[Co\ edta]^{2-}$	+0.13	1	II → III					[33]
$[Co\ pdta]^-$	+0.12	1	III → II					[33]
$[Co\ pdta]^{2-}$	+0.12	1	II → III					[33]
$[Co\ trdta]^-$	+0.05	1	III → II					[33]
$[Co\ trdta]^{2-}$	+0.05	1	II → III					[33]

[a]pim = 2-(2-pyridyl)-imidazolin; pic = 2-picolylamine; tren = triethylenetetramine; pdta = propylenediaminetetraacetate; trdta = trimethylenediaminetetraacetate; dg = dimethylglyoximate.

[b]It is impossible to determine the $E_{1/2}$ value exactly due to the hindrance of anodic waves, but it is in the positive region of potential.

TABLE 2.2.14. Half-Wave Potentials for the Monocyano Cobalt(III) Complexes of the trans-$[Co(CN)X(NH_3)_4]$ Type in Aqueous Media (25° C) [70][a]

Compound	$E_{1/2}$ (V vs SCE) Co(III) → Co(II)	$E_{1/2}$ (V vs SCE) Co(II) → Co(metal)	Supporting electrolyte
trans-$[Co(CN)(NO_2)(NH_3)_4]ClO_4 \cdot 1/2\ H_2O$	-0.57	-1.17	0.5 M Na_2SO_4
	-0.44	-1.06	HAc + NaAc buffer (pH = 4.1)
trans-$[Co(CN)(NO)(NH_3)_4]Cl \cdot 1/2\ H_2O$	-0.52	-1.12	0.5 M Na_2SO_4
trans-$[Co(CN)py(NH_3)_4](ClO_4)_2$	-0.46	-1.12	0.5 M Na_2SO_4
	-0.45	-1.05	HAc + NaAc buffer (pH = 4.1)
trans-$[Co(CN)(NH_2CH_3)(NH_3)_4](ClO_4)_2$	-0.21	-1.04	0.5 M Na_2SO_4
	-0.39	-1.12	HAc + NaAc buffer (pH = 4.1)
trans-$[Co(CN)(NH_2C_2H_5)(NH_3)_4](ClO_4)_2$	-0.47	-1.17	0.5 M Na_2SO_4
	-0.30	-1.02	HAc + NaAc buffer (pH = 4.1)
$[Co(CN)(NH_3)_5](ClO_4)_2 \cdot 1/2\ H_2O$	-0.36	-1.05	0.5 M Na_2SO_4
	-0.35		HAc + NaAc buffer (pH = 4.1)

(continued)

TABLE 2.2.14 (continued)

Compound	$E_{1/2}$ (V vs SCE) Co(III) → Co(II)	$E_{1/2}$ (V vs SCE) Co(II) → Co(metal)	Supporting electrolyte
trans-$[Co(CN)(SO_3)(NH_3)_4].2H_2O$	-0.63	-1.10	0.5 M Na_2SO_4
	-0.55	-1.00	HAc + NaAc buffer (pH = 4.1)
trans-$[Co(CN)(CO_3)(NH_3)_4].H_2O$	-0.46	-1.12	0.5 M Na_2SO_4
	-0.43	-1.10	HAc + NaAc buffer (pH = 4.1)
trans-$[Co(CN)(OH_2)(NH_3)_4]Cl_2$	-0.35	-1.05	0.5 M Na_2SO_4
	-0.31	-1.06	HAc + NaAc buffer (pH = 4.1)
trans-$[Co(CN)(N_3)(NH_3)_4]N_3.H_2O$	-0.52	-1.15	0.5 M Na_2SO_4
	-0.27	-1.06	HAc + NaAc buffer (pH = 4.1)
trans-$[Co(CN)I(NH_3)_4]I$	-0.24	-1.06	0.5 M Na_2SO_4
	~-0.13	~-1.13	HAc + NaAc buffer (pH = 4.1)
trans-$[Co(CN)(NCS)(NH_3)_4]ClO_4.3/2\ H_2O$	-0.23	-1.06	0.5 M Na_2SO_4
	-0.17	~-1.10	HAc + NaAc buffer (pH = 4.1)

[a]The compounds of this series, trans-$[Co(CN)(S_2O_3)(NH_3)_4].2H_2O$ and trans-$[Co(CN)(P{\equiv}\phi_3)(NH_3)_4]I$, are insoluble in aqueous solutions; $P{\equiv}\phi_3$ = triphenylphosphine.

TABLE 2.2.15. Half-Wave Potentials for the Dicyano and Related Complexes of the $[Co(CN)_2N_4]$ Type in Aqueous Media (25°C)[a]

Compound	First step (change of valence) (V)	Second step (change of valence) (V)	Third step (change of valence) (V)	Background salt	Refs.
trans-$[Co(CN)_2(NH_3)_4]NO_3 \cdot H_2O$	-0.62 (3 → 2)	-1.11 (2 → 0)		0.5 M Na_2SO_4	[114]
	-0.51 (3 → 2)	-1.02 (2 → 0)		0.5 M Na_2SO_4 + 1 M NH_3	
	-0.52 (3 → 2)	-1.04 (2 → 0)		0.5 M Na_2SO_4 + 0.1 M NH_3	
trans-$[Co(CN)_2en(NH_3)_2]Cl \cdot H_2O$	-0.71 (3 → 2)	-1.12 (2 → 0)		0.5 M Na_2SO_4	[103]
	-0.94 (3 → 2)				
	-0.74 (3 → 2)	-1.10 (2 → 1)	-1.17 (2 → 0)	0.5 M Na_2SO_4 + 1 M NH_3	
	-0.79 (3 → 2)	-1.09 (2 → 1)	max (2 → 0)	0.5 M Na_2SO_4 + 1 M en	
trans-$[Co(CN)_2en_2]NO_3$	-0.75 (3 → 2)	-1.16 (2 → 0)		0.5 M Na_2SO_4	[103]
	-0.94 (3 → 2)				
	-0.87 (3 → 2)	-1.20 (2 → 1)	max (2 → 0)	0.5 M Na_2SO_4 + 1 M en	
	-0.78 (3 → 2)	-1.31 (2 → 1)	max (1 → 0)	0.5 M Na_2SO_4 + 1 M en + 0.1 M KCN	
cis-$[Co(CN)_2en_2]NO_3$	-0.95 (3 → 2)	-1.38 (2 → 0)		0.5 M Na_2SO_4	[101]
	-0.86 (3 → 2)	-1.18 (2 → 1)	max (2 → 0)	0.5 M Na_2SO_4 + 1 M en + 0.00084% Triton X 100	
	-0.78 (3 → 2)	-1.33 (2 → 1)	max (2 → 0)	0.45 M Na_2SO_4 + 1 M en + 0.1 M KCN	[101]

(continued)

TABLE 2.2.15 (continued)

Compound	First step (change of valence) (V)	Second step (change of valence) (V)	Third step (change of valence) (V)	Background salt	Refs.
$K_4[Co^{II}(CN)_4$-en-$Co^{II}(CN)_4].H_2O$	-0.31 (3 → 2)	-1.23 (2 → 1)	-1.57 (1 → 0)	0.5 M Na_2SO_4 + 1.5 M en + 0.00084% Triton X 100	[101]
cis-$[Co(CN)_2pn_2]ClO_4$	-0.88 (3 → 2)	-1.19 (2 → 0)		0.5 M Na_2SO_4	[103]
cis-$[Co(CN)_2tn_2]ClO_4$	-0.65 (3 → 2)	-1.12 (2 → 0)		0.5 M Na_2SO_4	
cis-$[Co(CN)_2dip_2]NO_3.7H_2O$	-0.43 (3 → 2)	-0.87 (2 → 1)	-1.23 (1 → 0)	0.5 M Na_2SO_4	[114], [354]
	-0.43 (3 → 2)	-0.92 (2 → 1)	-1.25 (1 → 0)	1 M LiCl + 30% methanol	[114], [354]
	-0.43 (3 → 2)	-0.95 (2 → 1)	-1.26 (1 → 0)	1 M LiCl + 30% methanol + 1 mM dip	[354]
cis-$[Co(CN)_2phen_2]NO_3.6H_2O$	-0.39 (3 → 2)	-1.19 (2 → 0)		0.5 M Na_2SO_4 (25°C)	[114], [354]
	-0.32 (3 → 2)	-1.27 (2 → 0)		0.1 M $[(CH_3)_4N]Cl$ + 99.7% methanol (25°C)	
	-0.32 (3 → 2)	-0.91 (2 → 1)	-1.26 (2 → 0)	0.1 M $[(CH_3)_4N]Cl$ + 99.7% methanol (50°C)	
1,10-phenanthroline	-1.37 (current-rise)			0.1 M $[(CH_3)_4N]Cl$ + 99.7% methanol (25°C)	
trans-$K[Co(CN)_2(dgH)_2].3/2\ H_2O$	-1.10_5 (3 → 1)			0.5 M Na_2SO_4	[115]

[a] tn = trimethylenediamine; pn = propylenediamine; dg^{2-} = dimethylglyoximate ion; phen = phenanthroline. Voltage unit: V vs SCE.

TABLE 2.2.16. Half-Wave Potentials for Mixed Cyano Cobalt(III) Complexes in Aqueous Media (25° C)

Compound	First step (change of valence) (V)	Second step (change of valence) (V)	Background salt	Refs.
trans-$K[Co(OH_2)_2(CN)_4]\cdot 3/4\ H_2O$	-1.03 (3 → 2)	-1.41 (2 → 1)	0.5 M Na_2SO_4	[116], [356]
trans-$Na_5[Co(SO_3)_2(CN)_4]\cdot 3H_2O$	-1.21 (3 → 1)	No reduction to metal	0.5 M Na_2SO_4	[116], [117]
	-1.22 (3 → 1)	No reduction to metal	0.5 M Na_2SO_3	[356]
cis-$Na_2K_3[Co(SO_3)_2(CN)_4]\cdot 2.5H_2O$ (119)	-1.32 (3 → 1)	-1.58 (1 → 0)	0.5 M Na_2SO_4	[103], [356]
	-1.38 (3 → 1)	-1.75 (1 → 0)	0.5 M Na_2SO_3	[103], [356]
cis-$Na[Co(CN)_4en]\cdot 3.5\ H_2O$	-1.17 (3 → 0)		0.5 M Na_2SO_4	[118]
	-1.36_4 (3 → 1)	-1.51 (1 → 0)	0.5 M Na_2SO_4 + 2 M en	[118]
	-1.12 (3 → 1)	-1.39 (1 → 0)	0.5 M Na_2SO_4 + 1 M en	[116]
$K_4[Co(CN)_4\text{-en-}Co(CN)_4]\cdot H_2O$	-0.32 (2 → 3)	-1.23 (2 → 1)	0.5 M Na_2SO_4 + 1.5 M en + 0.00084% Triton X 100	
trans-$K[Co(CN)_4(NH_3)_2]\cdot H_2O$	-0.78 (3 → 2)	-1.05 (2 → 1)	0.5 M Na_2SO_4	[103]
	-0.98 (3 → 2)	-1.37 (2 → 1)	$NH_4Cl + NH_3$ buffer, $\mu = 1$ M (pH = 11.0)	

(continued)

TABLE 2.2.16 (continued)

Compound	First step (change of valence) (V)	Second step (change of valence) (V)	Background salt	Refs.
$K_3[CoCl(CN)_5]$	-0.86 (3 → 2)	-1.20 (2 → 1)	3 M KCl + 0.0032% Tween 80	[120], [122]
	-1.01 (3 → 2)	-1.34 (2 → 1)	0.5 M Na_2SO_4	[121], [355]
$K_3[CoBr(CN)_5]$	-0.71$_5$ (3 → 2)	-1.18 (2 → 1)	3 M KBr + 0.0032% Tween 80	[120], [355]
	-0.85 (3 → 2)	-1.26 (2 → 1)	0.5 M Na_2SO_4	[121], [355]
$K_3[CoI(CN)_5]$	-0.71 (3 → 2)	-1.13 (2 → 1)	3 M KI + 0.0032% Tween 80	[120], [355]
	-0.80 (3 → 2)	-1.30 (2 → 1)	0.5 M Na_2SO_4	[121], [355]
$K_3[Co(SCN)(CN)_5]$	-0.78 (3 → 2)	-1.23 (2 → 1)	1 M KCl	[120], [355]
	-0.85 (3 → 2)	-1.27 (2 → 1)	0.5 M Na_2SO_4	[121], [355]
$K_3[Co(N_3)(CN)_5] \cdot 2H_2O$	-1.37$_3$ (3 → 1)		1 M KCl	[120], [355]
	-1.31 (3 → 1)		0.5 M Na_2SO_4	[121], [355]
$K_4[Co(S_2O_3)(CN)_5]$	-1.29 (3 → 1)		1 M KCl	[120], [355]
	-1.38 (3 → 1)		0.5 M Na_2SO_4	[121], [355]
$K_3[Co(NO_2)(CN)_5]$	-1.33$_5$ (3 → 1)		1 M KCl	[120], [355]
	-1.44 (3 → 1)		0.5 M Na_2SO_4	[121], [355]

$K_4[Co(SO_3)(CN)_5] \cdot 3H_2O$	-1.53_5 (3 → 1)	1 M KCl	[120], [355]
	-1.56 (3 → 1)	0.5 M Na_2SO_4	[121], [355]
$K_2[Co^{III}(OH_2)(CN)_5]$	-1.45 (3 → 1)	1 M KCN	[123]
	-1.45 (3 → 1)	1 M KCl	[124]
	-1.65 (3 → 1)	pH = 12.5, 0°C	[125]
	-0.68 (3 → 2)	2 M KCl pH = 6.3	[126]
	-0.90 (3 → 2)	2 M KCl pH = 9	[126]
$K_3[Co^{II}(CN)_5]$	-1.30 (2 → 1)	1 M KCN	[123]
	-1.30 (2 → 1)	1 M KCl	[124]
	-1.59 (2 → 1)	pH = 12.5, 0°C	[125]
$K_6[(NC)_5Co^{III}OOCo^{III}(CN)_5] \cdot H_2O$	-1.66 (3 → 1)	pH = 12.5, 0°C	[125]
$[CoCl(CN)_5]^{3-}$	-0.82 (3 → 2)	2 M KCl pH = 9	[126]
	-0.63 (3 → 2)	2 M KCl pH = 6.3	[126]
$[CoBr(CN)_5]^{3-}$	-0.81 (3 → 2)	2 M KCl pH = 9	[126]
	-0.61 (3 → 2)	2 M KCl pH = 6.3	[126]
$[CoI(CN)_5]^{3-}$	-0.53 (3 → 2)	2 M KCl pH = 9	[126]
	-0.53 (3 → 2)	2 M KCl pH = 6.3	[126]
$K_3[Co^{III}(CN)_6]$	No reduction (0 to -2.0 V)	1 M KCl	[123]

TABLE 2.2.17. Half-Wave Potentials for the Co(III) and Co(II) Complexes of Schiff Bases in Aqueous and Nonaqueous Media[a]

Compound	First step (change of valence) (V)	Second step (change of valence) (V)	Supporting electrolyte	Refs.
$[Co(OH_2)_2(salen)]ClO_4$	+0.34 (3 → 2)	-0.555 (reduction of the ligand)	0.2 N HCl, H_2O (60%) - methanol (40%)	[127]
	+0.130 (3 → 2)	-1.150 (2 → 1)	0.25 M $LiNO_3$, 95% ethanol	[127]
$[Co(P{\equiv}\phi_3)(OH_2)(salen)]ClO_4$	-0.050 (3 → 2)	-0.67 (catalytic hydrogen wave)	0.2 N HCl, H_2O (60%) - methanol (40%)	[127]
	-0.035 (3 → 2)	-1.135 (2 → 1)	0.25 M $LiNO_3$, 95% ethanol	[127]
$[Co(P{\equiv}bu_3)(OH_2)(salen)]ClO_4$	-0.05 (3 → 2)	-0.575 (reduction of the ligand)	0.2 N HCl, H_2O (60%) - methanol (40%)	[127]
	-0.325 (3 → 2)	-1.140 (2 → 1)	0.25 M $LiNO_3$, 95% ethanol	[127]
$[Co(CH_3)(OH_2)(salen)]$	-1.34 (3 → 1)		0.25 M LiCl, 95% ethanol	[127]
$[Co^{III}(P{\equiv}\phi_3(OH_2)(bae)]ClO_4$	-0.0 (3 → 2)	-0.71 (catalytic hydrogen evolution)	0.2 N HCl, H_2O (60%) - methanol (40%)	[127]
	+0.070 (3 → 2)	-1.520 (2 → 1)	0.25 M $LiNO_3$, 95% ethanol	[127]
$[Co(P{\equiv}bu_3)(OH_2)(bae)]ClO_4$	-0.05 (3 → 2)	-0.71 (catalytic hydrogen evolution)	0.2 N HCl, H_2O (60%) - methanol (40%)	[127]
	-0.385 (3 → 2)	-1.520 (2 → 1)	0.25 M $LiNO_3$, 95% ethanol	[127]
$[Co(CH_3)(OH_2)(bae)]$	-1.60 (3 → 1)	Catalytic hydrogen evolution	0.25 M LiCl, 95% ethanol	[127]
$[Co(py)_2(bae)]Cl$	-0.82	-1.38	HAc + NaAc buffer	[128]
$[Co(aniline)_2(bae)]Cl$	-0.82	-1.35	HAc + NaAc buffer	[128]
$[Co(p\text{-}Cl\text{-}C_6H_4NH_2)_2(bae)]Cl$	-0.82	-1.34	HAc + NaAc buffer	[128]

[Co(p-$CH_3C_6H_4NH_2)_2$(bae)]Cl	-0.82	-1.3	HAc + NaAc buffer	[128]
[Co^{II}(o-HO-C_6H_4CH=N-R)]				
R = m-Cl-C_6H_4	-1.211 (2 → 1)	-1.45	0.1 M KCl, 80% ethylene-glycol-water soln	[129]
R = phenyl, ϕ	-1.238 (2 → 1)	-1.49	0.1 M KCl, 80% ethylene-glycol-water soln	[129]
R = m-$C_6H_4(CH_3)$	-1.225 (2 → 1)	-1.47	0.1 M KCl, 80% ethylene-glycol-water soln	[129]
R = p-$C_6H_4(CH_3)$	-1.237 (2 → 1)	-1.47	0.1 M KCl, 80% ethylene-glycol-water soln	[129]
R = cyclo-C_6H_{11}	-1.34 (2 → 1)		0.1 M KCl, 80% ethylene-glycol-water soln	[129]
R = p-ClC_6H_4	-1.260 (2 → 1)	-1.46	0.1 M KCl, 80% ethylene-glycol-water soln	[129]
	-1.190 (2 → 1)	-1.43	Satd LiCl, 80% ethylene-glycol-water soln	[129]
R = o-$C_6H_4OCH_3$	-1.280 (2 → 1)	-1.48	0.1 M KCl, 80% ethylene-glycol-water soln	[129]
	-1.240 (2 → 1)	-1.49	Satd LiCl, 80% ethylene-glycol-water soln	[129]
R = p-$C_6H_4OCH_3$	-1.299 (2 → 1)	-1.50	0.1 M KCl, 80% ethylene-glycol-water soln	[129]
	-1.250 (2 → 1)	-1.49	Satd LiCl, 80% ethylene-glycol-water soln	[129]
R = phenyl, ϕ	-1.220 (2 → 1)	-1.44	Satd LiCl, 80% ethylene-glycol-water soln	[129]

[a] salen = bis(salicylaldehyde)ethylenediimine; bae = bis(acetylacetone)ethylenediimine; py = pyridine; P≡bu_3 = tributylphosphine; P≡ϕ_3 = tripheylphosphine; ϕ = phenyl.

TABLE 2.2.18. Half-Wave Potentials for the π-Complexes of Cobalt in Aqueous and Nonaqueous Media[a]

Compound	First step (change of valence) (V)	Second step (change of valence) (V)	Supporting electrolyte	Refs.
$[Co^{III}(\pi\text{-}C_5H_5)_2]Cl$ (cobalticinium chloride)	-1.18 (3 → 2)		1 M LiCl aq soln	[130]
	-1.03 (3 → 2) vs 4 M NaCl calomel		1 M LiCl, ethanol (80%)-H_2O (20%)	[130]
	-0.98 (3 → 2)		1 M LiCl, ethanol (96%)-H_2O (4%)	[130]
$[Co^{III}(\pi\text{-}C_5H_5)_2]I_3$	-1.11 (3 → 2)		0.2 M $NaClO_4$, formamide	[131]
$[Co^{III}(\pi\text{-}C_5H_5)_2]ClO_4$	-1.16 (3 → 2)		0.1 M $NaClO_4$ aq soln (pH = 6.2)	[132]
$[Co^{III}(\pi\text{-}C_9H_7)_2]ClO_4$ [bis(indenyl)cobalt(III)perchlorate]	-0.71 (3 → 2)		0.2 M $NaClO_4$, formamide	[131]
	-0.53 (3 → 2)		0.2 M $NaClO_4$, N,N'-dimethylformamide	[131]
	-0.6 (3 → 2) vs the std calomel electrode		0.1 M $NaClO_4$ aq soln	[133]
$[(\pi\text{-}C_5H_5)Co(CO)_2]$	-2.5 (1-electron redn) vs 10^{-3} M Ag/$AgClO_4$	-3.2	0.1 M $[(bu)_4N]ClO_4$, $CH_3OCH_2CH_2OCH_3$ (dimethoxyethane)	[134]
$[(\pi\text{-}C_5H_5)Co(S_2C_4F_6)]$(red) $S_2C_4F_6$: 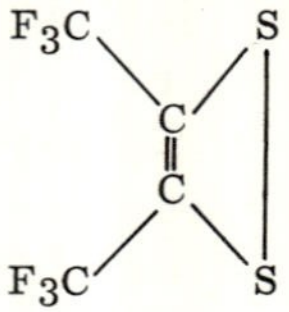	-1.1 (1-electron redn) vs 10^{-3} M Ag/$AgClO_4$	-2.9	0.1 M $[(bu)_4N]ClO_4$, $CH_3OCH_2CH_2OCH_3$ (dimethoxyethane)	[134]
$[(\pi\text{-}C_5H_5)Co(CO)]_3$	-1.6 (1-electron redn) vs 10^{-3} M Ag/$AgClO_4$		0.1 M $[(bu)_4N]ClO_4$, dimethoxyethane	[135]
$Co_2(CO)_6(\phi C\equiv C\phi)$	-1.6 (1-electron redn) vs 10^{-3} M Ag/$AgClO_4$		0.1 M $[(bu)_4N]ClO_4$, dimethoxyethane	[135]

$ClC(Co(CO)_3)_3$	-1.1 (1-electron redn) vs 10^{-3} M $Ag/AgClO_4$	0.1 M $[(bu)_4N]ClO_4$, dimethoxyethane	[135]
$Co_2(CO)_7(C_4HO_2)$	-1.2 (1-electron redn) vs 10^{-3} M $Ag/AgClO_4$	0.1 M $[(bu)_4N]ClO_4$, dimethoxyethane	[135]
$[Co(CO)_4]_2$	-0.9 (2-electron redn)(0 0) vs 10^{-3} M $Ag/AgClO_4$	0.1 M $[(bu)_4N]ClO_4$, CH_2Cl_2	[136]
$[Co(CO)_4(Sn\equiv\phi_3)]$	-1.6 (1-electron redn) vs 10^{-3} M $Ag/AgClO_4$	0.1 M $[(bu)_4N]ClO_4$, dimethoxyethane	[136]
$[Co(CO)_4Sn\equiv(CH_3)_3]$	-1.6 (1-electron redn) vs 10^{-3} M $Ag/AgClO_4$	0.1 M $[(bu)_4N]ClO_4$, dimethoxyethane	[136]
$[Co(CO)_3NO]$	-1.07 (1-electron redn) vs AgCl/Ag satd $[(C_2H_5)_4N]Cl$ in AN	0.1 M $[(C_2H_5)_4N]ClO_4$, CH_3CN, acetonitrile (AN)	[137]
$[Co(CO)_2(NO)Sb\equiv\phi_3]$	-1.18 (1-electron redn) vs AgCl/Ag satd $[(C_2H_5)_4N]ClO_4$ in AN	0.1 M $[(C_2H_5)_4N]ClO_4$, AN	[138]
$[Co(CO)_2(NO)As\equiv\phi_3]$	-1.27 (1-electron redn) vs AgCl/Ag	0.1 M $[(C_2H_5)_4N]ClO_4$, AN	[138]
$[Co(CO)_2(NO)P\equiv\phi_3]$	-1.46 (1-electron redn) vs AgCl/Ag	0.1 M $[(C_2H_5)_4N]ClO_4$, AN	[138]
$[Co(CO)_2(NO)]_2$diphos	-1.53 (1-electron redn) vs AgCl/Ag	0.1 M $[(C_2H_5)_4N]ClO_4$, AN	[138]
$[Cl(CO)(NO)(Sb\equiv\phi_3)_2]$	-1.65 (1-electron redn) vs AgCl/Ag	0.1 M $[(C_2H_5)_4N]ClO_4$, AN	[138]
$[Co(CO)(NO)(As\equiv\phi_3)_2]$	-1.70 (1-electron redn) vs AgCl/Ag	0.1 M $[(C_2H_5)_4N]ClO_4$, AN	[138]
$[Co(CO)(NO)(P\equiv\phi_3)_2]$	-1.89 (1-electron redn) vs AgCl/Ag	0.1 M $[(C_2H_5)_4N]ClO_4$, AN	[138]

(continued)

TABLE 2.2.18 (continued)

Compound	First step (change of valence) (V)	Second step (change of valence) (V)	Supporting electrolyte	Refs.
[Co(CO)(NO)diphos]	-2.02 (1-electron redn) vs AgCl/Ag		0.1 M $[(C_2H_5)_4N]ClO_4$, AN	[138]
NO (nitrosyl)	-0.96 vs AgCl/Ag		0.1 M $[(C_2H_5)_4N]ClO_4$, AN	[138]
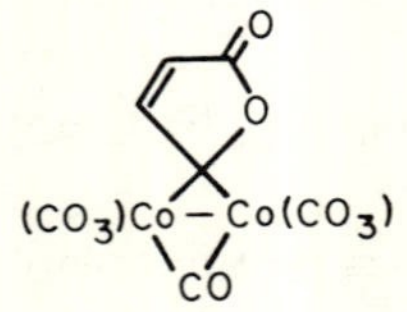	-1.3 (1-electron redn) vs 10^{-3} M Ag/$AgClO_4$		0.1 M $[(bu)_4N]ClO_4$, dimethoxyethane	[139]
	-2.9 (1-electron redn) vs 10^{-3} M Ag/$AgClO_4$		0.1 M $[(bu)_4N]ClO_4$, dimethoxyethane	[139]
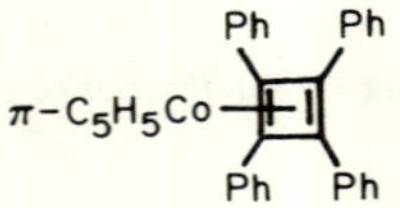	-1.6 (1-electron redn) vs 10^{-3} M Ag/$AgClO_4$	-2.1	0.1 M $[(bu)_4N]ClO_4$, dimethoxyethane	[139]
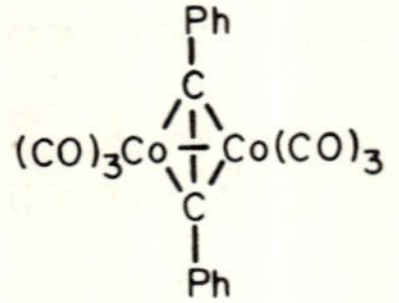				
$[(\pi\text{-}C_5H_5)CoP\equiv\phi_2]_2$	-0.3	-2.6 (1-electron redn) vs 10^{-3} M Ag/$AgClO_4$	0.1 M $[(bu)_4N]ClO_4$, dimethoxyethane	[140]
$[(\pi\text{-}C_5H_5)CoSCH_3]_2$	-0.5 (1-electron oxidn) vs 10^{-3} M Ag/$AgClO_4$	-2.4 (1-electron redn)	0.1 M $[(bu)_4N]ClO_4$, dimethoxyethane	[140]

[a] Ph (phenyl); ϕ (phenyl); Sb$\equiv\phi_3$ (triphenylstibine); As $\equiv\phi_3$ (triphenylarsine); P $\equiv\phi_3$ (triphenylphosphine); diphos (ϕ_2 =P—CH_2—CH_2—P =ϕ_2); π-C_5H_5 (cyclopentadienyl).

TABLE 2.2.19. Half-Wave Potentials for Cobalt Complexes with Ligands of π-Bonding Character in Aqueous and Nonaqueous Media[a]

Compound	$E_{1/2}$ (change of valence) (V)	$E_{1/2}$ (change of valence) (V)	$E_{1/2}$ (change of valence) (V)	$E_{1/2}$ (change of valence) (V)	Supporting electrolyte	Refs.
$[Co\ phen_3]^{3+}$	+0.09 (3→2)				1 M Na_2SO_4 + 0.03% gelatin	[109]
$[Co\ phen_3](ClO_4)_3 \cdot 2H_2O$	+0.10 (3→2)	-0.33 (pre-wave)	-1.0 (2→1)		1 M Na_2SO_4 + 0.03% gelatin	[141], [142]
$[Co\ phen_3]Cl_3 \cdot 7H_2O$	+0.08 (3→2)	-1.00_5 (2→1)	-1.34 (1→0)		0.5 M Na_2SO_4	[81], [82]
	+	-1.62 (2→1)			1 M LiCl, ethanol (99.7%)	[81], [82]
	+	-0.83 (2→1)			1 M LiCl + 0.1 M phen, ethanol (99.7%)	[81], [82]
$[Co^{II}phen_3](ClO_4)_2$	+0.10 (2→3)	-1.0 (2→1)			1 M Na_2SO_4 aq soln	[141]
$[Co\ phen_3](ClO_4)_3 \cdot 2H_2O$	+ (3→2)	-0.88 (2→1)	-1.58 (1→-1)	-1.89 (-1→-3)	0.1 M $[(C_2H_5)_4N]ClO_4$, DMSO (100%)	[65]
1,10-Phenanthroline		-1.83 (at which a current-rise starts)			0.1 M $[(C_2H_5)_4N]ClO_4$, DMSO (100%)	
$[Co\ bathophen_3]Cl_3$		-0.80 (2→1)	-1.11 (1→0)		1 M LiCl, ethanol (99.7%)	[82]
$[Co\ dip_3]^{3+}$	+0.05 (3→2)				1 M Na_2SO_4 aq soln	[109]
$[Co\ dip_3](ClO_4)_3 \cdot 3H_2O$	-0.14 (3→2)	-1.32 (2→1)			0.1 M KNO_3 + 0.05 M dip aq soln	[80]
	+0.09 (3→2)	-1.16 (2→1)			0.1 M $NaNO_3$	[79]
	+0.20 (3→2)				0.5 M LiCl + 0.1 M dip, 80% ethanol	[79]
	+0.235 (3→2)				0.05 M $[(C_2H_5)_4N]ClO_4$, CH_3CN(AN)	[39]

(continued)

TABLE 2.2.19 (continued)

Compound	$E_{1/2}$ (change of valence) (V)	$E_{1/2}$ (change of valence) (V)	$E_{1/2}$ (change of valence) (V)	$E_{1/2}$ (change of valence) (V)	Supporting electrolyte	Refs.
[Co dip_3]$(ClO_4)_2$	+0.241 (2→3)	(2→1)	-1.632 (1→-1)		0.05 M [$(C_2H_5)_4N$]ClO_4, CH_3CN(AN) + dip	[39]
[Co dip_3]$(ClO_4)_3 \cdot 3H_2O$	+0.203 (3→2)	+1.019 (2→1)	-1.505 (1→-1)	-2.132 (dip redn)	0.1 M [$(C_2H_5)_4$]ClO_4, DMSO (100%)	[103]
	+(3→2)	-1.10_4 (2→1)			1 M KCl aq soln	[143]
		-1.33 (2→1)			0.1 M KNO_3 aq soln	[80]
		-0.91 (2→1)			0.5 M LiCl + 1 M dip, ethanol, VPE	[79]
	+0.27 (3→2)				0.5 M LiCl + 0.1 M dip, 80% ethanol, VPE	[79]
cis-[Co gly dip_2]Cl_2	-0.14 (3→2)	~-1.4 (2→1)			1 M KCl aq soln	[143]
cis-[Co ala dip_2]Cl_2	-0.20 (3→2)	~-1.4 (2→1)			1 M KCl	[143]
cis-[Co tyrosine dip_2]Cl_2	-0.19 (3→2)	-1.36 (2→1)			1 M KCl	[143]
cis-[Coleuc dip_2]Cl_2	~-0.25 (3→2)	-1.37_8 (2→1)			1 M KCl	[143]
cis-[Co(OCOC(NO)CH_3) dip_2]Cl_2	-0.37_3(3→2)	-1.39_5 (2→1)			1 M KCl + 0.0032% Tween 80	[143]
2,2'-Dipyridyl (dip)		-2.05 (at which a current-rise starts)			0.1 M [$(C_2H_5)_4N$]ClO_4, DMSO (100%)	
		-1.47 (the current-rise potential)			1 M LiCl, ethanol (99.7%)	

[a] gly = glycinate; ala = alaninate; leuc = 1-leucinate; bathophen = 4,7-diphenyl-1,10-phenanthroline.

TABLE 2.2.20. Half-Wave Potentials for the Oximato Cobalt Complexes in Aqueous and Nonaqueous Media[a]

Compound	$E_{1/2}$ (change of valence) (V)	$E_{1/2}$ (change of valence) (V)	$E_{1/2}$ (change of valence) (V)	Supporting electrolyte	Refs.
trans-$[Co(OH_2)_2(dgH)_2]NO_3.H_2O$	+0.310 (3→2)	-1.600 (ligand redn)		0.2 N HCl, H_2O (60%)-methanol (40%)	[127]
	+0.075 (3→2)	-1.01 (2→1)		0.25 M $LiNO_3$, 95% ethanol	[127]
trans-$[Co(OH_2)_2(dgH)_2]$ $OCOCH_3$ (anhydrous)	+ (3→2)	-0.75 (2→1)	-0.99 (1→0)	0.5 M Na_2SO_4 aq soln	[144]
trans-$[Co(OH_2)_2(dgH)_2]ClO_4$	-0.62 (3→2)	-0.85 (2→1)		0.1 M$[(C_2H_5)_4N]ClO_4$, DMSO	[145]
trans-$[Co(P \equiv \phi_3)(OH_2)(dgH)_2]$-$ClO_4$	-0.040 (3→2)	-0.67 (che)[b]		0.2 M HCl, H_2O (60%)-methanol (40%)	[127]
	-0.150 (3→2)	-0.720 (2→1)		0.25 M $LiNO_3$, 95% ethanol	[127]
trans-$[Co(P \equiv \phi_3)_2(dgH)_2]ClO_4$	-0.075 (3→2)	-0.65 (che)		0.2 N HCl, H_2O (60%)-methanol (40%)	[127]
	-0.100 (3→2)	-0.710 (2→1)		0.25 M $LiNO_3$, 95% ethanol	[127]
trans-$[CoCl(P \equiv \phi_3)(dgH)_2]$	-0.050 (3→2)	-0.70 (che)		0.25 N HCl, H_2O (60%)-methanol (40%)	[127]
	-0.260 (3→2)	-0.715 (2→1)		0.25 M $LiNO_3$, 95% ethanol	[127]
trans-$[Co(P \equiv bu_3)(OH_2)(dgH)_2]$ NO_3	-0.385 (3→2)	-1.000 (ligand redn)		0.25 N HCl, H_2O (60%)-methanol (40%)	[127]
trans-$[Co(P \equiv bu_3)(OH_2)(dgH)_2]$ ClO_4	-0.525 (3→2)	-0.752 (2→1)		0.25 M $LiNO_3$, 95% ethanol	[127]
trans-$[Co(P \equiv bu_3)_2(dgH)_2]ClO_4$	-0.730 (3→2)			0.25 M LiCl, 95% ethanol	[127]
trans-$[Co(P \equiv bu_3)_2(dgH)_2]NO_3$	-0.585 (3→2)	-1.020 (ligand redn)		0.25 N HCl, H_2O (60%)-methanol (40%)	[127]

(continued)

TABLE 2.2.20 (continued)

Compound	$E_{1/2}$ (change of valence) (V)	$E_{1/2}$ (change of valence) (V)	$E_{1/2}$ (change of valence) (V)	Supporting electrolyte	Refs.
trans-[$CoCl(P\equiv bu_3)(dgH)_2$]	-0.625 (3→2)	-0.765 (2→1)		0.25 M $LiNO_3$, 95% ethanol	[127]
	+0.05 (3→2) vs Ag/AgCl standard electrode	-0.75 (2→1)	-0.82 (1→0)	$LiClO_4$, H_2O (50%)-ethanol (50%)	[146]
trans-[$Co(CH_3)(P\equiv bu_3)(dgH)_2$]	-0.90 (3→2)	Ligand redn		0.25 N HCl, H_2O (60%)-methanol (40%)	[127]
	-1.59 (3→2)	che		0.25 M LiCl, 95% ethanol	[127]
	-2.12 (3→1) vs Ag/0.10 M $AgNO_3$ electrode			CH_3CN	[147]
trans-[$Co(CH_3)(P\equiv \phi_3)(dgH)_2$]	-0.62 (3→2)	che		0.25 N HCl, H_2O (60%)-methanol (40%)	[127]
trans-[$Co(CH_3)(OH_2)(dgH)_2$]	-1.26 (3→2)			0.1 M K_2SO_4 aq soln	[148]
	-1.7 (3→2) vs Ag/0.10 M $AgNO_3$ electrode	-2.42 (2→1)		CH_3CN(AN)	[147]
trans-[$Co(CH_3)$(nico)$(dgH)_2$]	-1.28 (3→2)			0.1 M K_2SO_4 aq soln	[148]
trans-[$Co(CH_3)$(tolu)$(dgH)_2$]	-1.28			0.1 M K_2SO_4 aq soln	[148]
trans-[$Co(CH_3)$py$(dgH)_2$]	-1.31			0.1 M K_2SO_4 aq soln	[148]
	-1.75 (3→2) vs Ag/0.10 M $AgNO_3$ electrode	-2.44 (2→1)	-3.01 (redn of py)	CH_3CN	[147]
trans-[$Co(CH_3)$(pico)$(dgH)_2$]	-1.33			0.1 M K_2SO_4 aq soln	[148]
trans-[$Co(CH_3)$(imd)$(dgH)_2$]	-1.40			0.1 M K_2SO_4	[148]

trans-[Co(CH_3)py(dgB_2F_4)]	-1.3 (3→2) vs Ag/0.10 M $AgNO_3$ electrode	-2.39 (2→1)	-3.08 (redn of py)	CH_3CH	[147]
trans-[Co(CN)py$(dgH)_2$]	-1.19			0.1 M K_2SO_4	[148]
	-1.1 (3→2) vs Ag/0.10 M $AgNO_3$ electrode	-1.86 (2→1)	-3.08 (redn of py)	CH_3CN	[147]
trans-[Co($C_6H_5CH_2$)py$(dgH)_2$]	-1.20			0.1 M K_2SO_4	[148]
trans-[Co($CH_3CH(OH)CH_2$) py$(dgH)_2$]	-1.24			0.1 M K_2SO_4	[148]
trans-[Co($HOCH_2CH(OH)CH_2$)-py$(dgH)_2$]	-1.26			0.1 M K_2SO_4	[148]
trans-[Co($HOCH_2CH(CH_3)_2$)-py$(dgH)_2$]	-1.27			0.1 M K_2SO_4	[148]
trans-[CoCl(py)$(dgH)_2$]	-1.28			0.1 M K_2SO_4	[148]
trans-[Co($CH_3CH_2CH_2$)py$(dgH)_2$]	-1.30			0.1 M K_2SO_4	[148]
trans-[Co($CH_3CH(CH_3)_2$)-py$(dgH)_2$]	-1.30			0.1 M K_2SO_4	[148]
trans-[Co(C_2H_5)py$(dgH)_2$]	-1.31			0.1 M K_2SO_4	[148]
trans-[Co($CH_3CH_2CH_2CH_2$)-py$(dgH)_2$]	-1.80 (3→2) vs Ag/0.10 M $AgNO_3$ electrode	-2.44 (2→1)	-3.05 (redn of py)	CH_3CN	[147]
trans-[Co$(py)_2(dgH)_2$]$^+$	-0.325 vs Hg pool	-1.020	-1.263	0.1 M KCl	[149]

(continued)

TABLE 2.2.20 (continued)

Compound	$E_{1/2}$ (change of valence) (V)	$E_{1/2}$ (change of valence) (V)	$E_{1/2}$ (change of valence) (V)	Supporting electrolyte	Refs.
trans-[$Co(py)_2(dgH)_2$]NO_3	-0.18 (3→2) vs Ag/AgCl standard electrode	-0.84 (2→1)	-1.14 (1→0)	$LiClO_4$, H_2O-ethanol	[146]
	-0.22 (3→2)	-0.88 (2→1)	-1.18 (1→0)	1 M KCl aq soln	[115]
trans-[$CoCl(py)(dgH)_2$]	-1.28			0.1 M K_2SO_4 aq soln	[148]
	0.10 (3→2) vs Ag/AgCl standard electrode	-1.0 (2→1)	-1.27 (1→0)	$LiClO_4$, H_2O-ethanol	[148]
trans-[$Co(p\text{-}N(CH_3)_2\text{-}an)_2(dgH)_2$]$^+$	-0.391 vs Hg pool	-1.035	-1.260	0.1 M KCl aq soln	[149]
trans-[$Co(p\text{-}CH_3O\text{-}an)_2(dgH)_2$]$^+$	-0.373 vs Hg pool	-1.040	-1.280	0.1 M KCl	[149]
trans-[$Co(an)_2(dgH)_2$]$^+$	-0.364 vs Hg pool	-1.055	-1.280	0.1 M KCl	[149]
trans-[$Co(4\text{-}NH_2\text{-}py)_2(dgH)_2$]$^+$	-0.462 vs Hg pool	-1.040	-1.285	0.1 M KCl	[149]
trans-[$Co(4\text{-}CH_3\text{-}py)_2(dgH)_2$]$^+$	-0.310 vs Hg pool	-1.010	-1.250	0.1 M KCl	[149]
trans-[$Co(3\text{-}CH_3\text{-}py)_2(dgH)_2$]$^+$	-0.305 vs Hg pool	-1.020	-1.280	0.1 M KCl	[149]
trans-[$Co(4\text{-}CN\text{-}py)_2(dgH)_2$]$^+$	-0.152 vs Hg pool	-1.020	-1.280	0.1 M KCl	[149]
trans-[$Co(N(CH_3)_3)Cl(dgH)_2$]	+0.05 (3→2) vs Ag/AgCl standard electrode	-0.9 (2→1)	-1.30 (1→0)	$LiClO_4$, H_2O-ethanol (1:1)	[146]

trans-[Co($NH_2C_2H_5$)$_2$(dgH)$_2$]Cl	-0.66 (3→2)	-1.08 (2→1)	-1.25 (1→0)	1 M KCl aq soln	[115]
trans-[Co(copoly-am-vpy)(OH)-(dgH)$_2$] (mol wt = 1400)	-1.15			0.1 M K_2SO_4 aq soln	[148]
trans-[Co(copoly-am-vpy)(OH)-(dgH)$_2$] (mol wt = 3000)	-1.16			0.1 M K_2SO_4 aq soln	[148]
trans-[CoCl(OH_2)(dgH)$_2$]		-1.22		0.1 M K_2SO_4	[148]
trans-[Co(OH)(OH_2)(dgH)$_2$]		-1.26		0.1 M K_2SO_4	[148]
trans-[Co(dgH)$_2$(NH_3)$_2$]Cl. 5H_2O	-0.65 (3→2)	-1.045 (2→1)	-1.272 (1→0)	1 M KCl aq soln	[115]
	-0.60 (3→2)	-1.03 (2→1)	-1.23 (1→0)	0.5 M Na_2SO_4	[144]
	-0.78 (3→2)	-0.94 (2→1)	-1.27 (1→0)	0.5 M Na_2SO_4 + 1 M NH_3	[144]
trans-[Co(dgH)$_2$(NH_3)$_2$]ClO_4	-0.74 (3→1)			0.1 M [(C_2H_5)$_4$N]ClO_4, DMSO (100%)	[150]
cis-[Co(dgH)$_2$(NH_3)$_2$]ClO_4	-0.82 (3→2)	-2.04 (2→0)		0.1 M [(C_2H_5)$_4$N]ClO_4, DMSO (100%)	[150]
trans-K[Co(CN)$_2$(dgH)$_2$]. 3/2H_2O	-1.105 (3→1)			0.5 M Na_2SO_4 aq soln	[115], [122]
	-1.45 (3→1)			0.1 M [(C_2H_5)$_4$N]ClO_4, DMSO (100%)	[145]
trans-Na[Co(NO_2)$_2$(dgH)$_2$]. H_2O	-0.35 (3→2)	-1.1 (ligand redn)		1 M KCl aq soln	[115]
trans-[Co(NO_2)(OH_2)(dgH)$_2$]	-0.155 (3→2)	-1.1 (ligand redn)		1 M KCl	[115]
trans-H[CoCl_2(dgH)$_2$]	+ (3→2)	-0.724 (2→1)	-1.1 (ligand redn)	1 M KCl	[115]
	+ (3→2)	-0.81 (2→1)		0.1 M [(C_2H_5)$_4$N]ClO_4	[150]
trans-[Co(CN)(NH_3)(dgH)$_2$]. 1/2 H_2O	-0.65 (3→2)	-0.80 (2→1)	-1.05 (1→0)	0.5 M Na_2SO_4 aq soln	[151]
	-1.02 (3→1)			0.1 M [(C_2H_5)$_4$N]ClO_4, DMSO (100%)	[145]

(continued)

TABLE 2.2.20 (continued)

Compound	$E_{1/2}$ (change of valence) (V)	$E_{1/2}$ (change of valence) (V)	$E_{1/2}$ (change of valence) (V)	Supporting electrolyte	Refs.
trans-[CoCl(NH_3)($dgH)_2$]	-0.58 (3→2)	-1.03 (2→1)	-1.20 (1→0)	0.5 M Na_2SO_4 aq soln	[144]
	-0.58 (3→2)	-0.81 (2→1)		0.1 M [$(C_2H_5)_4N$]ClO_4, DMSO	[150]
trans-[CoBr(NH_3)($dgH)_2$]	-0.58 (3→2)	-1.03 (2→1)	-1.20 (1→0)	0.5 M Na_2SO_4	[144]
	-0.57 (3→2)	-0.81 (2→1)		0.1 M [$(C_2H_5)_4N$]ClO_4, DMSO	[145]
trans-[CoI(NH_3)($dgH)_2$]	-0.59 (3→2)	-1.01 (2→1)	-1.23 (1→0)	0.5 M Na_2SO_4	[144]
	-0.56 (3→2)	-0.81 (2→1)		0.1 M [$(C_2H_5)_4N$]ClO_4, DMSO	[145]
trans-[CoF(NH_3)($dgH)_2$]	-0.59 (3→2)	-0.81 (2→1)		0.1 M [$(C_2H_5)_4N$]ClO_4, DMSO	[150]
trans-[Co(OH_2)($dgH)_2$(NH_3)]-$OCOCH_3$	-0.58 (3→2)	-1.03 (2→1)	-1.22 (1→0)	0.5 M Na_2SO_4 aq soln	[144]
	-0.78 (3→2)	-0.94 (2→1)	-1.25 (1→0)	0.5 M Na_2SO_4 + 1 M NH_3	[144]
trans-[Co(OH_2)($dgH)_2$(NH_3)]ClO_4	-0.65 (3→2)	-0.84 (2→1)		0.1 M [$(C_2H_5)_4N$]ClO_4, DMSO	[145]
trans-[Co(NCS)($dgH)_2$(NH_3)]	-0.77 (3→1)			0.1 M [$(C_2H_5)_4N$]ClO_4, DMSO	[103]
trans-H[Co($NCS)_2$($dgH)_2$]	~-0.64 (3→2)	-1.32 (2→1)		0.1 M [$(C_2H_5)_4N$]ClO_4, DMSO	[103]
trans-$(NH_4)_2$[Co(CN)($dgH)_2$-(SO_3)]. $4H_2O$	-1.21 (3→2)	-1.57 (2→1)		0.5 M Na_2SO_4 aq soln	[103]
trans-Na[Co($NO_2)_2$($dgH)_2$]. H_2O	-0.95 (3→1)	-2.03 (Na^I→Na^0)		0.1 M [$(C_2H_5)_4N$]ClO_4, DMSO	[103]
[Co($dgH)_3$]. 5/2 H_2O	-0.81 (3→2)	-2.11 (2→0)		0.1 M [$(C_2H_5)_4N$]ClO_4, DMSO	[150]
	-1.04 (3→2)			0.5 M LiCl, 50% ethanol satd with dg^{2-}	[115]
cis-[Co($dgH)_2$en]ClO_4. 1/3 H_2O	-0.87 (3→2)	-2.29 (2→0)		0.1 M [$(C_2H_5)_4N$]ClO_4, DMSO	[150]
trans-[Co(p-CH_3O-an$)_2$($megH)_2$]$^+$	-0.364 vs Hg pool	-0.988	-1.418	0.1 M KCl aq soln	[149]

trans-[Co(p-CH_3O-an)$_2$(nioxH)$_2$]$^+$	-0.353 vs Hg pool	-0.770	-1.035	0.1 M KCl	[149]
trans-[Co(p-CH_3O-an)$_2$(bzdH)$_2$]$^+$	-0.160 vs Hg pool	-0.712	-0.920	0.1 M KCl	[149]
trans-[Co(py)$_2$(megH)$_2$]$^+$	-0.280	-0.965	-1.385	0.1 M KCl	[149]
trans-[Co(py)$_2$(nioxH)$_2$]$^+$	-0.291 vs Hg pool	-0.870	-1.155	0.1 M KCl	[149]
trans-[Co(py)$_2$(bzdH)$_2$]$^+$	-0.145 vs Hg pool	-0.783	-0.920	0.1 M KCl	[149]
trans-[Co(py)$_2$(frdH)$_2$]$^+$	-0.035 vs Hg pool	-0.733	-0.950	0.1 M KCl	[149]
trans-[Co(p-CH_3O-an)(frdH)$_2$]$^+$	-0.100 vs Hg pool	-0.745	-0.965	0.1 M KCl	[149]
Vitamin B_{12} (cyanocobalamine)	-1.12 (3→1)			0.25 M LiCl, 95% ethanol	[127]
	-1.09			0.1 M K_2SO_4 aq soln	[148]
	-1.26	-2.27		H_2O (20%)-CH_3CN (80%) vs Ag/0.10 M $AgNO_3$	[147]
	-1.07 (3→1)			$LiClO_4$, H_2O-ethanol (1:1) vs Ag/AgCl	[146]
	-1.11 (3→1)			0.100 M K_2SO_4 aq soln	[107]
	-1.12 (3→1)			0.1 M K_2SO_4 aq soln	[151]
	-1.11			0.1 M [$(CH_3)_4$N]Cl aq soln	[151]
	-1.15 (3→1)			0.1 M KNO_3 aq soln	[151]
	-1.19			[$(CH_3)_4$N]Cl-[$(CH_3)_4$N]OH	[151]

(continued)

TABLE 2.2.20 (continued)

Compound	$E_{1/2}$ (change of valence) (V)	$E_{1/2}$ (change of valence) (V)	$E_{1/2}$ (change of valence) (V)	Supporting electrolyte	Refs.
Vitamin B_{12} (cyanocobalamine)				pH = 5.2	[151]
	-1.11			pH = 6.9	[151]
	-1.13			pH = 8.1	[151]
	-1.10			pH = 9.4	[151]
	-1.10			pH = 11.7	[151]
Methylcobalamin	-1.20			0.1 M K_2SO_4	[148]
Vitamin B_{12a} (aquocobalamin)	-0.04 ($B_{12a} \rightarrow B_{12r}$) ($3 \rightarrow 2$)	-1.02 ($B_{12r} \rightarrow B_{12s}$) ($2 \rightarrow 1$)			[80], [152], [107]
	+0.01 ($3 \rightarrow 2$)	-0.97 ($2 \rightarrow 1$)		$LiClO_4$, H_2O-ethanol (1:1) vs Ag/AgCl	[146]
Vitamin B_{12r}[Co(II)]	-0.04 ($2 \rightarrow 3$) (anodic)	-0.95 ($2 \rightarrow 1$)	-1.55 (catalytic wave)	0.1 M K_2SO_4 aq soln	[107]
(hydrogenation of vitamin B_{12a} for 6 hr)	-0.04 ($2 \rightarrow 3$)	-0.94 ($2 \rightarrow 1$)		0.1 M K_2SO_4	[107]
Vitamin B_{12} coenzyme	-1.6			NaOH, H_3BO_3 (pH = 10.0)	[153]
	-1.2			NaAc, HAc (pH = 4.75)	[153]
	-1.36			K_2SO_4 etc. (pH = 6.9)	[153]
	-1.43 ($3 \rightarrow 1$)			0.1 M K_2SO_4 + KOH (pH = 11.0)	[153]
	-1.53 ($3 \rightarrow 1$)			0.1 M K_2SO_4 + KOH (pH = 11.4)	[153]

[a]bu = butyl; nico = nicotinamide; tolu = p-toluidine; py = pyridine; pico = δ-picoline; imd = imidazole; $dg_2B_2F_4$ = cyclic oxime boronic esters; an = aniline; copoly-am-vpy = a copolymer of acrylamide and 4-vinylpyridine; megH = 2, 3-pentanedionedioxime; nioxH = cyclohexanedionedioxime; bzdH = α-benzyldioxime; frdH = α-furyldioxime. Vitamin B_{12} (cyanocobalamine) is

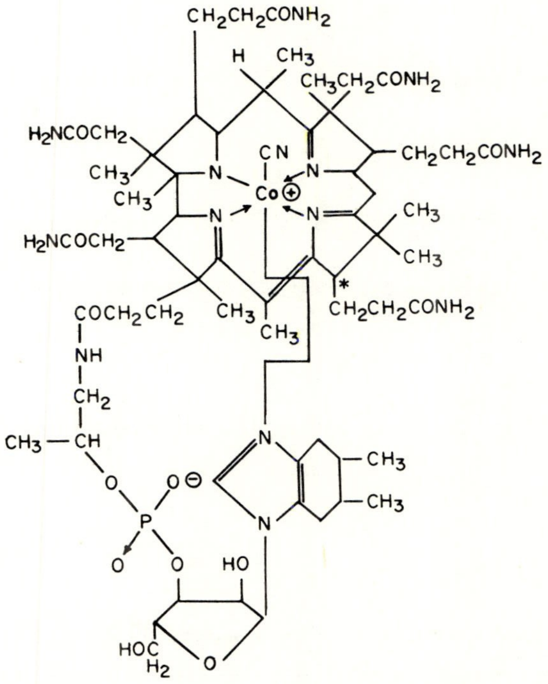

[b]che = catalytic hydrogen evolution.

TABLE 2.2.21. Half-Wave Potentials for the Cyclic Amine Cobalt Complexes in Nonaqueous Media[a]

Compound	$E_{1/2}$ (change of valence) (V)	$E_{1/2}$ (change of valence) (V)	$E_{1/2}$ (change of valence) (V)	Supporting electrolyte	Refs.
$[Co^{III}$(trans-tetramine)-$(CH_3CN)_2](ClO_4)_3$	+0.087 (3→2)	-2.028 (2→1)	-2.52 (1→0)	0.10 M $[(C_2H_5)_4N]ClO_4$, AN vs Ag/0.10 M $AgNO_3$	[154]
	+0.021 (3→2)	-2.037 (2→1)	(E_p, voltammetric data at a platinum sphere electrode, pse)	0.10 M $[(C_2H_5)_4N]ClO_4$, AN vs Ag/0.10 M $AgNO_3$	[154]
$[Co^{II}$(trans-tetramine)$](ClO_4)_2$	+0.078 (2→3)	-2.028 (2→1)	-2.52 (1→0)	0.10 M $[(C_2H_5)_4N]ClO_4$, AN vs Ag/0.10 M $AgNO_3$	[154]
	+0.42 (2→3)	-2.037 (2→1)	(E_p, voltammetric data)	0.10 M $[(C_2H_5)_4N]ClO_4$, AN vs Ag/0.10 M $AgNO_3$	[154]
$[Co^{I}$(trans-tetramine)$](BF_4)$	+0.078 (2→3)	(1→2)	-2.52 (1→0)	0.10 M $[(C_2H_5)_4N]ClO_4$, AN vs Ag/0.10 M $AgNO_3$	[154]
$[Co^{III}$(trans-diene)$(CH_3CN)_2]$-$(ClO_4)_3$	+0.129 (3→2)	-1.76 (2→1)	-2.7 (1→0)	0.10 M $[(C_2H_5)_4N]ClO_4$, AN vs Ag/0.10 M $AgNO_3$	[154]
	+0.010 (3→2)	-1.693 (2→1)	(E_p, voltammetric data)	0.10 M $[(C_2H_5)_4N]ClO_4$, AN vs Ag/0.10 M $AgNO_3$	[154]
$[Co^{II}$(trans-diene)$](ClO_4)_2$	+0.126 (2→3)	-1.76 (2→1)	-2.7 (1→0)	0.10 M $[(C_2H_5)_4N]ClO_4$, AN vs Ag/0.10 M $AgNO_3$	[154]
	+0.240 (2→3)	(E_p, voltammetric data)		0.10 M $[(C_2H_5)_4N]ClO_4$, AN vs Ag/0.10 M $AgNO_3$	[154]
$[Co^{I}$(trans-diene)$]ClO_4$	+0.126 (2→3)	-1.62 (1→2)	-2.7 (1→0)	0.10 M $[(C_2H_5)_4N]ClO_4$, AN vs Ag/0.10 M $AgNO_3$	[154]
	+0.240 (2→3)	-1.658 (1→2)	(E_p, voltammetric data)	0.10 M $[(C_2H_5)_4N]ClO_4$, AN vs Ag/0.10 M $AgNO_3$	[154]

$[Co^{II}tpp]$	+1.42 (ligand oxid)	+1.19 (ligand oxid)	+0.52 (2→3)	0.1 M $[(n\text{-}bu)_4N]ClO_4$, benzonitrile, Pt disk electrode	[155]
H_2tpp	+1.20 (ligand oxid)	+1.00 (ligand oxid)		0.1 M $[(n\text{-}bu)_4N]ClO_4$, benzonitrile, Pt disk electrode	
$[Co^{II}tpp)$	+1.26 (anodic oxid)	+1.06 (ligand oxid)	+0.32 (2→3)	0.1 M $LiClO_4$, butyronitrile, rotating platinum electrode	[156]
H_2tpp	+1.12 (ligand oxid)	+0.97 (ligand oxid)		0.1 M $LiClO_4$, butyronitrile, rotating platinum electrode	
$[Co^{II}tpp]$		-0.82 (2→1)	-1.87	0.1 M $[(n\text{-}C_3H_7)_4N]ClO_4$, DMSO	[157]
H_2tpp	-0.70 (pre-wave)	-1.05 (cath-odic)	-1.47 (cath-odic)	0.1 M $[(n\text{-}C_3H_7)_4N]ClO_4$, DMSO	
$[Co^{II}pc]$		+0.77 (2→3)		0.1 M $[(n\text{-}bu)_4N]ClO_4$, 1-chloro-naphthalene, cyclic voltammetry	[155]
H_2pc		+1.10 (ligand oxid)		0.1 M $[(n\text{-}bu)_4N]ClO_4$, 1-chloro-naphthalene, cyclic voltammetry	
$Na_4[Co^{II}pts]$	-0.547 (2→1)	-1.346 (1→0)		0.100 M $[(C_2H_5)_4N]ClO_4$, DMSO, DME	[158]
	+1.09 (2-elec-tron oxid)			0.100 M $[(C_2H_5)_4N]ClO_4$, DMSO, ro-tating platinum electrode	[158]
	+0.455 (2→3)		-1.355	0.100 M $[(C_2H_5)_4N]ClO_4$, DMSO, ro-tating platinum electrode	
Na_4H_2pts		-0.525	-0.970	0.100 M $[(C_2H_5)_4N]ClO_4$, DMSO, DME	[158]
	+0.90	-0.530	-0.980	0.100 M $[(C_2H_5)_4N]ClO_4$, DMSO, DME	
$[Co^{II}etioI]$	+1.18 (anodic)	+0.87 (anodic)	+0.30 (2→3)	0.1 M $LiClO_4$, butyronitrile, rotating platinum electrode	[156]
H_2etioI		+0.77 (anodic)		0.1 M $LiClO_4$, butyronitrile, rotating platinum electrode	

(continued)

TABLE 2.2.21 (continued)

Compound	$E_{1/2}$ (change of valence) (V)	$E_{1/2}$ (change of valence) (V)	$E_{1/2}$ (change of valence) (V)	Supporting electrolyte	Refs.
[Co^{II}etioI]		-1.04	-1.57	0.1 M [$(C_3H_7)_4N$]ClO_4, dimethylformamide	[157]
H_2etioI	-0.92 (prewave)	-1.34		0.1 M [$(C_3H_7)_4N$]ClO_4, dimethylformamide	
[Co^{II}(deuteropp)]	+0.94 (anodic)	+0.26 (anodic)		0.1 M $LiClO_4$, butyronitrile, rotating platinum electrode	[156]
H_2deuteropp	+1.33 (anodic)	+0.76 (anodic)		0.1 M $LiClO_4$, butyronitrile, rotating platinum electrode	

[a]tpp = tetraphenylporphyrin; pc = phthalocyamine; pts = tetrasulfonated phthalocyanine; etioI = etioporphyrin I; deuteropp = deuteroporphyrin-IX-dimethylester;

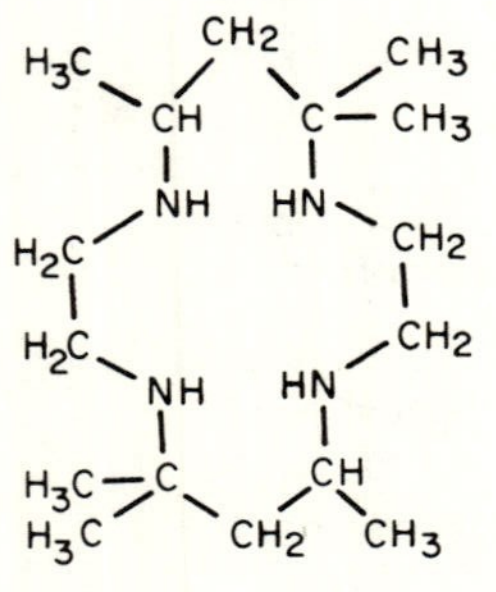

trans-TETRAMINE

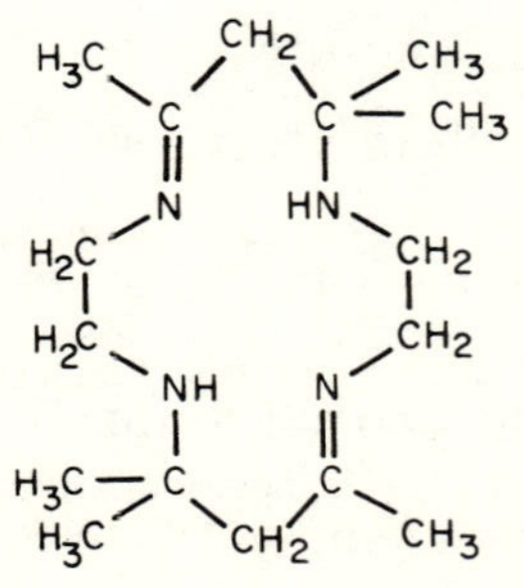

trans-DIENE

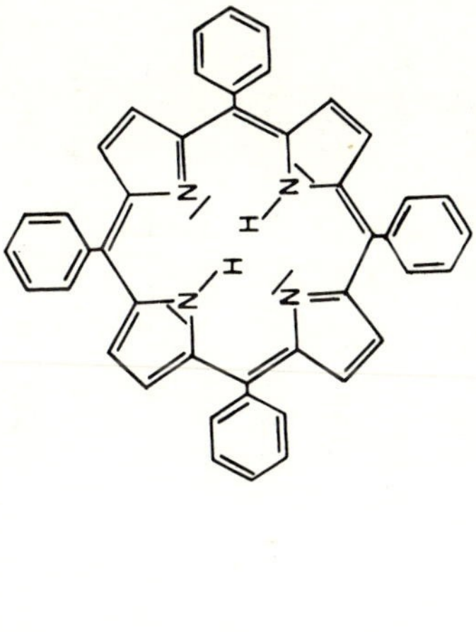

Tetraphenylporphine

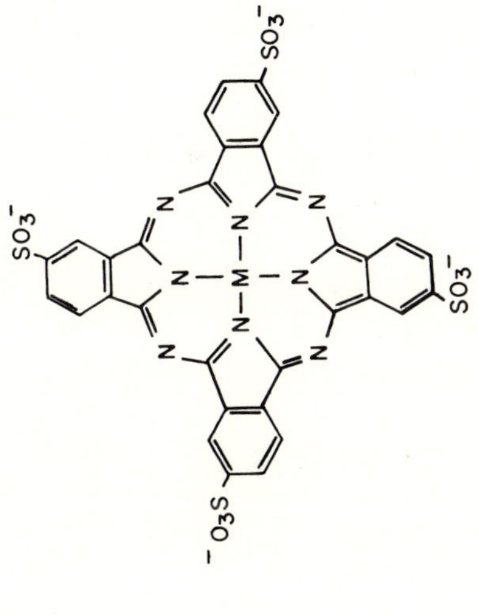

Tetrasulfonated metal phthalocyanine

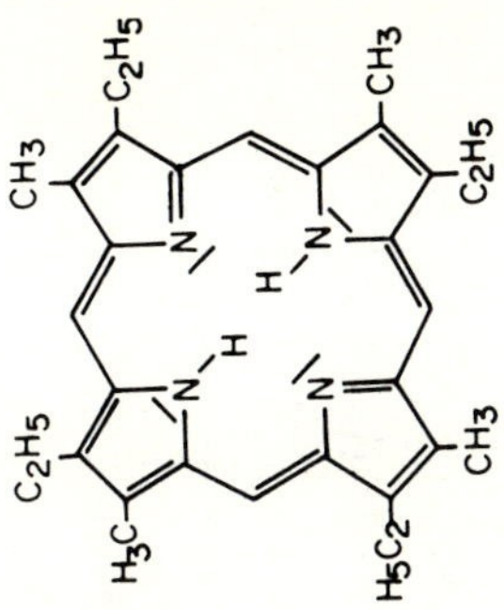

Atioporphyrine I

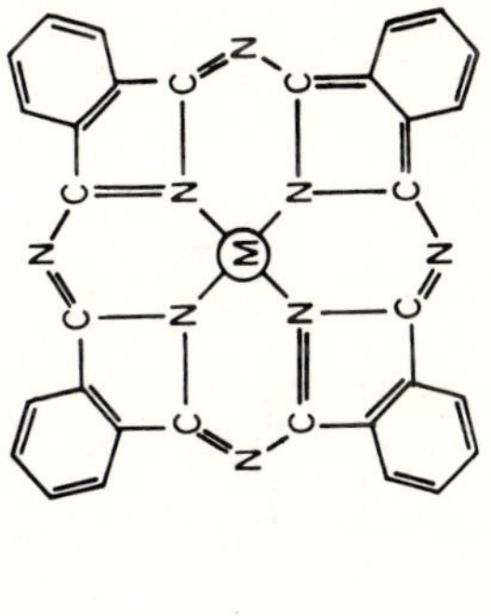

Phthalocyanine

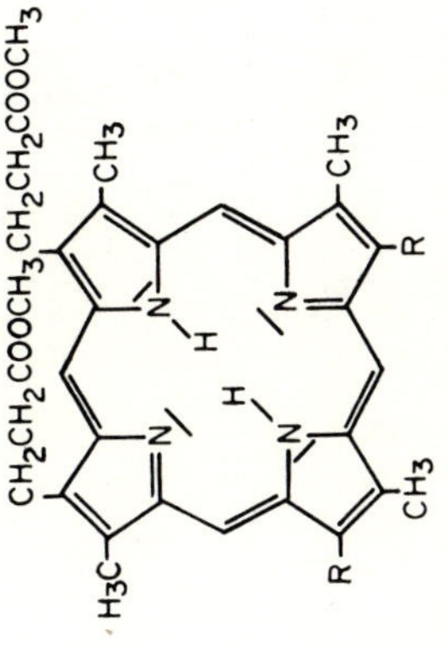

Deuteroporphyrin - Ix - Dimethylester (R=H)

TABLE 2.2.22. Half-Wave Potentials for the Monocyano Cobalt (III) Complexes of the trans-$[Co(CN)X(NH_3)_4]$ Type in Dimethyl Sulfoxide (25°C)

Compound	$E_{1/2}$ (V vs SCE)		Supporting electrolyte
	Co(III) → Co(II)	Co(II) → Co(metal)	
trans-$[Co(CN)(P\equiv\phi_3)(NH_3)_4]I$	~−0.13		
trans-$[Co(CN)(NO_2)(NH_3)_4]ClO_4 \cdot 1/2\ H_2O$	−0.44	−1.56	0.1 M $[(C_2H_5)_4N]ClO_4$, DMSO (100%)
trans-$[Co(CN)(NO)(NH_3)_4]Cl \cdot 1/2\ H_2O$	−0.43	−1.56	0.1 M $[(C_2H_5)_4N]ClO_4$, DMSO (100%)
trans-$[Co(CN)py(NH_3)_4](ClO_4)_2$	−0.38	−1.55	0.1 M $[(C_2H_5)_4N]ClO_4$, DMSO (100%)
trans-$[Co(CN)(NH_2CH_3)(NH_3)_4](ClO_4)_2$	−0.35	−1.52	0.1 M $[(C_2H_5)_4N]ClO_4$, DMSO (100%)
trans-$[Co(CN)(NH_2C_2H_5)(NH_3)_4](ClO_4)_2$	−0.45	−1.54	0.1 M $[(C_2H_5)_4N]ClO_4$, DMSO (100%)
trans-$[Co(CN)(NH_3)(NH_3)_4](ClO_4)_2$	−0.60	−1.57	0.1 M $[(C_2H_5)_4N]ClO_4$, DMSO (100%)
trans-$[Co(CN)(SO_3)(NH_3)_4] \cdot 2H_2O$[a]	Maximum	Maximum	0.1 M $[(C_2H_5)_4N]ClO_4$, DMSO (100%)
trans-$[Co(CN)(OH_2)(NH_3)_4](ClO_4)_2 \cdot H_2O$	−0.36	−1.55	0.1 M $[(C_2H_5)_4N]ClO_4$, DMSO (100%)
trans-$[Co(CN)(N_3)(NH_3)_4]N_3 \cdot H_2O$	−0.41	−1.55	0.1 M $[(C_2H_5)_4N]ClO_4$, DMSO (100%)
trans-$[Co(CN)I(NH_3)_4]I$	~−0.13[b]	−1.50	0.1 M $[(C_2H_5)_4N]ClO_4$, DMSO (100%)
trans-$[Co(CN)(NCS)(NH_3)_4]ClO_4 \cdot 3/2\ H_2O$	−0.46	Maximum	0.1 M $[(C_2H_5)_4N]ClO_4$, DMSO (100%)
$[Co(NH_3)_6](ClO_4)_3$	−0.43	−1.32	0.1 M $[(C_2H_5)_4N]ClO_4$, DMSO (100%)

[a] The two complexes of the same type, trans-$[Co(CN)(S_2O_3)(NH_3)_4] \cdot 2H_2O$ and trans-$[Co(CN)(CO_3)(NH_3)_4] \cdot H_2O$, are insoluble in DMSO; the former is also insoluble in water.

[b] Hindrance of anodic wave for I^- ions.

TABLE 2.2.23. Half-Wave Potentials for the Mixed Cyano Cobalt(III) Complexes in Dimethyl Sulfoxide (25°C)[a, b]

Compound	$E_{1/2}$ (change of valence) (V)	$E_{1/2}$ (change of valence) (V)	$E_{1/2}$ (change of valence) (V)	Conditions	Refs.
trans-$[Co(CN)_2(NH_3)_4]ClO_4 \cdot H_2O$	-0.71 (3→2)	-1.06 (2→1)	-1.57 (1→0)	0.1 M $[(C_2H_5)_4N]ClO_4$, DMSO (100%)	[114]
cis-$[Co(CN)_2en_2]ClO_4$	-0.95 (3→2)	-1.48 (2→1)	-1.99 (1→0)	0.1 M $[(C_2H_5)_4N]ClO_4$, DMSO (100%)	[114], [101]
trans-$[Co(CN)_2en(NH_3)_2]ClO_4$	-0.96 (3→2)	-1.51 (2→1)	-1.99 (1→0)	0.1 M $[(C_2H_5)_4N]ClO_4$, DMSO (100%)	[103]
trans-$[Co(CN)_2en_2]ClO_4$	-0.97 (3→2)	-1.52 (2→1)	-1.99 (1→0)	0.1 M $[(C_2H_5)_4N]ClO_4$, DMSO (100%)	[114]
cis-$[Co(CN)_2dip_2]NO_3 \cdot 7H_2O$	-0.48 (3→2)	-1.16 (2→1)	-1.96 (1→0)	0.1 M $[(C_2H_5)_4N]ClO_4$, DMSO (100%)	[114]
cis-$[Co(CN)_2phen_2]NO_3 \cdot 6H_2O$	-0.36 (3→2)	-1.28 (2→1)	-1.94 (1→0)	0.1 M $[(C_2H_5)_4N]ClO_4$, DMSO (100%)	[114]
trans-$Na[Co(CN)_2(dgH)_2] \cdot 2H_2O$	-1.105 (3→2)	-2.06 ($Na^I \rightarrow Na^0$)		0.1 M $[(C_2H_5)_4N]ClO_4$, DMSO (100%)	[115], [122]
cis-$[Co(CN)_2pn_2]ClO_4$	-0.97 (3→2)	-1.54 (2→1)	-1.90 (1→0)	0.1 M $[(C_2H_5)_4N]ClO_4$, DMSO (100%)	[103]
cis-$[Co(CN)_2tn_2]ClO_4$	-0.84 (3→2)	-1.42 (2→1)	-1.89 (1→0)	0.1 M $[(C_2H_5)_4N]ClO_4$, DMSO (100%)	[103]
cis-$[Co(CN)_2tren]ClO_4$	-1.06 (3→2)	-1.72 (2→1)	-1.93 (1→0)	0.1 M $[(C_2H_5)_4N]ClO_4$, DMSO (100%)	[103]
fac-$[Co(CN)_3dien] \cdot H_2O$	-1.37 (3→2)	-1.83 (2→1)		0.1 M $[(C_2H_5)_4N]ClO_4$, DMSO (100%)	[159]
mer-$[Co(CN)_3dien] \cdot 3H_2O$	-1.09 (3→2)	-1.74 (2→1)		0.1 M $[(C_2H_5)_4N]ClO_4$, DMSO (100%)	[159]
cis-$Na[Co(CN)_4en] \cdot 7/2\ H_2O$	-1.59 (3→2)	-1.78 (2→1)	-2.06 ($Na^I \rightarrow Na^0$)	0.1 M $[(C_2H_5)_4N]ClO_4$, DMSO (100%)	[159], [116]
trans-$Na[Co(P\equiv\phi_3)_2(CN)_4] \cdot 3H_2O$	-0.98 (3→1)		-2.06 ($Na^I \rightarrow Na^0$)	0.1 M $[(C_2H_5)_4N]ClO_4$, DMSO (100%)	[160]
trans-$Na[Co(As\equiv\phi_3)_2(CN)_4] \cdot 3H_2O$	-0.58 (3→2)	-1.06 (2→1)	-2.06 ($Na^I \rightarrow Na^0$)	0.1 M $[(C_2H_5)_4N]ClO_4$, DMSO (100%)	[160]

[a]tren = triethylenetetramine; pn = i-propylenediamine; tn = trimethylenediamine; dg^{2-} = dimethylglyoximate ion; dip = 2, 2'-dipyridyl; phen = 1, 10-phenanthroline; $P\equiv\phi_3$(triphenylphosphine); $As\equiv\phi_3$(triphenylarsine).

[b]The cyanide complexes with inorganic ligands are generally insoluble in aprotic dipolar solvents: e.g., the trans-$Na_5[Co(SO_3)_2(CN)_4] \cdot 3H_2O$, cis-$Na_2K_3[Co(SO_3)_2(CN)_4] \cdot 5/2\ H_2O$, trans-$K[Co(OH_2)_2(CN)_4] \cdot 3/4\ H_2O$, trans-$K[Co(CN)_4(NH_3)_2] \cdot H_2O$, and all other pentacyanocobaltate(III) complexes treated here were quite insoluble in DMSO or in AN.

2.3. VOLTAMMETRIC CHARACTERISTICS

Recently, the new voltammetric techniques such as current-controlled oscillopolarography (Heyrovský-Forejt type) [161], cyclic voltammetry, controlled-current or -potential electrolytic studies, and investigations of anodic oxidation using a Kalousek commutator [66] have attracted much attention in the field of electrochemistry of cobalt as powerful tools, along with classical or conventional polarographic and chronopotentiometric studies, for probing the redox processes of the Co(III)/Co(II) and Co(II)/Co(I) couples in solution. This is partly because a number of the electrode processes of an inert-inert type with π-electron systems have recently been discovered for π-bonded cobalt complexes, especially in organic solvents, and partly because the inertness or rigidity of the cobalt complexes in organic solvents made it possible to follow structurally the fate of the electrolyzed cobalt complexes in solution, not only in the reduction process but also in the counter process of oxidation with auxiliary instrumental methods of measurements such as ESR, NMR, and magnetic susceptibility. Such a trend can be recognized in several recent reports of cyclic amine complexes [162], Schiff base complexes [127, 163], and other similar organocobalt complexes in nonaqueous solvents, as summarized in Tables 2.2.20 and 2.2.21. Typical results can be observed in the voltammetric behavior of cobalt(II) and cobalt(III) complexes with macrocyclic rings, i.e., in the oxidation of cobalt(II) complexes with tetraphenylporphyrins (tpp) and phthalocyanine (pc) as determined by cyclic voltammetry [155]. The $[Co^{II}tpp]$ complex and its products, obtained by its controlled-potential electrochemical oxidation, were studied by ESR. The first oxidation occurred at the central cobalt(II) atom in the vicinity of +0.52 V (vs SCE) and all subsequent oxidations occurred at the tpp ligand. The potentials of central cobalt(II) oxidation showed a linear dependence on the third ionization potential of the ion, whereas the ligand oxidation potentials were approximately independent of the cobalt(II) ion. This distinguishes between the cobalt oxidation change and ligand oxidation. The cyclic voltammogram distinctly showed three one-electron reversible oxidation steps:

$$[Co^{II}tpp] \xrightarrow{+0.52\ V} [Co^{III}tpp]^{+} \xrightarrow{+1.19\ V} [Co^{III}tpp]^{2+} \xrightarrow{+1.42\ V} [Co^{III}tpp]^{3+}$$

in the butyronitrile containing 0.1 M $LiClO_4$ at a platinum disk electrode (PDE). The saturated aqueous calomel electrode (SCE) was separated from the solution by a fritted glass disk and a Luggin capillary. Cyclic potential sweep rates ranged from 1 to 150 V/min. The neutral original species is paramagnetic, with an effective spin of 1/2 compatible with a square-planar $3d^7$ configuration, and the first oxidation product, $[Co^{III}tpp]^+$, is very stable and showed no ESR spectrum, which is consistent with a square-planar $3d^6$ configuration. This change from a $3d^7$ to a $3d^6$ configuration implies cobalt(II) oxidation rather than ligand oxidation. The lifetime of the second oxidation product, $[Co^{III}tpp]^{2+}$ (paramagnetic), was found to be 7.5 hr by observing ESR spectrum, and the third oxidation species, $[Co^{III}tpp]^{3+}$, was diamagnetic [164]. Further, it was suggested that the $[Co^{III}tpp]^+$ cation can more readily interact with the electrolyte anions than the neutral $[Co^{II}tpp]$ complex.

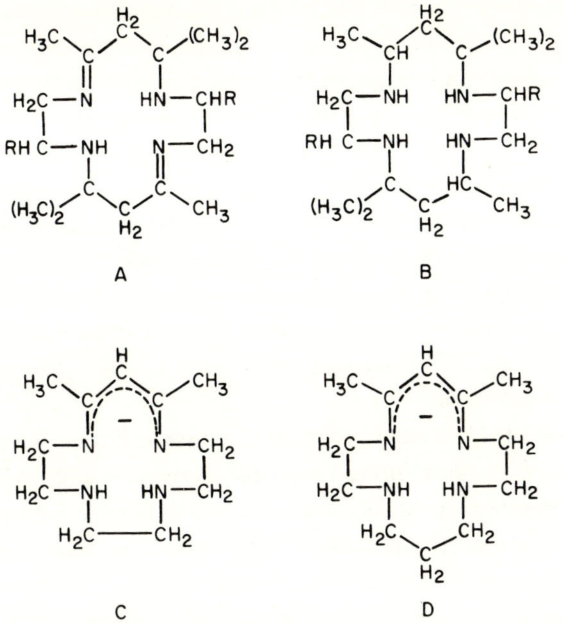

FIG. 2.3.1. Macrocyclic ligands: (A) R = H, 5, 7, 7, 12, 14, 14-hexamethyl-1, 4, 8, 11-tetraazacyclotetradeca-4, 11-diene(trans[14]diene); R = CH_3, 2, 5, 7, 7, 9, 12, 14, 14-octamethyl-1, 4, 8, 11-tetraazacyclotetradeca-4, 11-diene ($(CH_3)_2$-trans[14]diene); (B) R = H, 5, 7, 7, 12, 14, 14-hexamethyl-1, 4, 8, 11-tetraazacyclotetradecane (teta or tetb); R = CH_3, 2, 5, 7, 7, 9, 12, 14, 14-octamethyl-1, 4, 8, 11-tetraazacyclotetradecane ($(CH_3)_2$-teta); (C) 11, 13-dimethyl-1, 4, 7, 10-tetraazacyclotrideca-10, 12-diene (A[13]T); (D) 12, 14-dimethyl-1, 4, 8, 11-tetraazacyclotetradeca-11, 13-diene (A[14]T).

On the other hand, the free ligand H_2tpp gave two one-electron irreversible oxidation steps (see Table 2.2.21):

$$H_2\text{tpp} \xrightarrow{+1.00\ V} [H_2\text{tpp}]^+ \xrightarrow{+1.20\ V} [H_2\text{tpp}]^{2+}$$

An estimate of the lifetime of the oxidation products from the irreversible cyclic voltammogram showed values less than 10^{-2} sec, in contrast to the value of 16 min measured from the intensity decay of the ESR signal. This was interpreted as implying either a decomposition mechanism different at the electrode than in the solution or that their ESR spectra was of a decomposition product rather than the $[H_2\text{tpp}]^+$ cation radical itself.

Rillema, Endicott, and Papaconstantinou [162] studied the $[Co^{II}L]^{2+}$ and $[Co^{III}LX_2]^{n+}$ complexes in which cobalt is "trapped" or surrounded with four nitrogen donors of a macrocyclic ligand (of the Curtis type), as denoted in Fig. 2.3.1. These cobalt(II) and cobalt(III) complexes undergo one-electron oxidation or reduction in AN (acetonitrile) containing 0.1 M $[(C_2H_5)_4N]ClO_4$ at a rotating platinum electrode (RPE). The redox couples $[Co^{II}L]^{2+}/[Co^{I}L]^+$ and $[Co^{III}LX_2]^{n+}/[Co^{II}L]^{2+}$ were quasi-reversible to irreversible. That is, the cyclic voltammograms of several $[Co^{III}LX_2]$ complexes indicate that the Co(III) → Co(II) reductions are not quite reversible. Furthermore, the peak height may be somewhat smaller for the anodic sweep than for the cathodic sweep, indicating a slow chemical reaction following reduction (electron transfer). Nevertheless, the voltammetric behavior is sufficiently reversible to argue strongly that the $[Co^{III}LX_2]^{n+} \rightleftarrows [Co^{II}LX_2]$ couples are involved in each half-cycle; the ligands X^- are not rapidly replaced by solvent molecules. The $E_{1/2}$ values depend very strongly on the ligand X^-; in fact, one may take the oxidation peaks (~+0.5 V) for the $[Co^{II}L]^{2+}$ complexes as an indication of the values of $E_{1/2}$ expected for the solvolyzed cobalt(III) complexes (i.e., for $[CoL(an)_2]^{3+}$). The half-wave potential for the oxidation of $[Co^{I}L]^+$ and $[Co^{II}L]^{2+}$ parallels the variation of the ionization potential of the cobalt, while the $E_{1/2}$ values for $[Co^{III}LX_2]^{n+}$ $[Co^{II}LX_2]^{(n-1)+}$ vary with the spectrally determined ligand field strength of X in a largely predictable manner.

Table 2.3.1 shows the half-wave potentials from cyclic voltammetry at the RPE for reduction and oxidation of the cobalt(II) and cobalt(III) complexes of the Curtis type in AN.

The cobalt(III) complexes in which the ligand X has relatively low ligand field strength are reduced at relatively positive potentials. This is the qualitative trend one would expect for variation in E° (standard reduction or oxidation potential). Rillema et al. [162] modified the approach of Rock [3] to allow a semiquantitative comparison of $E_{1/2}$ values with ligand field parameters. The following equation may hold for a series of very similar complexes under the assumption that variations in ΔH° arise largely from variation in ligand field splitting energies:

$$\Delta G^\circ = -FE^\circ \doteq (-2.40\Delta + 2P) - (-1.80\Delta' + P') + T\Delta S \quad (2.3.1)$$

where the cobalt(II) species formed was assumed to be spin paired. Here, Rillema et al. [162] have made rather drastic assumptions: 1) $2\Delta' = \Delta$ for the same ligands; 2) for complexes of the type trans-$[CoL\,X_2]$, the total ligand field splitting parameter Δ is simply related to the splitting parameters for octahedral complexes with ligands X and L, respectively; thus $\Delta = 1/3\Delta_X + 2/3\Delta_L$; and 3) $E_{1/2} \doteq E^\circ$ + constant. With these assumptions and setting $\Delta = 10D_q{}^Z$, Eq. (2.3.1) may be put in the form where C is a collection of terms insensitive to changes in ligand field strength along the Z axis:

$$D_q{}^Z = 1.6E_{1/2} + C \quad (2.3.2)$$

Figure 2.3.2 shows the plot of $D_q{}^Z$ vs $E_{1/2}$ for complexes of the $[Co\ trans[14]diene\ X_2]^+$ and the $[Co(teta)X_2]^+$ types in AN; the slope of a plot is $1.7 \pm 0.2\ cm^{-1}\ mV^{-1}$. It is to be noted that a similar plot of $E_{1/2}$ vs $D_q{}^Z$ for the $[Co^{III}X(NH_3)_5]$ complexes is consistent with the theoretical slope of $3.2\ cm^{-1}\ mV^{-1}$ [75].

The voltammetric oxidation of a variety of cobalt complexes with porphyrins has also been studied in butyronitrile, and the intermediates were shown to be stable [156]. Felton, Dolphin, Borg, and Fajer [165] found that the cobalt(II) complex with octaethylporphyrin, $[Co^{II}oep]$, in chloroform undergoes two distinct one-electron oxidations when treated with bromine. The first step, which required 0.5 ± 0.02 mole of bromine, brought about the oxidation of Co(II) to Co(III), giving $[Co^{III}oep]Br$, which had been isolated, recrystallized

TABLE 2.3.1. Half-Wave Potentials from Cyclic Voltammetry and Polarographic Measurements at the Rotating Platinum Electrode for Reduction and Oxidation of Macrocyclic Cobalt Complexes in Acetonitrile[a]

	Reduction			Oxidation	
	Cyclic voltammetry		Polarographic	Cyclic voltammetry	Polarographic
Compound	$E_{1/2}$ (HDE)	$E_{1/2}$ (SPE)	$E_{1/2}$ (RPE)	$E_{1/2}$ (SPE)	$E_{1/2}$ (RPE)
$[Co^{II}trans[14]diene]^{2+}$	-1.59 (2→1)	-1.39 (2→1)	-1.40 (2→1)	+0.72 (2→3)	
	-1.19 (1→2)	-1.36 (1→2)		+0.40 (3→2)	
$[Co^{II}trans[14]diene(OH_2)_2]^{2+}$	-1.55 (2→1)	-1.40 (2→1)	-1.42 (2→1)	+0.94 (2→3)	+1.21 (2→3)
	-1.29 (1→2)	-1.39 (1→2)		+0.31 (3→2)	
$[Co^{II}trans[14]diene(py)_2]^{2+}$	-1.46 (2→1)	-1.42 (2→1)	+1.43 (2→1)	+1.06 (2→3)	+1.44 (2→3)
	-1.25 (1→2)	-1.41 (1→2)		+0.25 (3→2)	
$[Co^{II}(teta)]^{2+}$	-1.90 (2→1)	-1.76 (2→1)		+0.51 (2→3)	
	-1.66 (1→2)	-1.63 (1→2)		+0.37 (3→2)	
$[Co^{II}(CH_3)_2trans[14]diene]^{2+}$	-1.37 (2→1)	-1.36 (2→1)	-1.33 (2→1)	+0.75 (2→3)	
	-1.24 (1→2)	-1.36 (1→2)		+0.53 (3→2)	
$[Co^{III}trans[14]diene(Br)_2]^{+}$		+0.17 (3→2)	+0.10 (3→2)	-1.45 (2→1)	
		+0.17 (2→3)		-1.45 (1→2)	
$[Co^{III}trans[14]diene(Cl)_2]^{+}$		-0.20 (3→2)	-0.20 (3→2)	-1.55 (2→1)	-1.80 (2→1)
		-0.17 (2→3)		-1.47 (1→2)	
$[Co^{III}trans[14]diene(N_3)_2]^{+}$	-0.42 (3→2)	-0.45 (3→2)		-1.73[b] (2→1)	
	-0.42 (2→3)	-0.42 (2→3)		-1.64[b] (1→2)	
$[Co^{III}trans[14]diene(OH_2)_2]^{3+}$		+0.32 (3→2)		-1.53 (2→1)	
		-0.23 (2→3)		-1.50 (1→2)	
$[Co^{III}trans[14]diene(NCS)_2]^{+}$		+0.44 (3→2)	-0.45 (3→2)	-1.46 (2→1)	-1.48 (2→1)
		+0.59 (2→3)		-1.45 (1→2)	

(continued)

TABLE 2.3.1 (continued)

Compound	Reduction: Cyclic voltammetry $E_{1/2}$ (HDE)	Reduction: Cyclic voltammetry $E_{1/2}$ (SPE)	Reduction: Polarographic $E_{1/2}$ (RPE)	Oxidation: Cyclic voltammetry $E_{1/2}$ (SPE)	Oxidation: Polarographic $E_{1/2}$ (RPE)
$[Co^{III}trans[14]diene(OH)(OCOCH_3)]^+$	-0.58 (3→2)	-0.60 (3→2)		-1.78 (2→1)	
	-0.44 (2→3)	-0.55 (2→3)		-1.70 (1→2)	
$[Co^{III}trans[14]diene(NO_2)_2]^+$	-0.40 (3→2)	-0.48 (3→2)		-1.66 (2→1)	
	-0.38 (2→3)			-1.64 (1→2)	
$[Co^{III}trans[14]diene(CN)_2]^+$	-1.10 (3→2)	-1.10 (3→2)	-1.09 (3→2)	-1.78 (2→1)	-1.80 (2→1)
	-1.10 (2→3)	-1.10 (2→3)		-1.68 (1→2)	
$[Co^{III}trans[14]diene(NH_3)_2]^{3+}$		-0.91 (3→2)		-1.45 (2→1)	
		-0.91 (2→3)		-1.43 (1→2)	
$[Co^{III}(teta)Cl_2]^+$		-0.14 (3→2)	-0.08 (3→2)	-1.94[b] (2→1)	
		-0.14 (2→3)		(1→2)	
$[Co^{III}(teta)Br_2]^+$		+0.06 (3→2)	+0.05 (3→2)		
		+0.06 (2→3)			
$[Co^{III}(teta)(CN)_2]^+$	-1.00 (3→2)	-1.00 (3→2)			
	-1.00 (2→3)	-1.00 (2→3)			
$[Co^{III}(tetb)Cl_2]^+$		-0.17 (3→2)	-0.05 (3→2)		
		-0.17 (2→3)			

[a]Numerals in parentheses indicate the oxidation change of cobalt.

[b]Values were determined at the HDE (a hanging drop mercury electrode); voltage unit, V vs SCE. The electrode system consisted of a saturated calomel reference electrode (SDE), a Brinkmann Instruments hanging drop mercury electrode (HDE) or a rotating platinum electrode (RPE) (or a stationary platinum electrode (SPE) for cyclic voltammetry), and a platinum wire functioned as the third electrode.

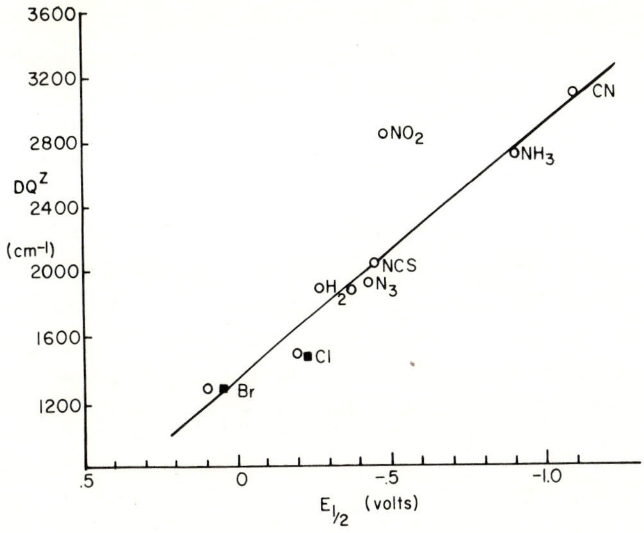

FIG. 2.3.2. Ligand field stabilization energy of X, D_q^2, vs the half-wave potential, $E_{1/2}$: (O) Co(trans[14]diene)X_2^+; (■) Co(teta)X_2^+.

from methylene chloride-ligroin, and identified. Further oxidation of the [Co^{III}oep]Br complex used 0.51 ± 0.02 mole of bromine and yielded a green species formulated as [Co^{III}oep]Br_2. This formulation is supported by the quantitative conversion of the dication according to

$$[Co^{III}oep]Br_2 + [Co^{II}oep] \rightarrow 2[Co^{III}oep]Br$$

This result of chemical oxidation is quite consistent with that of voltammetric oxidation obtained for the similar species [Co^{II}tpp] in butyronitrile [155]. These two-stage oxidations could also be reached by electrolysis in CH_2Cl_2. Similarly, the stable oxidized products of the [Co^{II}tpp] complex have been prepared and identified by these techniques [165].

As for the reduction processes, Felton and Linschitz [157] have examined the [Co^{II}tpp] and [Co^{II}etioI] complexes at the DME in DMF (dimethylformamide) containing 0.1 M tetra-n-propylammonium perchlorate as a background electrolyte, where tpp and etioI represent tetraphenylporphyrin and etioporphyrin, respectively, as denoted in Table 2.2.21. The [Co^{II}tpp] complex gave two well-defined waves of equal height and slope corresponding to one-electron reductions, while the polarogram of the [Co^{II}etioI] complex showed that the second wave at -1.57 V (vs SCE) was half the height of the first wave, which was accompanied by a maximum, independent of dye concentration. In the course of attempting to assign the waves to specific reduction processes, Felton et al. [157] have recognized that the nature of the reduction waves for the cobalt complex is sharply different from that of the reduction waves for metalloporphyrins involving Mg, Zn, Cd, Cu, Ni, Pb, and Sn in the magnitude of half-wave potentials, suggesting that the site of primary reduction for the [Co^{II}tpp] complex may be the cobalt(II) ion rather than the ring. For the cobalt complex, Zerner and Gouterman [166] find the lowest available level to be the half-filled cobalt $a_{1g}(d_{Z^2})$ orbital, which lies well below the lowest porphyrin $e_g(\pi)$ orbital. This finding was invoked to explain that preferential reduction of the cobalt in this complex may be reasonable.

Oscillopolarographic or voltammetric techniques were utilized mainly for elucidating the redox systems of π-bonded cobalt complexes, but the other aspect of the voltammetry relates to the entire field of electrochemistry and electroanalytical chemistry of cobalt. For example, the technique of anodic stripping voltammetry is well suited, in terms of sensitivity and convenience, for the determination of a trace quantity of cobalt in alloys, especially by using a solid electrode [167]. In fact, a number of papers have been published concerning the voltammetric determination of cobalt, together with voltammetric data. Since it is not possible for us to cover the entire field, suffice it to mention that excellent reviews are annually issued concerning the voltammetric behavior of cobalt [168-171].

3. KINETIC PARAMETERS AND DOUBLE-LAYER PROPERTIES

3.1. KINETIC PARAMETERS

In earlier years, Eyring, Marker, and Kwoh [172] and Tanaka and Tamamushi [173-176] independently applied absolute rate theory and the Nernst diffusion layer concept to the problem of irreversible electrode processes. Since then, Smutek [177], Delahay [178], and Evans and Hush [179] have independently developed the rigorous theoretical treatments involving the case of semi-infinite linear diffusion. Koutecky [180] solved this problem for the polarographic case.

The last decade has seen tremendous progress in evaluating kinetic parameters by such techniques as stationary electrode polarography [181], potentiostatic methods [182-186], and those which use polarographic instantaneous current [187-191], relaxation methods [192], and double-pulse galvanostatic methods [193-195], each capable of estimating the kinetic parameters of rapid heterogeneous electron-transfer reactions. All of these theoretical treatments and techniques are based on time-dependent transient phenomena.

In contrast to these methods, there are the other theoretical treatments and techniques for studies of heterogeneous electron-transfer reactions in hydrodynamic systems, but these methods have so far received relatively little attention because of the success of the time-dependent methods and partly because of the difficulties involved in mathematical calculations [196].

One of the merits for the latter is time-independence, which permits steady-state measurements, as opposed to the transient, time-dependent measurements of the former that require high-response electronics in the readout circuits. The other merit is that the steady-state feature of the hydrodynamic systems eliminates the capacitor current which often limits the sensitivity of the time-dependent methods, although several improvements and corrections for the structure of the electric double layer and charging have been made [197, 198]. Hydrodynamic steady-state measurements are quite suitable for studies of slow electrode reactions of totally irreversible systems or moderately rapid reactions of quasi-reversible systems, the group to which most systems of Co(III)/Co(II) couples in aqueous solutions belong. Non steady-state measurements, such as potentiostatic techniques, are suited to studies of fast electrode reactions of systems such as $[Co\ edta]^-/[Co\ edta]^{2-}$ and $[Co\ cydta]^-/[Co\ cydta]^{2-}$ couples [34, 199].

In this section we will describe, as one example, the procedures for evaluating the kinetic parameters for the slow electroreductions of pentacyanocobaltate(III) complexes on the basis of the temperature dependence of current-potential curves, and the effect of the double-layer structure on the kinetic parameters will be presented and discussed by correcting them in a simple way according to the original concept of Frumkin [200, 201].

The pentacyanocobaltate(III) complexes of the $[Co^{III}(CN)_5X]$ type have been studied by polarographic and voltammetric methods by several investigators [120, 123, 124, 126, 202-207]. As described in Section 4, the slow electroreduction processes of pentacyanocobalt(III) complexes can be divided into two groups. One group, the chloro-, bromo-, iodo-, thiocyanato-, and aquo-pentacyanocobalt(III) complexes, give rise to well-defined two one-electron reduction waves of Co(III) → Co(II) → Co(I), while the other group, the nitro-, azido-, thiosulfato-, sulfito-, and nitrosyl-pentacyano complexes, show a one step two-electron reduction wave in neutral, unbuffered, noncomplexing media. The corresponding electrode processes have been shown to be [206, 355]:

$$2[Co^{III}(CN)_5X]^{(2+n)-} \xrightarrow{e} 2[Co^{II}(CN)_5X]^{(3+n)-} \xrightarrow{e} 2[Co^{I}(CN)_5X]^{(4+n)-} \longrightarrow$$

$$2[Co^{I}(CN)_5]^{4-} + X^{n-} \longrightarrow [Co^{I}(CN)_5\text{-H-}Co^{I}(CN)_5]^{7-},$$

where the unidentate ligand, X, denotes the ion I^-, Br^-, Cl^-, SCN^-, NO_2^-, N_3^-, $S_2O_3^{2-}$, SO_3^{2-}, OH^-, or NO^-, or the molecule NH_3 or OH_2. Thus the primary product of the $[Co^I(CN)_5]^{4-}$ ion is a very strong base which extracts a hydrogen ion per 2 moles of the cobalt(I) complexes from water, and the final product of the binuclear cobalt(I) complex is very stable toward electrochemical oxidation in neutral solutions; neither further reduction to the cobalt(0) state nor anodic oxidation from the cobalt(I) to the cobalt(III) or to the cobalt(II) state was observed over the potential range from the positive potential at which dissolution of mercury occurs to -2.00 V (vs SCE) in a 0.5-$\underline{M}$ Na_2SO_4 aqueous solution. Therefore the reduction processes of Co(III)→Co(I) and of Co(III)→Co(II)→Co(I) are concluded to be totally irreversible systems since the anodic oxidation of the cobalt(I) state can never occur electrochemically under these conditions. Chemically, however, oxidation of the pentacyanocobaltate(I) complex occurs very rapidly and vigorously in water to the cobalt(III) state, evolving hydrogen and suggesting it to be a powerful reducing agent [208, 209]. Hence it can be understood that the electrochemical stability and the chemical stability of the pentacyanocobaltate(I) complex are quite different from each other with respect to its significance toward oxidation. The half-wave potentials of the reduction steps are given in Table 2.2.16.

The transfer coefficient for the cathodic reduction (α_C), the rate constant (k_C^o) at the potential where the electrode potential equals zero against a normal hydrogen electrode (NHE), and the heat of activation (ΔH) were evaluated for totally irreversible processes of Co(III) → Co(I) and of Co(III) → Co(II) → Co(I), respectively, in the solution with a unit ionic strength (0.5 M Na_2SO_4). The diffusion coefficient (D_0) required for the calculation of the rate constant was estimated from the Ilkovič equation using the data of the diffusion-controlled current, i_d, and is shown in Tables 3.1.1 and 3.1.2 as examples. From the $\log[i/(i_d - i)]$ vs E (the potential) plot, which is shown in Fig. 3.1.1 as an example, the cathodic transfer coefficient multiplied by the number of electrons involved in the rate-determining step ($\alpha_C n$) was evaluated and shown in Tables 3.1.3 and 3.1.4. The apparent rate constant of the cathodic reduction was computed according to

$$\alpha_C nFE_{1/2} = (RT)2.3 \log(k_C{}^a f_0 D^{1/2}) + 1.15 \log t - 0.12 \tag{3.1.1}$$

where $E_{1/2}$ is the half-wave potential in V vs NHE, $k_C{}^a$ is the apparent rate constant of the cathodic reduction at 0 V vs NHE in centemeters/second, f_0 is the activity coefficient of the oxidant in the solution (f_0 was assumed to be unity in this estimation), t is the duration time of electrolysis which corresponds to the drop life of the DME, and the other symbols have their usual significance [189, 210]. The apparent heat of activation, $\Delta H_C{}^a$, for the cathodic reduction was determined from the temperature dependence of the rate constant at 5, 15, 25, 35, and 45°C, the correlation of which is indicated in Tables 3.1.5 and 3.1.6. The $\Delta H_C{}^a$ values appearing in Tables 3.1.7 to 3.1.10 are those obtained from $\Delta H_C{}^{a*}$.

$$d \ln k_C{}^a/dT = \Delta H_C{}^{a*}/RT^2 \tag{3.1.2}$$

or

$$\log k_C{}^a = \text{const} - \Delta H_C{}^{a*}/RT \tag{3.1.3}$$

where $\Delta H_C{}^{a*}$ represents the heat of activation at a constant volume and $\Delta H_C{}^a$ that at a constant pressure. As one example, an Arrhenius plot is given in Fig. 3.1.2, from which the slope of the straight-line (tan θ) was obtained as 3330. Hence, $\Delta H_C{}^{a*}$ = (3330)2.303R = 15.24 kcal/mole. The kinetic parameters thus obtained are shown in Tables 3.1.7 to 3.1.11 for the bromo-, sulfono-, thiosulfato-, and azido-pentacyanocobaltate(III) complexes in a 0.5-M Na_2SO_4 aqueous solution.

TABLE 3.1.1. The Estimation of Diffusion Coefficient, D_0, for the Process Co(III) → Co(I) of the $K_4[Co(S_2O_3)(CN)_5]$ Complex in 0.5 M Na_2SO_4

Temperature (°C)	i_d (μA)	D_0 (cm/sec$^{1/2}$)	D_0 (cm^2/sec)
5	19.3	1.80×10^{-3}	3.24×10^{-6}
15	21.7	2.30×10^{-3}	4.12×10^{-6}
25	26.0	2.43×10^{-3}	5.90×10^{-6}
35	29.1	2.72×10^{-3}	7.40×10^{-6}
45	32.2	3.01×10^{-3}	9.06×10^{-6}

[a] 5×10^{-3} M complex in 0.5 M Na_2SO_4; m = 1.76 mg/sec, t = 4.2 sec at open circuit.

TABLE 3.1.2. The Estimation of Diffusion Coefficient, D_0, for the First and the Second Step of the $K_3[CoBr(CN)_5]$ Complex in 0.5 M Na_2SO_4[a]

Temperature (°C)	First step, Co(III) → Co(II)			Second step, Co(II) → Co(I)		
	i_d (μA)	D_0	D_0 (cm^2/sec)	i_d (μA)	D_0	D_0 (cm^2/sec)
5	9.5	1.68×10^{-3}	2.72×10^{-6}	9.8	1.72×10^{-3}	2.95×10^{-6}
15	10.9	1.93×10^{-3}	3.72×10^{-6}	11.0	1.99×10^{-3}	3.96×10^{-6}
25	11.9	2.12×10^{-3}	4.49×10^{-6}	12.3	2.06×10^{-3}	4.33×10^{-6}
30	13.0	2.31×10^{-3}	5.65×10^{-6}	13.1	2.39×10^{-3}	5.71×10^{-6}
45	14.1	2.50×10^{-3}	6.45×10^{-6}	14.3	2.60×10^{-3}	6.75×10^{-6}

[a] 5×10^{-3} M complex ion in 0.5 M Na_2SO_4 aq soln; m = 1.76 mg/sec, t = 4.2 sec at open circuit.

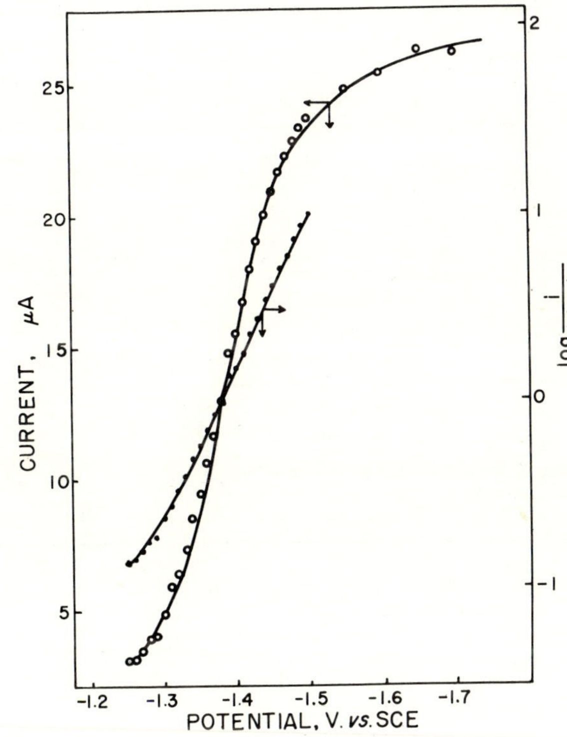

FIG. 3.1.1. The $\log[i/(i_d-i)]$ vs E plot of the reduction process, Co(III) → Co(I), for the $K_4[Co(S_2O_3)(CN)_5]$ complex. Concentration of the complex, 5×10^{-3} M in 0.5 M Na_2SO_4 at 25.0° C; m = 1.76 mg/sec; t = 4.2 sec.

TABLE 3.1.3. The Determination of Transfer Coefficient for the Cathodic Reduction Co(III) → Co(I) of the $K_4[Co(S_2O_3)(CN)_5]$ Complex[a]

Temperature (°C)	tan θ	$\alpha_C n$	$E_{1/2}$ (V vs SCE)
5	104	0.50	-1.380
15	111	0.49	-1.380
25	116	0.49	-1.379
35	122	0.48	-1.378
45	123	0.49	-1.371

[a] 5×10^{-3} M complex ion in 0.5 M Na_2SO_4; m = 1.76 mg/sec, t = 4.2 sec.

TABLE 3.1.4. The Determination of Transfer Coefficient for the Cathodic Reduction Steps of Co(III) → Co(II) and of Co(II) → Co(I) of the $K_3[CoBr(CN)_5]$ Complex in 0.5 M Na_2SO_4

	First step			Second step		
Temperature (°C)	tan θ	$\alpha_C n$	$E_{1/2}$ (V vs SCE)	tan θ	$\alpha_C n$	$E_{1/2}$ (V vs SCE)
5	119	0.44	-0.893	102	0.51	-1.259
15	125	0.44	-0.873	108	0.51	-1.267
25	126	0.45	-0.849	106	0.50	-1.256
30	133	0.43	-0.847	139	0.42	-1.250
45	140	0.42	-0.853	208	0.28	-1.242

TABLE 3.1.5. Apparent Rate Constants for the Co(III) → Co(I) Process of $K_4[Co(CN)_5S_2O_3]$ (5×10^{-3} M) in 0.5 M Na_2SO_4

Temperature (°C)	k_C^a (cm/sec)	$\log k_C^a$
5	8.38×10^{-14}	-13.077
15	2.27×10^{-13}	-12.645
25	5.55×10^{-13}	-12.256
35	1.42×10^{-12}	-11.847
45	2.49×10^{-12}	-11.603

TABLE 3.1.6. Apparent Rate Constants for the Co(III) → Co(II) and Co(II) → Co(I) Processes of $K_3[Co(CN)_5Br]$ (5×10^{-3} M) in 0.5 M Na_2SO_4

Temperature (°C)	First step ($Co^{III} \rightarrow Co^{II}$)		Second step ($Co^{II} \rightarrow Co^{I}$)	
	k_C^a (cm/sec)	log k_C^a	k_C^a (cm/sec)	log k_C^a
5	7.59×10^{-9}	-8.120	5.55×10^{-13}	-12.256
15	1.84×10^{-8}	-7.736	1.04×10^{-12}	-11.983
25	2.93×10^{-8}	-7.466	3.33×10^{-12}	-11.478
30	5.65×10^{-8}	-7.248	a	a
35	9.04×10^{-8}	-7.044	a	a

[a]The evaluation of k_C^a was impossible because of the unreliable value of $\alpha_C n$.

TABLE 3.1.7. Kinetic Parameters for Totally Irreversible Process Co(III) → Co(I) for the $K_4[Co(S_2O_3)(CN)_5]$ Complex in a 0.5-M Na_2SO_4 Solution[a]

Temperature, °C / °K	5 / 278	15 / 288	25 / 298	35 / 308	45 / 318
$E_{1/2}$ (V vs SCE)	-1.380	-1.380	-1.379	-1.378	-1.371
$E_{1/2}$ (V vs NHE)	-1.122	-1.129	-1.133	-1.149	-1.139
i_d (μA)	19.3	21.7	26.0	29.1	32.2
$\alpha_C n$	0.50	0.49	0.49	0.48	0.49
D_0 (cm^2/sec)	3.24×10^{-6}	4.12×10^{-6}	5.90×10^{-6}	7.40×10^{-6}	9.06×10^{-6}
log k_C^a	-13.08	-12.65	-12.26	-11.85	-11.60
k_C^a (cm/sec)	8.38×10^{-14}	2.27×10^{-13}	5.55×10^{-12}	1.42×10^{-12}	2.49×10^{-12}

[a]$(\Delta H_C^a)_{E=0}$ = 14.7 kcal/mole at 25°C; ΔH_D = 4.5 kcal/mole at 25°C.

TABLE 3.1.8. Kinetic Parameters of Totally Irreversible Process Co(III) → Co(II) for the $K_3[CoBr(CN)_5]$ Complex in a 0.5-M Na_2SO_4 Solution[a]

Temperature, °C / °K	5 / 278	15 / 288	25 / 298	30 / 303	35 / 308
$E_{1/2}$ (V vs SCE)	-0.893	-0.873	-0.849	-0.847	-0.853
$E_{1/2}$ (V vs NHE)	-0.635	-0.621	-0.603	-0.605	-0.614
i_d (μA)	9.52	10.85	11.90	13.00	14.10
$\alpha_C n$	0.44	0.44	0.45	0.43	0.42
D_0 (cm^2/sec)	2.72×10^{-6}	3.72×10^{-6}	4.49×10^{-6}	5.65×10^{-6}	6.45×10^{-6}
$\log k_C{}^a$	-8.12	-7.74	-7.45	-7.25	-7.04
$k_C{}^a$ (cm/sec)	7.59×10^{-9}	1.84×10^{-8}	2.93×10^{-8}	5.65×10^{-8}	9.04×10^{-8}

[a] $(\Delta H_C{}^a)_{E=0}$ = 12.7 kcal/mole at 25°C; ΔH_D = 4.3 kcal/mole at 25°C.

TABLE 3.1.9. Kinetic Parameters of Totally Irreversible Process Co(II) → Co(I) for the $K_3[CoBr(CN)_5]$ Complex in a 0.5-M Na_2SO_4 Solution[a]

Temperature, °C / °K	5 / 278	15 / 288	25 / 298	30 / 303	35 / 308
$E_{1/2}$ (V vs SCE)	-1.259	-1.267	-1.256	-1.250	-1.242
$E_{1/2}$ (V vs NHE)	-1.001	-1.015	-1.010	-1.008	-1.003
i_d (μA)	9.80	11.00	12.30	13.10	14.30
$\alpha_C n$	0.51	0.51	0.50	0.42	0.28
D_0 (cm^a/sec)	2.95×10^{-6}	3.96×10^{-6}	4.33×10^{-6}	5.71×10^{-6}	6.75×10^{-6}
$\log k_C{}^a$	-12.26	-11.98	-11.48	-	-
$k_C{}^a$ (cm/sec)	5.55×10^{-13}	1.04×10^{-12}	3.33×10^{-12}	-	-

[a] $(\Delta H_C{}^a)_{E=0}$ = ~11.8 kcal/mole at 25°C; ΔH_D = 4.3 kcal/mole at 25°C.

TABLE 3.1.10. Kinetic Parameters of Totally Irreversible Process Co(III) → Co(I) for the $K_4[Co(SO_3)(CN)_5].3H_2O$ Complex in a 0.5-M Na_2SO_4 Solution[a]

Temperature, °C / °K	10 / 283	15 / 288	25 / 298	35 / 308	45 / 318
$E_{1/2}$ (V vs SCE)	-1.563	-1.562	-1.560	-1.556	-1.547
$E_{1/2}$ (V vs NHE)	-1.306	-1.309	-1.314	-1.318	-1.315
i_d (μA)	18.0	20.0	22.8	26.3	29.2
$\alpha_C n$	0.58	0.58	0.59	0.58	0.60
D_0 (cm²/sec)	2.72×10^{-6}	3.39×10^{-6}	4.37×10^{-6}	5.81×10^{-6}	7.18×10^{-6}
log k_C^a	-16.57	-16.27	-16.07	-15.56	-15.32
k_C^a (cm/sec)	2.70×10^{-17}	5.38×10^{-17}	9.52×10^{-17}	2.76×10^{-16}	4.79×10^{-16}

[a] $(\Delta H_C^a)_{E=0}$ = 15.7 kcal/mole at 25°C; ΔH_D = 4.9 kcal/mole at 25°C.

TABLE 3.1.11. Kinetic Parameters of Totally Irreversible Process Co(III) → Co(I) for the $K_3[Co(N_3)(CN)_5].2H_2O$ Complex in a 0.5-M Na_2SO_4 Solution[a]

Temperature, °C / °K	5 / 278	15 / 288	25 / 298	35 / 308	45 / 318
$E_{1/2}$ (V vs SCE)	-1.314	-1.310	-1.310	-1.306	-1.296
$E_{1/2}$ (V vs NHE)	-1.056	-1.058	-1.064	-1.067	-1.064
i_d (μA)	15.0	17.1	20.0	21.8	26.0
$\alpha_C n$	0.39	0.40	0.39	0.39	0.39
D_0 (cm²/sec)	1.88×10^{-6}	2.43×10^{-6}	3.35×10^{-6}	4.00×10^{-6}	5.71×10^{-6}
log k_C^a	-10.64	-10.27	-10.00	-9.73	-9.44
k_C^a (cm/sec)	2.29×10^{-11}	5.38×10^{-11}	1.01×10^{-10}	1.87×10^{-10}	3.61×10^{-10}

[a] $(\Delta H_C^a)_{E=0}$ = 11.6 kcal/mole at 25°C; ΔH_D = 4.5 kcal/mole at 25°C.

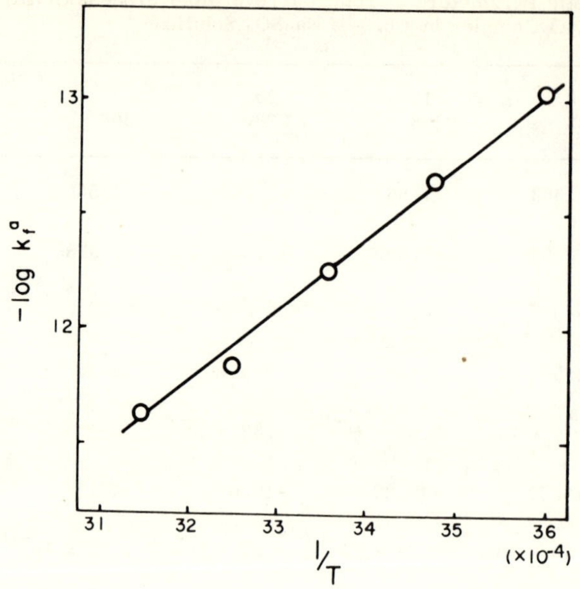

FIG. 3.1.2. Temperature dependence of $k_C{}^a$ for the cathodic reduction of the $K_4[Co(S_2O_3)(CN)_5]$ complex in a 0.5-M Na_2SO_4 solution.

In general, the rate constant k_C for the cathodic reduction is expressed by

$$k_C = k_C^\circ \exp[-(\alpha_C nF/RT)\Delta\phi] \tag{3.1.4}$$

with

$$k_C^\circ = k_C^\prime \exp[-\Delta G_C^\circ/RT] \tag{3.1.5}$$

where $\Delta\phi$ is the Galvani potential difference between the interphases of the electrode and the bulk of the solution; k_C° and ΔG_C° are the rate constant and the true heat of activation at $\Delta\phi = 0$, respectively; $k_C^\prime$ is the constant; and the other symbols have their usual significance. The situation for anodic oxidation is quite the same as that in cathodic reduction. The $\Delta\phi$ value equals the difference of inner potentials between the two phases of the electrode and the bulk of solution:

$$\Delta\phi = \phi_M - \phi_\sigma \tag{3.1.6}$$

where ϕ_M and ϕ_σ represent the inner potential of the phase of the electrode and of the bulk of the solution, respectively. The $\Delta\phi$ (Galvani potential difference) between the two phases with different chemical compositions is essentially an unknowable and unmeasurable value, as has already been pointed out by Gibbs and Guggenheim [211, 212]. Accordingly, from a practical point of view, Eq. (3.1.4) is better written as the approximate Eq. (3.1.8) involving a practically measurable value, E, which is the potential difference between the working electrode and the reference electrode. The E is expressed approximately by

$$E = \Delta\phi - \Delta\phi_r \tag{3.1.7}$$

where $\Delta\phi_r$ denotes the Galvani potential difference between the interphases of the electrode and the solution in the reference electrode.

Using E instead of $\Delta\phi$, Eq. (3.1.4) is rewritten as

$$k_C = k_C{}^a \exp\left[\frac{\alpha_C nF}{RT} E\right] \tag{3.1.8}$$

with

$$k_C{}^a = k_C^\circ \exp\left[\frac{\alpha_C nF}{RT} \Delta\phi_r\right] \tag{3.1.9}$$

At the standard redox potential, E°, where the rate constant of the cathodic reduction equals that of the anodic oxidation ($k_C = k_A$), the standard rate constant k_S does not depend on the nature of the reference electrode used, while the cathodic rate constant k_C° or the anodic rate constant k_A° at the potential where the electrode potential equals zero against a reference electrode (E = 0) is dependent on the reference electrode. The k_S cannot be determined for totally irreversible systems whose standard potentials are unknown.

Further, on the assumption that $\Delta\phi$ can be kept constant experimentally, the temperature dependence of k_C can be written as

$$\left(\frac{\partial \ln k_C}{\partial T}\right)_{\Delta\phi} = \frac{\partial \ln k_C{}^a}{\partial T} - \frac{\alpha_C nF}{RT^2}\Delta\phi \tag{3.1.10}$$

with

$$\frac{\partial \ln k_C{}^a}{\partial T} = \frac{\Delta H_C}{RT^2} \tag{3.1.11}$$

where ΔH_C is the true heat of activation when $\Delta\phi$ equals zero. Hence ΔH_C cannot be determined since $\Delta\phi$ is an unknowable value, i.e., the true heat of activation, ΔH_C, characteristic of the electron-transfer proper, can be estimated only under the conditions where the Galvani potential difference of the electrode/solution interphases equals zero ($\Delta\phi = 0$). From the practical point of view, therefore, the apparent heat of activation, $\Delta H_C{}^a$ is evaluated from the temperature dependence of k_C under conditions where E instead of $\Delta\phi$, is approximately kept constant. Then the following equations hold for the temperature dependence of k_C from Eqs. (3.1.8) and (3.1.9):

$$\left(\frac{\partial \ln k_C}{\partial T}\right)_E \equiv (\Delta H_C{}^a)_E = \frac{\partial \ln k_C{}^a}{\partial T} + \frac{\alpha_C nF}{RT^2} E \tag{3.1.12}$$

with

$$\frac{\partial \ln k_C{}^a}{\partial T} = \frac{\Delta H_C}{RT^2} - \frac{\alpha_C nF}{RT}\frac{\partial\Delta\phi_r}{\partial T} + \frac{\alpha_C nF}{RT^2}\Delta\phi_r \tag{3.1.13}$$

Hence

$$(\Delta H_C{}^a)_E = \Delta H_C - \alpha_C nFT\frac{\partial\Delta\phi_r}{\partial T} + \alpha_C nF\Delta\phi_r + \alpha_C nFE \tag{3.1.14}$$

Assuming that the Gibbs-Helmholtz equation holds for the potential $\Delta\phi_r$ of the reference electrode,

$$\Delta H_r = nFT\frac{\partial\Delta\phi_r}{\partial T} - nF\Delta\phi_r \qquad (3.1.15)$$

where ΔH_r implies the heat of activation with respect to the electrode reaction taking place at the reference electrode. Therefore $(\Delta H_C{}^a)_{E=0}$ can be expressed by Eq. (3.1.16) when the condition E = 0 is fulfilled:

$$(\Delta H_C{}^a)_{E=0} = \Delta H_C - \alpha_C\Delta H_r \qquad (3.1.16)$$

Since this equation involves the term $(\alpha\Delta H_r)$ of the heat of reaction dependent on the nature of the reference electrode used in addition to the true heat of activation (ΔH_C), the apparent heat of activation, $\Delta H_C{}^a$, is not necessarily characteristic of the electron-transfer proper. The ΔH_r value is still an essentially unknowable and unmeasurable value, as is the $\Delta\phi$ value. Nernst proposed the convention for equilibrium electrode potentials that the ΔH_r value of the NHE (normal hydrogen electrode) be 0 V. It is worthwhile to note that this assumption is very convenient for comparison of the temperature dependence of equilibrium electrode potentials, but it does not have any theoretical advantages for estimating and comparing the heats of activation for the electron-transfer processes. The above discussions are based on the excellent review concerning the heat of activation by Tamamushi [213].

Thus the true heat of activation is essentially an unknown quantity. Nevertheless, the term $-\alpha\Delta H_r$ may be regarded as being relatively constant for a series of very similar cobalt complexes, leading to the hypotheses that the apparent heat of activation $(\Delta H_C{}^a)_{E=0}$ is essentially proportional to the true heat of activation $(\Delta H_C)_{\Delta\phi=0}$ under the above conditions. As a matter of fact, comparison of the apparent heats of activation is possible for a series of cobalt complexes, provided that the same reference electrode is used in a series of experiments and that the cathodic transfer coefficients α_C have similar values. Taking account of these conditions, Vlček has succeeded in verifying that a linear relation holds for a series of the $[CoX(NH_3)_5]$, $[CoX_2(NH_3)_4]$, and $[CoX(CN)_5]$ complexes between the apparent heat of activation and the excitation energy expressed in terms of the absorption maxima ν_{max} [126, 203, 214, 215]. Table 3.1.12 shows the apparent heat of activation and the other kinetic parameters obtained for the cathodic reduction Co(III) → Co(II) for a series of pentammine- and tetrammine-cobalt(III) complexes under the same experimental conditions [214].

On the other hand, there is another steady-state treatment based on the temperature dependence of the exchange current density i_0 of the system involved. The apparent heat of activation $\Delta H_S{}^a$ at the standard potential is obtained from the temperature dependence of the standard rate constant k_S, the rate constant at the standard equilibrium state $(k_C = k_A = k_S)$. In other words, this is a method of following variations of k_S with temperature, the system being in an equilibrium state under constant overpotential. The apparent heat of activation $\Delta H_S{}^a$ at the standard electrode potential E_e° is defined as

TABLE 3.1.12. Apparent Heat of Activation and Other Kinetic Parameters of the Cathodic Reduction Co(III) → Co(II) at the Dropping Mercury Electrode. Solution: 0.14 M $HClO_4$ + 1.26 M $NaClO_4$. Reference Electrode: 0.1 N Calomel Electrode[a]

Complex	$k_C^a/\sqrt{D^b}$	$(\Delta H_C^a)_{E=0}$ (kcal/mole)	α_C
$[Co(NH_3)_6]^{3+}$	2.1×10^{-4}	12.6	0.67
$[Co(OH_2)(NH_3)_5]^{3+}$	2.6×10^{-2}	9.5	0.56
$[Co(NO_3)(NH_3)_5]^{2+}$	1.7×10^{-1}	9.3	0.53
$[Co(NO_2)(NH_3)_5]^{2+}$	1×10^{-3}	15.2	0.59
$[CoF(NH_3)_5]^{2+}$	4.2×10^{-4}	11.2	0.51
$[CoF(NH_3)_5]^{2+}$ [c]	1.3×10^{-3}	6.8	0.51
$[Co(fo)(NH_3)_5]^{2+}$	4.8×10^{-4}	7.5	0.63
$[Co(ac)(NH_3)_5]^{2+}$	9.5×10^{-4}	8.3	0.65
$[Co(po)(NH_3)_5]^{2+}$	9.5×10^{-4}	8.5	0.64
$[CoN_3(NH_3)_5]^{2+}$	1.1×10^{-1}	10.5	0.55
$[Co(ONO)(NH_3)_5]^{2+}$	7×10^{-4}	15	0.5
$[Co(oxH)(NH_3)_5]^{2+}$	9.2×10^{-4}	12.7	0.59
$[Co(fuH)(NH_3)_5]^{2+}$ [c]	6.7×10^{-4}	12.5	0.73
$[Co(maH)(NH_3)_5]^{2+}$ [c]	1.3×10^{-3}	12.5	0.76
$[Co(fuCH_3)(NH_3)_5]^{2+}$ [c]	6.8×10^{-4}	12.0	0.71
$[Co(SO_4)(NH_3)_5]^{+}$	3.8×10^{-3}	9.3	0.49
cis-$[Co(OH_2)_2(NH_3)_4]^{3+}$	2.8	8.0	0.55
trans-$[Co(OH_2)_2(NH_3)_4]^{3+}$	8.5×10^{-1}	7.5	0.52
cis-$[Co(NO_2)_2(NH_3)_4]^{+}$	2.3×10^{-2}	14.2	0.54
trans-$[Co(NO_2)_2(NH_3)_4]^{+}$	1.9×10^{-4}	15.5	0.6
cis-$[Co(ac)_2(NH_3)_4]^{+}$ [c]	2.5×10^{-1}	6.0	0.55
trans-$[Co(ac)_2(NH_3)_4]^{+}$ [c]	3.3×10^{-1}	4.8	0.52

[a] ac = acetate; po = propionate; fu = fumarate; ma = maleate.

[b] D = the diffusion coefficient for the cathodic process.

[c] Measured in 1.4 M $HClO_4$.

$$\Delta H_S{}^a = RT^2 \frac{\partial \ln k_S}{\partial T} \tag{3.1.17}$$

or

$$\Delta H_S{}^a = \Delta H_C + \alpha_C nF\Delta\phi_e{}^a - \alpha_C nFT \frac{\partial \Delta\phi_e{}^a}{\partial T} \tag{3.1.18}$$

where $\Delta\phi_e{}^a$ is the standard electrode potential of the system in question (not the value measured against a reference electrode). Further, at the standard equilibrium potential E_e° the overall faradaic current is equal to zero, and the concentrations of the depolarizers of the reduced and oxidized forms at the electrode surface, C_R or C_O, respectively, are equal to those in the bulk of the solution, C° ($C_R = C_O = C^\circ$). Therefore, the exchange current density j_0 can be written as follows by assuming that $\alpha_C + \alpha_A = 1$:

$$j_0 = nFk_S(C_O)^{\alpha_C}(C_R)^{\alpha_A} \tag{3.1.19}$$

Accordingly, at the standard equilibrium conditions ($C_O = C_R = C^\circ = 1$ M), the standard exchange current density $^\circ j$ is written as

$$^\circ j = nFC^\circ k_S \tag{3.1.20}$$

This relation between $^\circ j$ and k_S indicates that variations with temperature of k_S can be followed or monitored by measuring the exchange current density.

It is clear from Eq. (3.1.18) that $\Delta H_S{}^a$ at the standard potential is by no means identical with the true heat of activation, ΔH_C or ΔH_A, although it is characteristic of the electron-transfer process considered. The practical applications of this treatment can be seen in the voltammetric work of Bartelt et al. [222, 224-229]. Tamamushi [213] also suggested the idea of adopting "the potential at zero-charge of the metal" or "the null-point of metal" [216] for the purpose of evaluating the true heat of activation, and he recommended that the electrode potential according to the ϕ-scale seems to be preferable to that referred to a reference electrode, such as a hydrogen electrode for the kinetic study of the electrode processes, because the former seems likely to be related directly to the charge of the working electrode. This concept was applied in practice to the study of corrosion phenomena, and its use in electrode kinetic studies has been discussed by Antropov [217-219].

Kinetic parameters for the reversible electrode process of $[Co^{III}edta]^-/[Co^{II}edta]^{2-}$ and $[Co^{III}cydta]^-/[Co^{II}cydta]^{2-}$ couples at the DME have been determined by the potentiostatic method in solutions of ionic strength 0.5 ($NaNO_3$) containing 0.1 M acetate buffer (pH = 5.0), where $cydta^{4-}$ denotes the 1,2-cyclohexanediamine-tetraacetate ion [34, 199]. The numerical calculations and analysis were carried out on the basis of the general equation for the anodic wave [197] with a NEAC 2230 electronic computer. Methods of programming for the analysis and the results are given in Refs. 34 and 198. The diffusion coefficients of EDTA-cobalt complexes were determined from the measurements of the limiting diffusion current density in the above supporting electrolyte at 25°C:

$D_O = 6.9 \times 10^{-6}$ cm²/sec for $[Co\ edta]^-$
$D_R = 5.3 \times 10^{-6}$ cm²/sec for $[Co\ edta]^{2-}$

and are in good agreement with those obtained by the other methods [40, 220]. The kinetic parameters of the $[Co\ edta]^-/[Co\ edta]^{2-}$ couple, which include the measured and the corrected value for the effect of double-layer structure, were compared with those of $[Cr\ edta]^-/[Cr\ edta]^{2-}$ [221] and are given in Table 3.1.13.

TABLE 3.1.13. Kinetic Parameters of the Electrode Reaction of $[Co\ edta]^-/[Co\ edta]^{2-}$ and $[Cr\ edta]^-/[Co\ edta]^{2-}$ at 25°C

Parameters[a]	$[Co\ edta]^-/[Co\ edta]^{2-}$	$[Cr\ edta]^-/[Cr\ edta]^{2-}$
Supporting electrolyte	0.1 M acetate buffer + 0.4 M $NaNO_3$ (pH = 5.0)	0.1 M acetate buffer + 0.4 M NaCl (pH = 5.0)
$(E^\circ)_{ms}$ (V vs SCE)	+0.135	-1.220
$(E^\circ)_{corr}$ (V vs SCE)[b]	-	-1.220
$(k_S)_{ms}$ (cm/sec)	2.9×10^{-2}	2.1×10^{-1}
$(k_S)_{corr\ a}$ (cm/sec)[c]	5.8×10^{-2}	-
$(k_S)_{corr\ c}$ (cm/sec)[c]	-	1.7×10
$(\alpha_C)_{ms}$	0.51	0.58
$(\alpha_C)_{corr}$[d]	-	0.54
$(\alpha_A)_{ms}$	0.50	0.39
$(\alpha_A)_{corr}$[d]	-	0.43
D_O (cm^2/sec)	6.88×10^{-6}	6.33×10^{-6}
D_R (cm^2/sec)	5.34×10^{-6}	5.63×10^{-6}
$\Delta\phi$ (V)	-0.012	-0.073

[a]The subscript ms denotes the value of measured parameter, and the subscripts O and R indicate the oxidant and the reductant, respectively.

[b]Corrected standard potential defined by Eq. (33) in Ref. 197.

[c]Corrected standard rate constant expressed in terms of concentration for the anodic and the cathodic process, respectively [cf. Eqs. (37) and (38) in Ref. 197].

[d]Corrected transfer coefficient defined by Eqs. (44) and (45) in Ref. 197.

Considerable differences were recognized in comparing the corrected standard rate constant of Cr(III)-Cr(II) EDTA with those of Co(III)-Co(II) EDTA. The rate of the electron transfer of Cr(III)-Cr(II) EDTA is much larger than that of Co(III)-Co(II) EDTA. The effect of the double-layer structure on the kinetic parameters is more striking in the case of the electrode reaction of Cr(III)-EDTA/Cr(II)-EDTA than that of Co(III)-EDTA/Co(II)-EDTA, the result of which is consistent with the prediction that the influence of the double-layer structure would be less on the slow electrode reaction of cobalt complexes than on the fast one. In other words, the electrode process with an inherently fast electron transfer is apparently observed as a slower one because of the repulsion taking place between the discharged species and the electrode. Moreover, since the electrode reaction of Co(III)-EDTA/Co(II)-EDTA takes place at the positive region of the electrocapillary curve of mercury, it may be strongly affected by the electrical double layer because of the magnitude of its rate constant.

Bartelt and Landazury [222] have determined the exchange current density j_0 of the redox system $[Co(NH_3)_6]^{2+}/[Co(NH_3)_6]^{3+}$ by measuring the polarization resistance at the equilibrium potential E_e at the rotating platinum disk electrode (RDE). The concentration-dependence of j_0 gave the cathodic and the anodic transfer coefficients (see Eq. 3.1.20). At higher concentrations of the $[Co(NH_3)_6]^{3+}$ ion there is a deviation from the linear relation between log j_0 and log concentration of this complex. This is caused by coverage of the electrode by the aquated or hydrolysis product of the cobalt(III) complex; hence this phenomenon depends on the NH_4Cl concentration and the temperature. The anodic formation of the $[Co(NH_3)_6]^{3+}$ complex (the regeneration of the former species) favors the coverage of the electrode by the hydrolysis or aquated product (probably $[Co(OH)_2(NH_3)_4]^+$ ion, as suggested in Section 4), causing an anomalously low anodic limiting current. The dependence of j_0 on the ammonia concentration corresponds to the dependence of the $[Co(NH_3)_6]^{2+}$ ion concentration on ammonia content and shows that the hexammine complex alone takes part in the redox reaction. An increase of NH_4Cl and KCl concentration causes a lowering of j_0 as a result of the formation of the $[Co(NH_3)_6]^{3+}$-Cl^- ion pair and Cl^- adsorption on the electrode. The heterogeneous rate constant (5.25×10^{-8} cm/sec in 6.8 M NH_3 + 1 M NH_4Cl) derived is in good agreement with the literature value for the homogeneous rate constant [223] according to the relation proposed by Marcus. The apparent heat of activation was estimated to be 7.5 kcal/mole from the dependence of j_0 on temperature, which is illustrated in Fig. 3.1.3.

Bartelt and Skilandat [224] have determined the kinetic parameters of the reversible $[Co\ en_3]^{3+}/[Co\ en_3]^{2+}$ couple at the RDE by the similar method of following the variation of exchange current density near equilibrium. Evidence was found for a hydrolysis process which caused partial coverage of the platinum electrode surface, and hence, a decrease in the exchange current density. By varying the concentration of the free ligand (en) in excess, it was possible to show that only those ions of similar stoichiometry take part in the electron exchange reaction while ions such as $[Co(OH)_4en]^-$ and $[Co(OH)_2en_2]^+$ are electrochemically inactive. Additional evidence for the formation of outer-sphere complexes with anions (i.e., "super complexes") was provided by investigations of several authors [230, 231]. The effect of the anions on the rate constant has been examined with the results given in Table 3.1.14.

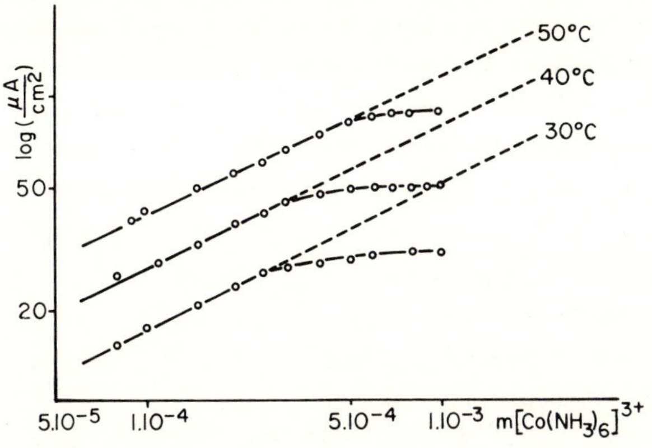

FIG. 3.1.3. Temperature-dependence of j_0. (Data from Ref. 222.)

TABLE 3.1.14. The Effect of Foreign Anions on Rate Constant of the $[Co\ en_3]^{3+}/[Co\ en_3]^{2+}$ Couple in Aqueous Media

Redox system; concn of background salt, 1 M	Rate constant, k_S (cm/sec) (25°C)
$[Co\ en_3]^{2+}/[Co\ en_3]^{3+}$ ClO_4^-	$(2.9 \pm 0.2) \times 10^{-2}$
$[Co\ en_3]^{2+}/[Co\ en_3]^{3+}$ Cl^-	$(2.1 \pm 0.1) \times 10^{-2}$
$[Co\ en_3]^{2+}/[Co\ en_3]^{3+}$ Br^-	$(0.62 \pm 0.02) \times 10^{-2}$
$[Co\ en_3]^{2+}/[Co\ en_3]^{3+}$ SO_4^{2-}	$(1.64 \pm 0.2) \times 10^{-2}$
$[Co\ en_3]^{2+}/[Co\ en_3]^{3+}$ $S_2O_3^{2-}$	$(0.46 \pm 0.03) \times 10^{-2}$

Among the anions, the perchlorate ion exhibited the smallest tendency of forming ion pairs; this observation is consistent with the results obtained by Stranks [230] and Larson [231, 232]. The rate constant of the $[Co\ en_3]^{2+}/[Co\ en_3]^{3+}$ system was 2.1×10^{-2} cm/sec at 25°C in the 1-M KCl solution containing 0.5 M ethylenediamine. For the homogeneous redox reaction of the $[Co\ en_3]^{2+}/[Co\ en_3]^{3+}$ system, Dwyer and Sargeson [233] have estimated the homogeneous rate constant, k_{homo}, to be 7.7×10^{-5} 1/mole/sec under the similar conditions. A comparison of R_S and k_{homo} was made according to the theory of Marcus [234];

$$(k_{homo}/10^{11})^{1/2} \approx k_S/10^4$$

and the following values were obtained:

$$(k_{homo}/10^{11})^{1/2} = 2.78 \times 10^{-8}; \qquad k_S/10^4 = 2.1 \times 10^{-6}$$

On comparing the homogeneous and the heterogeneous (electrochemical) rate constants, it was found that the heterogeneous rate constant k_s is much larger than the Marcus theory [234] predicts. Bartelt and Skilandat [224] have discussed this fact and the fractional reaction order (0.7) for the $[Co\ en_3]^{3+}$ ion. On the other hand, Laitinen and Randles [236] have determined the cathodic rate constant of the $[Co\ en_3]^{3+}/[Co\ en_3]^{2+}$ system at the dropping mercury electrode (DME) to be 9×10^{-2} cm/sec. The agreement between these values may be satisfactory.

Bartelt [225] has also determined the kinetic parameters of the $[Co\ chn_3]^{2+}/[Co\ chn_3]^{3+}$ system at the RDE by measuring the exchange current density in 1 M KCl solution containing 0.1 M chn (chn represents 1,2-cyclohexanediamine). The redox potential was -0.515 ± 0.002 V (vs SCE) in the same aqueous solution (25°C). In comparison with earlier redox systems investigated by Bartelt et al., the $[Co\ chn_3]^{2+}/[Co\ chn_3]^{3+}$ system has a lower exchange current density; the reproducibility of measurements near the equilibrium potential ($\eta = 5$ mV) was unsatisfactory. The kinetic parameters of this system are given in Table 3.1.15. Cobalt(II) as well as cobalt(III) complexes were found to influence the charge-transfer reaction by partially covering the platinum electrode. The coverage of the platinum electrode (θ) is relatively high ($\theta = 80\%$) in the presence of the redox system $[Co\ chn_3]^{2+}/[Co\ chn_3]^{3+}$, 0.1 M chn, due to the complexing agent and hydrolysis products of the $[Co\ chn_3]^{3+}$ ion. The coverage was determined from the measurements in the limiting current region [235] using the equation

$$j_0 = nFk_s(C_O)^{\alpha_C}(C_R)^{\alpha_A}(1 - \theta)$$

The coverage obtained during variation of the cobalt(II) and cobalt(III) concentrations agreed well with this equation. A calculation of the coverage from the value of the exchange current density of a similar redox system also gave satisfactory agreement for the degree of coverage.

Bartelt and Prügel [229, 237] and Bartelt and Skilandat [226] have determined by similar methods the kinetic parameters and the degree of coverage at a platinum electrode for the redox system $[Co\ dien_2]^{2+}/[Co\ dien_2]^{3+}$, $[Co\ pn_3]^{2+}/[Co\ pn_3]^{3+}$ and $[Co\ bn_3]^{2+}/[Co\ bn_3]^{3+}$. These results are summarized in Table 3.1.15 (dien, pn and bn denote diethylenetriamine, 1,2-propanediamine, and 2,3-butanediamine, respectively).

Klatt and Blaedel [238] have theoretically derived current-potential equations for the quasi-reversible and totally irreversible heterogeneous charge transfer reaction at a tubular platinum electrode (Fig. 3.1.4) and experimentally confirmed the validity of the theoretical equations for the rate-controlled heterogeneous electron-transfer reaction occurring at a tubular electrode of the $[Co(NH_3)_6]Cl_3$ complex in solutions of 1 M NH_4Cl and 1 M NH_3. The observed and calculated current-potential curve for the hexamminecobalt(III) ion is shown in Fig. 3.1.5 and the kinetic parameters evaluated are given in Table 3.1.16.

TABLE 3.1.15. Kinetic Parameters Evaluated from the Electrode Reaction of Cobalt Complexes in Aqueous Media[a]

Reaction	Medium	Temp (°C)	Electrode	Rate constant[b, c]	α	Heat of activation	Method	Refs.
$[Co(OH_2)_6]^{2+} + 2e \rightarrow Co$				$j_0 = 8 \times 10^{-7}$	0.5			[242]
$[Co(OH_2)_6]^{2+} + 2e \rightarrow Co$	1 M KCl		DME	$\log k^{\circ}_C = -9.7$	$\alpha_C n = 0.38$			[243]
$[Co(OH_2)_6]^{2+} + 2e \rightarrow Co$	1 mM(Co), $NaClO_4$ + 0.005% gelatin	30	DME	3.2×10^{-12}	$\alpha_C n = 0.96$		dc pol	[244]
				$k^{\circ}_C = 2.5 \times 10^{-16}$				
	NaCl + 0.005% gelatin	30	DME	1.2×10^{-12}	$\alpha_C n = 0.93$		dc pol	[244]
				$k^{\circ}_C = 8.0 \times 10^{-17}$	$\alpha_C n = 0.93$		dc pol	[244]
	$[(CH_3)_4N]Br$ + 0.005% gelatin	30		3.5×10^{-12}	$\alpha_C n = 0.81$		dc pol	
				$k^{\circ}_C = 4.0 \times 10^{-16}$	$\alpha_C n = 0.81$		dc pol	[244]
	Na_2SO_4 + 0.005% gelatin	30	DME	4.0×10^{-12}	$\alpha_C n = 0.88$		dc pol	[244]
$[Co(OH_2)_6]^{2+} + 2e \rightarrow Co$	1 mM Co^{2+}, 0.1 M $NaClO_4$, 40% (vol%) AN	30	DME	$k^{\circ}_C = 1.4 \times 10^{-10}$	$\alpha_C n = 0.54$		dc pol	[244]
	0.1 M $NaClO_4$, 60% AN	30	DME	$k^{\circ}_C = 4.5 \times 10^{-9}$	$\alpha_C n = 0.42$		dc pol	[244]
	0.1 M $NaClO_4$, 80% AN	30	DME	$k^{\circ}_C = 7.0 \times 10^{-13}$	$\alpha_C n = 0.81$		dc pol	[244]
	0.1 M $NaClO_4$, 100% AN	30	DME	$k^{\circ}_C = 2.8 \times 10^{-6}$	$\alpha_C n = 0.66$		dc pol	[244]

(continued)

TABLE 3.1.15[a] (continued)

Reaction	Medium	Temp (°C)	Electrode	Rate constant[b, c]	α	Heat of activation	Method	Refs.
$[Co(OH_2)_6]^{2+} + 2e \rightarrow Co$	1 m$\underline{M}$ Co^{2+}, 0.1 $\underline{M}$ $NaClO_4$, 20% (vol%) DMF	30	DME	$k_C^o = 1.0 \times 10^{-14}$	$\alpha_C n = 0.81$		dc pol	[244]
	0.1 $\underline{M}$ $NaClO_4$, 40% DMF	30	DME	$k_C^o = 3.5 \times 10^{-17}$	$\alpha_C n = 0.66$		dc pol	[244]
	0.1 $\underline{M}$ $NaClO_4$, 60% DMF	30	DME	$k_C^o = 3.2 \times 10^{-16}$	$\alpha_C n = 0.81$		dc pol	[244]
$[Co(OH_2)_6]^{2+} + 2e \rightarrow Co$	0.1 $\underline{M}$ $NaClO_4$, 80% DMF	30	DME	$k_C^o = 3.5 \times 10^{-15}$	$\alpha_C n = 0.87$		dc pol	[244]
	0.1 $\underline{M}$ $NaClO_4$, 100% DMF	30	DME	$k_C^o = 5.0 \times 10^{-13}$	$\alpha_C n = 0.70$		dc pol	[244]
$[Co(OH_2)_6]^{2+} + 2e \rightarrow Co$	1 m$\underline{M}$ Co^{2+}, 0.1 $\underline{M}$ $NaClO_4$, 0% (vol%) FA	30	DME	$k_C^o = 2.5 \times 10^{-16}$	$\alpha_C n = 0.9$		dc pol	[244]
	0.1 $\underline{M}$ $NaClO_4$, 20% FA	30	DME	$k_C^o = 1.0 \times 10^{-14}$	$\alpha_C n = 0.84$		dc pol	[244]
	0.1 $\underline{M}$ $NaClO_4$, 40% FA	30	DME	$k_C^o = 5.6 \times 10^{-15}$	$\alpha_C n = 0.84$		dc pol	[244]
	0.1 $\underline{M}$ $NaClO_4$, 60% FA	30	DME	$k_C^o = 4.5 \times 10^{-15}$	$\alpha_C n = 0.80$		dc pol	[244]
	0.1 $\underline{M}$ $NaClO_4$, 80% FA	30	DME	$k_C^o = 1.4 \times 10^{-14}$	$\alpha_C n = 0.79$		dc pol	[244]
	0.1 $\underline{M}$ $NaClO_4$, 100% FA	30	DME	$k_C^o = 2.8 \times 10^{-15}$	$\alpha_C n = 0.84$		dc pol	[244]
$[Co(NH_3)_6]^{3+} + e \rightleftarrows [Co(NH_3)_6]^{2+}$	1 $\underline{M}$ NH_3 + 1 $\underline{M}$ NH_4Cl	25	Pt (RDE)	$k_C^o = 4.5 \times 10^{-6}$	$\alpha_C n = 0.65$		Voltammetry	[241]
	1 $\underline{M}$ NH_3 + 1 $\underline{M}$ NH_4Cl	25	Pt (TE)	$k_C^o = 6 \times 10^{-6}$	$\alpha_C n = 0.61$		Voltammetry	[238]

	1 M NH_3 + 1 M NH_4NO_3	25	DME	$k^\circ_C = 3.1 \times 10^{-5}$	$\alpha_C n = 0.75$		dc pol	[92]
	0.14 M $HClO_4$ + 1.26 M $NaClO_4$			$k^\circ_C = 5.2 \times 10^{-7}$	$\alpha_C n = 0.67$		dc pol	[214]
	7 M NH_3 + 0.1 M NH_4Cl + 1.0 M NH_4Cl (xM)	25	Pt (RDE)	$(1.33 \pm 0.2) \times 10^3$ (x = 0.2 M)	$\alpha_A n = 0.41 \pm 0.02$	7.5 kcal/mole	Voltammetry	[227]
				$j_0 = 270 \pm 20$ $\mu A/cm^2$ (x = 0.2 M)				
				$j_0 = 460$ $\mu A/cm^2$ (x = 0 M)				
$[Co(NH_3)_6]^{3+} + e \rightleftarrows [Co(NH_3)_6]^{2+}$	1 mM $[Co(NH_3)_6]^{3+}$, 1 mM $[Co(NH_3)_6]^{2+}$, 7 M NH_3 + xM NH_4Cl	25	Pt (RDE)	5.25×10^{-4} (x = 1 M)	$\alpha_A n = 0.42$, $\alpha_C n = 0.58$		Voltammetry	[222]
	7 M NH_3 + 0.2 M NH_4Cl	25	Pt (RDE)	1.33×10^{-3}	$\alpha_A n = 0.41$, $\alpha_C n = 0.59$		Voltammetry	[222]
$[Co(NH_3)_6]^{3+} + e \rightleftarrows [Co(NH_3)_6]^{2+}$	1 M $NaNO_2$	25	DME	$k^\circ_C = [4.5 \times 10^{-3}]$	$\alpha_C n = 0.69 \pm 0.03$	7.8 kcal/mole	dc pol	[95]
$[Co(NH_3)_6]^{3+} + e \rightleftarrows [Co(NH_3)_6]^{2+}$	7 M NH_3 + (x → 0)M NH_4Cl 5 mM Co^{2+}, 1 mM $[Co(NH_3)_6]^{3+}$, 7 M NH_3 + xM $(NH_4)_2SO_4$	25	Pt (RDE)	2.3×10^{-3}	$\alpha_A n = 0.41$, $\alpha_C n = 0.59$		Voltammetry	[222]
	x = 1 M	25	Pt (RDE)	2.1×10^{-4}	$\alpha_A n = 0.41$, $\alpha_C n = 0.59$		Voltammetry	[222]
	x = 0.2 M	25	Pt (RDE)	7.2×10^{-4}	$\alpha_A n = 0.41$, $\alpha_C n = 0.59$		Voltammetry	[222]

(continued)

TABLE 3.1.15[a] (continued)

Reaction	Medium	Temp (°C)	Elec-trode	Rate constant[b, c]	α	Heat of activa-tion	Method	Refs
$[Co\ phen_3]^{3+} + e \rightleftarrows [Co\ phen_3]^{2+}$	1 m$\underline{M}$ Co^{3+}, 1 m$\underline{M}$ CO^{2+}, 1 $\underline{M}$ KCl + 4 m$\underline{M}$ phen	25	Pt (RDE)	4.8×10^{-2}	$\alpha_A n$ = 0.77 ± 0.02 $\alpha_C n$ = 0.23 ± 0.02		Voltam-metry	[228]
$[Co\ dien_2]^{3+} + e \rightleftarrows [Co\ dien_2]^{2+}$	1 $\underline{M}$ KCl + 0.1 $\underline{M}$ dien	25	Pt (RDE)	7.7×10^{-2}	$\alpha_A n = 0.77$, $\alpha_C n = 0.23$		Voltam-metry	[229]
$[Co\ en_3]^{3+} + e \rightleftarrows [Co\ en_3]^{2+}$		25	DME	9×10^{-2}			dc pol	[236]
	1 $\underline{M}$ KCl + 0.1 $\underline{M}$ en	25	Pt (RDE)	2.69×10^{-2}	$\alpha_A n = 0.77$, $\alpha_C n = 0.23$	10 kcal/mole	Voltam-metry	[224]
$[Co\ pn_3]^{3+} + e \rightleftarrows [Co\ pn_3]^{2+}$	1 $\underline{M}$ KCl + 0.1 $\underline{M}$ pn	25	Pt (RDE)	1.4×10^{-2}	$\alpha_A n = 0.77$, $\alpha_C n = 0.23$		Voltam-metry	[226]
$[Co\ bn_3]^{3+} + e \rightleftarrows [Co\ bn_3]^{2+}$	1 $\underline{M}$ KCl + 0.1 $\underline{M}$ bn	25	Pt (RDE)	1.04×10^{-2}	$\alpha_A n = 0.77$, $\alpha_C n = 0.23$		Voltam-metry	[226]
$[Co\ chn_3]^{3+} + e \rightleftarrows [Co\ chn_3]^{2+}$	1 $\underline{M}$ KCl + 0.1 $\underline{M}$ chn	25	Pt (RDE)	0.65×10^{-2}	$\alpha_A n = 0.77$, $\alpha_C n = 0.23$		Voltam-metry	[225]
$[Co\ edta]^- + e \rightleftarrows [Co\ edta]^{2-}$	0.4 $\underline{M}$ $NaNO_3$ + 0.1 $\underline{M}$ acetate buffer (pH = 5.0)	25	DME	2.9×10^{-2}	$\alpha_A n = 0.49$, $\alpha_C n = 0.52$		ps	[197]
				5.8×10^{-2}	$\alpha_A n = [0.49]$, $\alpha_C n = [0.52]$			
$[Co(OH_2)_6]^{2+} + 2e \rightarrow Co$	0.1 $\underline{M}$ K_2SO_4	25	DME		$\alpha_C n = 0.37$		dc pol	[245]

trans-[Co(CN)(dgH)$_2$py] + e → trans-[Co^{II}(CN)(dgH)$_2$py]$^-$	0.1 M K_2SO_4	25	DME	$\alpha_C n = 0.33$	dc pol	[245]
cyanocobalamin [Co(III)] + e → cyanocobalamin [Co(II)]	0.1 M K_2SO_4	25	DME	$\alpha_C n = 0.41$	dc pol	[245]
methylcobalamin [Co(III)] + e → methylcobalamin [Co(II)]	0.1 M K_2SO_4	25	DME	$\alpha_C n = 0.25$	dc pol	[245]
trans-[Co(X)(dgH)$_2$(py)] or trans-[CoR(dgH)$_2$(py)] + e → trans-[Co(X)(dgH)$_2$(py)]$^-$ or trans-[CoR(dgH)$_2$(py)]$^-$	0.1 M K_2SO_4	25	DME			[245]
X or R = CN	0.1 M K_2SO_4	25	DME	$\alpha_C n = 0.33$	dc pol	[245]
$C_6H_5CH_2$	0.1 M K_2SO_4	25	DME	$\alpha_C n = 0.74$	dc pol	[245]
$CH_3CH(OH)CH_2$	0.1 M K_2SO_4	25	DME	$\alpha_C n = 0.35$	dc pol	[245]
$HOCH_2CH(OH)CH_2$	0.1 M K_2SO_4	25	DME	$\alpha_C n = 0.27$	dc pol	[245]
$HOCH_2CH(CH_3)$	0.1 M K_2SO_4	25	DME	$\alpha_C n = 0.28$	dc pol	[245]
Cl	0.1 M K_2SO_4	25	DME	$\alpha_C n = 0.31$	dc pol	[245]
CH_3CH_2CH2	0.1 M K_2SO_4	25	DME	$\alpha_C n = 0.33$	dc pol	[245]
$CH_3CH(CH_3)$	0.1 M K_2SO_4	25	DME	$\alpha_C n = 0.31$	dc pol	[245]
CH_3	0.1 M K_2SO_4	25	DME	$\alpha_C n = 0.33$	dc pol	[245]
CH_3CH_2	0.1 M K_2SO_4	25	DME	$\alpha_C n = 0.32$	dc pol	[245]

(continued)

TABLE 3.1.15[a] (continued)

Reaction	Medium	Temp (°C)	Electrode	Rate constant[b, c]	α	Heat of activation	Method	Refs.
trans-[Co(OH)$(dgH)_2$(copoly-AM-VPy]: mol wt = 1400	0.1 M K_2SO_4	25	DME		$\alpha_C n = 0.70$		dc pol	[245]
mol wt = 3000		25	DME		$\alpha_C n = 0.60$		dc pol	[245]
trans-$[CoCl(dgH)_2(OH_2)]$ + e → $[CoCl(dgH)_2(OH_2)]^-$	0.1 M K_2SO_4	25	DME		$\alpha_C n = 0.36$		dc pol	[245]
trans-$[Co(OH)(dgH)_2(OH_2)]$ + e → $[Co(OH)(dgH)_2(OH_2)]^-$	0.1 M K_2SO_4	25	DME		$\alpha_C n = 0.25$		dc pol	[245]
$[Co(OH_2)(NH_3)_5]^{3+}$ + e → $[Co(OH_2)(NH_3)_5]^{2+}$	1 M HNO_3 + $NaNO_3$ + Th^{4+} ($\mu = 2$)	0	DME	$k^\circ_C = 4 \times 10^{-5}$		dc pol	dc pol	[246]
$[CoF(NH_3)_5]^{2+}$ + e → $[CoF(NH_3)_5]^+$	1 M HNO_3 + $NaNO_3$ + Th^{4+} or La^{3+} ($\mu = 2$)	0	DME	$k^\circ = 1.4 \times 10^{-6}$[d]			dc pol	[246]
$[Co(SO_4)NH_3)_5]^+$ + e → $[Co(SO_4)(NH_3)_5]$	1 M HNO_3 + $NaNO_3$ + Th^{4+} or La^{3+} ($\mu = 2$)	0	DME	$k^\circ = 9.55 \times 10^{-6}$[d]			dc pol	[246]
$[Co(NO_2)(NH_3)_5]^{2+}$ + e → $[Co(NO_2)(NH_3)_5]^+$	1 M $NaNO_2$	25	DME	$k^\circ_C = 2.5 \times 10^{-3}$ = $[1.3 \times 10^{-2}]$	$\alpha_C n$ = 0.66 ± 0.03	8.4 kcal/mole	dc pol	[95]
trans-$[Co(NO_2)_2(NH_3)_4]^+$ + e → trans-$[Co(NO_2)_2(NH_3)_4]$	1 M $NaNO_2$	25	DME	$k^\circ_C = 2.9 \times 10^{-3}$ = $[5.7 \times 10^{-3}]$	$\alpha_C n$ = 0.65 ± 0.03	7.3 kcal/mole	dc pol	[95]
cis-$[Co(NO_2)_2(NH_3)_4]^+$ + e → cis-$[Co(NO_2)_2(NH_3)_4]$	1 M $NaNO_2$	25	DME	$k^\circ_C = 7.1 \times 10^{-2}$ = $[1.1 \times 10^{-1}]$	$\alpha_C n$ = 0.58 ± 0.03	7.1 kcal/mole	dc pol	[95]
$[Co(NO_2)_3(NH_3)_3]$ + e → $[Co(NO_2)_3(NH_3)_3]$	1 M $NaNO_2$	25	DME	$k^\circ_C = 6.3 \times 10^{-2}$ = $[1.9 \times 10^{-2}]$	$\alpha_C n$ = 0.66 ± 0.03	6.7 kcal/mole	dc pol	[95]

trans-$Na_5[Co(SO_3)_2(CN)_4].3H_2O$ + 2e → trans-$[Co^I(SO_3)_2(CN)_4]^{7-}$	0.5 M Na_2SO_3	15	DME	$k_C^\circ = 2.67 \times 10^{-13}$	$\alpha_C n = 0.55$		dc pol	[116]
	0.5 M Na_2SO_3	20	DME	$k_C^\circ = 3.32 \times 10^{-13}$	$\alpha_C n = 0.56$		dc pol	[116]
	0.5 M Na_2SO_3	25	DME	$k_C^\circ = 5.29 \times 10^{-13}$	$\alpha_C n = 0.53$	4.7 kcal/mole (at 25°C)	dc pol	[116]
	0.5 M Na_2SO_3	30	DME	$k_C^\circ = 6.56 \times 10^{-13}$	$\alpha_C n = 0.57$		dc pol	[116]
	0.5 M Na_2SO_3	35	DME	$k_C^\circ = 9.12 \times 10^{-13}$	$\alpha_C n = 0.61$		dc pol	[116]

[a] AN = acetonitrile; DMF = dimethylformamide; FA = formamide; phen = 1,10-phenanthroline; dien = diethylenetriamine, $NH_2CH_2CH_2NHCH_2CH_2NH_2$; pn = 1,2-propanediamine; bn = 2,3-butanediamine; chn = 1,2-cyclohexanediamine; dg^{2-} = dimethylgloximate ion, $CH_3C(NO)C(NO)CH_3$; py = pyridine; R = alkyl group; α_A = anodic transfer coefficient; α_C = cathodic transfer coefficient; ps = a potentiostatic method; values in brackets are corrected for the double layer effect. RDE = rotating platinum disk electrode; TE = tubular platinum electrode.

[b] Standard rate constant in cm/sec unless otherwise stated.

[c] k_C° = cathodic rate constant in cm/sec at E = 0.

[d] The cathodic rate constant in the presence of Th^{4+} and La^{3+} is represented by the equation $k = k^\circ + {}_Mk^\circ[M^{n+}]$, where k° is the rate constant in a solution without polyvalent cations. The rate constant is related to the potential of a 0.1 M calomel electrode at 0°C.

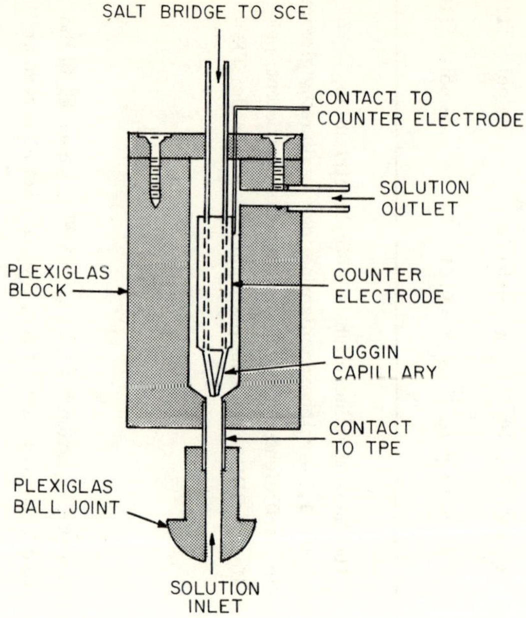

FIG. 3.1.4. Electrode assembly. (Data from Ref. 238.)

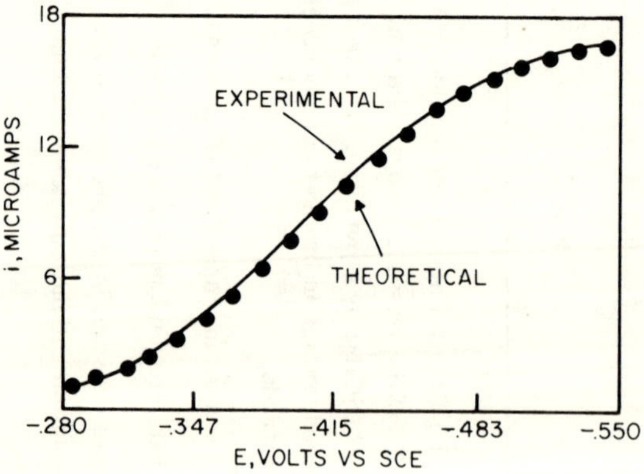

FIG. 3.1.5. Observed and calculated current-potential curve for the $[Co(NH_3)_6]Cl_3$ system. $C_O = 2.00 \times 10^{-4}$ M, V = 5.72 ml/min, $k_S = 6 \times 10^{-6}$ cm/sec, $\alpha_C = 0.61$, and $D_O = 6.2 \times 10^{-6}$ cm^2/sec. (Data from Ref. 238.)

TABLE 3.1.16. [$Co(NH_3)_6Cl_3$ System][a]

V (ml/min)	$E_{1/2}$ (V vs SCE)	$E_{3/4} - E_{1/4}$ (mV)	$0.0618/\alpha n_a$ (V)	Log $(k_a/D_0^{2/3})$	$i_d/V^{1/3}$
1.43	-0.389	-93	0.096	-1.91	9.42
2.74	-0.399	-95	0.099	-1.93	9.43
5.72	-0.403	-96	0.101	-1.86	9.73
7.36	-0.405	-92	0.096	-1.84	9.39
10.0	-0.411	-93	0.097	-1.86	9.49

[a]2.00×10^{-4} F Co $(NH_3)_6Cl_3$ in 1 F NH_3 - 1 F NH_4Cl.

The diffusion coefficient of $[Co(NH_3)_6]^{3+}$ ion in the aqueous solution containing 1 M NH_3 and 1 M NH_4Cl was calculated from the i (current) values at various flow rates (ml/min) and concentrations to give 6.2×10^{-6} cm^2/sec with a standard deviation of 0.1×10^{-6} cm^2/sec. Klatt and Blaedel [238] recommend this voltammetric treatment of evaluating the kinetic parameters in hydrodynamic electrochemical systems.

Still more recently, Suzuki [239, 240] has promoted hydrodynamic voltammetric studies with the convection electrode system in order to evaluate the kinetic parameters of totally irreversible systems and applied this method to the totally irreversible process, Co(III) → Co(II), of the $[Co(NH_3)_6]Cl_3$ complex in the same solution used by Klatt and Blaedel [238]. The parameters obtained at the RDE ($\alpha_C = 0.61$; $k_C{}^a = 6 \times 10^{-6}$ cm/sec) were in good agreement with those measured at a tubular platinum electrode, suggesting that the theoretical treatment is valid for evaluation of kinetic parameters. The results are listed in Table 3.1.17 [241].

TABLE 3.1.17. Kinetic Parameters for the $[Co(NH_3)_6]Cl_3$ Complex in 1 M NH_3 and 1 M NH_4Cl at 25°C

Tangential velocity v (cm/sec)	α_C	k_C° ($\times 10^{-6}$ cm/sec)
94.2	0.67 ± 0.01	5.0 ± 0.4
125.6	0.65 ± 0.02	5.5 ± 0.8
157.0	0.64 ± 0.02	3.5 ± 0.5
188.4	0.65 ± 0.02	4.1 ± 0.2
Average	0.65	4.5

In a 0.5-M Na_2SO_4 aqueous solution the bis(dimethylglyoximato)-cobalt(III) complexes in which ammonia and cyanide ligands are bound at the axial fifth and the sixth positions have been found to be irreversibly reduced in one or three steps to the cobalt(I) or cobalt(0) state at the DME [115]:

$$\text{trans-}[Co^{III}(dgH)_2(NH_3)_2]^+ \xrightarrow{e} \text{trans-}[Co^{II}(dgH)_2(NH_3)_2] \xrightarrow{e} \text{trans-}[Co^{I}(dgH)_2(NH_3)_2]^-$$

$$\xrightarrow{e} [Co^0(dgH)_2]^{2-} + 2\ NH_4^+$$

$$\text{trans-}[Co^{III}(CN)_2(dgH)_2]^- \xrightarrow{2e} \text{trans-}[Co^{I}(CN)_2(dgH)_2]^{3-}$$

On the other hand, in DMSO (100%) containing 0.1 M $[(C_2H_5)_4N]ClO_4$, the dicyano- and the diammine cobalt(III) complexes are irreversibly reduced in one step to the cobalt(I) state at the DME and no further reduction to the metal occurs over the potential range of 0 – -2.70 V (vs SCE) [145]:

$$\text{trans-}[Co^{III}(dgH)_2(NH_3)_2]^+ \xrightarrow{2e} \text{trans-}[Co^{I}(dgH)_2(NH_3)_2]^-$$

$$\text{trans-}[Co^{III}(dgH)_2(CN)_2]^- \xrightarrow{2e} \text{trans-}[Co^{I}(dgH)_2(CN)_2]^{3-} \quad \text{(in DMSO)}$$

Similar methods have been used for the pentacyanocobaltate(III) complexes, and the kinetic parameters for the cathodic electrode reduction have been evaluated and discussed [145]. Those results are summarized in Tables 3.1.18 to 3.1.23.

TABLE 3.1.18. Kinetic Parameters of the Co(III) → Co(I) Process for the trans-$K[Co(CN)_2(dgH)_2]\cdot 3/2\ H_2O$ Complex in 0.5 M Na_2SO_4 Solution[a]

Temperature (°K)	283.0 (10°C)	288.0 (15°C)	293.0 (20°C)	298.0 (25°C)	303.0 (30°C)
$E_{1/2}$ (V vs NHE)	-0.863	-0.861	-0.860	-0.859	-0.862
i_d (μA)	4.21	5.03	5.32	5.72	6.19
$\alpha_C n$	0.80	0.75	0.69	0.69	0.72
D_0 (cm²/sec)	0.83×10^{-6}	4.58×10^{-6}	5.95×10^{-6}	7.29×10^{-6}	8.80×10^{-6}
$\log k_C^\circ$	-13.93	-13.68	-13.42	-13.17	-13.06
k_C^a (cm/sec)	1.00×10^{-14}	2.09×10^{-14}	3.80×10^{-14}	8.51×10^{-14}	8.71×10^{-14}

[a]Complex concentration, 1 mM. k_C^a = cathodic rate constant at E = 0 (V vs NHE). ΔH_C^a = 21.7 kcal/mole at 25°C. ΔH_D = 6.6 kcal/mole (the heat of mass transfer) at 25°C.

TABLE 3.1.19. Kinetic Parameters of the Co(III) → Co(II) Process for the trans-$[Co^{III}(dgH)_2(NH_3)_2]Cl.5H_2O$ Complex in 0.5 M Na_2SO_4 Aqueous Solution[a]

Temperature (°K)	288.0 (15°C)	293.0 (20°C)	298.0 (25°C)	303.0 (30°C)	308.0 (35°C)
$E_{1/2}$ (V vs NHE)	-0.353	-0.339	-0.335	-0.324	-0.312
$\alpha_C n$	0.72	0.69	0.72	0.73	0.70
D_0 (cm²/sec)	2.22×10^{-6}	2.96×10^{-6}	3.31×10^{-6}	4.16×10^{-6}	4.93×10^{-6}
log k_C^a	-7.48	-7.16	-7.02	-6.78	-6.54
k_C^a (cm/sec)	3.30×10^{-8}	6.86×10^{-8}	9.49×10^{-8}	1.68×10^{-8}	2.90×10^{-7}

[a] ΔH_C^a = 13.6 kcal/mole at 25°C. ΔH_D = 7.2 kcal/mole at 25°C.

TABLE 3.1.20. Kinetic Parameters of the Co(III) → Co(I) Process for the trans-$K[Co(CN)_2(dgH)_2].3/2\ H_2O$ Complex in the DMSO Containing 0.1 M $[(C_2H_5)_4N]ClO_4$[a]

Temperature (°K)	293.0 (20°C)	298.0 (25°C)	303.0 (30°C)	308.0 (35°C)	313.0 (40°C)
$E_{1/2}$ (V vs NHE)	-1.243	-1.239	-1.239	-1.237	-1.237
i_d (μA)	1.24	1.40	1.51	1.76	1.84
$\alpha_C n$	0.85	0.82	0.82	0.81	0.85
D_0 (cm²/sec)	1.46×10^{-6}	1.85×10^{-6}	2.16×10^{-6}	2.92×10^{-6}	3.20×10^{-6}
log k_C^a	-20.89	-20.49	-20.17	-19.80	-19.48
k_C^a	1.29×10^{-21}	3.25×10^{-21}	6.81×10^{-21}	1.60×10^{-20}	3.29×10^{-20}

[a] ΔH_C^a = 21.9 kcal/mole at 25°C. ΔH_D = 7.0 kcal/mole at 25°C.

TABLE 3.1.21. Kinetic Parameters of the Co(III) → Co(I) Process for the trans-$[Co^{III}(dgH)_2(NH_3)_2]ClO_4$ (anhydrous) in the DMSO Containing 0.1 M $[(C_2H_5)_4N]ClO_4$[a]

Temperature (°K)	293.0 (20°C)	298.0 (25°C)	303.0 (30°C)	308.0 (35°C)	313.0 (40°C)
$E_{1/2}$ (V vs NHE)	-0.504	-0.487	-0.471	-0.470	-0.412
i_d (μA)	1.20	1.26	1.42	1.57	1.72
$\alpha_C n$	0.43	0.44	0.43	0.42	0.38
D_0 (cm²/sec)	1.42×10^{-6}	1.56×10^{-6}	1.96×10^{-6}	2.40×10^{-6}	2.89×10^{-6}
log k_C^a	-6.71	-6.51	-6.30	-6.13	-5.87
k_C^a	1.97×10^{-7}	3.11×10^{-7}	5.06×10^{-7}	7.47×10^{-7}	1.34×10^{-6}

[a] ΔH_C^a = 14.1 kcal/mole at 25°C. ΔH_D = 7.5 kcal/mole at 25°C.

TABLE 3.1.22. Comparison of the Kinetic Parameters for the First Step of Cathodic Reduction in Aqueous Media at 25°C[a]

Complex	Change of valence	k_C[a] (cm/sec)	ΔH_C[a] (kcal/mole)
trans-K[Co(CN)$_2$(dgH)$_2$]. 3/2 H_2O	Co(III) → Co(I)	8.51×10^{-14}	21.7
trans-[Co(dgH)$_2$(NH$_3$)$_2$]Cl. 5H_2O	Co(III) → Co(II)	9.49×10^{-8}	13.6

[a]0.5 M Na_2SO_4 aq soln.

TABLE 3.1.23. Comparison of the Kinetic Parameters for the Co(III) → Co(I) Process of Bis(dimethylglyoximato)cobalt(III) Complexes in the Dimethyl Sulfoxide at 25°C[a]

Complex	k_C[a] (cm/sec)	ΔH_C[a] (kcal/mole)
trans-K[Co(CN)$_2$(dgH)$_2$]. 3/2 H_2O	3.25×10^{-21}	21.9
trans-[Co(CN)(dgH)$_2$(NH$_3$)]. 1/2 H_2O	4.75×10^{-11}	19.1
trans-[Co(dgH)$_2$(NH$_3$)$_2$]ClO$_4$ (anhydrous)	3.11×10^{-7}	14.1

[a]DMSO (100%) solution containing 0.1 M [(C_2H_5)$_4$N]ClO$_4$.

3.2. DOUBLE-LAYER PROPERTIES

The polarographic reduction of $[Co^{II}(OH_2)_6]^{2+}$ ion in solutions of not too concentrated, noncomplexing background salts, e.g., perchlorate, results typically in one irreversible two-electron wave with a well-defined diffusion plateau. Perchlorate is used because it has the smallest coordinating tendency among a number of anions.

In contrast, at sufficiently high ionic strength, the limiting current becomes appreciably lower and becomes dependent on both the ϕ-potential of the double layer and the molar concentration [j] of the supporting electrolyte. That is, by increasing the concentration of the supporting electrolyte, a kinetically-controlled feature in the mass transfer is observed [247, 248]. This behavior can be explained by the fact that with increasing electrolyte concentration, the diffuse part of the double layer becomes more and more compressed and, at last, the electrode potential E becomes equal to that in the bulk of the solution; this is caused by the variations of the equivalent half-thickness of the diffuse part of the double layer $1/\kappa$ (Debye-Hückel length) with the molar concentration [j] and the charge Z of the supporting electrolyte. This concept was first described by Gierst [249]. According to him, the $[Co(OH)_2)_6]^{2+}$ ion should be partially dehydrated before the electron transfer process proceeds (probably by an S_N1 reaction mechanism). The rate of dehydration increases on raising the negative ϕ-potential, where the term ϕ is the "rational potential," measured with reference to the null charge point observed in the absence of specific adsorption. In any case, the retarding or accelerating effect of foreign ions has been interpreted as due to the change in the double-layer structure.

On the other hand, the reduction mechanism of cobalt(II) complexes in the presence of complexing ligand is strongly dependent on the nature of the ligand and on the stability of the primary product, which means that the electrode process proceeds either in two separate steps to give the cobalt(I) complex as a relatively stable product of the first step or in a single two-electron reduction process. In this case the effect of the double-layer structure on the rate of the electrode reactions has been noted frequently. The same effect was found for the cathodic reduction of the $[Co(NH_3)_6]Cl_3$ complex by Kolthoff and Khalafalla [250]. Abnormal phenomena involving exhibition of a "current-dip" on part of a limiting diffusion plateau in noncomplexing media was fully explained in terms of the variation of the ϕ-potential. Such an effect of the double layer on the rate of electrode reactions has so far received several different interpretations. According to Frumkin [200], the actual surface concentration of the complex has to be corrected by a Boltzmann term representative of the electrostatic work involved in the approach of the particles to the electrode. On the other hand, in Levich's opinion [251], the essential role of the diffuse potential barrier which surrounds the electrode is to fix a limit to the translation speed of the penetrating particles. As a matter of fact, however, in the case of a moderate or slow reaction rate, in which the thickness of the diffusion layer is much larger than that of the diffuse double layer, the effect of the double-layer structure on the kinetic parameters can be corrected [252] in a simple way according to the original concept of Frumkin [200]. In fact, many cobalt(III) complexes can be satisfactorily treated with this procedure. In the cases of inherently fast electrode processes, though for only a very few cobalt complexes, the structure of the double layer must be taken into account in the boundary value problem [253, 254]. The examples for this case are very rare as far as the cobalt complexes are concerned.

To illustrate the effect of the double-layer structure on kinetic parameters, one example will be presented concerning the slow electroreduction of pentacyanocobaltate(III) complexes, the uncorrected parameters of which are given in Tables 3.1.7 to 3.1.11.

3.2.1. Frumkin's Correction for Double-Layer Structure Effects

In the negative potential region, cations are crowded in the vicinity of the mercury electrode surface, and this distribution of the cations (K^+ or Na^+) results in a potential difference $\Delta\phi$ in the diffuse double layer with reference to the potential in the bulk of the solution. This causes the true heat of activation to be larger than the apparent one. On the other hand, the

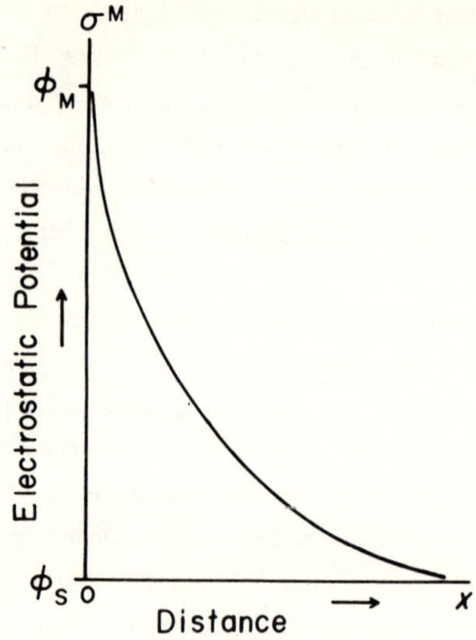

FIG. 3.2.1. The Gouy-Chapman model of the double layer. ϕ_M is the electrostatic potential of the mercury surface, and ϕ_S is the electrostatic potential in the bulk of the solution.

repulsion of the pentacyanocobaltate(III) anion by the negatively charged electrode causes a decrease in the concentration of the complex at the surface of the electrode (i.e., at the pre-electrode layer) with reference to that in the bulk of the solution, so that the correction of the rate constant with respect to these two factors is expressed by

$$k_C{}^a = k_C^\circ \exp[(\alpha_C - Z_0)F\Delta\phi/RT] \tag{3.2.1}$$

where k_C° is the true rate constant for the cathodic reduction, Z_0 the ionic charge of the depolarizer, and the other symbols have their usual significance. For evaluation of $\Delta\phi$, the differential capacity of the electrode/solution interface was measured and plotted against the variation of potential at 5, 15, 25, 35, and 45°C, respectively, where $\Delta\phi$ is the Galvani potential difference between the electrode and the bulk of the solution, as shown in Fig. 3.2.1 ($\Delta\phi = \phi_2 - \phi_S$), where ϕ_2 (or ψ_d) is the potential in the outer Helmholtz layer or the "d" plane of the double layer (see p. 156). The plot diagram of the measured differential capacity potential for the thiosulfato complex is given as an example in Fig. 3.2.2. The surface charge density 2^M ($\mu C/cm^2$) of the mercury electrode was evaluated from dividing q by A, in which the value of q had been estimated experimentally by a graphical integration of the plotted curve in Fig. 3.2.2 where q is the total surface charge of the DME and A is the total surface area (in cm^2) of the mercury drop expressed by

$$A = 4\pi\,(3mt/4\pi d)^{2/3} \tag{3.2.2}$$

where d is the density of mercury and m is the flow rate of mercury (mg/sec). The integration was made for the region between the potential of the capillary maximum (Lippmann's po-

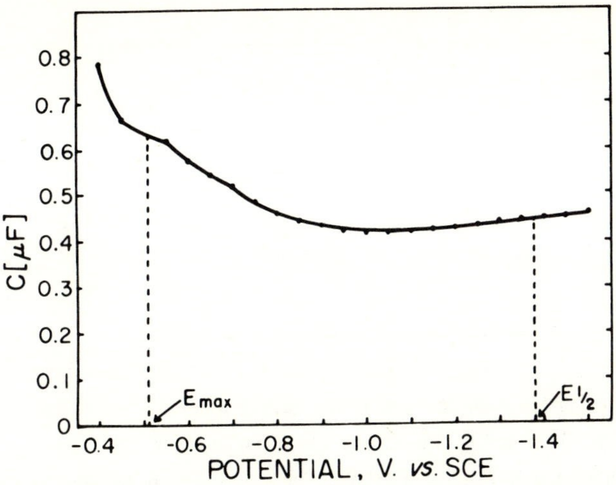

FIG. 3.2.2. Differential capacity of the double layer vs potential obtained for the $K_4[Co(S_2O_3)(CN)_5]$ complex (5×10^{-3} M) in a 0.5-M Na_2SO_4 aqueous solution; m = 1.37 mg/sec; t = 4.0 sec.

TABLE 3.2.1. The Estimate of Surface Charge Density for $[Co(S_2O_3)(CN)_5]^{4-}$ [a]

Temperature (°C)	q (μC)	A (cm^2)	q^M ($\mu C/cm^2$)
15	0.4169	0.02643	15.77
25	0.4061	0.02646	15.34
35	0.4054	0.02649	15.30
45	0.3915	0.02653	14.76

[a]Conditions: 5×10^{-3} M complex ion in 0.5 M Na_2SO_4.

tential, E_{max}) and $E_{1/2}$, as indicated in Fig. 3.2.2. Table 3.2.1 shows the numerical data obtained for the $K_4[Co(S_2O_3)(CN)_5]$ complex (5×10^{-3} M) in 0.5 M Na_2SO_4. A mechanical "drop-knocker" was used to keep the magnitude of the mercury drop time independent of applied potential. The value of $\Delta\phi$ can be evaluated from Eq. (3.2.3), which is based on the Gouy-Chapman theory by using the numerical value of q^M computed above:

$$q^M = \pm[RT\varepsilon/2\pi\sum_i C_i^\circ\{\exp(-Z_iF\Delta\phi/RT) - 1\}]^{1/2} \quad (3.2.3)$$

where C_i° represents the concentration of ions (i) (moles/cm^3) in the bulk of the solution, ε is the dielectric constant of the solvent (water in this case), and the other symbols have the usual significance. The dielectric constant ε of water is given in Table 3.2.2.

TABLE 3.2.2. Dielectric Constant of Water

Temperature (°C)	ε	esu ($\times 10^9$)
5	86.04	9.56
15	82.22	9.14
25	78.54	8.73
35	75.00	8.33
45	71.59	7.95

The concentration of the ions present in the solution of the $K_4[Co(S_2O_3)(CN)_5]$ complex (5×10^{-3} M complex ion in 0.5 M Na_2SO_4) is (in mole/cm^2): K^+, 2×10^{-5}; SO_4^{2-}; 5×10^{-4}, Na^+, 10^{-3}; and $[Co(S_2O_3)(CN)_5]^{4-}$, 5×10^{-6}. These values can be used in Eq. (3.2.3) to calculate q^M:

$$q^M = [RT\varepsilon/2\pi](2 \times 10^{-5})\left\{\exp(-F\Delta\phi/RT) - 1\right\} + (5 \times 10^{-6})\left\{\exp(4F\Delta\phi/RT) - 1\right\} + 10^{-3}\left\{\exp(-F\Delta\phi/RT) - 1\right\} + 5 \times 10^{-4}\left\{\exp(2F\Delta\phi/RT) - 1\right\}^{1/2} \qquad (3.2.4)$$

The numerical calculations, however, are rather tedious and require the use of a digital computer, especially since the supporting electrolyte used (0.5 M Na_2SO_4) is not a 1:1 electrolyte. Calculation of $\Delta\phi$ was carried out by a trial and error method, that is, a value of $\Delta\phi$ is assumed and q^M calculated using Eq. (3.2.4). The value of $\Delta\phi$ for which the calculated value of q^M agrees with the measured value of q^M is then the true $\Delta\phi$ value. For example, from Eq. (3.2.4), for $\Delta\phi = 34.0$, $q^M = 15.31$; for $\Delta\phi = 34.1$, $q^M = 15.33$; and for $\Delta\phi = 34.2$, $q^M = 15.46$, where the dielectric constant ε of water is 78.54 at 25°C. Since the measured value of q^M is 15.34 at 25°C (Table 3.2.1), $\Delta\phi = 34.1$ mV. Therefore, the true rate constant k_C° corrected for the effect of the double layer can be evaluated by placing this value ($\Delta\phi = 34.1$ mV) in the Eq. (3.2.1). The results are given in Table 3.2.3. The true rate constant is about a hundredth of the apparent one, while the slope of the straight line of the "Arrhenius plot" is scarcely influenced by the electrical double-layer correction. The true heat of activation ΔH_C° obtained from the temperature dependence of k_C° is 14.74 kcal/mole at 25°C, while the apparent heat of activation $\Delta H_C{}^a$ obtained from that of $k_C{}^a$ is 14.66 kcal/mole at 25°C. The good agreement between these suggests that the effect of the double-layer structure on the heat of activation is negligible. On the other hand, the double layer exerts a remarkable influence on the frequency factor of the cathodic electrode reduction; the intercept of the line on the vertical axis is log $PZ_a = -0.5$ for $k_C{}^a$, but log $PZ_0 = -2.5$ for k_C° (when Z_a and Z_0 are frequency factors representing the numbers of collisions).

In general, the heat of activation thus obtained involves the heat of activation for diffusion (mass transfer). Therefore, the true heat of activation for the electron transfer has to be

TABLE 3.2.3. The True and Apparent Rate Constant for the Process Co(III) → Co(I) of the $K_4[Co(S_2O_3)(CN)_5]$ Complex[a]

Temperature (°C)	$\Delta\phi$ (mV)	$k_C{}^a$ (cm/sec)	k_C° (cm/sec)
15	33.3	2.27×10^{-13}	2.10×10^{-15}
25	34.1	5.55×10^{-13}	5.39×10^{-15}
35	35.3	1.42×10^{-12}	1.38×10^{-14}
45	35.8	2.49×10^{-12}	2.61×10^{-14}

[a]Conditions: 5×10^{-3} M complex ion in 0.5 M Na_2SO_4.

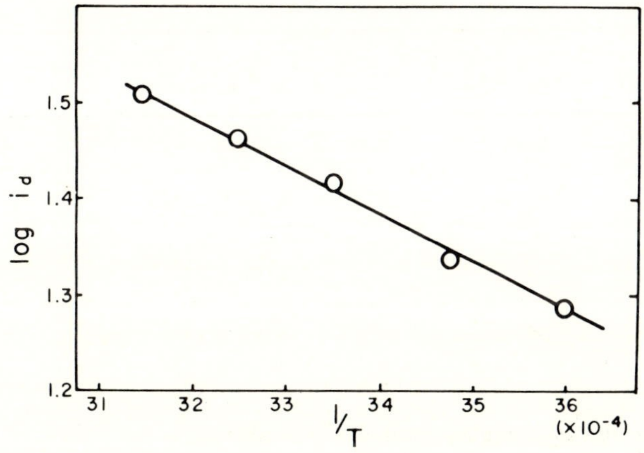

FIG. 3.2.3. The log i_d vs 1/T plot of estimating ΔH_D for the $K_4[Co(S_2O_3)(CN)_5]$ complex (5×10^{-3} M) in 0.5 M Na_2SO_4.

subtracted from the ΔH_C° value by the ΔH_D value (the heat of activation for diffusion), which could be evaluated from

$$\log(i_d/m^{2/3}t^{1/6}) = \log(0.627nFCD_O{}^{1/2}) - 1/2\ \Delta H_D/2.303RT \tag{3.2.5}$$

The slope of the $\log(i_d)$ vs 1/T plot gives the heat of activation for diffusion [255]. All the values of ΔH_D for the pentacyanocobaltate(III) complexes examined fall within the range of 4.3 to 4.9 kcal/mole at 25°C. Figure 3.2.3 shows one example of the Arrhenius plot for estimating the value of ΔH_D for the $K_4[Co(S_2O_3)(CN)_5]$ complex (5×10^{-3} M) in 0.5 M Na_2SO_4. The value of ΔH_D was calculated to be $\Delta H_D = 500 \times 2 \times 2.303\ R = 4.5$ (kcal/mole) from the slope (tan Θ = 500) in Fig. 3.2.3 according to Eq. (3.2.5). Therefore the true heat of activation for the electron transfer ΔH°_{Ct} can be evaluated from

$$\Delta H^\circ_{Ct} = \Delta H^\circ_C - \Delta H_D - \alpha_C nF\Delta\phi \quad (3.2.6)$$

As to the totally irreversible process for the $K_4[Co(S_2O_3)(CN)_5]$ complex,

$$\Delta H^\circ_{Ct} = \Delta H^\circ_C - \Delta H_D - \alpha_C nF$$

$$= 9.8 \text{ (kcal/mole at } 25^\circ C)$$

The results obtained for the other pentacyano complexes by similar procedures are tabulated in Tables 3.2.4 and 3.2.5.

TABLE 3.2.4. The True and Apparent Rate Constant for the Process Co(III) → Co(I) of the $K_3[Co(N_3)(CN)_5].2H_2O$ Complex[a]

Temperature (°C)	$\Delta\phi$ (mV)	$k_C{}^a$ (cm/sec)	k°_C (cm/sec)
5	35.1	2.29×10^{-11}	1.61×10^{-13}
15	36.7	5.38×10^{-11}	3.53×10^{-13}
25	37.5	1.01×10^{-10}	9.00×10^{-13}
35	37.8	1.87×10^{-10}	1.49×10^{-12}

[a]Conditions: 5×10^{-3} M complex ion in 0.5 M Na_2SO_4.

TABLE 3.2.5. The True and Apparent Rate Constant for the Process Co(III) → Co(I) of the $K_4[Co(SO_3)(CN)_5].3H_2O$ Complex[a]

Temperature (°C)	$\Delta\phi$ (mV)	$k_C{}^a$ (cm/sec)	k°_C (cm/sec)
10	38.9	2.70×10^{-17}	1.81×10^{-20}
15	39.1	5.38×10^{-17}	3.27×10^{-20}
25	40.5	8.52×10^{-17}	6.12×10^{-20}
35	41.4	2.76×10^{-16}	2.18×10^{-19}

[a]Conditions: 5×10^{-3} M complex ion in 0.5 M Na_2SO_4.

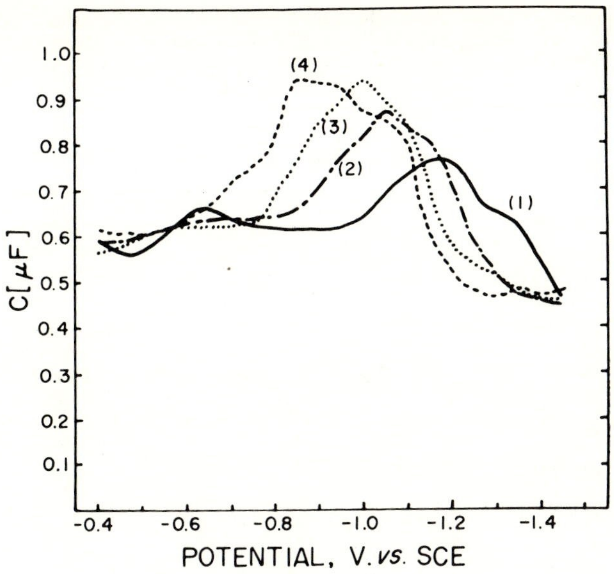

FIG. 3.2.4. Temperature dependence of the differential capacity-potential curve for the $K_3[Co\,Br(CN)_5]$ complex (5×10^{-3} M) in 0.5 M Na_2SO_4. Temperatures in °C: (1) 5, (2) 15, (3) 25, and (4) 45; m = 1.37 mg/sec; t = 4.0 sec.

Figure 3.2.4 shows the effect of temperature on the differential capacity-potential curve measured for the bromo-pentacyano complex. The remarkable variations of the differential capacity with temperature suggest that specific adsorption of the complex molecule can take place on the mercury surface, probably by formation of a Hg..... Br—$Co(CN)_5$ bridge. Therefore, Frumkin's correction for the double-layer effect could not be applied to estimate the apparent rate constant for the bromo complex. The situation is almost the same for all the halogeno-, isothiocyanato-, and thiocyanato-pentacyanocobaltate(III) complexes. Strong adsorption of the halogeno and thiocyanato complexes has also been observed for the i-t curves. These facts [207] lead to the conclusion that specific adsorption on the mercury surface is caused by the strong affinity of halides and the sulfur of the isothiocyanate ligand in these pentacyano complexes. The heat absorbed upon the liberation of ammonia gas (1 mole) from the $[Co(NH_3)_6]Cl_3$ complex (a solid phase) was estimated to be about 14 to 16 kcal/mole by means of differential thermogravimetric analysis [256], suggesting that the bonding energy of the σ-bond, such as Co^{III}—N, is more or less of this order. Inspection of Table 3.2.6 shows that the heat of activation for pentacyano complexes is much smaller than expected from this value, implying that bond-breaking of the Co^{III}—X bond could not take place during reduction, although the ligand X is finally released from the coordination sphere at the Co(I) state, as has been verified previously [206]. It is noteworthy that the mechanism of electron-transfer reactions is closely associated with the sixth ligand, X, since the $[Co^{III}(CN)_6]^{3-}$ ion cannot be reduced polarographically at the DME in neutral unbuffered solutions. In addition, no

TABLE 3.2.6. Kinetic Parameters of the Totally Irreversible Process Co(III) → Co(I) for the Thiosulfato, Sulfito, and Azido Complexes, and of the Process Co(III) → Co(II) → Co(I) for the Bromo Complex at 25°C[a]

	$K_4[Co(S_2O_3)(CN)_5]$	$K_4[Co(SO_3)(CN)_5]\cdot 3H_2O$	$K_3[Co(N_3)(CN)_5]\cdot 2H_2O$	$K_3[Co(Br)(CN)_5]$	
				First wave	Second wave
$E_{1/2}$ (V vs SCE)	-1.379	-1.560	-1.310	-0.847	-1.256
i_d (μA)	26.0	22.8	20.0	11.9	12.3
$\alpha_C n$	0.49	0.59	0.39	0.45	0.50
D_O (cm²/sec)	5.90×10^{-6}	4.37×10^{-6}	4.00×10^{-6}	4.49×10^{-6}	4.33×10^{-6}
k_C^a (cm/sec)	5.55×10^{-13}	9.52×10^{-17}	1.87×10^{-10}	2.93×10^{-8}	3.33×10^{-12}
k_C° (cm/sec)	5.39×10^{-15}	6.12×10^{-20}	9.00×10^{-13}	-	-
ΔH_C^a (kcal/mole)	14.7	15.7	11.6	12.7	~11.8
ΔH_C° (kcal/mole)	14.7	15.4	12.1	-	-
ΔH_D (kcal/mole)	4.5	4.9	4.5	4.3	4.3
ΔH_{Ct}° (kcal/mole)	9.8	9.9	7.2	-	-

[a]Conditions: 5×10^{-3} M complex ion in 0.5 M Na_2SO_4 (25°C); m = 1.37 mg/sec, t = 4.0 sec.

oxidation from the Co(I) to the Co(III) state occurs polarographically except for the case of a highly alkaline solution. As tentatively proposed previously [257], the mercury electrode (DME) can be regarded as a kind of electron donor (or a ligand), the coordinating ability of which increases with increasing potential. With this consideration, it is possible that electron transfer from the mercury atom to the 3d shells of the cobalt(III) occurs by three different mechanisms: 1) those in which Co(III) forms a bridging activated complex, [Hg..... X—Co(CN)$_5$]; 2) those in which Co(III) passes through the seven-coordinated activated complex, [Hg.....CoX(CN)$_5$], by an S_N2 mechanism; and 3) those in which Co(III) reacts via the five-coordinated activated complex, [Hg.....Co(CN)$_5$], by an S_N1 reaction mechanism. From the unexpectedly small values for the heat of activation (7.2 to 9.9 kcal/mole), one can presume that the S_N2 mechanism operates in the electron-transfer reaction of sulfito-, thiosulfato-, and azido-pentacyanocobaltate(III) complexes. The mercury atom (DME) would take its place with the cyanide and the X ligands at the closest approach to the electrode, giving an activated complex, [Hg.....CoX(CN)$_5$], with a bond loosening. Above all, the azide group N≡N≡N⁻ with π electron systems conducts electrons more readily. This may be one of the reasons why the heat of activation (ΔH_{Ct}° = 7.2 kcal/mole at 25°C) is so small.

On the other hand, the halogeno- and thiocyanato-pentacyanocobaltate(III) complexes give a double wave in which the halide and thiocyanate ligands are capable of offering a bridging path for the electron transfer and are probably reduced by a bridging mechanism, partly because of the strong affinity of halides or the sulfur atom of thiocyanate and partly because of the property of the thiocyanate ligand of readily conducting electrons. Bond rupture of the Co^{III}—X scarcely seems possible from the apparent heat of activation ($\Delta H_C{}^a$ = 10 to 13 kcal/mole at 25°C). Further speculation, however, seems unwise until a more complete discussion concerning entropy changes in these systems becomes possible.

4. ELECTROCHEMICAL STUDIES

4.1. POLAROGRAPHIC CHARACTERISTICS

Most early electrochemical studies of cobalt were mainly for analytical purposes; simultaneous determinations of cobalt, nickel, copper, and so on were shown to be possible by using complexing agents. For these purposes the formation of cobalt complex ions was usually achieved merely by mixing the hexaquocobalt(II) ion and a complexing agent at room temperature. Consequently, most cobalt complexes treated in earlier years were of a "substitution-labile" type toward solvolysis with a few exceptions for complexes with ligands such as 2,2'-dipyridyl (π-bonding ligands). Therefore, these solutions consist of a mixture of labile cobalt(II) complex species in equilibrium with free ligands in aqueous media.

It is well known generally that the cobalt(II) ion surrounded by ligands such as ammonia tends to convert into the cobalt(III) ion through oxidation, provided that the strength of its ligand field is large enough to cause some splitting of the 3d energy levels of the cobalt. Most of the resultant cobalt(III) complexes are of an "inert" type toward solvolysis. Usually the dissociation of cobalt(III) complexes need not be taken into account in noncomplexing aqueous media except for a partial aquation of some ligands in mixed ligand Co(III) complexes such as halides. In contrast, Co(II) complexes of a high-spin type dissociate rapidly into the aquated species in noncomplexing media. Therefore, the first step of Co(III) → Co(II) → Co(metal) is related only to the original structure of the inert low-spin Co(III) complexes, and the second step is not related to any of the structures of cobalt complexes in noncomplexing media (the reduction of inert-labile type).

The modern trend in the field of polarography of cobalt is to establish the relationship between the structures or bonding types of cobalt complexes and their electrode processes; this is because cobalt complexes, among transition metal complexes, are large in number and exhibit substantial variety in their bonding nature from compound to compound. In these studies, nonaqueous polarography is now attracting special attention and promises to provide clues for the elucidation of electron transfer mechanisms of the electrode reactions and their correlation to the electronic configuration of a great variety of mixed ligand complexes. The most striking feature of nonaqueous polarography is that, by adopting aprotic nonaqueous solvents, the Co(III) complexes, which would otherwise lead to a loss of ligands in the lower oxidation states of cobalt, are reduced in stepwise fashion with a complete retention of the original configuration. For example, the pathway Co(III) → Co(II) → Co(I) → Co(0) → Co(-I) takes place without a loss of ligands in aprotic solvents ($[Co\ phen_3]^{3+}$ in DMSO; see Table 2.2.19), although the appropriate valence orbitals are, of course, delocalized over the entire complex molecule and Co(-I) should be interpreted only as a change in the formal oxidation state of the central cobalt. The electrode processes of the completely "inert-inert" type or of the completely "inert-labile" type are the ones in which it is easiest and simplest to determine the assignment of the polarographic waves. For either extreme case these consist of the complex remaining structurally intact or, alternatively, undergoing a rapid dissociative equilibrium upon reduction. Moreover, the complete retention of the structure in nonaqueous media makes it possible to relate the structures in both the oxidized and the reduced forms to the redox reaction of the electrode processes. For this reason cyclic voltammetry or controlled-current oscillopolarography (Heyrovský–Forejt type) are powerful methods not only for the examination of the reversibility of the electrode reaction, but also for the identification of the cobalt species responsible for the conventional polarographic waves. Here the interpretation is facilitated because the cathodic reductions can be observed and compared with the anodic oxidations in the same experiment. The aprotic solvent may not only play the role of stabilizing the lower oxidation states of cobalt but may also simplify the electrode process by converting the labile cobalt complexes into inert ones, (where the terms "labile" and "inert" are used with regard to the "lability" of ligands bound to the cobalt toward solvolysis, i.e., the ligand exchange reactions with solvent molecules). In this respect the lability used here has a significance somewhat different from that defined by Taube [68] on the basis of the magnitude of the rate for the ligand substitution reactions with the same kind of free ligands present in solution. The terms "substitution-inert" and "substitution-labile," as defined by Taube [68] are independent of the nature of the medium. However, the inertness or lability of the cobalt complexes used here is with respect to solvolysis and is dependent on the nature of the solvent. In conclusion, the simplified electrode process obtained by converting labile complexes into inert-type ones with nonaqueous solvents can be regarded as representing an ideal model of the more complicated electrode processes in water, where a pathway similar to that in the nonaqueous solvent would operate if the preceding or subsequent chemical reactions, such as aquation, dissociation, and rearrangement of ligands, had been completely prevented or inhibited throughout the course of reduction.

4.1.1. Hexamminecobalt(III) Complex

The ammine cobalt(III) complexes have been investigated most thoroughly and have received a great deal of attention. Above all, the behavior of the hexamminecobalt(III) ion is

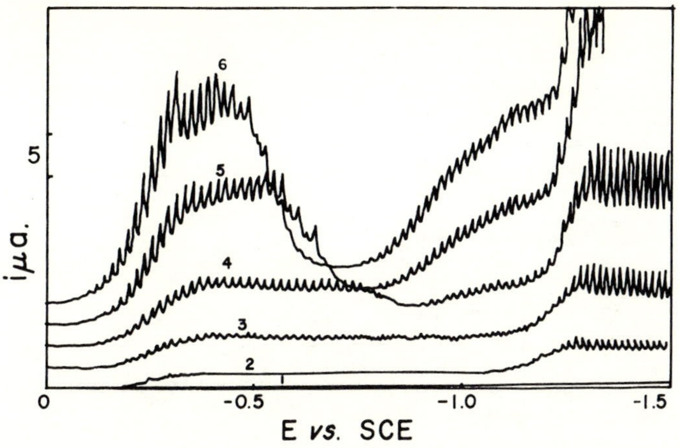

FIG. 4.1.1. Polarograms obtained in 0.1 M $NaClO_4$ with various concentrations of the $[Co(NH_3)_6]Cl_3$ complex: (1) 0 m<u>M</u>, (2) 0.5 m<u>M</u>, (3) 1 m<u>M</u>, (4) 2 m<u>M</u>, (5) 5 m<u>M</u>, and (6) 10 m<u>M</u>. (Data from Ref. 93.)

of prime importance since it is typical of the electrode reactions of ammonia complexes and their analogs with similar donor molecules.

The $[Co^{III}(NH_3)_6]^{3+}$ ion gives a well-defined wave of two steps in noncomplexing media. The first step corresponds to the one-electron reduction to Co(II), and the second to the two-electron reduction to the metal. The first step in the cathodic reduction produces the hexamminecobalt(II) ion which in turn decomposes, more or less, depending on the nature of the medium. In aqueous media without free ligands, the following electrode reaction scheme was proposed by Kolthoff and Khalafalla [250, 258, 259]:

$$\underset{\text{inert}}{[Co^{III}(NH_3)_6]^{3+}} \xrightarrow{e} \underset{\text{labile}}{[Co^{II}(NH_3)_6]^{2+}} \longrightarrow [Co^{II}(OH)(NH_3)_4]OH_{ppt} + 2NH_4^+$$

$$\downarrow$$

$$[Co^{II}(OH)(NH_3)_4]^+$$

$$\downarrow$$

$$[Co^{II}(OH_2)_6]^{2+}, \text{ etc.}$$

$$\downarrow 2e$$

$$\text{Co(metal)}$$

Laitinen et al. [91-93] recognized that upon the reduction of Co(III) to Co(II) in neutral unbuffered medium a so-called "current-dip" occurs in the neighborhood of -0.8 V (vs SCE) due to the adsorption of the $[Co^{II}(OH)(NH_3)_4]OH$ precipitate [250] at the DME. Figure 4.1.1 shows the current-dip observed for various concentrations of the $[Co(NH_3)_6]^{3+}$ ion in a 0.1-<u>M</u> $NaClO_4$ aqueous solution. The dip in the diffusion current was attributed to a film of the

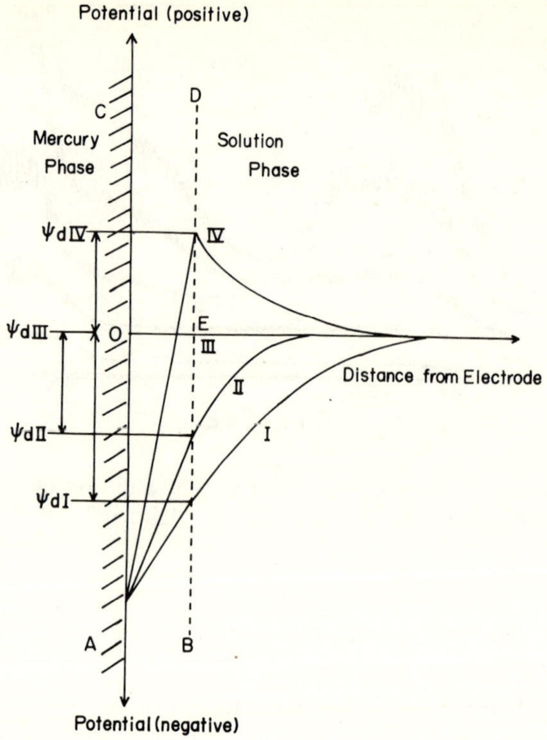

FIG. 4.1.2. Potential profile within the double layer and the variation of the ϕ_2 (or ψ_d) potential with increasing supporting electrolyte concentration (Curve I, II, and III) and in the presence of cationic surfactants (Curve IV). (Data from Ref. 69.)

positively charged $[Co^{II}(OH)(NH_3)_4]^+$ ion which firmly adhered to the mercury surface on the negative side of the isoelectric point:

$$[Co^{II}(NH_3)_6]^{2+} \rightarrow [Co^{II}(OH)_2(NH_3)_4] \rightarrow [Co(OH)(NH_3)_4]^+ + OH^-$$

In order to account for the fact that the minimum disappears at high supporting electrolyte concentrations, Kolthoff and Khalafalla [69] took into consideration the effect of the ϕ_2 (or ψ_d) potential (the potential in the outer Helmholtz layer or the "d" plane of the double layer) on the adsorption of the secondary product of the $[Co^{II}(OH)(NH_3)_4]^+$ ion at the negatively charged mercury surface. Figure 4.1.2 shows a schematic diagram of the double layer at potentials more negative than the potential of zero charge on mercury. With increasing electrolyte concentration the diffuse part of the double layer becomes more and more compressed (Curves II and III) and is at last completely suppressed, whereupon the potential at E is equal to that in the bulk of the solution; the minimum disappears. The adsorption of the $[Co^{II}(OH)$-$(NH_3)_4]^+$ ion may change the ϕ_2 potential from its normal negative value to less negative values. This may cause a retarding effect on the rate of the reduction of the $[Co(NH_3)_6]^{3+}$ ion and thus bring about a minimum in polarograms.

Strongly adsorbed cations (e.g., tetrabutylammonium ion) or uncharged surfactants (e.g., gelatin) eliminate the dip. When they are present in a concentration of 2 x 10^{-5} M, no dip is observed. Under these conditions the double layer may become completely suppressed (Curve III in Fig. 4.1.2) or the ϕ_2 potential may become positive (Curve IV). The presence of such adsorbed particles increases the overvoltage and decreases the rate of electroreduction of the $[Co(NH_3)_6]^{3+}$ ion.

The same interpretation accounts for the decelerating effect of anionic surfactants like sulfate ion on the reduction of Co(III) to the Co(II) state. With increasing sulfate ion concentration, $E_{1/2}$ shifts markedly to more negative values. This is attributed to the effects of both the change in the double-layer structure and ion pair formation with the complex cation ("super-complex") [260, 261].

Contrary to this, the iodide ion shifts the $E_{1/2}$ in a positive direction, a marked effect being observed when the iodide concentration is as small as 10^{-6} M. Adsorbed iodide eliminates the decelerating effect of cationic surfactants. This is attributed to the effect of a film of adsorbed mercury(I) iodide on the ϕ_2 potential [259].

The adherence of the $[Co^{II}(OH)(NH_3)_4]OH$ precipitate to the mercury surface was also noted on the instantaneous current-time (i-t) curves in the potential range in which the dip occurs [69]. Figure 4.1.3 exemplifies the abnormal shape of i-t curves observed at the minimum. The oscillations observed in the abnormal curves are undoubtedly connected with the time involved in the formation and adherence of the secondary reduction product of the $[Co(NH_3)_6]^{3+}$ ion at the expanding surface of the growing mercury drop. In the experiments in Fig. 4.1.3 the drop time was about 7 sec at the minimum. The shape of the curve at -0.8 V (vs SCE) remained unchanged when the drop time was varied between 6 and 17 sec.

The i-t curves of the $[Co(NH_3)_6]Cl_3$ complex have been frequently examined as an example of totally irreversible systems in acidic solutions. From the theory of i-t curves in the presence of surface-active substances, Kůta, Weber, and Koutecky [262] derived equations for the decrease of mean current and for the displacement of $E_{1/2}$ to more negative values under the assumption that the electrode process does not proceed on that part of the mercury electrode surface which is covered with the surface-active substance. Experiments with uncharged surfactants, viz., eosin, polyvinyl alcohol, camphor, and thymol in the case of the totally irreversible process, $[Co(NH_3)_6]^{3+} \rightarrow [Co(NH_3)_6]^{2+}$ in 0.1 M H_2SO_4 were in very good agreement with theory. Further, Kůta and Smoler [263] studied the effect of temperature on the electrode reactions of the one-electron process, $[Co(NH_3)_6]^{3+} \rightarrow [Co(NH_3)_6]^{2+}$, inhibited by some cations in acidic solutions (0.1 M H_2SO_4). Tribenzylamine, tetrabutylammonium sulfate and Triton X 45 were used as inhibitors. From the analysis of i-t curves, it was concluded that the time (t) necessary for the complete coverage of the electrode decreases with rising temperature. The rate constant of the electrode reaction on the covered electrode surface increases with rising temperature, and the

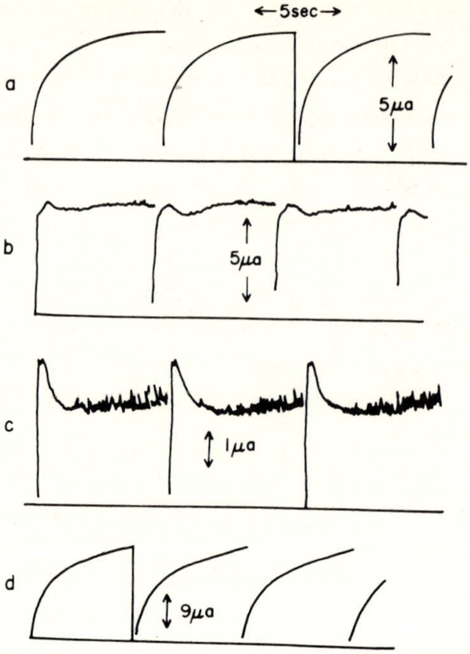

FIG. 4.1.3. Current-time curves in 0.1 M $NaClO_4$. Concentration of the complex, 2 mM. Potential (V vs SCE): (a) -0.5, (b) -0.7, (c) -0.8, and (d) -1.4. (Data from Ref. 69.)

temperature dependence of it follows the Arrhenius equation. So far as the rate constant increases appreciably with temperature, the inhibited polarographic wave is practically temperature independent.

The i-t curve analyses were also made for the $[Co(NH_3)_6]^{3+}$ ion in 0.5 M K_2SO_4 - 0.1 M H_2SO_4 in the presence of acridine hydrochloride [264]. The addition of 4.23×10^{-5} M acridine hydrochloride causes the current to decrease continuously to finite but not to zero values within 4 sec of the drop life due to the adsorption. Thus the one-electron reduction process of the $[Co(NH_3)_6]^{3+}$ ion in acidic solutions has often been used as a typical instance of totally irreversible systems for checking experimentally the theoretical equations; no current dip was, of course, observable in acidic solutions.

Similar types of current-time curves are usually observable not only for the cases of the aquated precipitates of the secondary product, but also for the cases where the primary reduction product is insoluble in water. For example, the cis-dicyanobis(2,2'-dipyridyl)-cobalt(III) complex gives a normal wave of three steps at a concentration of less than 10^{-3} M in 0.5 M Na_2SO_4 (25°C), in which each step corresponds to the stepwise reduction of Co(III) → Co(II) → Co(I) → Co(0). With increasing complex concentration, the first step ($Co^{III} \rightarrow Co^{II}$) becomes split, with the appearance of two or more steps due to the depression of the current. Figure 4.1.4 shows the effect of the complex concentration on the current-potential curves

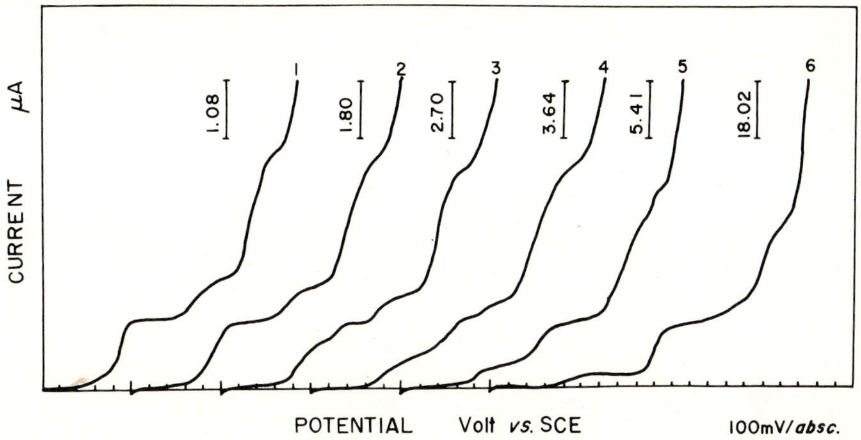

FIG. 4.1.4. The concentration dependence of the current-potential curve for the cis-$[Co(CN)_2dip_2]NO_3 \cdot 7H_2O$ complex obtained in 0.5 M Na_2SO_4 (25°C). Concentration (M): (1) 0.6×10^{-3}, (2) 1.0×10^{-3}, (3) 1.5×10^{-3}, (4) 2.0×10^{-3}, (5) 3.0×10^{-3}, and (6) 10^{-2}. Recorded from 0.0 V vs SCE.

of the cis-$[Co^{III}(CN)_2dip_2]NO_3 \cdot 7H_2O$ complex in 0.5 M Na_2SO_4 at 25°C. The normal first step of the one-electron reduction (Curves 1 and 2) gradually changes into multisteps (Curves 3, 4, and 5) with increasing complex concentration, and become suppressed almost completely (Curve 6) as the concentration reaches 10^{-2} M. This is attributed to the interference of the insoluble product, $[Co^{II}(CN)_2dip_2]$, accumulated on the mercury surface with the current passing through the DME in the potential range of -0.4 to -0.6 V (vs SCE), at which point the Co(III) to Co(II) reduction begins to occur. Thus the prevention of the current is closely correlated with the concentration of the depolarizer in solution. This conclusion is reinforced by the fact that the abnormal shaped i-t curves are always obtained only under conditions where the depression of current occurs at the first diffusion plateau. Figure 4.1.5 shows one example of i-t curves measured on single-drops of the DME at various fixed potentials. At concentrations higher than 2×10^{-3} M, the abnormal curve (2) is observed at -0.55 V (vs SCE). Fairly large oscillations on the curve suggest that the insoluble substance of the primary product prevents the current from passing through the mercury surface. This postulate is further reinforced by observations concerning temperature and solvent effects on the solubility of the reduction product in solution. These two effects on the current-potential curves can clearly be demonstrated by selecting experimental conditions suitable for detecting a slight increase in the solubility of the $[Co^{II}(CN)_2dip_2]$ complex formed at the DME. Figure 4.1.6 shows that the deformed wave is improved successively and regularly from left to right with rising temperature, and the depressed current is restored again slightly but regularly up to the normal situation. This should be expected from a linear dependence of current on temperature.

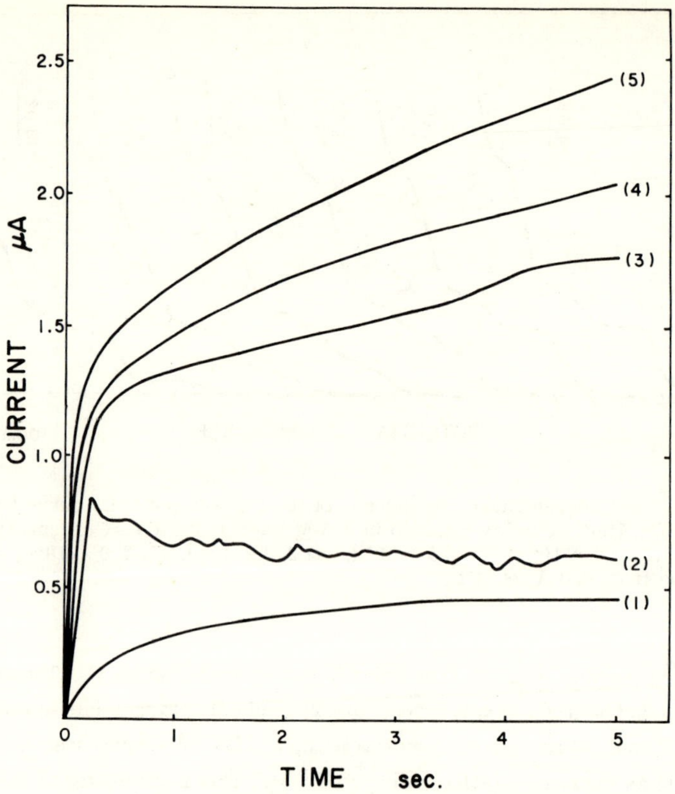

FIG. 4.1.5. Current-time curves obtained for the cis-[Co(CN)$_2$dip$_2$]NO$_3$. 7H$_2$O complex (2 x 10^{-3} M) in 0.5 M Na_2SO_4 (25°C). Fixed potential (V vs SCE): (1) -0.40, (2) -0.55, (3) -0.75, (4) -0.85, and (5) -1.00. The current in Curves (4) and (5) is diffusion controlled.

The addition of methanol to the aqueous solution also improves the wave shape and restores the current to the original state of displaying a diffusion-controlled wave, as shown in Fig. 4.1.7. Conclusively, the primary reduction product, [Co^{II}(CN)$_2$dip$_2$], is more soluble in methanol than in water. We have frequently encountered similar phenomena in nonaqueous media because most of degraded inorganic complex anions have lower solubilities in aprotic solvents than has the original complex [354].

In DMSO the cis-[Co(CN)$_2$dip$_2$]NO$_3$. 7H$_2$O complex gives a normal wave of three steps, Co(III) → Co(II) → Co(I) → Co(0). The height of each step always has a ratio of approximately 1:1:1, but in aqueous and methanolic solutions a rearrangement of cyanides occurs during reduction, yielding the tetracyanocobaltate(II) complex, [Co^{II}(CN)$_4$dip]$^{2-}$. The former process in DMSO corresponds to the step-by-step reduction of the "inert-inert" type, whose original configuration remains intact throughout reduction, while the latter is accompanied by the rearrangement of ligands due to the change in the lability of cyanides in aqueous and

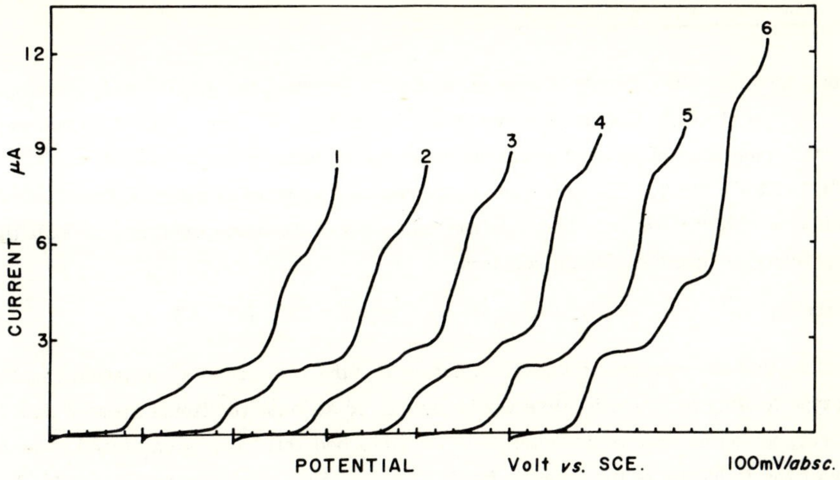

FIG. 4.1.6. The effect of temperature on current-potential curves for the cis-$[Co(CN)_2dip_2]NO_3 \cdot 7H_2O$ complex (1.5×10^{-3} M) in 0.5 M Na_2SO_4. Temperature (°C): (1) 5, (2) 10, (3) 20, (4) 25, (5) 30, and (6) 50.

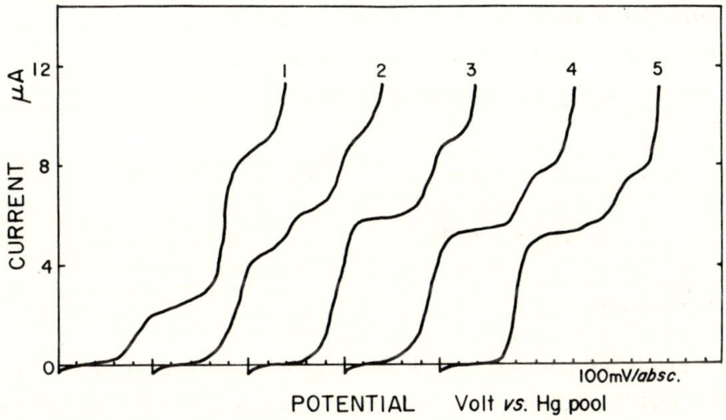

FIG. 4.1.7. The effect of methanol on the current-potential curves obtained for the cis-$[Co(CN)_2dip_2]NO_3 \cdot 7H_2O$ complex (5×10^{-3} M) in 1 M LiCl H_2O-methanolic solution (25° C). Methanol in vol%: (1) 0, (2) 5, (3) 10, (4) 30, and (5) 50, recorded from 0 V vs Hg pool.

methanolic solutions. There is, in general, a tendency toward forming the symmetrical complex in which the central cobalt is surrounded with bonding of the same nature as far as possible. Consequently, the mixed liganded complex is difficult to form in the labile state of the complex. Conversely, it is therefore not strange that the rearrangement of cyanides occurs at the labile state of the Co(II) complex on the pathway of reduction (see Tables 2.2.15 and 2.2.23) [265].

4.1.2. The Other Amminecobalt(III) Complexes

Vlček and Kůta [266] reported several instances in which the single, well-developed wave for the process Co(III) → Co(II), observed with $[CoY(NH_3)_5]^{n+}$ ions, splits into two waves when small quantities of halide ion are added to the supporting electrolyte (Y = NO_3^-, H_2O, CH_3COO^-, NO_2^-, etc.). The new wave (prewave) that appears at more positive potentials upon addition of halide was attributed to the reduction of the halo-pentamminecobalt(III) ion formed through a ligand exchange reaction:

$$[CoY(NH_3)_5]^{n+} + X^- \rightarrow [CoX(NH_3)_5]^{2+} + Y^{0,-} \quad (X = Cl^-, Br^-, I^-) \qquad (4.1.1)$$

They stated that the prewave is kinetic in nature and the rate of the substitution reaction of such a type leading to a kinetic wave would have to be at least 10^7 times greater than that of the corresponding homogeneous reaction in the adsorbed state and is catalyzed by adsorption under the influence of the electric field. Tanaka, Sato, and Tamamushi [267], however, found that such a prewave is never observed for the aquopentamminechromium(III) complex under the same conditions in the presence of halide ion although the same phenomenon may be expected to be observed with chromium(III) complexes. Following these findings they investigated the influence of anions on the cathodic wave of aquopentamminecobalt(III) ion in more detail and concluded that the current of prewave rises from the anodic background and becomes cathodic when the potential reaches the value where reduction of the chemically formed Hg_2X_2 can proceed according to

$$[Co(OH_2)(NH_3)_5]^{3+} + Hg + X^- + nH_3O^+ \rightarrow [Co(OH_2)_{n+1}(NH_3)_{5-n}]^{2+} + \frac{1}{2}Hg_2X_2 + nNH_4^+ \qquad (4.1.2)$$

$$Hg_2X_2 + 2e \xrightarrow{\text{fast}} 2Hg + X^- \qquad (4.1.3)$$

That is, the prewave is not the reduction of the halopentamminecobalt(III) ion formed by substitution Reaction (4.1.1), but the reduction wave of mercury(I) halide formed by the oxidation-reduction Reaction (4.1.2). Independently, Anson and Chang [268] reached the same conclusion concerning the prewave that appears when halide ion is added to the solution of the $[Co(OH_2)(NH_3)_5]^{3+}$ ion. The dependence of the current-potential curves on the ionic strength and composition of the supporting electrolyte revealed that either the chemical oxidation-reduction Reaction (4.1.2) or the electrochemical reduction (4.1.3) of Hg_2X_2 controls the rate of the electrode reaction responsible for the prewave, depending upon the electrode potential. They also attempted to account for the observed current-potential behavior in terms of the changes in diffuse layer potential ϕ_2 between bromide-containing and bromide-free perchlorate supporting electrolytes, since the addition and adsorption of halide anion decreases the charge of the positively charged mercury surface. A mechanism somewhat similar to Reactions (4.1.2) and (4.1.3) was previously suggested by Laitinen, Frank, and Kivalo [93] to account for the effects of added halide on the polarographic behavior of the $[Co(NH_3)_6]^{3+}$ ion.

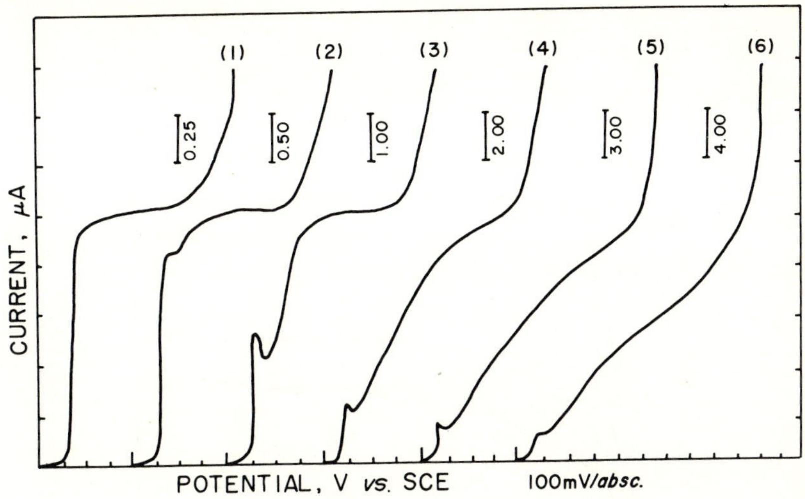

FIG. 4.1.8. The effect of the complex concentration on the current-potential curve of $[Co(CN)(NH_3)_5](ClO_4)_2 \cdot 1/2\ H_2O$ in 0.5 M Na_2SO_4 (25°C). Concentration (M): (1) 5×10^{-4}, (2) 10^{-3}, (3) 2×10^{-3}, (4) 4×10^{-3}, (5) 6×10^{-3}, and (6) 8×10^{-3}. Recorded from -0.20 V vs SCE.

For the reduction of the cyanide substituted complex, $[Co^{III}(CN)(NH_3)_5]Cl_2$, the formation of the $[Co(CN)(NH_3)_4]OH$ precipitate should analogously be predicted to occur more rapidly and far more easily than that of the $[Co^{II}(OH)(NH_3)_4]OH$ precipitate upon the reduction of $[Co(NH_3)_6]^{3+}$, since it necessitates nothing but the liberation of only one ammonia ligand from the sphere of coordination:

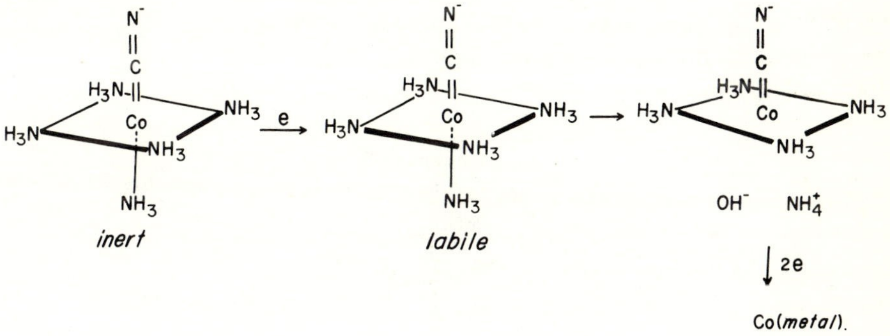

Figure 4.1.8 illustrates the effect of the complex concentration on the current-potential curves of the $[Co(CN)(NH_3)_5]^{2+}$ ion over the range between 5×10^{-4} and 8×10^{-3} M in 0.5 M Na_2SO_4. The depression in current becomes more noticeable with increasing concentration in the potential range of the current-rise at which the reduction to the Co(II) state begins and not in the more negative potential region of the diffusion-controlled plateau for the case of the hexamminecobalt(III) ion, suggesting that a secondary reaction involving release of

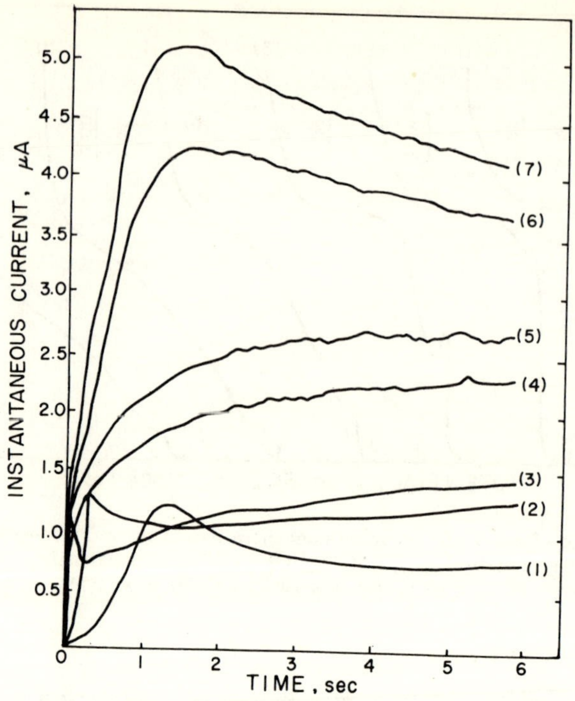

FIG. 4.1.9. The i-t curves measured for the $[Co(CN)(NH_3)_5](ClO_4)_2 \cdot 1/2\ H_2O$ complex (10^{-2} M) in 0.5 M Na_2SO_4 (25°C). Fixed potential (V vs SCE): (1) -0.30, (2) -0.35, (3) -0.40, (4) -0.45, (5) -0.50, (6) -0.65, and (7) -0.80.

ammonia proceeds more readily. From evidence presented later, the ammonia ligand at a trans-position with regard to the cyanide is activated and liberated from the coordination sphere when the Co(III) is reduced to the Co(II) state, giving a five-coordinate complex, $[Co^{II}(CN)(NH_3)_4]OH$:

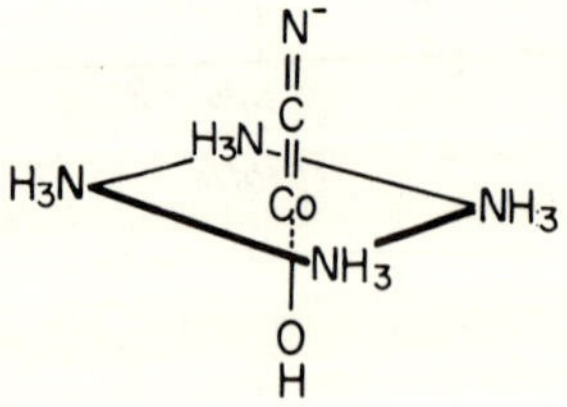

A marked dependence of the current depression on the complex concentration has been observed for i-t curves during the life of a single drop of the DME. Small oscillations in the i-t curves, similar to the case of the hexamminecobalt(III) ion, are also noticed whenever the depression of current occurs on polarograms. Figure 4.1.9 exemplifies the abnormal

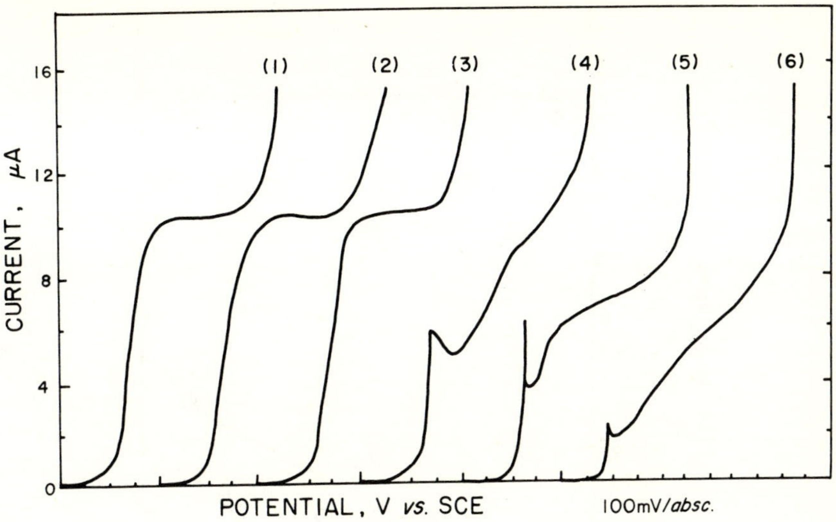

FIG. 4.1.10. The effect of pH on the current-potential curve of the $[Co(CN)(NH_3)_5](ClO_4)_2 \cdot 1/2\ H_2O$ complex (4×10^{-3} $\underline{M}$) in NaAc + HAc buffer solutions ($\mu = 1$ $\underline{M}$). The pH values are: (1) 3.56, (2) 4.52, (3) 5.52, (4) 5.75, (5) 7.20, and (6) 9.23. Recorded from -0.10 V vs SCE.

shape of i-t curves which may be attributed to the deposition of the $[Co^{II}(CN)(NH_3)_4]OH$ precipitate which is insoluble in water. As anticipated, this hydroxide would be soluble in acidic solutions; indeed, the effect of pH on the current-potential curves shows the depression of current disappears completely as soon as the pH decreases across a neutral point, as indicated in Fig. 4.1.10.

In DMSO, a quite similar process via the $[Co^{II}(CN)(NH_3)_4]^+$ ion operates in the cathodic reduction as shown from studies of Co(III) complexes of the trans-$[Co(CN)X(NH_3)_4]$ type both in DMSO and in aqueous solutions, where the ligand, X, denotes the ion NO_2^-, NO^-, N_3^-, I^-, NCS^-, $S_2O_3^{2-}$, CO_3^{2-}, or SO_3^{2-} or the molecule $NH_2C_2H_5$, NH_2CH_3, NH_3, py (pyridine), OH_2, or $P{\equiv}\phi_3$ (triphenylphosphine). The only difference in polarographic behavior between DMSO and aqueous media is that further aquation of the $[Co^{II}(CN)(NH_3)_4]^+$ ion proceeds before the two-electron reduction to the metal in the second step. In DMSO (100%) containing 0.1 $\underline{M}$ $[(C_2H_5)_4N]ClO_4$, the $E_{1/2}$ of the first step, Co(III) → Co(II), follows the order of the spectrochemical series of the ligand, X, whereas those of the second step, Co(II) → Co(metal), exhibit almost the same value (-1.55 V vs SCE), irrespective of the kind of the ligand, X, suggesting that the cobalt(II) species responsible for the second step is the $[Co^{II}(CN)(NH_3)_4]^+$ ion without the ligand, X (see Table 2.2.22). Figure 4.1.11

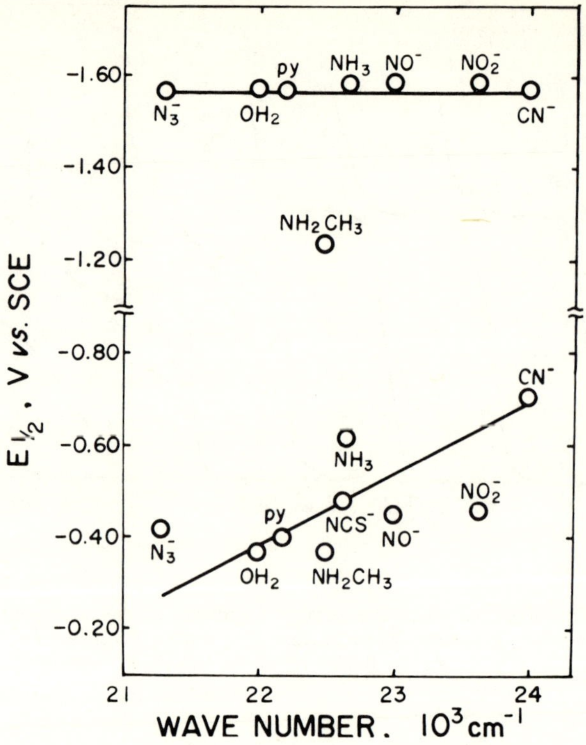

FIG. 4.1.11. The relationship between $E_{1/2}$ and ν_{max} both for the first and the second step of the trans-$[Co(CN)X(NH_3)_4]$ complexes (10^{-3} M) in the DMSO containing 0.1 M $[(C_2H_5)_4N]ClO_4$ at 25°C.

shows the plot of $E_{1/2}$ vs ν_{max} (the absorption maximum of the first d-d band) in wave number. Therefore, the main processes of the electrode reaction in DMSO are:

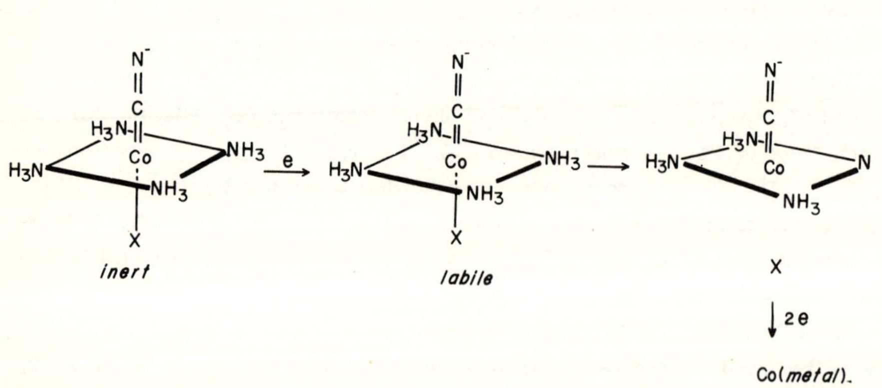

The liberation of the ligand, X, from the Co(II) complex was further substantiated by the observation that the $E_{1/2}$ of the anodic waves are all nearly the same (+0.03 V vs SCE)

within the experimental error, independent of the kind of the ligand, X, indicating that the Co(II) species responsible for the anodic oxidation, Co(II) → Co(III), have the same formula, $[Co^{II}(CN)(NH_3)_4]^+$ (see Table 4.1.1).

TABLE 4.1.1. The Half-Wave Potentials of Anodic Waves Obtained for the Kalousek Polarograms of the trans-$[Co(CN)X(NH_3)_4]$ Complexes[a]

Ligand, X for $[Co(CN)X(NH_3)_4]$ complexes	$E_{1/2}$ of anodic wave Co(II) → Co(III) (V vs SCE)
NO_2^-	$+0.03_7$
OH_2	+0.03
$NH_2C_2H_5$	+0.026
NCS^-	+0.03
py	+0.03
NH_3	+0.08
N_3^-	$+0.03_8$

[a]DMSO (100%) containing 0.1 M $[(C_2H_5)_4N]ClO_4$ (25°C).

Figure 4.1.12 shows typical Kalousek polarograms of trans-$[Co(CN)(NO_2)(NH_3)_4]ClO_4 \cdot 1/2\ H_2O$ at fixed potentials of -1.00 and -1.70 V (vs SCE) at which the current reaches limiting diffusion plateaus for the first and the second step, suggesting the formation of Co(II) complex and of metallic cobalt, respectively. Thus the sixth ligand, X, is liberated from the Co(II) complex during reduction in DMSO. The counter bond, Co—X, with regard to the cyanide seems most likely to be loosened relative to the strong bond, $Co{=}C{=}N^-$. Such a trans-effect of cyanide is explained in terms of the π-bonding character of cyanide. The back donation causes withdrawal of electrons toward the axial cobalt-cyanide linkage which causes an increased residual positive charge of the cobalt ion. This results in a decrease of the electron density on the donor atom of the axial counter ligand, X, with concomitant weakening of the Co—X bond. A few exceptions, however, are found for complexes with $X = CN^-$, NH_3, and $P{\equiv}\phi_3$ in DMSO. These complexes undergo reduction to the metal in the second step with a retention of the six-coordinate formula in DMSO. The discrepancy of the $E_{1/2}$ (+0.08 V) for the anodic wave of $[Co(CN)(NH_3)_5]^+$ with those (+0.03 ± 0.01 V) of the other indicates no liberation of ammonia for the complex with $X = NH_3$. The addition of a small quantity of water to the DMSO (100%) solution, however, immediately causes not only the liberation of the ligand, X, but also of ammonia ligands to yield further aquated Co(II) species; the anodic wave due to the Co(II) complex with ligands could no longer be

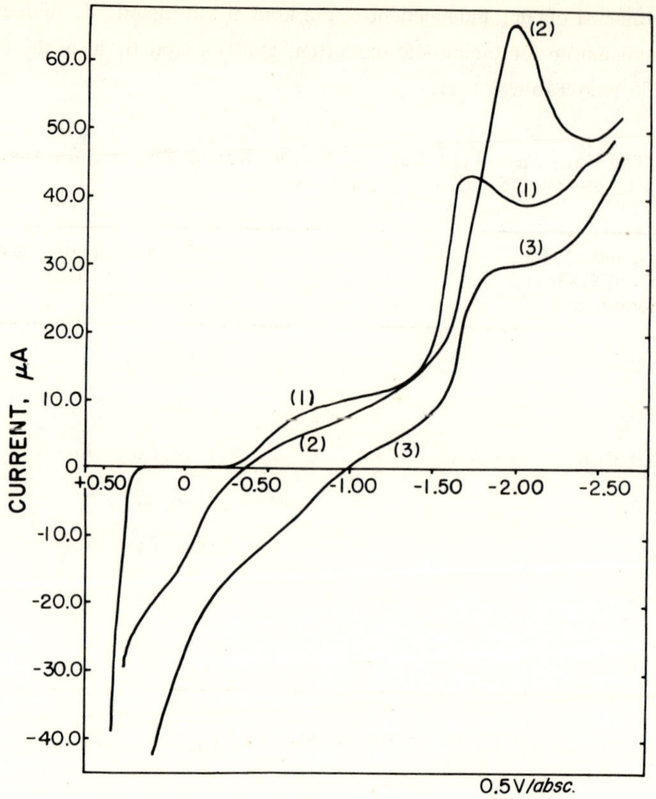

FIG. 4.1.12. Direct current and Kalousek polarograms of trans-[Co(CN)$(NO_2)(NH_3)_4$]-ClO_4.1/2 H_2O (5 × 10^{-3} M) in DMSO containing 0.1 M [$(C_2H_5)_4N$]ClO_4 (25°C). (1) Conventional dc polarogram; (2) commutated wave at the fixed potential (V vs SCE) of -1.00; (3) the same at -1.70.

observed by using a Kalousek commutator [66]. In aqueous media these ammine cobalt(III) complexes are considered to follow a course of reduction similar to that in DMSO, at least at the beginning of the electrode pathway, but different in the respect that in water further aquation proceeds. This subsequent chemical reaction causes the electrode process to be more complicated in aqueous media. This is the reason why nonaqueous polarography is of prime importance in studies of cobalt complexes; the electrode pathway (an inert-inert type) in aprotic nonaqueous solvents can be regarded as a model of the course which most cobalt complexes would tend to take in aqueous media at the beginning of the electrode reaction. Accordingly, the situation in aqueous media is almost the same as in DMSO for the relationship between $E_{1/2}$ and ν_{max}. The $E_{1/2}$ for the first step, but not for the second step, correlates with the structure of the Co(III) complex as given in Figs. 4.1.13 and 4.1.14 (see Table 2.2.14).

There are several other papers concerned with the electrochemistry of the ammine cobalt(III) complexes. For instance, Holtzclaw [269] attempted to differentiate the cis- and

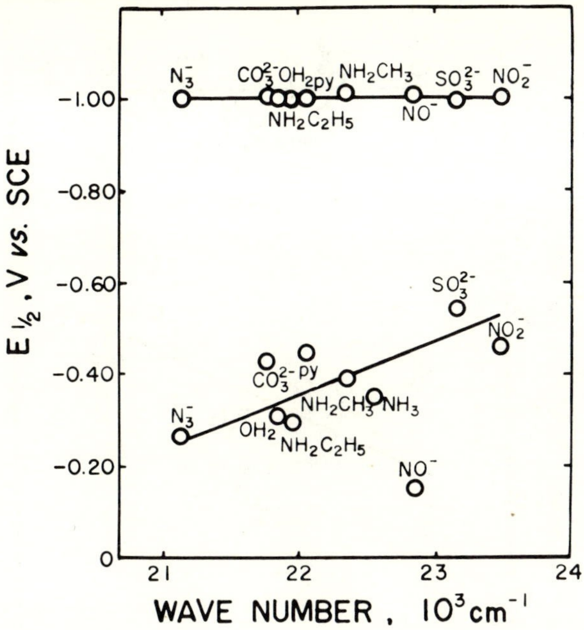

FIG. 4.1.13. Relationship between $E_{1/2}$ and ν_{max} for the trans-[Co(CN)X(NH_3)$_4$] complexes in NaAc + HAc buffer solutions (μ = 1 M, pH = 4.1).

trans- isomers of the [CoX$_2$(NH_3)$_4$] complexes from each other on the basis of the $E_{1/2}$ for the Co(III) → Co(II) process as early as in 1951. Yanai and Kuroda [96] studied the acidopentamminecobalt(III) complexes polarographically and showed that the $E_{1/2}$ of the first step shifts toward a more negative potential with an increase in the basisity of the acid ligand in the order CH_3COO^- < CH_2ClCOO^- < $CHCl_2COO^-$ < CCl_3COO^-. Ikeuchi [270] has recently determined the diffusion coefficient for the Co(III) complexes of the nitro-ammine series in aqueous solutions of various ionic strengths. Yamashita and Imai [271] reported the photochemical reaction rate of Co(III) complexes of the nitro-ammine series by polarographic i-t curves. Malik and Aslam [272, 273] have determined the stability constant of [Co(amino acid)(NH_3)$_5$] and [Co(amino acid)(NH_3)$_4$] complexes polarographically.

4.1.3. Ammine Cobalt(III) Complexes

As mentioned in Section 1, although the electron-transfer reaction of the [Co en$_3$]$^{3+}$/[Co en$_3$]$^{2+}$ system is faster than the diffusion rate ($k_c^a > 10^{-2}$ cm/sec), "irreversible" slopes for the plots of log[(i_d - i)/i] against potential are obtained [273-275]. For this reason the [Co en$_3$]$^{3+}$/[Co en$_3$]$^{2+}$ system has often been mistaken as an irreversible or a quasi-reversible process. This system, however, is actually a reversible process which

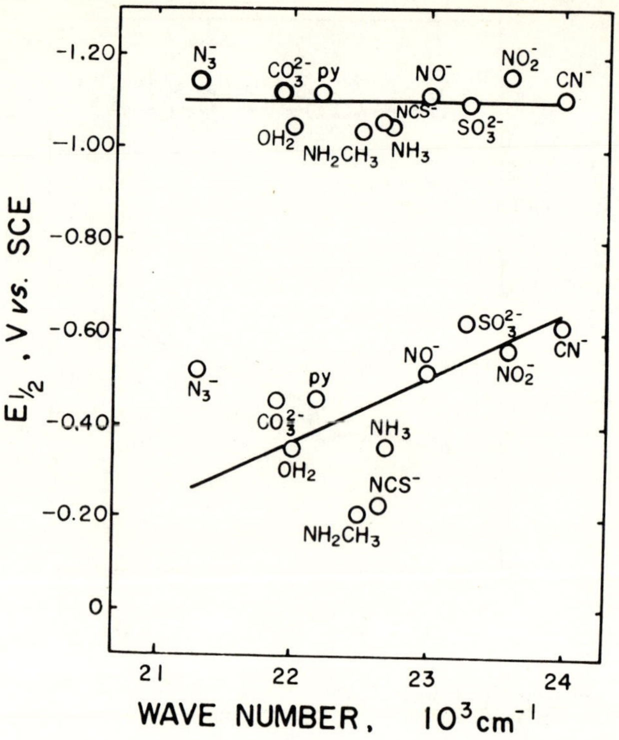

FIG. 4.1.14. Relationship between $E_{1/2}$ and ν_{max} for the trans-$[Co(CN)X(NH_3)_4]$ complexes in a 0.5-M Na_2SO_4 unbuffered solution (25°C).

gives a polarographically irreversible one electron reduction wave even in the presence of an excess of ethylenediamine [89]. This type of polarographic process belongs to a group in which the primary product of the electron-transfer reaction is labile toward solvolysis (i.e., aquation) and undergoes subsequent chemical aquations in aqueous solutions with excess ethylenediamines [276]. The $[Co\ en_3]^{3+}/[Co\ en_3]^{2+}$ couple in an excess of ethylenediamine has been shown to be reversible potentiometrically by Bjerrum [1] and polarographically by Laitinen and Grieb [87]. The $[Co\ en_3]^{3+}/[Co\ en_3]^{2+}$ couple, however, shows irreversible log plot slopes in the absence of ethylenediamine although the electron-transfer reaction is reversible as shown by alternating current polarography [92, 284]. Konrad and Vlček [105] confirmed that the corrected cathodic half-wave potential is identical with the redox potential measured at a platinum electrode. They found that when the equilibrium reaction of $[Co\ en_3]^{2+}$ with water shifts toward the aquated Co(II) species, the anodic limiting current has a kinetic character. The slowest reaction in the consecutive complex formation is the attachment of the third ethylenediamine molecule to the $[Co\ en_2(OH_2)]$ complex. The redox potential of the system was found to depend upon the concentration of free ethylenediamine present and not to depend on the pH of the solution (except insofar as the pH determines the concentration of free ethylenediamine).

A reversible process of this type was recently found for the cis-$[Co(CN)_2tren]^+$/cis-$[Co(CN)_2tren]$ system in aqueous solutions with free ligands (tren) in excess and in DMSO (where the tetradentate ligand "tren" is triethylenetetramine). That is, the cathodic and the anodic half-wave potentials were identical with the redox potential (-1.062 V vs SCE) in the DMSO containing 0.1 M $[(C_2H_5)_4N]ClO_4$, and the slopes of log plot were 59.8 ± 0.05 mV. On the other hand, the cathodic and the anodic half-wave potentials were -0.877 V (vs SCE) at 25°C in 0.5 M Na_2SO_4 aqueous solutions with 0.021 to 0.084 M tren in excess [103].

As to the polarographic behavior of the amine cobalt(III) complexes, several papers have been published. For instance, Henney, Holtzclaw, and Larson [98] studied the aquated bis (ethylenediamine)cobalt(III) complexes, such as $[Co(OH_2)(NH_3)en_2]^{3+}$, $[Co(OH_2)(NCS)en_2]^{2+}$ and $[Co(OH)(OH_2)en_2]^{2+}$ ion by polarography in connection with an investigation of solvolysis reactions following the electron-transfer process. Mason and White [278] reported on the effect of pH and free ligand concentration (NO_2^-) on the polarography of trans-$[Co(NO_2)_2en_2]^+$. Hargens, Min, and Henney [279] similarly investigated the reduction of the bis(ethylenediamine)cobalt(III) complexes with sulfite ligands (oxygen coordinated). The difference in behavior between cis- and trans-isomers was distinctly demonstrated. All of the reductions were hindered by the polarized character of the ions. Laitinen and Randles [236] and Sherwood and Laitinen [280] found anomalous impedance behavior of the $[Co\ en_3]^{3+}/[Co\ en_3]^{2+}$ and $[Co\ dien_2]^{3+}/[Co\ dien_2]^{2+}$ systems at the DME over the range 200 Hz to 20 KHz (where the tridentate ligand "dien" is diethylenetriamine). The results were interpreted in terms of specific adsorption of the reactants at a mercury surface [281]. The Senda-Delahay model [254] was found inapplicable but the Sluyters-Rehbach-Delahay model [282] yielded results consistent with the adsorption of a single species (the oxidant), a large charge-separation capacitance, and a rapid charge-transfer reaction. Fischerova, Dracka, and Meloun [283] derived the theoretical equation for the kinetics of a first-order chemical reaction of the reduction product, $[Co\ en_3]^{2+}$, in the presence of ethylenediamine and applied it to the $[Co\ en_3]^{3+}/[Co\ en_3]^{2+}$ system. The homogeneous rate constant of the following chemical reaction was calculated from the polarographic current (1.85×10^{-7} l/mole at 20°C). Several other reports on the ethylenediamine and triethylenetetramine cobalt complexes are found in the literature [285-291].

4.1.4. Dicyano Cobalt(III) Complexes

The cobalt(III) complexes of the $[Co(CN)_2N_4]$ type have been found to undergo the stepwise reduction Co(III) → Co(II) → Co(I) → Co(0) at the DME in aprotic solvents such as DMSO [where N_4 represents four nitrogen donors, which may come from one tetradentate ligand;

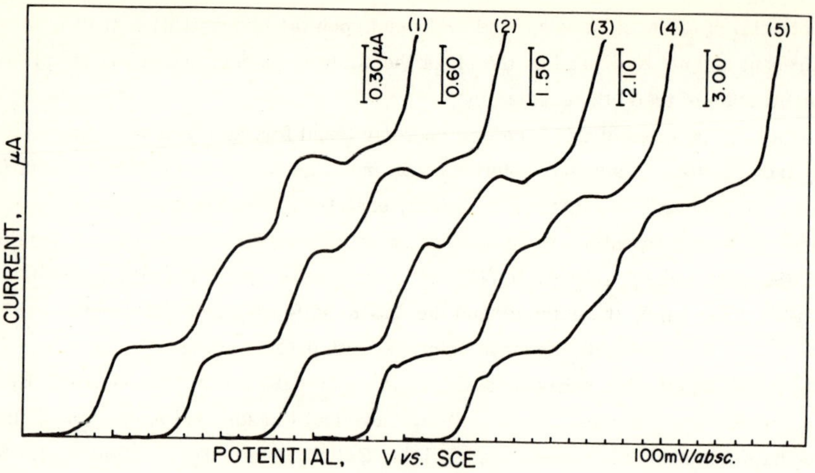

FIG. 4.1.15. The effect of the complex concentration on the current-potential curve for trans-$[Co(CN)_2en_2]NO_3$ in DMSO containing 0.1 M $[(C_2H_5)_4N]ClO_4$. Concentration (M at 25°C): (1) 0.52×10^{-3}, (2) 1.04×10^{-3}, (3) 2.61×10^{-3}, (4) 3.65×10^{-3}, and (5) 5.21×10^{-3}.

tren (triethylenetetramine), from two chelate ligands, en, pn, tn, dip, or phen; or from four ammonia ligands] [114]. Figures 4.1.15 and 4.1.16 show typical dc polarograms in which effects of the complex concentration on the current-potential curves are given for the dicyano Co(III) complexes with ethylenediamine (σ-bonding) and with 2, 2'-dipyridyl (π-bonding) in DMSO containing 0.1 M $[(C_2H_5)_4N]ClO_4$ (25°C). The scales of the current axes are given on each curve for the different complex concentrations for convenience of comparison.

Thus the dicyano Co(III) complexes give a wave of three steps in which each step corresponds to a gain of one electron. The structure of the original complex seems to remain intact throughout the reductions:

$$[Co^{III}(CN)_2en_2]^+ \xrightarrow{e} [Co^{II}(CN)_2en_2] \xrightarrow{e} [Co^{I}(CN)_2en_2]^- \xrightarrow{e} [Co^{0}(CN)_2en_2]^{2-} \quad \text{(in DMSO)}$$

Oscillopolarograms (dE/dt vs E) curves show three indentations on both the cathodic and anodic branches, implying that the reduction is of an inert-inert type and that the Co atom still retains ligands around itself. The metallic cobalt as a final product usually shows only one indentation, similar to that in aqueous media (see Fig. 4.1.25), on the anodic branch. The $E_{1/2}$ values and electrode processes for the $[Co(CN)_2N_4]$ complexes are summarized in Table 2.2.23. The $E_{1/2}$ of the first step for trans-$[Co(CN)_2en_2]^+$ lies at a more negative potential than that for the corresponding cis-isomer, so that the resultant cis-dicyano Co(II) complex is more stable toward electrochemical reduction than the corresponding trans-complex. The effect of temperature on current-potential curves provides the most useful data for judging the stabilities of degraded cobalt complexes in DMSO. For example, the

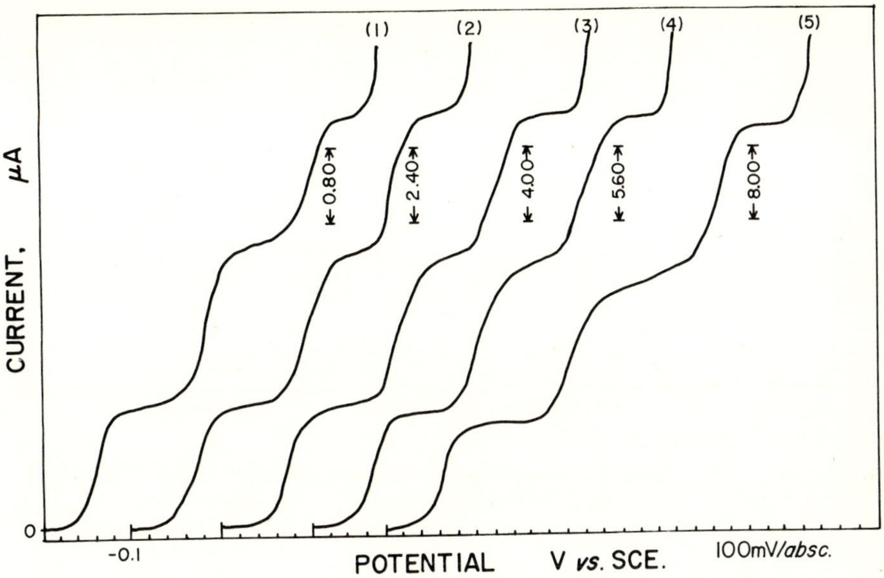

FIG. 4.1.16. The effect of the complex concentration on the current-potential curve for cis-[Co(CN)$_2$dip$_2$]NO$_3$.7H$_2$O in DMSO containing 0.1 M [$(C_2H_5)_4N$]ClO$_4$ (25°C). Concentration (M): (1) 10^{-3}, (2) 3 x 10^{-3}, (3) 3 x 10^{-3}, (4) 7 x 10^{-3}, and (5) 10^{-2}.

thermal dependence of current-potential curves for the above two complexes are given in Figs. 4.1.17 and 4.1.18. These reductions of an inert-inert type in nonaqueous solutions change immediately to those of an inert-labile type with the addition of water to the DMSO solution. The dicyano Co(II) complex of the labile type forms several aquated equilibrium mixtures in aqueous media, except for the cases of the complexes with 2,2'-dipyridyl (dip) and 1,10-phenanthroline (phen) which involve π-bonding. For instance, the cis-[Co(CN)$_2$en$_2$]$^+$ ion gives an irreversible one electron reduction wave, Co(III) → Co(II), in noncomplexing media [101]. The reduction of Co(II) to Co(metal) cannot be observed by conventional polarography, but can be seen by oscillopolarography, as shown in Fig. 4.1.25. The presence of ethylenediamine in excess makes it possible to prevent the aquation of the Co(II) complex to some extent, but it results in the overall rearrangement of cyanides in the labile Co(II) state, since the cobalt ion essentially tends to take the ligands of the same bonding character as far as possible. Hence the Co(II) complex is broken up into two kinds of more stable Co(II) species; one of the species is the binuclear tetracyanocobaltate(II) complex, [Co^{II}(CN)$_4$-en-Co^{II}(CN)$_4$]$^{4-}$, which is reduced to the Co(0) state via the Co(I) species (polarographically active), and the other is an equilibrium mixture of hydroxoethylenediamine Co(II) complexes which give no polarographic reduction wave, Co(II) → Co(metal), as described above (polarographically inactive). The further addition of free cyanide ions to this solution causes

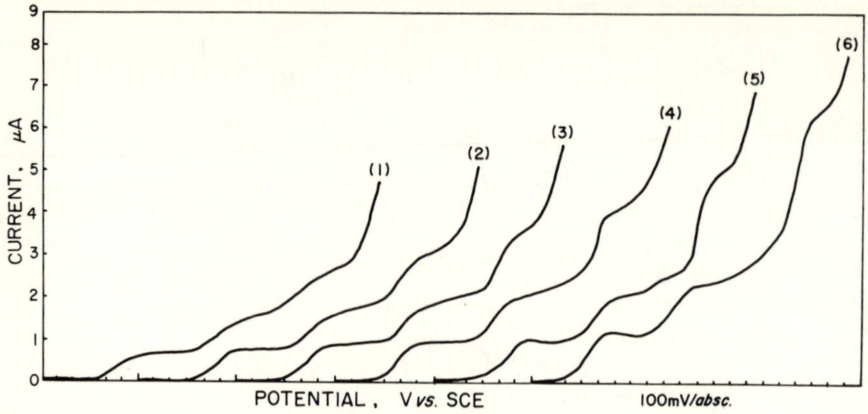

FIG. 4.1.17. The effect of temperature on current-potential curves for trans-$[Co(CN)_2en_2]NO_3$ (5.2×10^{-3} M) in DMSO containing 0.1 M $[(C_2H_5)_4N]ClO_4$. Temperature (°C): (1) 20, (2) 30, (3) 40, (4) 50, (5) 60, and (6) 70. Recorded from -0.50 V vs SCE.

conversion of all the inactive species into those of the former. As a result, the apparent 1/2 electron reduction wave for the second step becomes double in its wave height. The electrode processes in aqueous solutions with free ligands (en) in excess were concluded to be:

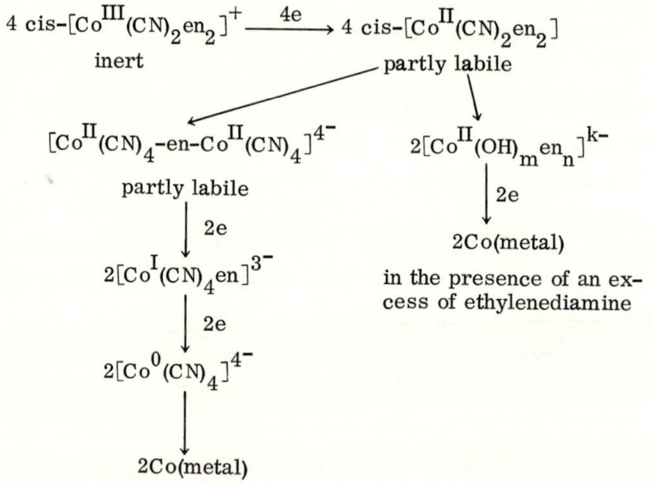

For the dE/dt vs E curves (in Fig. 4.1.25), three indentations are observable on the cathodic branch, corresponding to the above reductions of three steps in aqueous media with free ligands in excess. On the other hand, only one indentation is observed on the anodic branch, indicating that the final product is metallic cobalt without ligands.

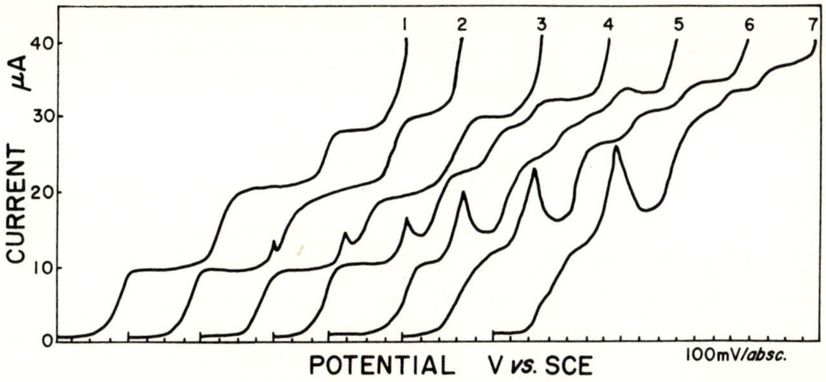

FIG. 4.1.18. The effect of temperature on current-potential curves for cis-$[Co(CN)_2dip_2]NO_3 \cdot 7H_2O$ (5×10^{-3} M) in DMSO containing 0.1 M $[(C_2H_5)_4N]ClO_4$. Temperature (°C): (1) 20, (2) 25, (3) 30, (4) 40, (5) 50, (6) 60, and (7) 70.

The presence of ethylenediamine in excess is essential for the existence of the $[Co^{II}(CN)_4\text{-en-}Co^{II}(CN)_4]^{4-}$ ion in aqueous media, but cyanide ions in excess are not necessarily needed; this Co(II) complex should, therefore, be regarded as "partly labile" toward solvolysis.

The presence of both CN^- ions and en in excess makes it possible to prevent completely the formation of aquated Co(II) species, $[Co^{II}(OH)_m en_n]^{k-}$, and to simplify further the electrode processes in aqueous media; the overall rearrangement of cyanides, leading to the formation of the binuclear tetracyanocobalt(II) complex can be detectable.

$$\underset{\text{inert}}{2\text{cis-}[Co^{III}(CN)_2en_2]^+} \xrightarrow{2e} \underset{\text{labile}}{2\text{cis-}[Co^{II}(CN)_2en_2]^\circ}$$

$$\downarrow$$

$$[Co^{II}(CN)_4\text{-en-}Co^{II}(CN)_4]^{4-}$$

$$\downarrow 2e$$

$$2[Co^{I}(CN)_4en]^{3-}$$

$$\downarrow 2e$$

$$2[Co^{0}(CN)_4]^{4-}$$

$$\downarrow$$

2Co(metal)

in the presence of both CN^- ions and en

The simplification of electrode processes, however, does not reach completeness with the inert-inert type in DMSO. From a comparison of the electrode processes of cis-$[Co(CN)_2en_2]^+$

ion in DMSO, in the presence of both CN^- ions and en, in the presence of en alone, and in a noncomplexing medium (0.5 M Na_2SO_4), it can clearly be demonstrated that the processes in DMSO are simplest and can be best understood. The behavior in DMSO provides information for interpreting the more complicated processes in aqueous media, since the pathway of an inert-inert type provides a model for the reactions of many cobalt complexes at least, provided that the high lability of the ligands bound to coordination sites can be fastened during reduction. Conversely, the cobalt complex tends to convert into more stable species through rearrangements or solvolysis of ligands, maintaining equilibrium with the coordinating ions in solution when it transits a labile state during the reduction pathway. The half-wave potentials and the electrode processes for the dicyano Co(III) complexes in aqueous media are summarized in Table 2.2.15.

4.1.5. Tricyano Cobalt(III) Complexes

Two possible steric isomers are considered: one is the meridional form of the Co(III) complex in which three cyanides are located at the coplanar coordination sites, and the other the facial form of the complex in which three cyanides lie at the nonplanar positions. In connection with the planarity of the Co(I) complex of electronic structure d^8, it is of prime importance whether three cyanides occupy the coplanar positions or not, since one would expect that the tricyano cobalt(I) complex with a mer-form is more stable than that with the corresponding fac-form upon reduction in DMSO. Indeed, we did find that the fac- and mer-$[Co^{III}(CN)_3dien]$ complexes give well-defined two-step waves at the DME in the DMSO (100%) containing 0.1 M tetraethylammonium perchlorate. The steps, each corresponding to an acceptance of one electron, represent the reduction of Co(III) to Co(II) and that of Co(II) to Co(I), respectively [159], No further reduction to Co(0) takes place over the potential range of 0 to -2.70 V (vs SCE). A linear dependence of the current upon the complex concentration was confirmed between 5×10^{-4} and 10^{-2} M for the fac-isomer, and between 5×10^{-4} and 3.5×10^{-3} M for the mer-isomer, since the solubility of the latter in DMSO is much smaller than that of the former. As predicted, a large difference was found in the half-wave potentials of the waves of the fac- and mer-isomers. The values for the first and the second steps of the fac-isomer lie at potentials more negative than those for the mer-isomer by 280 and 90 mV, respectively, indicating that the fac-Co(III) and Co(II) species are more stable toward the electrochemical reduction than are the mer ones. Such a large difference in $E_{1/2}$ has not previously been reported for steric isomers. This result can also be invoked to emphasize the larger stability of the tricyano Co(I) complex in the mer-form (see Table 2.2.23). The three cyanides located on coplanar coordination sites (mer-form) may contribute much more to the stability of the univalent cobalt than those in a nonplanar configuration do, as can be inferred from the electronic structure of the Co(I) complex. The net electrode reaction can be fully interpreted in terms of an inert-inert type reduction without any structural changes:

$$[Co^{III}(CN)_3dien] \xrightarrow{e} [Co^{II}(CN)_3dien]^- \xrightarrow{e} [Co^{I}(CN)_3dien]^{2-} \quad \text{(in DMSO)}$$

Here it should be strongly emphasized that the electrode processes of these tricyano Co(III) complexes in aqueous media are much more complicated than those of the dicyano Co(III) complexes, even in the presence of free ligands in excess where interpretation of polarograms was impossible due to the large maxima. This result also shows the advantage of adopting nonaqueous polarography to simplify the electrode process by converting a reduced labile complex into an inert one.

4.1.6. Tetracyano Cobalt(III) Complexes

The tetracyano Co(III) complexes not only give a great difference in half-wave potentials, but also quite a different electrode process depending upon whether the four cyanides lie at the coplanar coordinate positions or not. The trans-tetracyano Co(III) complex in which four cyanides are located on the coplanar positions undergoes direct or stepwise reduction to the Co(I) state and not to the Co(0) state, while the cis-tetracyano Co(III) complex in which four cyanides occupy nonplanar coordinate positions undergoes further reduction from the Co(III) to the Co(0) state via the Co(I) complex at the DME in aqueous media. Most of the tetracyanocobaltate(III) complexes with inorganic ligands are only slightly soluble in aprotic solvents such as DMSO, with a few exceptions, i.e., complexes with organic ligands such as en, $P \equiv \phi_3$ (triphenylphosphine), or $As \equiv \phi_3$ (triphenylarsine).

The trans-$Na_5[Co(SO_3)_2(CN)_4] \cdot 3H_2O$ complex gives a well-defined wave of one step, corresponding to a two-electron reduction from the Co(III) to the Co(I) state at the DME in 0.5 $\underline{M}$ Na_2SO_3; no further reduction to the Co(0) state occurs over the potential range of 0 to -2.0 V (vs SCE):

$$\text{trans-}[Co^{III}(OSO_2)_2(CN)_4]^{5-} \xrightarrow{2e} \text{trans-}[Co^{I}(OSO_2)_2(CN)_4]^{7-}$$

in the presence of SO_3^{2-} ions.

Figure 4.1.19 shows the temperature dependence of the current-potential curve for trans-$Na_5[Co(SO_3)_2(CN)_4] \cdot 3H_2O$ (5×10^{-3} $\underline{M}$) in 0.5 $\underline{M}$ Na_2SO_3. Below 40°C, a linear dependence of the diffusion current with temperature was confirmed, suggesting a high stability of the tetracyanocobaltate(I) complex. The presence of SO_3^{2-} ions in excess, however, is not necessarily needed for the direct reduction from the Co(III) to the Co(I) state to appear in dc polarography, but the aquation may occur when the disulfito complex goes through a labile Co(II) state:

$$\text{trans-}[Co^{III}(SO_3)_2(CN)_4]^{5-} \xrightarrow{e} \text{trans-}[Co^{II}(SO_3)_2(CN)_4]^{6-}$$
$$\text{partly labile} \searrow \qquad \downarrow e$$
$$\text{trans-}[Co^{II}(OH_2)_2(CN)_4]^{2-} \qquad \text{trans-}[Co^{I}(SO_3)_2(CN)_4]^{7-}$$
$$\downarrow e$$
$$\text{trans-}[Co^{I}(OH_2)_2(CN)_4]^{3-}$$

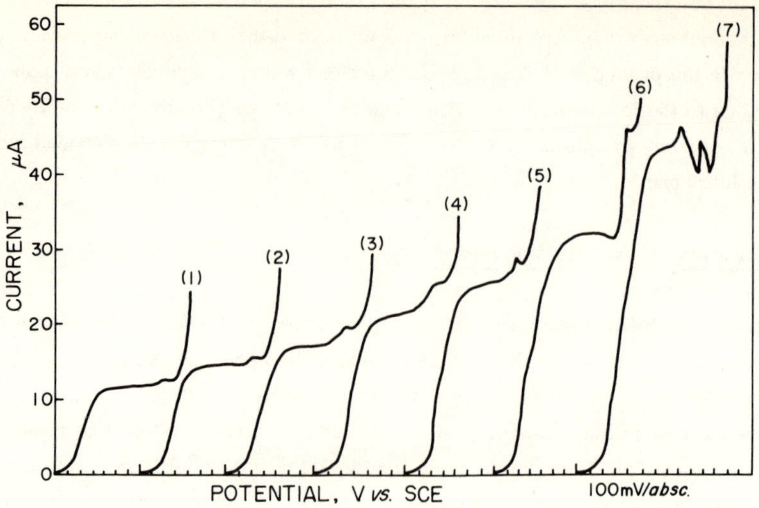

FIG. 4.1.19. The effect of temperature on the current-potential curve for trans-$Na_5[Co(SO_3)_2(CN)_4].3H_2O$ (5×10^{-3} M) in 0.5 M Na_2SO_3 (free ligands in excess). Temperature (°C): (1) 0, (2) 10, (3) 20, (4) 30, (5) 40, (6) 50, and (7) 60. Recorded from -1.00 V vs SCE.

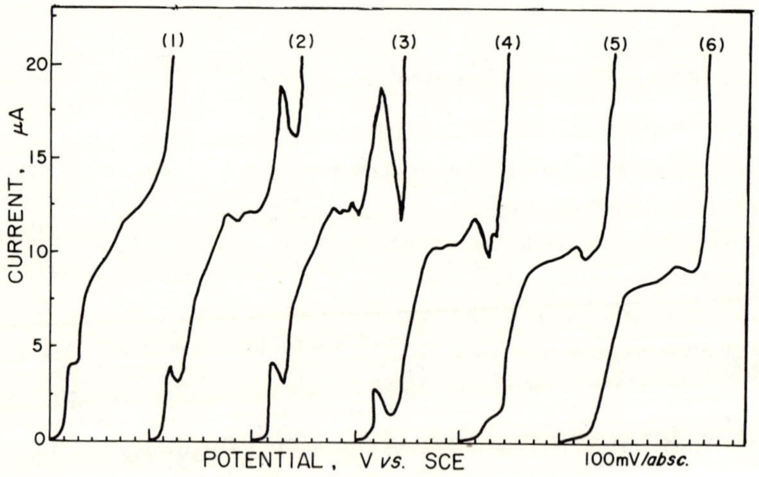

FIG. 4.1.20. The effect of sulfite ion concentration on the current-potential curve for trans-$Na_5[Co(SO_3)_2(CN)_4].3H_2O$ under a unit ionic strength ($Na_2SO_3 + Na_2SO_4$). Concentration of SO_3^{2-} ions (M): (1) 0, (2) 0.005, (3) 0.01, (4) 0.05, (5) 0.1, and (6) 0.5. Recorded from -0.90 V vs SCE.

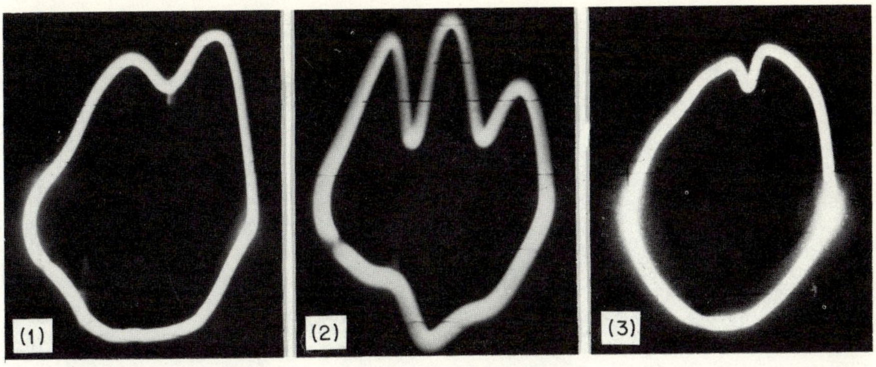

FIG. 4.1.21. Oscillopolarograms of dE/dt - E curve for mixed cyano Co(III) complexes (10^{-2} M). (1) trans-$Na_5[Co(SO_3)_2(CN)_4].3H_2O$ in 0.5 M Na_2SO_3 (free ligands in excess), (2) trans-$Na_5[Co(SO_3)_2(CN)_4].3H_2O$ in 0.5 M Na_2SO_4 (a noncomplexing medium), and (3) $K_4[Co(SO_3)(CN)_5].3H_2O$ in 0.5 M Na_2SO_4.

in the absence of SO_3^{2-} ions. Hence the electrode processes become more complicated in sulfate solutions than those in sulfite solutions. Clearer evidence for aquation is obtained by observation of the dE/dt vs E curves in the presence or absence of sulfite ions, as shown in Fig. 4.1.21. Only one cathodic indentation, Co(III) → Co(I), is observed in the presence of SO_3^{2-} ions, while two cathodic indentations, each corresponding to the reductions of Co(III) → Co(II) and of Co(II) → Co(I), are observable in the absence of SO_3^{2-}, the result of the latter being consistent with that for trans-$K[Co(OH_2)_2(CN)_4].3/4\ H_2O$ in 0.5 M Na_2SO_4 solutions. Figure 4.1.20 shows the effect of sulfite ion concentration on the current-potential curve for the trans-disulfito complex [116].

On the other hand, the cis-$Na_2K_3[Co(SO_3)_2(CN)_4].2.5H_2O$ complex [119] gives a wave of two steps in which each represents the reduction of Co(III) → Co(I) and of Co(I) → Co(0), respectively, in 0.5 M Na_2SO_3 or in 0.5 M Na_2SO_4 solutions (see Table 2.2.16). The cis-isomer is different from the corresponding trans-isomer in that further reduction from Co(I) to Co(0) takes place. The ratio of the wave height of the first step to that of the second is always approximately 2:1 since all the limiting currents are diffusion-controlled:

$$\text{cis-}[Co^{III}(SO_3)_2(CN)_4]^{5-} \xrightarrow{2e} \text{cis-}[Co^{I}(SO_3)_2(CN)_4]^{7-} \xrightarrow{e} \text{cis-}[Co^{0}(SO_3)_2(CN)_4]^{8-}$$

in the presence of SO_3^{2-} ions [356].

For the dE/dt vs E curves in the presence of SO_3^{2-} ions, two cathodic indentations, Co(III) → Co(I) → Co(0), are observable, but no anodic indentation exists for the cis-disulfito complex. This result, no anodic oxidation for the cis-isomer, was the same as that for the corresponding trans-isomer, suggesting that the cis-$[Co^0(SO_3)_2(CN)_4]^{8-}$ and

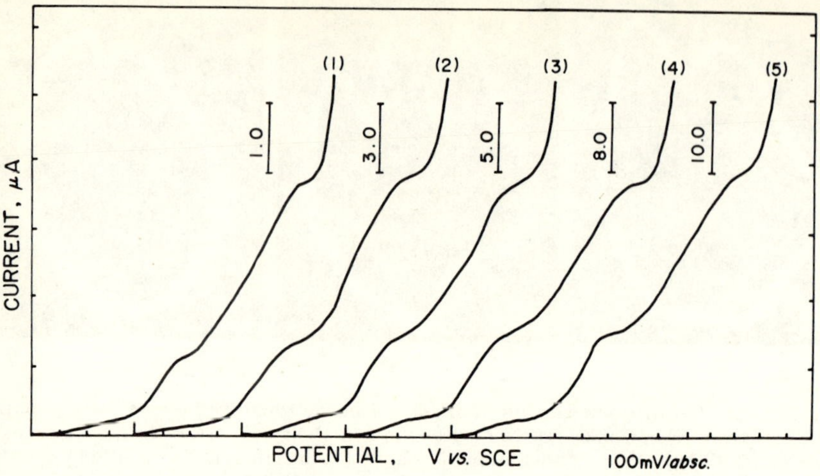

FIG. 4.1.22. The effect of the complex concentration on the current-potential curve for trans-K[$Co(OH_2)_2(CN)_4$].3/4 H_2O in 0.5 M Na_2SO_4 (25°C). Concentration (M): (1) 10^{-3}, (2) 3 x 10^{-3}, (3) 5 x 10^{-3}, (4) 8 x 10^{-3}, and (5) 10^{-2}. Recorded from -0.60 V vs SCE.

trans-[$Co^{I}(SO_3)_2(CN)_4]^{7-}$ ions are not electrochemically oxidized at the DME. Figure 4.1.21 shows the current-controlled oscillopolarograms obtained in the presence or in the absence of SO_3^{2-} ions in excess. These results are quite consistent with those obtained by using a Kalousek commutator, from which the cis-[$Co^{I}(SO_3)_2(CN)_4]^{7-}$ ion was also found not to be electrochemically oxidizable in the presence of SO_3^{2-} ions, since no anodic wave was observed.

In contrast to these results obtained in the presence of SO_3^{2-} ions, the anodic oxidations, Co(0) → Co(I) → Co(III), for the cis-isomer and Co(I) → Co(III) for the corresponding trans-isomer are observed in the absence of SO_3^{2-} ions both on the Kalousek polarograms and on the dE/dt vs E curves, suggesting that the aquated tetracyano Co(I) and Co(0) complexes are all oxidizable electrochemically at the DME.

The trans-K[$Co(OH_2)_2(CN)_4$].3/4 H_2O and trans-K[$Co(CN)_4(NH_3)_2$].H_2O complexes are reduced in two steps to the Co(I) state via the Co(II) complex in 0.5 M Na_2SO_4. A comparison of $E_{1/2}$ values reveals that the aquation of ammonia-containing species is not appreciable during reduction in 0.5 M Na_2SO_4 (see Table 2.2.16). Figure 4.1.22 shows one example of the variation with complex concentration of the current-potential curve for trans-[$Co(OH_2)_2(CN)_4]^{-}$ ion. The electrode processes are:

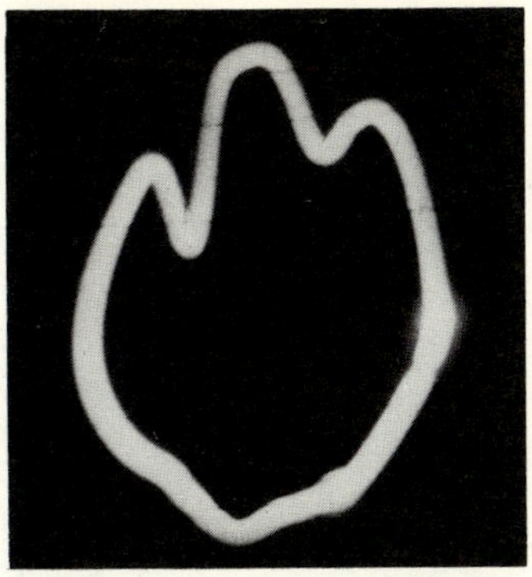

FIG. 4.1.23. The current-controlled oscillopolarograms of trans-K[$Co(OH_2)_2(CN)_4$]. 3/4 H_2O (5×10^{-3} M) in 0.5 M Na_2SO_4 (a noncomplexing medium).

$$\text{trans-}[Co^{III}(OH_2)_2(CN)_4]^- \xrightarrow{e} \text{trans-}[Co^{II}(OH_2)_2(CN)_4]^{2-} \xrightarrow{e} \text{trans-}[Co^{I}(OH_2)_2(CN)_4]^{3-}$$

$$\text{trans-}[Co^{III}(CN)_4(NH_3)_2]^- \xrightarrow{e} \text{trans-}[Co^{II}(CN)_4(NH_3)_2]^{2-}$$

$$\text{trans-}[Co^{II}(CN)_4(NH_3)_2]^{2-} \rightarrow \text{trans-}[Co^{II}(CN)_4(OH)_2]^{4-} + 2\ NH_4^+ \text{ etc.}$$

$$\text{trans-}[Co^{II}(CN)_4(OH)_2]^{4-} \xrightarrow{e} \text{trans-}[Co^{I}(CN)_4(OH)_2]^{5-}$$

$$\text{trans-}[Co^{II}(CN)_4(NH_3)_2]^{2-} \xrightarrow{e} \text{trans-}[Co^{I}(CN)_4(NH_3)_2]^{3-}$$

for the dE/dt vs E curves, two cathodic indentations, Co(III) → Co(II) → Co(I), are observed, but no anodic one was found for the diaquo complex, suggesting neither further reduction of the planar Co(I) complex, $[Co^I(CN)_4]^{3-}$ or $[Co^I(OH_2)_2(CN)_4]^{3-}$, to the Co(0) state nor anodic oxidation from the Co(I) to the Co(III) state occurs at the DME (see Fig. 4.1.23). Note that the anodic behavior of the trans-diaquo complex is quite different from that due to the aquated species of the trans-$Na_5[Co(SO_3)_2(CN)_4]$.$3H_2O$ complex in 0.5 M Na_2SO_4 since the anodic oxidation of the Co(I) complex is possible. Since the trans-K[$Co(CN)_4(NH_3)_2$].H_2O complex gives anodic indentations or waves of Co(I) → Co(II) → Co(III) on both the dE/dt vs E curve and Kalousek polarograms, the trans-K[$Co(OH_2)_2(CN)_4$].3/4 H_2O complex might take a dimeric configuration in solution.

The cis-Na[$Co(CN)_4en$].3.$5H_2O$ complex is soluble both in DMSO and in water. In the DMSO containing 0.1 M tetraethylammonium perchlorate, the cis-$[Co(CN)_4en]^-$ ion is reduced in two steps to the Co(I) state via the Co(II) complex at the DME; no further reduction to the Co(0) complex takes place at the DME in DMSO:

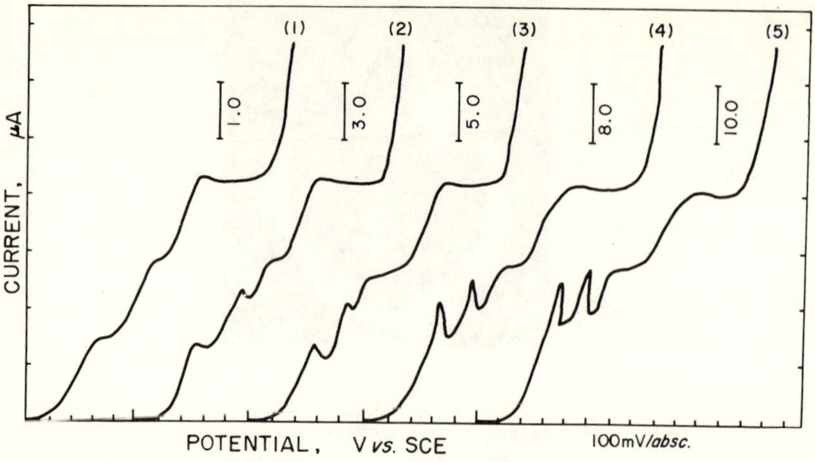

FIG. 4.1.24. The effect of complex concentrations on the current-potential curve for cis-Na[Co(CN)4en].3.5 H_2O in DMSO containing 0.1 M $[(C_2H_5)_4N]ClO_4$ (25°C). Complex concentration (M): (1) 10^{-3}, (2) 3×10^{-3}, (3) 5×10^{-3} (4) 8×10^{-3}, and (5) 10^{-2}. Recorded from -1.40 V vs SCE.

$$\text{cis-}[Co^{III}(CN)_4en]^- \xrightarrow{e} \text{cis-}[Co^{II}(CN)_4en]^{2-} \xrightarrow{e} \text{cis-}[Co^{I}(CN)_4en]^{3-}$$

Figure 4.1.24 illustrates the wave of three steps in which the last step is due to the reduction of sodium ion involved in the complex. The small maxima are attributed to the deposition of the Co(II) product having a smaller solubility in DMSO. No anodic oxidation was observed for the cis-$[Co^{I}(CN)_4en]^{3-}$ ion on the dE/dt vs E curves or on Kalousek polarograms in DMSO.

In contrast to the simple electrode processes of an inert-inert type in DMSO, more complicated processes are predicted in aqueous media due to the increase in the lability of the reduced Co(II) and Co(I) complexes. In fact, under the conditions where aquation is prevented to an extent, the following processes proceed at the DME:

$$2\text{cis-}[Co^{III}(CN)_4en]^- \xrightarrow{2e} [Co^{II}(CN)_4\text{-en-}Co^{II}(CN)_4]^{4-}$$

$$\downarrow 2e$$

$$[Co^{I}(CN)_4\text{-en-}Co^{I}(CN)_4]^{6-}$$

$$\downarrow 2e$$

$$2[Co^{0}(CN)_4]^{4-} + en$$

in the presence of 2 M en in excess

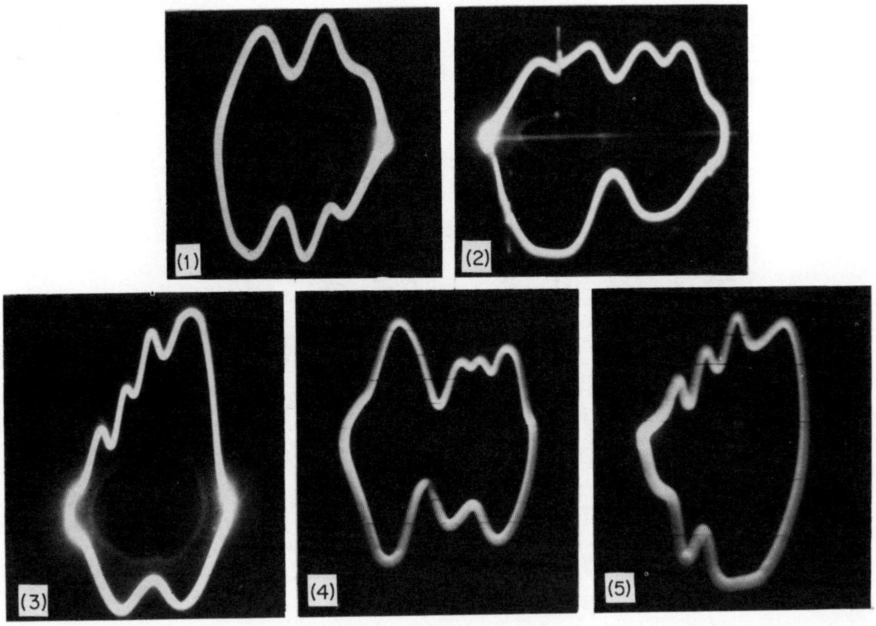

FIG. 4.1.25. Oscillopolarograms of Heyrovský-Forejt type for dicyano and tetracyano Co(III) complexes (10^{-2} M complex ion. (1) cis-[Co(CN)$_2$en$_2$]NO$_3$ in 0.5 M Na_2SO_4 alone, (2) cis-[Co(CN)$_2$en$_2$]NO$_3$ in 0.5 M Na_2SO_4 + 1 M en, (3) trans-[Co(CN)$_2$en$_2$]NO$_3$ in 0.5 M Na_2SO_4 alone, (4) cis-Na[Co(CN)$_4$en].3.5 H_2O in 0.5 M Na_2SO_4 alone, and (5) cis-Na[Co (CN)$_4$en].3.5 H_2O in 0.5 M Na_2SO_4 + 1 M en (25°C). Measured by Polaroscope P 576, Kovo Inc., Prague.

There exist three cathodic and anodic indentations due to a time delay or "Verweilung" on the charging and discharging current, indicating that the final product is a soluble Co(0) complex bonded with ligands, since metallic cobalt usually gives only one indentation due to the dissolution to Co(II) ions from the cobalt amalgam [118]. These observations are also fully consistent with those obtained for the K_4[Co^{II}(CN)$_4$-en-Co^{II} (CN)$_4$]. H_2O complex [277] under the same conditions (see Fig. 4.1.25).

In noncomplexing media, such as in 0.5 M Na_2SO_4, the wave of the three electron reduction, corresponding to the apparently direct reduction from the Co(III) to the Co(metal), was observed as shown in Fig. 4.1.26. Oscillopolarographic studies, however, revealed that the apparent direct reduction, Co(III) → Co(metal), actually takes place in two steps to the metal at the DME. These electrode processes of an inert-labile type involve so many aquated labile species and are so complicated that they can no longer be discussed in relation to their structures. Figure 4.1.26 shows current-potential curves measured under the same experimental conditions. The wave height is roughly proportional to the number of electrons participating in the electrode processes, the $E_{1/2}$ values of which are summarized in Tables 2.2.16 and

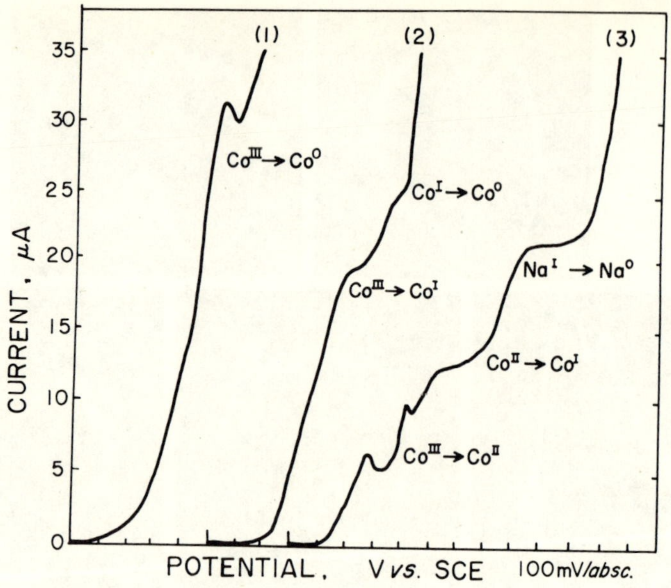

FIG. 4.1.26. Typical current-potential curves obtained in three sorts of supporting electrolytes under the same experimental conditions where the same complex concentration, the same capillary, etc., were used. (1) 0.5 M Na_2SO_4 only, recorded from -0.80 V; (2) 0.5 M Na_2SO_4 + 1 M en, recorded from -0.80 V; and (3) DMSO containing 0.1 M $[(C_2H_5)_4N]ClO_4$, recorded from -1.40 V (vs SCE).

2.2.23. From a comparison of the behavior in the three different supporting electrolytes, one can see the merit of adopting aprotic solvents for interpretation of more complicated processes in aqueous media. These results show that the tetracyano Co(III) complexes with a cis-configuration are reduced to the Co(0) state via the Co(I) complex, whereas those with a trans-configuration are reduced to the Co(I) state and not to the Co(0) complex polarographically in aqueous media.

4.1.7. Pentacyano Cobalt(III) Complexes

An outline of the electrode processes has been presented in Section 3.1 from the point of view of kinetics in connection with an estimate of kinetic parameters. They will be presented here from the viewpoint of their structures. As shown in Fig. 4.1.21, all the pentacyanocobaltate(III) complexes of the $[Co^{III}X(CN)_5]$ type give one or two cathodic indentations, corresponding to a one or two electron reduction in conventional polarography; no anodic indentations are found, indicating neither further reduction from the Co(I) to the Co(0) state nor anodic oxidation from the Co(I) to the Co(III) state occurs electrochemically at the DME [120, 122, 123], where the ligand, X, denotes the ion I^-, Br^-, Cl^-, SCN^-, N_3^-, NO_2^-, NO^-, SO_3^{2-},

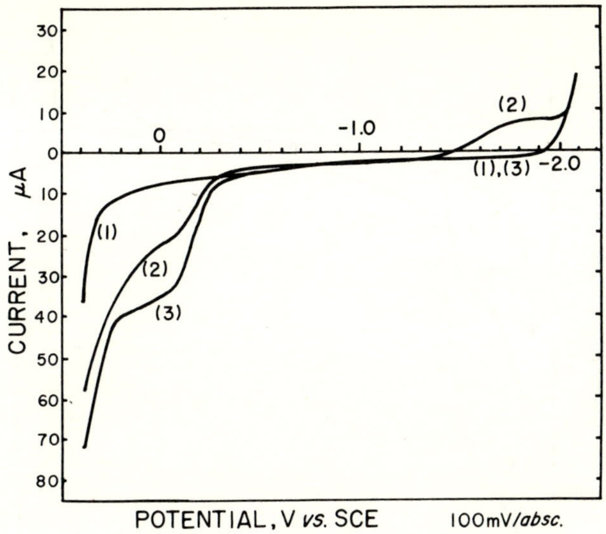

FIG. 4.1.27. Kalousek polarograms at the fixed potential of -1.60 V vs SCE obtained for (1) 0.5 M Na_2SO_4 alone, (2) 5 × 10^{-3} M $K_4[Co(S_2O_3)(CN)_5]$ dissolved in 0.5 M Na_2SO_4, and (3) 5 × 10^{-3} M $Na_2S_2O_3 \cdot 5H_2O$ dissolved in 0.5 M Na_2SO_4 aqueous solution (25°C).

$S_2O_3^{2-}$ or NCS^-, or the molecule OH_2. Similarly, no anodic waves are found in Kalousek polarograms for all the pentacyano Co(I) complexes in 0.5 M Na_2SO_4. This situation favors the detection of the anodic wave of free ligands, X^{m-}, liberated from the pentacyano complex during reduction. Figures 4.1.27 to 4.1.29 show typical Kalousek polarograms for the thiosulfato-, sulfito-, and bromopentacyano complexes which were measured by holding the DME at a fixed potential of -1.6 or -1.5 V (vs SCE) and switching to more positive potentials at a frequency of 11.8 Hz. Anodic waves due to free anions, such as $S_2O_3^{2-}$, SO_3^{2-} or Br^-, that are liberated from the complex are clearly observed since no anodic wave of the Co(I) complex exists, suggesting that the liberation of the sixth ligand, X, occurs during the reduction. The $E_{1/2}$ values for the anions released from the pentacyano complexes agree with that for the corresponding free anion of the potassium or sodium salt. The situation was the same for all the anions of the sixth ligand, X.

The composition and structure of the pentacyanocobaltate(I) complex formed by the reduction of pentacyano-cobaltate(II) and -cobaltate(III) complexes is still controversial, in spite of the number of studies on this subject [208, 209]. The $[Co^I(CN)_5H]^{3-}$ formula is now widely accepted and used as the best possible structure, since it was proposed by Griffith and Wilkinson [292] on the basis of the proton NMR spectrum. Although the $[Co^I(CN)_5]^{4-}$ ion is not electrochemically oxidizable to the Co(III) complex at the DME, chemically the Co(I) cyanide complex is rapidly oxidized, evolving hydrogen, to the $[Co^{III}(OH_2)(CN)_5]^{2-}$ ion in aqueous

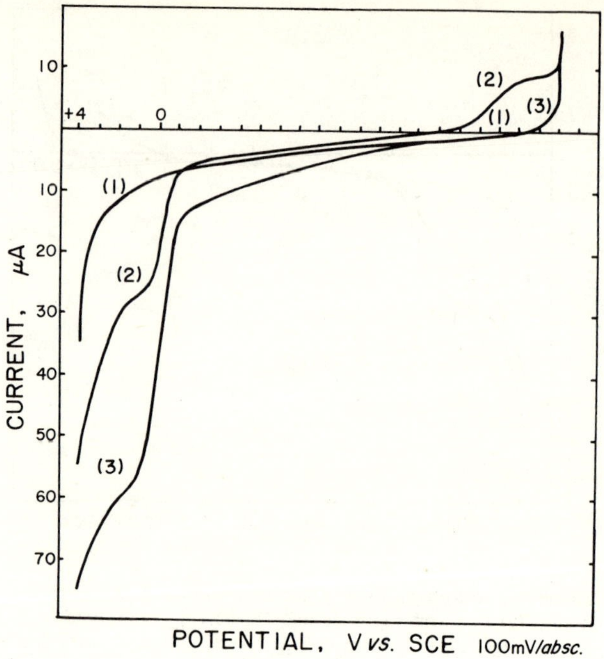

FIG. 4.1.28. Kalousek polarograms of $K_4[Co(SO_3)(CN)_5].3H_2O$ at the fixed potential of -1.60 V vs SCE obtained for 0.5 M Na_2SO_4 alone, (2) 5 x 10^{-3} M $K_4[Co(SO_3)(CN)_5].3H_2O$ dissolved in 0.5 M Na_2SO_4, and (3) 5 x 10^{-3} M Na_2SO_3 dissolved in 0.5 M Na_2SO_4 (25°C).

solutions. Hanzlík and Vlček [205] showed that the $[Co^I(CN)_5]^{4-}$ ion is a very strong base which extracts hydrogen ions from both proton donors and water molecules, and they speculate that for this reason the $[Co^I(CN)_5]^{4-}$ ion can exist only in solutions of very low proton activity. The pentacyano Co(I) complex can be obtained by reduction with $NaBH_4$ in deaerated alcohol in vacuo (10^{-7} Torr) [293]. The Co(I) complex usually tends to undergo a disproportionation reaction so that it is considered almost impossible to isolate it in a pure crystalline state from solution. Under these circumstances, the stoichiometric relation between $[Co^I(CN)_5]^{4-}$ and H^+ has not yet been determined as far as the electrode process is concerned. However, the findings that both of the polarographic reductions, Co(III) → Co(I) and $2H^+ \rightarrow H_2$, can take place simultaneously at -1.75 V (vs SCE) at the DME [120] suggested the use of the diffusion currents of both waves to follow the reaction between $[Co^I(CN)_5]^{4-}$ and H^+ by means of an amperometric titration. The result shows that the molar ratio of the $[Co^I(CN)_5]^{4-}$ ion to H^+ is always approximately 2:1, independent of the nature of the sixth ligand, X. That is, titrations of the parent complexes $K_m[Co(X)(CN)_5]$ were carried out with 0.2 N HCl under a stream of nitrogen at a fixed potential (-1.75 V vs SCE) at which the current reaches a limiting diffusion plateau, indicating the formation of the $[Co^I(CN)_5]^{4-}$ ion at the surface of the

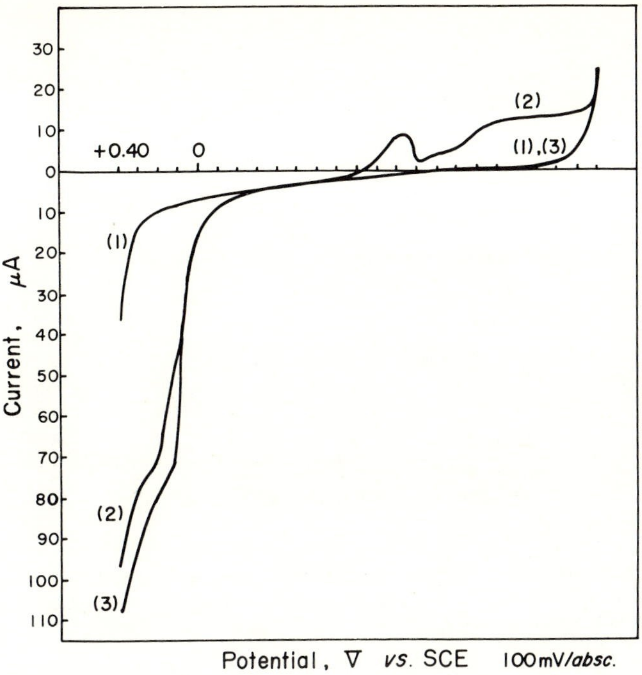

FIG. 4.1.29. Kalousek polarograms of $K_3[CoBr(CN)_5]$ at the fixed potential of -1.50 V vs SCE obtained for (1) 0.5 M Na_2SO_4 alone, (2) 5 x 10^{-3} M $K_3[CoBr(CN)_5]$ dissolved in 0.5 M Na_2SO_4, and (3) 5 x 10^{-3} M KBr dissolved in 0.5 M Na_2SO_4 (25°C).

DME. The neutral solution to be titrated was prepared by dissolving the yellow crystals of the complex at 10^{-2} M in a 0.5-M Na_2SO_4 solution which had been deaerated, in advance, by nitrogen. As one example, Fig. 4.1.30 illustrates the amperometric titration curve of the $[Co^{I}(CN)_5]^{4-}$ ion formed by the polarographic reduction of the $K_4[Co(S_2O_3)(CN)_5]$ complex (10^{-2} M) with the titrant 2 N HCl. At the beginning of titration the current [Co(III) → Co(I)] remains constant while H^+ ions are being consumed by the $[Co^{I}(CN)_5]^{4-}$ ion to be titrated, and then starts to increase after all the $[Co^{I}(CN)_5]^{4-}$ ions formed at the DME react with the H^+ ions of the titrant. The end point of the titration of $[Co^{I}(CN)_5]^{4-}$ with H^+ is indicated by the beginning of an increase in the reduction wave for the titrant, H^+ ions in excess; the new current corresponds to the reduction of excess hydrogen ions. The following reaction scheme is the most plausible:

$$2[Co^{III}(S_2O_3)(CN)_5]^{4-} \xrightarrow{4e} 2[Co^{I}(S_2O_3)(CN)_5]^{6-}$$

$$\downarrow$$

$$2[Co^{I}(CN)_5]^{4-} + 2S_2O_3^{2-}$$

$$\downarrow$$

$$[Co^{I}(CN)_5{-}H{-}Co^{I}(CN)_5]^{7-}$$

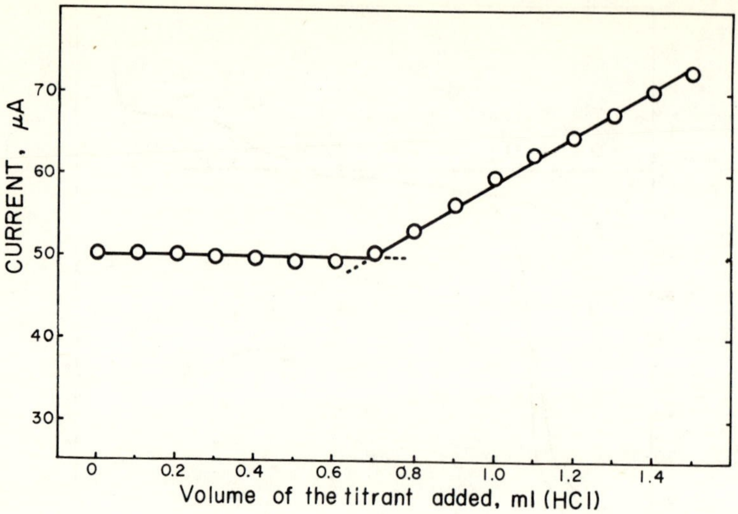

FIG. 4.1.30. Amperometric titration curve of the $K_4[Co(S_2O_3)(CN)_5]$ complex (35 ml) with 0.2 N HCl (f = 1.173) at -1.75 V (vs SCE), with a complex concentration of 10^{-2} M in 0.5 M Na_2SO_4 aqueous solution (25°C).

A blank test was carried out, in advance, individually for a 0.5-M Na_2SO_4 solution only and for the 0.5-M Na_2SO_4 solution containing 0.01 M $Na_2S_2O_3 \cdot 5H_2O$, 0.01 M Na_2SO_3, 0.01 M NaN_3, etc., but a straight line without a break was always obtained for each solution. The current ($2H^+ \rightarrow H_2$) increased linearly at -1.75 V vs SCE when 0.2 N HCl (0 to 1.5 ml) was added drop by drop to 35 ml of the above solution, indicating that no reaction of capturing H^+ ions takes place to a degree that will influence the titration curve between the anions, X^{m-}, and H^+ ions. Figure 4.1.31 exemplifies a blank of the supporting electrolyte alone.

Table 4.1.2 summarizes some of the results. Thus the stoichiometric relation between $[Co^I(CN)_5]^{4-}$ and H^+ is always 2:1, irrespective of the kind of the ligand, X, suggesting that the sixth ligand, X, is finally released from the pentacyano Co(I) complex, confirming the results obtained in the Kalousek polarograms. This result, coupled with evidence for the existence of a direct linkage between the metal and H^+ [292], also suggests that the hydridopentacyanocobaltate(I) complex takes a dimeric structure, $[Co^I(CN)_5\text{-H-}Co^I(CN)_5]^{7-}$, through a cobalt-hydrogen linkage [294].

In connection with this, similar amperometric titrations of the tetracyano Co(I) complex with 0.2 N HCl were carried out for several parent Co(III) complexes, trans-$K[Co(OH_2)_2(CN)_4] \cdot 3/4\ H_2O$, trans-$Na_5[Co(SO_3)_2(CN)_4] \cdot 3H_2O$, and cis-$Na_2K_3[Co(SO_3)_2(CN)_4] \cdot 2.5\ H_2O$, at a fixed potential at which the current reaches a limiting plateau, indicating the formation of the tetracyano Co(I) complex. No reactions, however, take place between the $[Co^I(CN)_4]^{3-}$ ion and H^+ at the DME in 0.5 M Na_2SO_4; a straight line without a break was always obtained

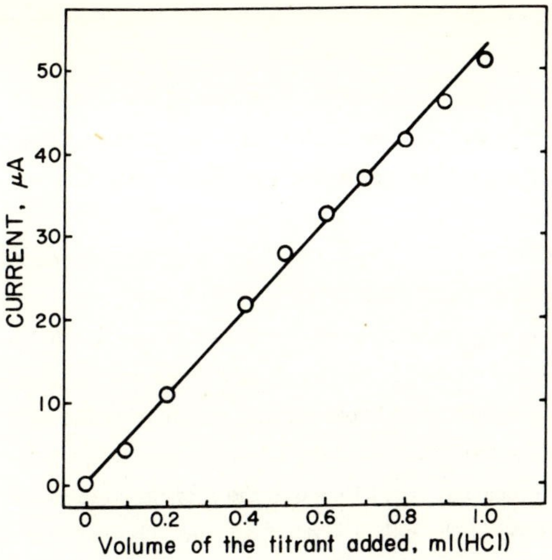

FIG. 4.1.31. Amperometric titration curve (blank test) for the background solution of 0.5 M Na_2SO_4 with 0.2 N HCl (f = 1.173) at the fixed potential of -1.75 V (vs SCE).

TABLE 4.1.2. Results of Amperometric Titrations of the $[Co^I(CN)_5]^{4-}$ Ion with 0.2 N HCl (25°C)[a]

Parent compound	Volume of HCl added at equiv point (ml)	Mole of HCl consumed at equiv point (x 10^4)	Ratio of $[Co^I(CN)_5]^{4-}/H^+$ at equiv point
$K_4[Co(S_2O_3)(CN)_5]$	0.69	1.62	2.16
$K_4[Co(SO_3)(CN)_5].3H_2O$	0.70	1.64	2.13
$K_3[CoN_3(CN)_5].2H_2O$	0.71	1.67	2.11
$K_3[CoBr(CN)_5]$	0.67	1.57	2.22

[a]Concentration of the complex: 10^{-2} M in 0.5 M Na_2SO_4 (35 ml). Titrant: 0.2 N HCl (F = 1.173).

for all the tetracyano complexes. This result led us to conclude that the tetracyano Co(I) complex, $[Co^I(CN)_4]^{3-}$ or $[Co^I(X)_2(CN)_4]^{m-}$, does not capture any protons from the solution, unlike the $[Co^I(CN)_5]^{4-}$ ion. Conversely, this fact can be invoked to emphasize that the strong basicity of the $[Co^I(CN)_5]^{4-}$ ion is due to the trans-effect of the cyanide located at the trans-position with regard to the sixth ligand, X. The Co—X bond may be strongly activated by interaction of the strong counter bond, Co= C = N$^-$. The vicinity of the Co(I) ion is extremely negatively charged, leading to liberation of the ligand, X, by repulsion. Accordingly, instead of the ligand, X, the H^+ ion would enter this electron-density rich position from the solution. Such a trans-effect of cyanide has already been observed for the trans-[Co(CN)X$(NH_3)_4$] complexes.

In highly alkaline solutions, the [$Co^I(CN)_5]^{4-}$/[$Co^{II}(CN)_5]^{3-}$ couple is reversible, with a formal electrode potential of -1.2 V (vs SCE) in 12 M NaOH [205]. The anodic wave of Kalousek polarograms, corresponding to the oxidation of the [$Co^I(CN)_5]^{4-}$ to the Co(III) state, first appears in 10 M NaOH for the $K_3[CoBr(CN)_5]$ and $K_4[Co(S_2O_3)(CN)_5]$ complexes. Hydroxide ion seems to contribute to breaking and eliminating the hydrogen bridge which shielded the sixth coordination position of the Co(I) ion against attack by the mercury electrode. Only under deprotonating conditions can the mercury electrode be accessible to the Co(I) ion from the direction of the naked sixth position. The reversibility of the Co(I)/Co(III) couple is high, the cathodic half-wave potential being close to the anodic one. In connection with the structures of the pentacyano Co(I) and Co(II) complexes, several other works can be cited [295-298].

4.1.8. Cobalt(III) Complexes with Ligands of π-Bonding Character

The Co(III) complexes with ligands of a π-bonding nature generally give stable reduced cobalt complexes of an inert type both in aqueous and nonaqueous media. In spite of such an inertness, a great difference is still found between the polarographic behavior in aqueous and nonaqueous solutions. For example, the bis(dimethylglyoximato)cobalt(III) complexes of the trans-[$CoX_2(dgH)_2$] type cited below are all reduced to the Co(I) state in DMSO; no further reduction to the Co(0) state occurs over the potential range of 0 to -2.70 V (vs SCE): H[$CoCl_2(dgH)_2$], Na[$Co(NO_2)_2(dgH)_2$]$\cdot H_2O$, H[$Co(NCS)_2(dgH)_2$], H[$Co(CN)(dgH)_2(Cl)$]$\cdot 4H_2O$, $(NH_4)_2$[$Co(CN)(dgH)_2(SO_3)$]$\cdot 4H_2O$, [$Co(CN)(dgH)_2(NH_3)$]$\cdot$1/2 H_2O, [$Co(dgH)_2(NH_3)_2$]ClO_4, [$Co(dgH)_2py(NH_3)$], [$Co(dgH)_2(OH_2)_2$]ClO_4, [$Co(dgH)_2(OH_2)(NH_3)$]$OCOCH_3$, [CoX$(dgH)_2$-(NH_3)] (X = I, Br, Cl, F, NCS), and Na[$Co(CN)_2(dgH)_2$]$\cdot 2H_2O$.

The most striking feature of this result is that the axially coordinated fifth and the sixth ligands, X_2, are liberated at the Co(I) state during reduction, except for the case of cyanide ligands. Therefore the contribution of the two axial ligands located above and below the planar [$Co(dgH)_2$] structure probably need not be considered in determining whether or not a stable Co(I) state can exist during reduction; i.e., the Co(I) complex takes a planar structure of the [$Co^I(dgH)_2$] type with a pair of hydrogen bonds, as indicated by

$$\text{trans-}[Co^{III}X_2(dgH)_2]^{n-} \xrightarrow{2e} [Co^I(dgH)_2]^- + 2X^{m-}$$

Figure 4.1.32 shows an example of current-potential curves for the dichloro complex in which two axial chloride ligands exhibit a strong interaction with the mercury electrode during

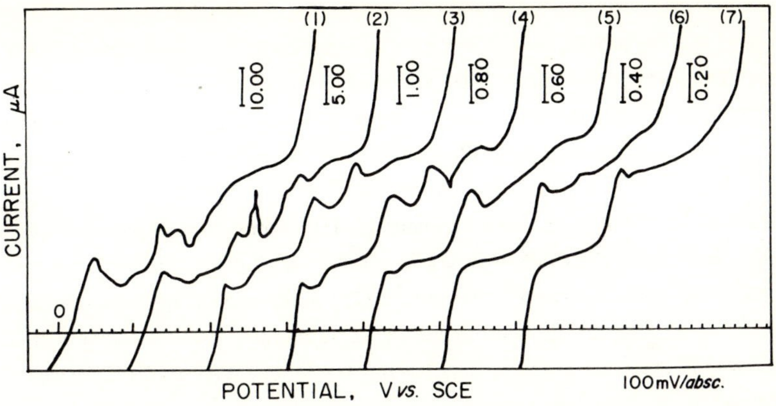

FIG. 4.1.32. The effect of the complex concentration on current-potential curves for trans-H[$CoCl_2(dgH)_2$] in DMSO containing 0.1 M [$(C_2H_5)_4N$]ClO_4 (25°C). Concentration (M): (1) 10^{-2}, (2) 5 x 10^{-3}, (3) 10^{-3}, (4) 8 x 10^{-4}, (5) 6 x 10^{-4}, and (7) 2 x 10^{-4}.

reduction in DMSO. An electron-transfer mechanism through a Hg—Cl—Co bridge may operate in the reduction of Co(III) → Co(II) → Co(I) in DMSO at the DME.

In aqueous solutions the polarographic behavior is dependent on the nature of the axial ligands, X_2. For example, the trans-Na[$Co(NO_2)_2(dgH)_2$]. H_2O complex is reduced to the metal through the Co(II) complex in 0.5 M Na_2SO_4 solutions with or without an excess of NO_2^- ions; no Co(I) state was found in aqueous media. The trans-[$CoX(dgH)_2(NH_3)$] complexes (X = I, Br, Cl, F, or H_2O) are reduced to the metal via the Co(II) and Co(I) complexes with aquation in 0.5 M Na_2SO_4 [144]. The trans-[$Co(dgH)_2(NH_3)_2$]Cl. $5H_2O$ complex undergoes a step-by-step reduction of Co(III) → Co(II) → Co(I) → Co(0), while the trans-Na[$Co(CN)_2(dgH)_2$]. $2H_2O$ complex is reduced in one step directly to the Co(I) state in aqueous solutions (see Table 2.2.20) [115, 122].

In contrast to this behavior of the trans-[$CoX_2(dgH)_2$] complexes, the tris(dimethylglyoximato)cobalt(III) complex, in which no planar structure [$Co(dgH)_2$] is involved, was found to be reduced to the metal via the Co(II) complex in DMSO containing 0.1 M [$(C_2H_5)_4N$]ClO_4 at the DME:

$$[Co^{III}(dgH)_3] \xrightarrow{e} [Co^{II}(dgH)_3]^- \xrightarrow{2e} Co(metal)$$

No cobalt(I) state was found for this reduction over the range of potential 0 to -2.7 V. This is consistent with the prediction that the [$Co^I(dgH)_2$]$^-$ anion with a planar structure could not be formed from the reduction of the cobalt(III) complexes with a cis-configuration, since it is unlikely that rearrangement of ligands, i.e., isomerization, takes place in DMSO. This result can also be invoked to emphasize the importance of the planarity of the bis(dimethylglyoximato)cobalt(I) complex (d^8) from the standpoint of the electronic structure. Several kinds of bis(dimethylglyoximato)cobalt(III) complexes with the cis-configuration cited below have been examined polarographically and found to be reduced to the metal through the Co(II) complex and not through the Co(I) state in DMSO or in methanolic solutions at the DME: cis-[Co(dgH)$_2$en]ClO$_4$.1/3 H$_2$O [299, 300], cis-[Co(dgH)$_2$(NH$_3$)$_2$]ClO$_4$, cis-K$_2$[Co(CO$_3$)dg(dgH)].4H$_2$O [103], and cis-K$_2$[Co(ox)dg(dgH)].4H$_2$O [301, 103] in methanolic solutions containing 1 M LiCl.

Thus the cis- vs trans-distinction has been clearly demonstrated polarographically with regard to the stability of the Co(I) complex in DMSO. A similar differentiation has already been shown for the cis- and trans-tetracyano Co(III) complexes in aqueous media in relation to the stability of the tetracyano Co(I) complex. This method of distinguishing steric isomers is more promising than that based on the interpretation and assignment of IR spectra [302].

Polarographic examination of the cis-bis(dimethylglyoximato)-cobalt(III) complexes in aqueous media has not been possible due to the presence of a large hydrogen wave, since solutions of [Co(dgH)$_3$].5/2 H$_2$O, cis-[Co(dgH)(dgH$_2$)en]Cl$_2$.2H$_2$O, and related complexes require a strong acidity (pH = 1 to 3). Only in highly alkaline solutions are the waves of Co(III) → Co(II) → Co(metal) observable. The ultimate formation of metallic cobalt for the cis-isomer could be verified by the occurrence of the reduction of free dimethylglyoximes [303].

These bis(dimethylglyoximato)cobalt(III) complexes are of fundamental importance as a model of vitamin B$_{12}$ (cyanocobalamin), together with the trans-[Co salen X$_2$] complexes [where "salen" denotes the bis(salicylaldehyde)ethylenediimine tetradentate ligand and its analogs]. Co(III) complexes with Schiff bases exhibited polarographic behavior quite similar to that of the oximato Co(III) complexes [127, 163]. Excellent reviews concerning the behavior of vitamin B$_{12}$ are found in the literature [106, 304]. The effect of the two axial ligands on the polarographic behavior for this kind of Co(III) complexes and vitamin B$_{12}$ has been studied extensively [146-149, 151-153, 305].

The tris(2,2'-dipyridyl)cobalt(III) complex was found for the first time to be reduced to the [Co dip$_3$]$^-$ anion in acetonitrile (AN) containing an excess of 2,2'-dipyridyl by Tanaka and Sato [39]:

$$[Co^{III}dip_3]^{3+} \xrightarrow[\text{reversibly}]{e} [Co^{II}dip_3]^{2+} \xrightarrow{e} [Co^{I}dip_3]^{+} \xrightarrow{2e} [Co^{-I}dip_3]^{-}$$

The [Co dip$_3$]$^-$ anion could exist as a stable species only in the presence of 2,2'-dipyridyl. Note also that the electrons added to the molecule are delocalized molecular orbitals in π-bonded systems over the entire molecule, and the Co(-I) state should be interpreted only as a formal oxidation state. More recently the [Co^{-I}dip$_3$]$^-$ and [Co^{-I}phen$_3$]$^-$ anions were found to exist as stable species even in the absence of free ligands in a DMSO (100%) solution [103]. Figure 4.1.33 shows the four-step wave in which the first three steps represent the reduction of Co(III) → Co(II) → Co(I) → Co(-I) and the last step the reduction of free 2,2'-dipyridyl liberated from the [Co^{-I}dip$_3$]$^-$ anion during reduction in DMSO [103].

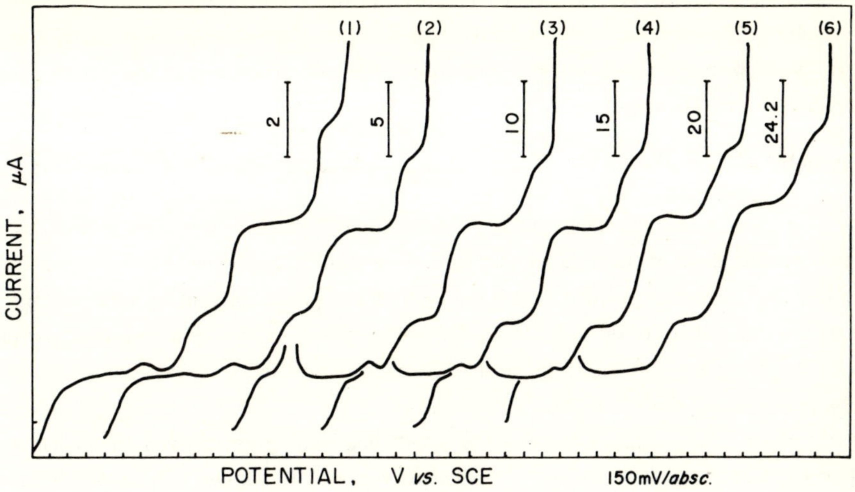

FIG. 4.1.33. The concentration-dependence of the current-potential curve for $[Co^{III}dip_3](ClO_4)_3 . 3H_2O$ in DMSO containing 0.1 M $[(C_2H_5)_4N]ClO_4$ (25°C). Complex concentration (M): (1) 8.26×10^{-4}, (2) 2.07×10^{-3}, (3) 4.13×10^{-3}, (4) 6.20×10^{-3}, (5) 8.26×10^{-3}, and (6) 10^{-2}. Recorded from +0.30 V vs SCE.

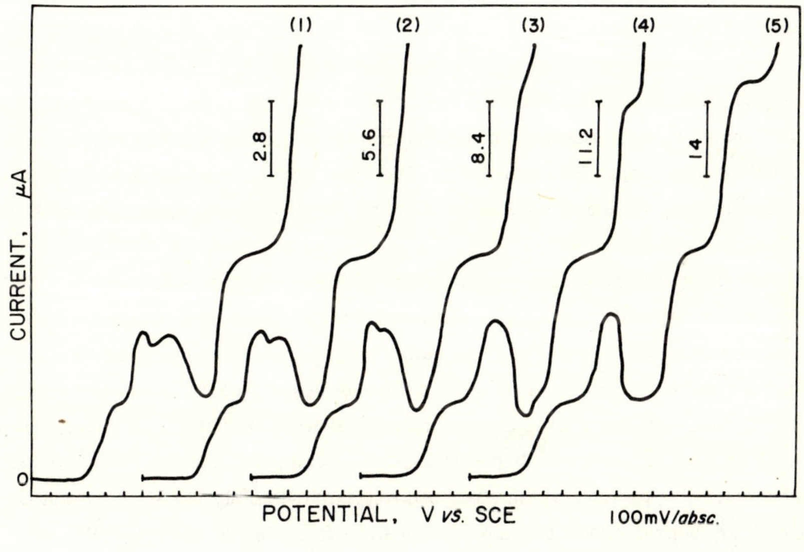

FIG. 4.1.34. The concentration dependence of the current-potential curve for $[Co^{II}phen_3](ClO_4)_2 . H_2O$ in DMSO containing 0.1 M $[(C_2H_5)_4N]ClO_4$ (25°C). Complex concentration (M): (1) 2×10^{-3}, (2) 4×10^{-3}, (3) 6×10^{-3}, (4) 8×10^{-3}, and (5) 10^{-2}. Recorded from -0.50 V vs SCE.

Figure 4.1.34 shows the three-step wave of the $[Co^{II}phen_3](ClO_4)_2 . H_2O$ complex, corresponding to the reduction of Co(II) → Co(I) → Co(-I) → Co(-III) in stepwise fashion in DMSO.

No subsequent reduction waves of 1,10-phenanthroline were observed over the potential range of this process. The presence of free ligands (phen) in a slight excess causes a maximum on the Co(II) → Co(I) process to disappear completely. The net processes of the Co(III) complex with 1,10-phenanthrolines are:

$$[Co^{III}phen_3]^{3+} \xrightarrow{e} [Co^{II}phen_3]^{2+} \xrightarrow{e} [Co^{I}phen_3]^{+} \xrightarrow{2e} [Co^{-I}phen_3]^{-} \xrightarrow{2e} [Co^{-III}phen_3]^{3-}$$

in DMSO (see Table 2.2.19). The existence of the Co(-III) anion is the most surprising feature of this result; these facts suggest that the use of DMSO is capable of simplifying the electrode processes to an even greater extent than AN as an aprotic solvent by retaining the inertness or the immobility of the ligands tightly in the sphere of the reduced cobalt complexes throughout the reduction pathways, since the presence of an excess of free ligands is no longer needed for maintaining the -1 species in DMSO.

The $[Co^{I}phen_3]ClO_4$ complex has been isolated successfully as a brown compound in vacuo (10^{-7} Torr) from an aqueous solution by reduction with $NaBH_4$, since it is insoluble in water [293]. The absorption spectrum measured in vacuo is given in Fig. 4.1.35. Polarographic evidence of a Co(I) complex with terpyridyl was given, and the rate constant for the Co(II) → Co(I) process was determined by Macero, Lovecchio, and Pace [306].

4.1.9. Cobalt(II) Complexes

Brdička [307, 308] first described the "prewave" due to the reduction of the $[Co^{II}(OH)(OH_2)_5]^{+}$ ion formed through the loss of ligands in noncomplexing aqueous media, especially in weakly alkaline solutions. Jain and Zutshi [309] and Verdier and Baptiste [310] have examined the $[Co^{II}(OH_2)_6]^{2+}$ ion in various supporting electrolytes in connection with the prewave.

On the other hand, the polarographic behavior of the so-called "salcomine" Co(II) complexes is of special interest because they exhibit the property of binding oxygen reversibly [311]. The mechanism of the reaction was studied polarographically for the first time for the bis(histidine) and bis(glycylglycinato)cobalt(II) complexes by Silvestroni and Illuminati [312, 313]. These Co(II) complexes are electrochemically oxidizable in a reversible manner to $[Co^{III}(histidine)_2]^{+}$ or $[Co^{III}glygly_2]^{+}$ ions at -0.2 and -0.5 V (vs SCE), respectively, in aqueous media. The product which absorbed oxygen reversibly has the formula $[X_2—Co—OO—Co—X_2]$, but the product of the irreversible oxidation is $[Co^{III}X_2(OH)]$ or $[Co^{III}X_2(OH)_2]^{-}$ without the peroxo bridge [314].

Catalytic hydrogen waves are observed with sulfhydryl-containing compounds in suitable buffers in the presence of Co(II); e.g., well-developed prewaves of Co(II) occur in the presence of cysteine, cystine, and similar compounds. Studies have been carried out extensively with cysteine ethyl ester, α-alanine, thioglycolic acid, β-mercaptopropionic acid, cystamine, and so on [315-320].

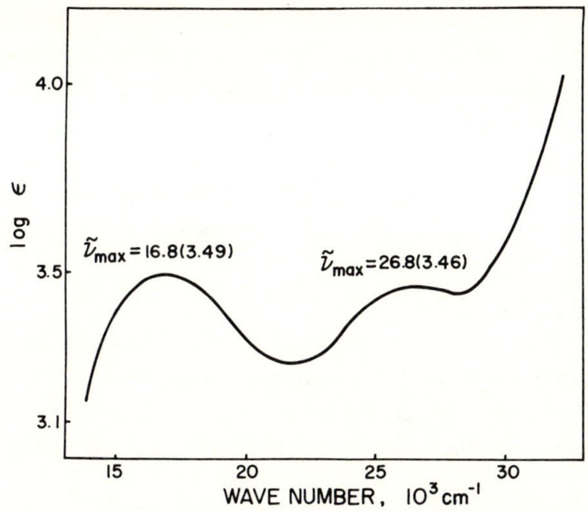

FIG. 4.1.35. The absorption spectrum of $[Co^{I}phen_3]ClO_4$ in methanol.

The double wave seen with Co(II) in a thiocyanate medium at the DME is of interest from the standpoint of adsorption [323, 324]. In this connection, attempts to differentiate the linkage isomers from one another have been carried out for the cathodic process, Co(III) → Co(II), in aqueous solutions. That is, the isothiocyanato- and thiocyanato-pentacyano complexes can clearly be distinguished from each other by i-t curves on a single drop of the DME. The $K_3[Co(NCS)(CN)_5]$ [325] and $K_3[Co(SCN)(CN)_5]$ complexes behave quite differentially in conventional and oscillographic polarography at the DME; the strong interaction or affinity of the exposed sulfur in the isothiocyanato complex with mercury electrode causes anomalous i-t curves in comparison with normal ones for the thiocyanato complex [103].

4.2. VOLTAMMETRIC CHARACTERISTICS

Vlček [321], using a vibrating platinum cathode, showed that the reduction of $[(NH_3)_5Co—O_2—Co(NH_3)_5]^{5+}$ occurs reversibly in ammoniacal solutions, suggesting that the relatively stable tervalent complex ion exists in the bulk solution in equilibrium with dissolved oxygen and the amminecobalt(II) species. However, Barnartt and Charles [322] showed that the $[(NH_3)_5Co—O_2—Co(NH_3)_5]^{5+}$ ion is reduced quantitatively to the $[Co^{II}(OH_2)_6]^{2+}$ ion in a one-electron reduction at a bright platinum cathode in H_2SO_4 solutions with simultaneous release of the peroxo oxygen as oxygen gas. The following cathodic reaction was established:

$$[(NH_3)_5Co—O_2—Co(NH_3)_5]^{5+} + 10H^+ + e \rightarrow O_2 + 2[Co^{II}(OH_2)_6]^{2+} + 10NH_4^+$$

Kimball and Kaska [326] have studied the cyclic voltammetry of pentakismethylisonitrile-cobalt(II) nitrate, $[Co^{II}(CH_3NC)_5](NO_3)_2$, at a platinum electrode, which underwent reduction to the $[Co^{I}(CH_3NC)_5]^+$ cation without the loss of ligands in a 1-M $NaNO_3$ aqueous solution. At the DME, however, the cathodic peak split into two peaks; the first peak corresponds to the reduction of the original Co(II) complex to the Co(I) state, and the second to the reduction of the newly formed species, $[Co^{II}(CH_3NC)_5]_2Hg^{2+}$, in contact with mercury in aqueous media. The following reduction of the second step was established:

$$[Co^{II}(CH_3NC)_5]_2Hg^{2+} \xrightarrow{e} Co{-}Hg + [Co^{I}(CH_3NC)_5]^+ + 5CH_3NC$$

Bond, Heath, and Martin [327] have examined the sulfur-bonded Co(III) and Co(II) complexes of the $[Co^{III}(sacsac)_3]$ and $[Co^{II}(sacsac)_2]$ (planar) type in the acetone containing 0.1 M $[(C_2H_5)_4N]ClO_4$ with Tast polarography and rapid scanning [where the abbreviation, "sacsac" (I) denotes dithioacetylacetone ligand, $S_2C_5H_7^-$]. The net processes are:

$$[Co^{III}(sacsac)_3]^0 \xrightarrow[-0.7\text{ V vs Ag/AgCl}]{2e} [Co^{I}(sacsac)_2]^- + sacsac^- \xrightarrow[\text{reversibly}]{e} [Co^0(sacsac)_2]^{2-}$$

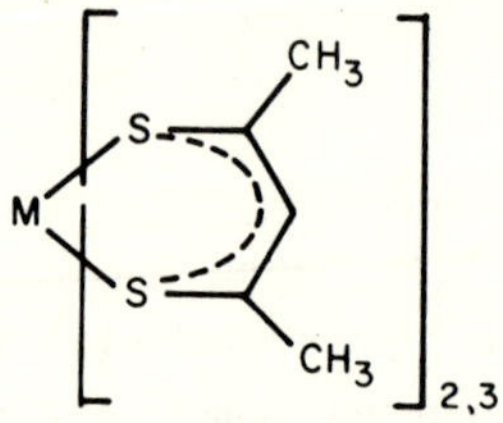

I

However, the current-potential curve was observed to change with time, even under stringent degassing procedures, which presumably is due to the partial rearrangement of ligands in the lower oxidation states. Similar time-dependent changes of the curves were frequently observed for vitamin B_{12} in AN, in DMSO, and in methanol, and for the Co(III) complexes of the [CoX(edta)] type in DMSO ($X = CN^-$, Cl^-, Br^-, NO_2^-, or OH^-). Such an organic solvent molecule may participate in promoting the rearrangement of ligands during reduction, since the curves did not change with time in aqueous media [103].

The cyclic voltammetry of vitamin B_{12} has recently been examined in aqueous media [328]. Both B_{12} and B_{12r} showed slow reduction to B_{12s}, whereas the B_{12s} exhibited a rapid one-electron oxidation to B_{12r}. A peak at -1.48 V (vs SCE) suggested an irreversible reduction which would be associated with catalytic hydrogen evolution. The catalytic hydrogen peak for a fresh solution is reversible, but irreversible for a solution of B_{12a}.

Lim and Anson [207] have studied the $[CoX(CN)_5]$ complexes with $X = I^-$, Br^-, Cl^-, N_3^-, SCN^-, OH_2, or CN^- by means of cyclic voltammetry. Unusually large and sharp "adsorption waves" appeared in cyclic voltammograms of $[Co^{II}(CN)_5]^{3-}$ and several Co(III) complexes at stationary mercury electrodes. The data were compatible in all cases with the adsorption of a coordinatively unsaturated Co(II) complex $[Co(CN)_4]^{2-}$ by means of a cobalt-mercury bond. The formation of the $[(NC)_5CoHgX]^{3-}$ and $[(NC)_5Co—Hg—Co(CN)_5]^{6-}$ anions has been verified [329]. Anson [335] also reported chronopotentiometric evidence for the adsorption of the reduction product of the ethylenedinitrilotetraacetatocobalt(III) complex at a platinum electrode. The quantitative determination of the adsorbed cobalt(III) complex of this kind at a platinum electrode was carried out by thin-layer electrochemistry [336].

The application of carbon paste electrodes to the cathodic reduction of cobalt complexes and anodic stripping voltammetry has been thoroughly investigated for practical use in cobalt analysis [167, 330, 331]. The reversible system of the $[Co\ edta]^-/[Co\ edta]^{2-}$ couple was found to become an irreversible one at the carbon paste electrode [332]. No catalytic anodic oxidation of the $[Co^{II}edta]^{2-}$ ion occurs, unlike the results obtained at a platinum electrode by Anson [333].

A precise study has been made of the current-time curves during discharge of Co(II) ions at the DME in the presence of noncomplexing indifferent electrolytes [334, 263]. The influence of various factors such as concentration, surface active agents, and oxygen was reported in detail.

Finally, the anodic oxidation of cobalt metal in alkaline solutions has received little attention in comparison with the large number of investigations of electrooxidation of the other metals. The reasons may be related to the small number of practical applications of cobalt in commercial products. As to the mechanism of anodic oxidation of cobalt, Behl and Toni [337] have studied cobalt metal in KOH with cyclic voltammetry at a stationary disk, rotating disk, and rotating ring-disk electrodes, while Benson, Briggs, and Wynne-Jones [338, 339] have carried out investigations with electrolytically prepared cobalt(II) hydroxide. Cowling and Riddiford [340] have recently summarized the scarce literature on the anodic oxidation of cobalt.

5. APPLIED ELECTROCHEMISTRY

5.1. ELECTROWINNING AND ELECTROREFINING

At present, the main sources of electrolytic cobalt is from the copper in the African Congo. The oxidized deposits, chiefly malachite, contain 0.2 to 0.3% Co, together with 5% Cu. At Kolwezi they are concentrated with hydrolyzed palm oil in a pulp dispersed with sodium carbonate and sodium silicate to 27% Cu and 1% Co. They are then sent to the Jadotville leaching and electrowinning plant. Since cobalt is not deposited in the copper electrolysis circuit, it accumulates in solution to an equilibrium where the quantity of cobalt eliminated with the washed gangue of the copper leach section is equal to the quantity of cobalt dissolved, so that all the soluble cobalt is found in the solution accompanying the gangue eliminated from the copper deposition cell after washing. Following the removal of copper by electrolysis at high current densities with vigorous air agitation, a lime treatment lowers the copper content to about 0.2 g/l. Further addition of a lime slurry in agitation machines now produces a cobalt precipitate which is the feed for the cobalt electrolysis cells (Co, 16 to 18%). Use of a pulp containing cobalt hydroxide in suspension and vigorous air agitation of the electrolyte maintain a high pH (about 5.8) in the electrolysis cells. The electrolyte usually contains 10 g/l Co. Power consumption is about 6.5 kWh/kg of cobalt at a current density of 500 A/m^2; the electrolyte temperature is 55 to 58°C. The cathode deposit normally contains the following: Co 92 to 94, Ni 0.25 to 0.30, Zn 1 to 1.8, Fe 0.05 to 0.1, Mn 0.1 to 0.3, Cu 0.01 to 0.02, and S 0.1 to 0.3%. These deposits, after being stripped from the cathode, are melted in an electric furnace under slightly oxidizing conditions and the product granulated into shot by pouring it into a stream of water. The final cobalt content reaches 99.1%. Refining electrolysis is carried out in diaphragm cells similar to those used in nickel refining [341].

Most of the so-called cobalt(II) salt reagents usually contain 2 to 3% nickel. To obtain pure, nickel-free cobalt, thermal decomposition of the cobalt(III) complex is adopted rather than a refining electrolysis; e.g., the decomposition of the hexamminecobalt(III) chloride with volatile ammonia ligands may be used to obtain nickel-free cobalt metal for academic uses.

5.2. ELECTRODEPOSITION AND ELECTROPLATING

Fink and Hutton [342] have reported that cathodic codeposits of cobalt-copper alloys are largely dependent on the cathode substrate, the pH, and the molar ratio of Co to Cu concentrations. Alberts, Wright, and Parker [343] examined the effects of ammonia on the deposition of cobalt from alkaline electroless nickel and cobalt plating baths. The rate of deposition of thin cobalt metal films on both metallic and nonmetallic substrates, and the magnetic and physical properties of such films have been shown to correlate closely with spectral and polarographic observations in ammonia buffer and nonbuffer systems. Clark and Holt [344] obtained bright deposits of Co—W alloys from a bath of cobalt(II) sulfate, sodium tungstate, and citric acid in a molar ratio of 1:1:1.5 at 70°C, a pH of 7, and a cathodic current density 15 A/dm^2 with anodes of cobalt-tungsten alloy, cobalt, tungsten, or an inert anode. Seaman, Myers, and Saxer [345] investigated anodic polarization of annealed cobalt, chromium, and nine Co—Cr alloys in hydrogen-saturated 1, 5, and 10 $\underline{N}$ H_2SO_4 solutions at 22°C using potentiostatic techniques. All specimens except pure cobalt and the 95Co-5Cr alloy showed an active-to-passive transition in all acid environments. Pure cobalt and the 95-5 alloy did exhibit marked "secondary" passivation in the transpassive region just prior to visible oxygen evolution. This "secondary" passivation was attributed to adsorbed oxygen. The potential width of the passive region for specimens exhibiting passive behavior was independent of acid concentration, but increased (especially for alloys containing 70 wt% or more chromium) with increasing chromium content. Pure cobalt and the 95-5 alloy exhibited visible oxygen evolution in the transpassive region. Corrosion potentials for cobalt and cobalt-chromium alloys were linear functions of pH over the pH range +1.21 to -1.04. dE_{corr}/d pH varied from -0.050 to -0.060 V, independent of composition. Passivation potentials E_p, and also linear functions of pH, shifted in the active potential direction with increasing chromium content and increasing pH, dE_p/d pH being from -0.050 to -0.070 V, independent of composition. The critical current density for passivation decreased as chromium was added to pure cobalt; the minimum value of I_0 for the Co-Cr system occurred at 30 wt% chromium. The passive current density also appeared to be minimal at the same chromium content [346, 347].

The cathodic overvoltage of cobalt is similar to that of nickel, but more definite [348]. With increasing current density, the overvoltage increases rapidly, which is attributed to the surface conditions of the substrate, since cleaning the surface restores the normal value of overvoltage. The cathodic or hydrogen overvoltage for a cobalt electrode in 1 N H_2SO_4 is 0.22 to 0.29 V at 20°C.

Cobalt becomes passive under certain conditions of electrolysis. A great number of papers regarding this subject have been published; there are some conflicting opinions and confusion because of the number of factors affecting the formation of oxide films of cobalt under conditions of passivity. Therefore, it suffices only to cite the most recent study, since other literature references are found therein. The passivation process for cobalt in aqueous solutions can be regarded as a competition between anodic oxide formation with water and the reaction that removes the oxide. Due to the low exchange current density for active dissolution, this competition is particularly favored with cobalt. Based on this concept, Ebersbach, Schwabe, and Ritter [349] have matched the experimental current-density/potential curves for the passivation process with calculated curves when corrected parameters are chosen. A satisfactory agreement was obtained between them. The computations show, moreover, that no normal passivation can occur at negative potentials when the "oxygen" passive layer is rapidly removed. In this case a salt layer formed directly on the metal can impede the anodic current to a considerable extent.

Cobalt amalgams have been prepared by electrolysis of cobalt(II) sulfate between a cobalt anode and a mercury cathode [350]. Both cubic and hexagonal cobalt were highly dispersed in mercury [350]. Hovsepian and Shain [351] determined the composition of the intermetallic compound between cobalt and zinc in mercury from anodic stationary electrode polarograms. Then the rate of formation of these intermetallic compounds was measured using a step-function controlled potential method [352]. The results indicated that the intermetallic compound is CoZn, in equilibrium with Co_2 and Zn_2 with $K = [Co_2][Zn_2]/[CoZn]^2 = 2$. Zinc was deposited on a previously prepared cobalt amalgam electrode, and then, in a second step, the unreacted zinc was reoxidized. The cobalt amalgam concentration was maintained at 0.5 M, and the pseudo-first-order rate constant for the disappearance of zinc was 5.6/sec.

5.3. ELECTROPOLISHING

The pioneer work on electrolytic polishing of cobalt was done by Tegart [353]. For example, electrolysis was carried out for 0.5 to 1.5 min at 20°C with a current density of 250 A/dm^2 and 8 to 9 V across the cell in a series circuit involving a stainless steel cathode and a bath of HCl containing 1/100 ethanol. The bluish-green anodic film is soluble in water, while a polished surface with a slight grain boundary delineation was obtained for the cathode.

REFERENCES

1. J. Bjerrum, Metal Ammine Formation in Aqueous Solution, Haase, Copenhagen, 1941, pp. 219-234.
2. D. A. Buckingham and A. M. Sargeson, in Chelating Agents and Metal Chelates (F. P. Dwyer and D. P. Mellor, eds.), Academic, New York, 1964, p. 242.
3. P. A. Rock, Inorg. Chem., 7, 837 (1968).
4. J. J. Kim and P. A. Rock, ibid., 8, 563 (1969).
5. W. M. Latimer, Oxidation Potentials — The Oxidation States of the Elements and Their Potentials in Aqueous Solutions, 2nd ed., Prentice-Hall, New York, 1952.
6. G. Charlot, Oxidation-Reduction Potentials, Pergamon, New York, 1958.
7. A. L. Rotinyan, V. L. Kheifets, and S. A. Nikolaeva, Zh. Neorg. Khim., 6, 21 (1961).
8. V. L. Kheifets, A. L. Rotinyan, and E. S. Kozich, Zh. Obshch. Khim., 29, 1052 (1959).
9. D. A. Johnson and A. G. Sharpe, J. Chem. Soc., 1964, 3490.
10. J. H. Baxendale and C. F. Wells, Trans. Faraday Soc., 53, 800 (1957).
11. L. H. Sutcliffe and J. R. Weber, ibid., 52, 1225 (1956); J. Inorg. Nucl. Chem., 12, 281 (1960).
12. G. Hargreaves and L. H. Sutcliffe, Trans. Faraday Soc., 51, 786 (1955).
13. D. H. Huchital, N. Sutin, and B. Warnqvist, Inorg. Chem., 6, 838 (1967).
14. A. A. Noyes, D. DeVault, C. D. Coryell, and T. J. Deahl, J. Amer. Chem. Soc., 59, 1326 (1937).
15. A. A. Noyes and T. J. Deahl, ibid., 59, 1337 (1937).
16. C. F. Wells, Discussions Faraday Soc., 46, 197 (1968).
17. B. Warnqvist, Inorg. Chem., 9, 682 (1970).
18. P. A. Rock, ibid., 4, 1667 (1965).

19. R. Farina and R. G. Wilkins, ibid., 7, 514 (1968).

20. L. Hin-Fat and W. C. E. Higginson, J. Chem. Soc., A, 1967, 298.

21. R. G. Yalman, Inorg. Chem., 1, 16 (1962).

22. J. W. Larson, P. Cerutti, H. K. Garber, and L. G. Hepler, J. Phys. Chem., 72, 2902 (1968).

23. R. N. Goldberg, R. G. Riddell, M. R. Wingard, H. P. Hopkins, C. A. Wulff, and L. G. Hepler, ibid., 70, 706 (1966).

24. L. H. Adami and E. G. King, U.S. Bur. Mines Rep. Invest., 6617 (1965).

25. K. K. Kelley and E. G. King, U.S. Bur. Mines Bull., 592 (1961).

26. D. D. Wagman, W. H. Evans, V. B. Parker, I. Halow, S. M. Bailey, and R. H. Schumm, Nat. Bur. Stand. Tech. Note, 370-3 (1968).

27. A. J. de Bethune and N. A. Swendeman Loud, Standard Aqueous Electrode Potentials and Temperature Coefficients at 25°C, Hampel, Skokie, Illinois, 1964.

28. A. B. Lamb and A. T. Larson, J. Amer. Chem. Soc., 42, 2024 (1920).

29. M. Pourbaix, Atlas of Electrochemical Equilibria in Aqueous Solutions, Pergamon, New York, 1966.

30. S. Jahn, Z. Anorg. Allgem. Chem., 60, 292 (1908).

31. H. Diebler and N. Sutin, J. Phys. Chem., 68, 174 (1964).

32. E. Paglia and C. Sironi, Gazz. Chim. Ital., 87, 1125 (1957).

33. N. Tanaka and H. Ogino, Bull. Chem. Soc. Japan, 38, 1054 (1965).

34. A. Yamada and N. Tanaka, Sci. Rep. Tohoku Univ. Ser. I, 52(2), 73 (1969).

35. N. Tanaka, Nippon Kagaku Zassi, 92, 919 (1971).

36. W. Hieber and W. Hübel, Z. Elektrochem., 57, 331 (1953).

37. H. M. Koepp, H. Wendt, and H. Strehlow, ibid., 64, 483 (1960).

38. G. Grube, ibid., 33, 389 (1927).

39. N. Tanaka and Y. Sato, Bull. Chem. Soc. Japan, 41, 2059 (1968).

40. Z. Samec and I. Němec, J. Electroanal. Chem., 31, 161 (1971).

41. L. G. Sillén and A. E. Martell, Stability Constants of Metal Ion Complexes (Special Publication No. 17), Chemical Society, London, 1964.

42. S. Taniewska-Osinska, Acta. Chim. Soc. Lodz., 7, 23 (1961).

43. P. Cescon, R. Marassi, V. Bartocci and M. Fiorani, J. Electroanal. Chem., 23, 255 (1969).

44. R. Hultgren, R. Orr, P. D. Anderson, and K. K. Kelley, Selected Values of Thermodynamic Properties of Metals and Alloys, Wiley, New York, 1963.

45. H. C. Gaur and H. L. Jindal, J. Electroanal. Chem., 23, 289 (1969).

46. Yu. K. Delimarskii and B. Markov, Electrochemistry of Fused Salts (English Translation), Sigma Press, Washington, D.C., 1961, p. 297.

47. H. C. Gaur and H. L. Jindal, Electrochim, Acta, 13, 835 (1968).

48. H. C. Gaur and W. K. Behl, in Proceedings of the First Australian Conference of Electrochemistry (A. Friend and F. Gutmann, eds.), Pergamon, London, 1964, p. 543.

49. W. K. Behl, Thesis, University of Delhi, 1962.

50. H. C. Gaur and W. K. Behl, Electrochim. Acta, 8, 107 (1963).

51. H. C. Gaur and W. K. Behl, Bull. Nat. Inst. Sci. India, 29, 179 (1965).

52. H. C. Gaur and H. L. Jindal, Indian. J. Chem., 4, 496 (1966).

53. H. A. Laitinen and C. H. Liu, J. Amer. Chem. Soc., 80, 1015 (1958).

54. S. N. Flengas and T. R. Ingraham, J. Electrochem. Soc., 106, 714 (1959).

55. W. J. Hamer, M. S. Malmberg, and B. Rubin, ibid., 103, 8 (1956).

56. J. A. N. Friend (ed.), Textbook of Inorganic Chemistry, Vol. X. The Metal Ammines, by M. M. J. Sutherland, Griffin, London, 1928.

57. Gmelins Handbuch der anorganischen Chemie, System No. 58, 8 Aufl., Kobalt, Teil A, Die Ammine des Kobalts(II), Verlag Chemie, Berlin, 1930.

58. Gmelins Handbuch der anorganischen Chemie, System No. 58, 8 Aufl., Kobalt, Teil B, Die Ammine des Kobalts(III), Verlag Chemie, Berlin, 1930.

59. Gmelins Handbuch der anorganischen Chemie, System No. 58, 8 Aulf., Kobalt, Teil B-Ergänzungsband, Verlag Chemie, Berlin, 1964.

60. R. S. Young, Cobalt, Its Chemistry, Metallurgy, and Uses (American Chemical Society Monograph), Reinhold, New York, 1960.

61. R. S. Young, The Analytical Chemistry of Cobalt, Pergamon, London, 1966.

62. I. V. Pyatiniskii, Analytical Chemistry of Cobalt (English Translation), Israel Program for Scientific Translations, Jerusalem, 1966.

63. N. Maki, Y. Shimura, and R. Tsuchida, Bull. Chem. Soc. Japan, 32, 23 (1959).

64. L. Malatesta and F. Bonati, Isocyanide Complexes of Metals, Wiley, London, 1969. Polarographic examinations have not yet been attempted of the cobalt complexes with isocyanides; a few reports, however, are found for the iron complexes.

65. N. Maki and M. Inoue, Ann. Rep. Fac. Eng. Shizuoka Univ., 21, 95 (1970).

66. M. Kalousek, Collect. Czech. Chem. Commun., 13, 105 (1948).

67. N. Maki, Bull. Chem. Soc. Japan, 44, 1447 (1971).

68. H. Taube, Chem. Rev., 50, 69 (1952).

69. I. M. Kolthoff and S. E. Khalafalla, in Modern Aspects of Polarography: A Tribute to Isamu Tachi (T. Kambara, ed.), Plenum, New York, 1966, pp. 11-25.

70. N. Maki and M. Inoue, Ann. Rep. Fac. Eng. Shizuoka Univ., 22, 137 (1971).

71. For a review, see S. Koide, J. Electrochem. Soc. Japan, 23, 484 (1955).

72. N. Maki, Y. Shimura, and R. Tsuchida, Bull. Chem. Soc. Japan, 30, 909 (1957).

73. N. Maki, Y. Shimura, and R. Tsuchida, ibid., 32, 150 (1959).

74. N. Maki, Y. Shimura, and R. Tsuchida, ibid., 32, 833 (1959).

75. A. A. Vlček, Discussions Faraday Soc., 26, 164 (1958).

76. J. Heyrovský and D. Ilkovič, Collect. Czech. Chem. Commun., 7, 198 (1935).

77. A. A. Vlček, Nature, 180, 573 (1957).

78. A. A. Vlček, Z. Elektrochem., 61, 1014 (1957).

79. A. A. Vlček, Z. Phys. Chem. Sonderheft (Internationales Polarographisches Kolloquium, Dresden), (1958) p. 143.

80. G. M. Waind and B. Martin, J. Inorg, Nucl. Chem., 8, 551 (1958); Proceedings of the 4th I.C.C.C. (Chemistry of the Coordination Compounds) (L. Cambi, ed.), Pergamon, London, 1957, pp. 551-556.

81. N. Maki, T. Hirano, and S. Musha, Bull. Chem. Soc. Japan, 36, 756 (1963).

82. N. Maki, T. Hirano, and S. Musha, Ann. Rep. Radiat. Center, 3, 26 (1962).

83. N. Tanaka, T. Tomita, and A. Yamada, Bull. Chem. Soc. Japan, 43, 2042 (1970).

84. N. Tanaka, J. Chem. Soc. Japan (in Japanese), 92, 919 (1971).

85. N. Maki, Bull. Chem. Soc. Japan, 42, 2275 (1969).

86. E.g., H. Taube, Electron-Transfer Reactions of Complex Ions in Solution, Academic, New York, 1970, pp. 28-34.

87. H. A. Laitinen and M. W. Grieb, J. Amer. Chem. Soc., 77, 5201 (1955).

88. P. Kivalo, ibid., 77, 2678 (1955).

89. J. Dolezal, Collect. Czech. Chem. Commun., 21, 113 (1956).

90. A. A. Vlček, Tables of Half-Wave Potentials of Inorganic Depolarizers (in Czechoslovakian and German), NCSAV, Prague, 1956; Reprinted (explanations added in Japanese), Polar. Soc. Japan, Kyoto, 1957, pp. 1-96.

91. H. A. Laitinen, J. C. Bailar, Jr., H. F. Holtzclaw, Jr., and J. V. Quagliano, J. Amer. Chem. Soc., 70, 2999 (1948).

92. H. A. Laitinen and P. Kivalo, ibid., 75, 2198 (1953).

93. H. A. Laitinen, A. J. Frank, and P. Kivalo, ibid., 75, 2865 (1953).

94. J. B. Willis, J. A. Friend, and P. Kivalo, ibid., 67, 1680 (1945).

95. S. Inoue and H. Imai, Rev. Polarogr. (Japan), 12, 163 (1964).

96. T. Yanai and K. Kuroda, J. Chem. Soc. Japan (in Japanese), 82, 1641 (1961).

97. H. A. Laitinen and M. W. Grieb, J. Amer. Chem. Soc., 77, 5201 (1955).

98. R. C. Henney, H. F. Holtzclaw, Jr., and R. C. Larson, J. Electroanal. Chem., 14, 435 (1967).

99. H. F. Holtzclaw, Jr. and D. P. Sheetz, J. Amer. Chem. Soc., 77, 5201 (1955).

100. H. F. Holtzclaw, Jr., J. Phys. Chem., 59, 300 (1955).

101. N. Maki, K. Yamamoto, H. Sunahara, and S. Sakuraba, Bull. Chem. Soc. Japan, 42, 3159 (1969).

102. N. Maki and S. Sakuraba, Ann. Rep. Fac. Eng. Shizuoka Univ., 20, 105 (1969).

103. N. Maki, Unpublished.

104. J. G. Mason and R. L. White, J. Electroanal. Chem., 8, 454 (1964).

105. D. Konrád and A. A. Vlček, Collect. Czech. Chem. Commun., 28, 808 (1963).

106. M. Brezina and P. Zuman, Polarography in Medicine, Biochemistry and Pharmacy, Interscience, New York, 1958, p. 400.

107. B. Jaselkis and H. Diehl, J. Amer. Chem. Soc., 76, 4345 (1954).

108. Cang Je-Sia, Thesis, Charles University, Prague, 1961.

109. K. Morinaga, Rev. Polarogr. (Japan), 14, 251 (1967).

110. E. Jacobsen and K. Schrøder, J. Phys. Chem., 66, 134 (1962).

111. R. D. Hargens, W. Min, and R. C. Henney, J. Electroanal. Chem., 26, 285 (1970).

112. N. Maki, Kagaku (Kyoto) (in Japanese), 14, 632 (1959).

113. C. N. Reilley, W. G. Scribner, and C. Temple, Anal. Chem., 28, 450 (1956).

114. N. Maki, Bull. Chem. Soc. Japan, 42, 3617 (1969).

115. N. Maki, Nature, 188, 227 (1960).

116. N. Maki and K. Yamamoto, Bull. Chem. Soc. Japan, 43, 2450 (1970).

117. N. Maki and R. Tsuchida, ibid., 37, 1233 (1964).

118. N. Maki and K. Okawa, J. Electroanal. Chem., 8, 262 (1964); Bull. Chem. Soc. Japan, 37, 1233 (1964).

119. H. Siebert, C. Siebert, and S. Thym, Z. Anorg. Allgem. Chem., 383, 165 (1971).

120. N. Maki, J. Fujita, and R. Tsuchida, Nature, 183, 458 (1959).

121. N. Maki, Ann. Rep. Fac. Eng. Shizuoka Univ., 22, 119 (1971).

122. N. Maki, Proceedings of the 3rd International Conference of Polarography, Vol. 1 (Southampton), (G. J. Hills, ed.), Macmillan, London, 1964, pp. 505-533.

123. D. N. Hume and I. M. Kolthoff, J. Amer. Chem. Soc., 71, 867 (1949).

124. M. Takahashi, M. Kuwahara, and M. Iguchi, Nippon Kagaku Zasshi, 91, 543 (1970).

125. T. Asai and T. Hara, Bull. Chem. Soc. Japan, 42, 3580 (1969).

126. A. A. Vlček, in Progress in Inorganic Chemistry, Vol. 5 (F. A. Cotton, ed.), Wiley (Interscience), New York, 1963.

127. G. Costa, A. Puxeddu, and G. Tauzher, Inorg. Nucl. Chem. Lett., 4, 319 (1968).

128. T. Kato and Y. Sunami, Ann. Rep. Fac. Sci. Eng. Kinki Univ., 2, 35 (1967).

129. J. R. Urwin and B. West, J. Chem. Soc., 1952, 4727.

130. A. A. Vlček, Collect. Czech. Chem. Commun., 30, 952 (1965).

131. H-S. Hsiung and G. H. Brown, J. Electrochem. Soc., 110, 1085 (1963).

132. J. A. Page and G. Wilkinson, J. Amer. Chem. Soc., 74, 6149 (1952).

133. P. L. Pauson and G. Wilkinson, ibid., 76, 2024 (1954).

134. R. E. Dessy, F. E. Stary, R. B. King, and M. Waldrop, ibid., 88, 471 (1966).

135. R. E. Dessy, R. B. King, and M. Waldrop, ibid., 88, 5112 (1966).

136. R. E. Dessy, P. M. Weissman, and R. L. Pohl, ibid., 88, 5117 (1966).

137. G. Piazza, A. Foffani, and G. Paliani, Z. Phys. Chem., Neue Folge, 60, 167 (1968).

138. G. Piazza, A. Foffani, and G. Paliani, ibid., Neue Folge, 60, 177 (1968).

139. R. E. Dessy and R. L. Pohl, J. Amer. Chem. Soc., 90, 1995 (1968).

140. R. E. Dessy, R. Kornmann, C. Smith, and R. Haytor, ibid., 90, 2001 (1968).

141. K. Morinaga, K. Nakano, S. Saito, and K. Nakamura, Bull. Chem. Soc. Japan, 39, 357 (1966).

142. K. Morinaga, K. Nakano, and K. Nakamura, Nippon Kagaku Zasshi, 84, 198 (1963).

143. N. Maki and H. Itatani, Bull. Chem. Soc. Japan, 36, 757 (1963).

144. N. Maki, Y. Ishiuchi, and S. Sakuraba, ibid., 42, 3166 (1969).

145. N. Maki, Ann. Rep. Fac. Eng. Shizuoka Univ., 20, 121 (1969).

146. G. N. Schrauzer, R. J. Windgassen, and J. Kohnle, Chem. Ber., 98, 3324 (1965).

147. G. N. Schrauzer and R. J. Windgassen, J. Amer. Chem. Soc., 88, 3738 (1966).

148. Y. Hohokabe and N. Yamazaki, Bull. Chem. Soc. Japan, 44, 1563 (1971).

149. Y. Yamada, I. Masuda, and K. Shinra, Nippon Kagaku Zasshi, 92, 707 (1971).

150. N. Maki, Bull. Chem. Soc. Japan, 44, 1447 (1971).

151. H. Diehl, R. R. Sealock, and J. I. Morrison, Iowa State Coll. J. Sci., 24, 433 (1950).

152. H. A. O. Hill, J. M. Pratt, and R. J. P. Williams, Chem. Ind., 1964, 197.

153. B. Kratochvil and H. Diehl, Talanta, 13, 1013 (1966).

154. J. Vasilevskis and D. C. Olson, Inorg. Chem., 10, 1228 (1971).

155. A. Wolberg and J. Manassen, J. Amer. Chem. Soc., 92, 2982 (1970).

156. A. Stanienda and G. Bibel, Z. Phys. Chem., 52, 254 (1967).

157. R. H. Felton and H. Linschitz, J. Amer. Chem. Soc., 88, 1113 (1966).

158. L. D. Rollmann and R. T. Iwamoto, ibid., 90, 1455 (1968).

159. N. Maki and K. Ohkawa, Bull. Chem. Soc. Japan, 44, 2005 (1971).

160. N. Maki and K. Ohshima, ibid., 43, 3970 (1970).

161. R. Kalvoda, Techniques of Oscillographic Polarography, 2nd ed., Elsevier, Amsterdam, 1965.

162. D. P. Rillema, J. F. Endicott, and E. Papaconstantinou, Inorg. Chem., 10, 1739 (1971).

163. G. Costa, G. Mestroni, A. Puxeddu, and E. Reisenhofer, J. Chem. Soc., 1970, 2870.

164. J. E. Falk, Porphyrins and Metalloporphyrins, Elsevier, New York, 1964.

165. R. H. Felton, D. Dolphin, D. C. Borg, and J. Fajer, J. Amer. Chem. Soc., 91, 196 (1969).

166. M. Zerner, M. Gouterman, and H. Kobayashi, Theor. Chim. Acta, 6, 363 (1966); Chem. Abstr., 66, 42136s (1967).

167. R. N. Adams, Electrochemistry at Solid Electrodes, Dekker, New York, 1969.

168. J. Milazzo, ed., Electroanalytical Abstracts.

169. E.g., R. S. Nicholson, Anal. Chem., 42, 130R (1970).

170. E.g., A. Patterson, Ann. Rev. Phys. Chem., 20, 91 (1969).

171. E.g., H. Matsuda, K. Umemoto, and T. Osa, Japan Anal. Ann. Rev., 20, 73R (1971).

172. H. Eyring, L. Marker, and T. C. Kwoh, J. Phys. Colloid Chem., 53, 1453 (1949).

173. N. Tanaka and R. Tamamushi, Sbornik Mezinarodniho Polarografickeho Sjezdu Praze (Proceedings of the 1st Congress of Polarography), Part 1, Prirodovedecke Vydavatelstvi, Prague, 1951, pp. 486-524.

174. N. Tanaka and R. Tamamushi, Bull. Chem. Soc. Japan, 22, 187 (1949).

175. R. Tamamushi and N. Tanaka, ibid., 22, 227 (1949).

176. R. Tamamushi and N. Tanaka, ibid., 23, 110 (1950).

177. M. Smutek, Collect. Czech. Chem. Commun., 18, 171 (1953).

178. P. Delahay, J. Amer. Chem. Soc., 73, 4944 (1951).

179. M. G. Evans and N. S. Hush, J. Chim. Phys., 49, C159 (1952).

180. J. Koutecký, Chem. Listy, 47, 323 (1953).

181. R. S. Nicholson and I. Shain, Anal. Chem., 36, 706 (1964).

182. S. P. Perone, ibid., 38, 1158 (1966).

183. S. Toshima, Y. Okinaka, and H. Okaniwa, J. Electrochem. Soc. Japan, 31, 854 (1963).

184. S. Toshima, Y. Okinaka, and H. Okaniwa, ibid., 33, 19 (1965).

185. S. Toshima, Y. Okinaka, and H. Okaniwa, Talanta, 11, 203 (1964).

186. W. Vielstich and H. Gerischer, Z. Phys. Chem., Neue Folge, 4, 10 (1955).

187. M. Matsuda, Z. Elektrochem., Ber. Bunsenges. Phys. Chem., 61, 489 (1957).

188. H. Matsuda, Z. Elektrochem., ibid., 62, 977 (1958).

189. H. Matsuda and Y. Ayabe, ibid., 63, 1164 (1959).

190. R. Tamamushi, K. Ishibashi, and N. Tanaka, Z. Phys. Chem., Neue Folge, 35, 117 (1962).

191. R. Tamamushi and N. Tanaka, ibid., Neue Folge, 39, 117 (1963).

192. P. Delahay and T. Berzins, J. Amer. Chem. Soc., 77, 6448 (1955).

193. H. Z. Gerischer, Phys. Chem., 10, 264 (1957).

194. H. Z. Gerischer, ibid., 14, 184 (1958).

195. P. Delahay, J. Amer. Chem. Soc., 81, 5077 (1959).

196. W. J. Blaedel and L. N. Klatt, Anal. Chem., 38, 879 (1966).

197. N. Tanaka and A. Yamada, Electrochim. Acta, 14, 491 (1969).

198. A. Yamada and N. Tanaka, Sci. Rep. Tohoku Univ., Ser. I, 53, 110 (1970).

199. N. Tanaka, M. Maeda, and A. Yamada, Proceedings of the 17th Symposium on Polarography (Japan), October 16, 1971, Chemical Society of Japan, Tokyo, p. 33.

200. A. N. Frumkin, Z. Elektrochem., 59, 807 (1955).

201. For a general review, see P. Delahay, Double Layer and Electrode Kinetics, Wiley (Interscience), New York, 1965.

202. H. Matschiner, Thesis, Polarographic Institute, Prague, 1962.

203. A. A. Vlček, Proceedings of the 7th I.C.C.C., Stockholm and Uppsala, June 25-29, 1962, L. G. Sillén, ed., Alinquist and Wiksell, Uppsala, Sweden, 1962, p. 285.

204. A. A. Vlček, "A Plenary Lecture," in The 8th International Conference on Coordination Chemistry (Austria), Butterworths, London, 1964, pp. 61-70.

205. J. Hanzlík and A. A. Vlček, Chem. Commun., 1969, 47.

206. N. Maki and Y. Ishiuchi, Bull. Chem. Soc. Japan, 44, 1721 (1971).

207. H. S. Lim and F. C. Anson, J. Electroanal. Chem., 31, 297 (1971).

208. For a general review, see J. Kwiatek, in Catalysis Reviews, Vol. 1 (H. Heinemann, ed.), Dekker, New York, 1968, pp. 38-152.

209. E.g., J. Kwiatek, I. L. Mador, and J. K. Seyler, in Reactions of Coordinated Ligands and Homogeneous Catalysis (Advances in Chemistry Series, No. 37) (D. H. Busch, ed.), American Chemical Society, Washington, D.C., 1963.

210. H. Matsuda and Y. Ayabe, Bull. Chem. Soc. Japan, 28, 422 (1955).

211. E. A. Guggenheim, Thermodynamics, North-Holland, Amsterdam, 1950.

212. R. Parsons, in Modern Aspects of Electrochemistry, Vol. 1 (J. O'M. Bockris and B. E. Conway, eds.), Butterworths, London, 1954.

213. R. Tamamushi, Rev. Polarogr. (Japan), 10, 1 (1962).

214. A. A. Vlček, Advances in Chemistry of Coordination Compounds [Proceedings of the 6th I.C.C.C. (Detroit)] (S. Kirschner, ed.), Macmillan, New York, 1961, pp. 590-603.

215. A. A. Vlček, Collect. Czech. Chem. Commun., 24, 3538 (1959).

216. A. N. Frumkin, Ergebn. Exakt. Naturw., 7, 235 (1928).

217. L. I. Antropov, J. Indian Chem. Soc., 35, 309 (1958).

218. L. I. Antropov and S. N. Banerjee, ibid., 35, 531 (1958).

219. L. I. Antropov and S. N. Banerjee, ibid., 36, 451 (1959).

220. T. Kitagawa and S. Tsushima, Rev. Polarogr. (Japan), 14, 17 (1966).

221. A. Yamada and N. Tanaka, Bull. Chem. Soc. Japan, 42, 1600 (1969).

222. H. Bartelt and S. Landazury, J. Electroanal. Chem., 22, 105 (1969).

223. N. S. Biradar, D. R. Stranks, and M. S. Vaidya, Trans. Faraday Soc., 58, 2421 (1962).

224. H. Bartelt and H. Skilandat, J. Electroanal. Chem., 23, 407 (1969).

225. H. Bartelt, ibid., 25, 79 (1970).

226. H. Bartelt and H. Skilandat, ibid., 24, 207 (1970).

227. H. Bartelt, Electrochim. Acta, 16, 307 (1971).

228. H. Bartelt, ibid., 16, 629 (1971).

229. H. Bartelt and M. Prügel, ibid., in press.

230. D. R. Stranks, Advances in Chemistry of Coordination Compounds [Proceedings of the 6th I.C.C.C., (Detroit)] (S. Kirschner, ed.), Macmillan, New York, 1961, p. 571.

231. R. Larson and J. Tobiasson, Acta Chem. Scand., 16, 1919 (1962).

232. R. Larson, ibid., 16, 2267 (1962).

233. F. P. Dwyer and A. M. Sargeson, J. Phys. Chem., 65, 1892 (1961).

234. R. A. Marcus, J. Chem. Phys., 43, 679 (1965).

235. H. Bartelt and M. Prügel, J. Electroanal. Chem., 29, 293 (1971).

236. H. A. Laitinen and J. E. B. Randles, Trans. Faraday Soc., 51, 54 (1955).

237. H. Bartelt and M. Prügel, J. Electroanal. Chem., 32, 309 (1971).

238. L. N. Klatt and W. J. Blaedel, Anal. Chem., 39, 1065 (1967).

239. J. Suzuki, Rev. Polarogr. (Japan), 15, 21 (1968).

240. J. Suzuki, Bull. Chem. Soc. Japan, 42, 3847 (1969).

241. J. Suzuki, ibid., 43, 755 (1970).

242. A. B. Sheinin, V. A. Zinov'ev, and V. L. Kheifets, Zh. Fiz. Khim., 35, 513 (1961); Chem. Abstr., 55, 17305a (1961).

243. H. Matsuda and Y. Ayabe, Z. Elektrochem., 59, 494 (1955).

244. J. N. Gaur and N. K. Goswami, Electrochim. Acta, 15, 519 (1970).

245. Y. Hohokabe and N. Yamazaki, Bull. Chem. Soc. Japan, 44, 1563 (1971).

246. A. A. Vlček, Nature, 197, 786 (1963).

247. E. Verdier and F. Rouelle, C. R. Acad. Sci., Paris, 259, 1856 (1964).

248. E. Verdier and F. Rouelle, J. Chim. Phys., 62, 297 (1965).

249. L. Gierst, in Transaction of the Symposium on Electrode Processes (E. Yeager, ed.), Wiley, New York, 1961, pp. 109-144.

250. I. M. Kolthoff and S. E. Khalafalla, Rev. Polarogr. (Japan), 11, 11 (1963).

251. B. Levich, Dokl. Akad. Nauk S.S.S.R., 67, 309 (1949).

252. M. Breiter, M. Kleinerman, and P. Delahay, J. Amer. Chem. Soc., 80, 5111 (1958).

253. H. Matsuda and P. Delahay, J. Phys. Chem., 64, 332 (1960).

254. M. Senda and P. Delahay, ibid., 65, 1580 (1961).

255. A. A. Vlček, Collect. Czech. Chem. Commun., 24, 3538 (1960).

256. W. Biltz, Z. Anorg. Allgem. Chem., 83, 188 (1913).

257. N. Maki, Rev. Polarogr. (Japan), 9, 203 (1961).

258. I. M. Kolthoff and S. E. Khalafalla, Inorg. Chem., 2, 133 (1963).

259. I. M. Kolthoff and S. E. Khalafalla, J. Amer. Chem. Soc., 85, 664 (1963).

260. K. Ogino and N. Tanaka, Bull. Chem. Soc. Japan, 40, 1119 (1967).

261. N. Tanaka, K. Ogino, and G. Sato, ibid., 39, 366 (1966).

262. J. Kůta, J. Weber, and J. Koutecky, Collect. Czech. Chem. Commun., 25, 2376 (1960).

263. J. Kůta and I. Smoler, J. Electroanal. Chem., 12, 535 (1966).
264. R. G. Barradas and F. M. Kimmerle, Anal. Lett., 1(3), 179 (1967).
265. S. Sakuraba, N. Maki, and T. Hamazaki, Ann. Rep. Fac. Eng. Shizuoka Univ., 19, 45 (1968).
266. A. A. Vlček and J. Kůta, Nature, 185, 95 (1960).
267. N. Tanaka, Y. Sato, and R. Tamamushi, Rev. Polarogr., 12, 127 (1964).
268. F. C. Anson and T-L. Chang, Inorg. Chem., 5, 2092 (1966).
269. H. F. Holtzclaw, Jr., J. Amer. Chem. Soc., 73, 1821 (1951).
270. H. Ikeuchi, J. Electroanal. Chem., 16, 405 (1968).
271. K. Yamashita and H. Imai, Nippon Kagaku Zasshi, 72, 101 (1969).
272. W. U. Malik and M. Aslam, Electrochim. Acta, 13, 263 (1968).
273. W. U. Malik and M. Aslam, ibid., 15, 689 (1970).
274. J. Tomes, Collect. Czech. Chem. Commun., 9, 12 (1937).
275. J. Tomes, ibid., 9, 81 (1937).
276. J. Tomes, ibid., 9, 150 (1937).
277. R. Ripan, A. Farcas, and O. Piringer, Z. Anorg. Allgem. Chem., 346, 211 (1966).
278. J. G. Mason and R. L. White, J. Electroanal. Chem., 8, 454 (1964).
279. R. D. Hargens, W. Min, and R. C. Henney, ibid., 26, 285 (1970).
280. P. J. Sherwood and H. A. Laitinen, J. Phys. Chem., 74, 1757 (1970).
281. H. A. Laitinen and L. M. Chambers, Anal. Chem., 36, 1881 (1964).
282. B. Timmer, M. Sluyters-Rehbach, and J. H. Sluyters, J. Electroanal. Chem., 15, 343 (1967).
283. E. Fischerova, O. Dracka, and M. Meloun, Collect. Czech. Chem. Commun., 33, 473 (1968).
284. P. Kivalo, J. Amer. Chem. Soc., 77, 2678 (1955).
285. J. Dolezal, Chem. Listy, 49, 1237 (1955).
286. E. H. Lyons, Jr., Z. Elektrochem., 59, 766 (1955).
287. V. Carunchio and L. Campanella, Ann. Chim. (Rome), 57, 1372 (1967); Chem. Abstr., 68, 45544a (1968).
288. C. Gh. Macarovici, F. Manok, and Cs. Varhelyi, Rev. Roum. Chim., 12, 279 (1967); Chem. Abstr., 67, 78483n (1967).
289. E. Jacobsen and K. Schrøder, J. Phys. Chem., 66, 134 (1962).
290. E. T. Verdier and J. Piro, J. Chim. Phys., Physicochim. Biol., 66, 812 (1969).
291. J. R. Urwin and B. West, J. Chem. Soc., 1952, 4727.
292. W. P. Griffith and G. Wilkinson, ibid., 1959, 2757.
293. N. Maki, M. Yamagami, and H. Itatani, J. Amer. Chem. Soc., 86, 514 (1964).
294. N. Maki and Y. Ishiuchi, Bull. Chem. Soc. Japan, 44, 1721 (1971).
295. K. M. Abubacker and W. U. Malik, Proc. Indian Acad. Sci., A50, 132 (1959).
296. M. Takahashi, M. Kuwahara, and M. Iguchi, Nippon Kagaku Zasshi, 91, 543 (1970).
297. A. A. Vlček and J. Hanzlík, Inorg. Chem., 6, 2053 (1967).
298. J. Hanzlík and A. A. Vlček, ibid., 8, 669 (1969).
299. N. Maki, Bull. Chem. Soc. Japan, 44, 2283 (1971).

300. A. V. Ablov, O. A. Bologa, and N. M. Samus, Inorg. Chem. (Russian), 11, 2632 (1966); Chem. Abstr., 66, 43277a (1966).

301. A. V. Ablov, N. M. Samus, and O. A. Bologa, ibid., (Russian), 14, 3320 (1969).

302. R. D. Gillard and G. Wilkinson, J. Chem. Soc., 1963, 6041.

303. M. Spritzer and L. Meites, Anal. Chim. Acta, 26, 58 (1962).

304. G. N. Schrautzer, Acct. Chem. Res., 1, 97 (1968).

305. P. K. Das, H. A. O. Hills, J. M. Pratt, and R. J. P. Williams, J. Chem. Soc., A, 1969, 1261.

306. D. J. Macero, F. V. Lovecchio, and S. J. Pace, Inorg. Chim. Acta, 3, 65 (1969).

307. R. Brdička, Collect. Czech. Chem. Commun., 2, 489 (1930).

308. R. Brdička, ibid., 3, 396 (1931).

309. D. S. Jain and K. Zutshi, J. Electroanal. Chem., 5, 389 (1963).

310. E. T. Verdier and M. G. Baptiste, ibid., 10, 42 (1965).

311. A. E. Martell and M. Calvin, The Chemistry of the Metal Chelate Compounds, Prentice-Hall, New York, 1952, p. 337.

312. P. Silvestroni and G. Illuminati, Ric. Sci., 28, 1211 (1958).

313. P. Silvestroni, ibid., 29, 301 (1959).

314. V. Caglioti, P. Silvestroni, and C. Furlani, J. Inorg. Nucl. Chem., 13, 95 (1960).

315. I. M. Kolthoff, P. Mader, and S. E. Khalafalla, J. Electroanal. Chem., 18, 315 (1968).

316. A. Calusaru, ibid., 15, 269 (1967).

317. P. Mader and I. M. Kolthoff, Anal. Chem., 41, 932 (1969).

318. A. Calusaru and V. Voicu, J. Electroanal. Chem., 20, 463 (1969).

319. P. Mader and I. M. Kolthoff, J. Polarogr. Soc., 14, 42 (1968).

320. M. Asthana, R. C. Kapoor, and M. L. Nigam, Electrochim. Acta, 11, 1587 (1966).

321. A. A. Vlček, Collect. Czech. Chem. Commun., 25, 3036 (1960).

322. S. Barnartt and R. G. Charles, J. Electrochem. Soc., 109, 333 (1962).

323. S. P. Perone and W. F. Gutknecht, Anal. Chem., 39, 892 (1967).

324. D. Kyriacou, ibid., 37, 1036 (1965).

325. I. Stotz, W. K. Wilmarth, and A. Haim, Inorg. Chem., 7, 1250 (1968).

326. M. E. Kimball and W. C. Kaska, Inorg. Nucl. Chem. Lett., 7, 119 (1971).

327. A. M. Bond, G. A. Heath, and R. L. Martin, Inorg. Chem., 10, 2026 (1971).

328. S. L. Tackett and J. W. Ide, J. Electroanal. Chem., 30, 510 (1971).

329. H. S. Lim and F. C. Anson, Inorg. Chem., 10, 103 (1971).

330. C. Olson and R. N. Adams, Anal. Chim. Acta, 29, 358 (1963).

331. T. Kitagawa, Rev. Polarogr. (Japan), 14, 1 (1966).

332. T. Kitagawa and S. Tsushima, ibid., 14, 17 (1966); Bull. Chem. Soc. Japan, 39, 636 (1966).

333. F. C. Anson, J. Electrochem. Soc., 110, 436 (1963).

334. E. T. Verdier, Collect. Czech. Chem. Commun., 32, 3500 (1967).

335. F. C. Anson, Anal. Chem., 36, 520 (1964).

336. A. T. Hubbard and F. C. Anson, ibid., 38, 1601 (1966).

337. W. K. Behl and J. E. Toni, J. Electroanal. Chem., 31, 63 (1971).

338. P. Benson, G. W. D. Briggs, and W. F. K. Wynne-Jones, Electrochim. Acta, 9, 281 (1964).

339. P. Benson, G. W. D. Briggs, and W. F. K. Wynne-Jones, ibid., 9, 275 (1964).

340. R. D. Cowling and A. C. Riddiford, ibid., 14, 981 (1969).

341. R. S. Young, ed., Cobalt, Its Chemistry, Metallurgy and Uses, Reinhold, New York, 1961, pp. 40-41.

342. C. G. Fink and J. L. Hutton, Trans. Electrochem. Soc., 85, 119 (1944).

343. G. S. Alberts, R. H. Wright, and C. C. Parker, J. Electrochem. Soc., 113, 687 (1966).

344. W. E. Clark and M. L. Holt, ibid., 94, 244 (1948).

345. G. T. Seaman, J. R. Myers, and R. K. Saxer, Electrochim. Acta, 12, 855 (1967).

346. W. Crow, J. R. Myers, and R. K. Saxer, Proceedings of the 3rd International Congress on Metallic Corrosion, Moscow, U.S.S.R., Butterworths, London, 1966.

347. Z. A. Iofa and W. Pao-Ming, Zh. Fiz. Khim., 36, 1395 (1962).

348. A. Murtazaev, ibid., 23, 1247 (1949); Chem. Abstr., 44, 1344 (1950).

349. U. Ebersbach, K. Schwabe, and K. Ritter, Electrochim. Acta, 12, 927 (1967).

350. N. Katoh, Nippon Kagaku Zasshi, 64, 1211 (1943); Chem. Abstr., 41, 3338 (1947).

351. B. K. Hovsepian and I. Shain, J. Electroanal. Chem., 14, 1 (1967).

352. W. M. Schwarz and I. Shain, J. Phys. Chem., 69, 30 (1965).

353. W. J. M. Tegart, The Electrolytic and Chemical Polishing of Metals in Research and Industry, Pergamon, London, 1956.

354. N. Maki, J. Electroanal. Chem., 51, 353 (1974).

355. N. Maki, Inorg. Chem., 13, 2180 (1974).

356. N. Maki, Chem. Lett., 5, 521 (1973).

Chapter III-3

NICKEL

ALEJANDRO JORGE ARVIA

and

DIONISIO POSADAS

Instituto de Investigaciones Fisicoquimicas Teóricas y Aplicadas
División Electroquímica
Universidad Nacional de La Plata
La Plata, Argentina

1. STANDARD AND FORMAL POTENTIALS

1.1. AQUEOUS SOLUTION

1.1.1. Electrode Potentials and Electromotive Force of Cells

The potential of the Ni/Ni(II) couple cannot be given with accuracy because a truly reversible equilibrium appears difficult to attain. Values of the standard electrode potential as given by different authors are in Table 1.1.1. Latimer [6] adopted the standard potential of the Ni/Ni(II) couple as equal to -0.250 V. However, the calculations of Larson, Cerutti, Garber, and Kepler [7] from the free energy of the reaction $Ni(c) = Ni^{2+}(aq) + 2e$ give the value -0.228 V, in good agreement with that reported by Carr and Bonilla [4]. The dispersion of standard potential values derived from electromotive force measurements [1-3, 5] is probably due either to oxygen interference or to the onset of mixed potentials instead of reversible potentials.

The real reversible potential of the Ni(II)/Ni(III) couple equal to 0.424 V (vs HgO electrode) has been determined by Conway and Bourgault [18] and Conway and Gileadi [13] at the intersection point of the anodic and cathodic potential/log(time) decay curves. From the open circuit potential of the positive plate of the alkali accumulator, Zedner [19-21], Foerster [22, 23], and Foerster and Krueger [24] obtained a value between 0.47 and 0.49 V (vs HgO electrode).

Table 1.1.1 contains potential values of electrode reactions involving different nickel oxygen-containing species. Most of the figures were calculated from thermal data. The interpretation of the experimental data is open to question because nickel oxide electrodes may involve nonstoichiometric oxidation states and consequently no equilibrium potentials could be recorded. They may correspond to "oxide/oxide" rather than metal/oxide redox systems. Hickling and Spice [25] measured the potential of the cell Pt/NiO(Ni)/NaOH(x<u>M</u>)/NHE and obtained a potential of 0.56 V for 1 <u>M</u> NaOH. After correcting for the hydrogen electrode in basic solutions, the potential is equal to 1.39 V. This figure lies close to the standard potential of the β-NiO(OH)/NiO_2 couple, according to Pourbaix, de Zoubov, and Deltombe [12], but as NiO_2 has never been isolated, under steady-state conditions the oxidation state of the metal should be related to Ni(III). Then the equilibrium involving nickel hydroxide in alkaline solutions is $\beta\text{-}NiO(OH) + H_2O + e = Ni(OH)_2 + OH^-$. Experimental electrode potentials of hydroxo nickel electrodes are assembled in Table 1.1.2. Table 1.1.3 refers to the electromotive force of galvanic cells with the participation of nickel-oxygen compounds. Data reported by Bourgault and Conway [18] and Conway and Gileadi [13] are considered the most reliable ones. Experimental and calculated electrode potentials of nickel sulfide electrodes are given in Table 1.1.4. Experimental data are subject to the same objections already reported for the other electrodes. Other electrochemical measurements assigned to systems at equilibrium are given in Gmelin's Handbook [35].

TABLE 1.1.1. Standard Electrode Potentials

Oxidation number change	Half-reaction	Std or formal potential (V)	Conditions	Refs.
$2 \rightarrow 0$	$Ni^{2+} + 2e = Ni$	-0.227 ± 0.002	$20^\circ C$, 1 $\underline{N}$ $NiSO_4$	[1]
		-0.2496	$25^\circ C$, 1 m Ni(II)	[2]
		-0.2480	1 m Ni(II) activity	[2]
		-0.231	1 $\underline{N}$ $NiSO_4$	[3]
		-0.232	Computed from isothermal measurements	[4]
		-0.246	Computed from exptl measurements	[5]
		-0.250	calcd	[6]
		-0.228	calcd from exptl thermal data	[7]
		$(dE^\circ/dT)_T = 0.93 \times 10^{-3}$; $(dE^\circ/dT)_i = 0.06 \times 10^{-3}$		[8]
		$(\Delta E/\Delta T) = -0.09 \times 10^{-3}$; 0.05 $\underline{M}$ $NiSO_4$		[2]
		$(\Delta E/\Delta T) = -0.03 \times 10^{-3}$; 0.005 $\underline{M}$ $NiSO_4$		[2]
	$NiO + 2H^+ + 2e = Ni + H_2O$	0.110	calcd	[9]
	$NiO.H_2O + 2H^+ + 2e = Ni + 2H_2O$	0.117	calcd	[10], [11]
	$NiO_{ads}.H_2O + 2H^+ + 2e = H_2O.Ni + H_2O$	0.130	calcd	[10], [11]
	✓ $Ni(OH)_2 + 2e = Ni + 2OH^-$	-0.72	calcd	[6], [8]
	$Ni(OH)_2 + 2H^+ + 2e = Ni + 2H_2O$	0.116	calcd	[9]

(continued)

TABLE 1.1.1 (continued)

Oxidation number change	Half-reaction	Std or formal potential (V)	Conditions	Refs.
	$HNiO_2^- + 3H^+ + 2e = Ni + 2H_2O$	-0.648	calcd	[9]
	$Ni(NH_3)_6^{2+} + 2e = Ni + 6NH_3(aq)$	-0.49	calcd	[6]
	$NiCO_3 + 2e = Ni + CO_3^{2-}$	-0.45	calcd solubility	[6]
2 → 1	$Ni(CN)_4^{2-} + e = Ni(CN)_4^{3-}$	-0.82	measured	[6]
2.67 → 0	$Ni_3O_4 + 4H_2O + 8e = 3Ni + 8OH^-$	-0.505	calcd	[10], [11]
	$Ni_3O_4 + 6H^+ + 6e = 3Ni + \frac{1}{2}O_2 + 3H_2O$	0.0	calcd	[12], [13]
	$Ni_3O_4 + 8H^+ + 8e = 3Ni + 4H_2O$	0.307	calcd	[10], [11]
2.67 → 2	$Ni_3O_4 + 2H_2O + 2e = 3HNiO_2 + H^+$	-0.718	calcd	[9]
	$Ni_3O_4 + 2H^+ + 2e = 3NiO + H_2O$	0.876 0.897	Hydrated oxide Unhydrated oxide	[9] [9]
	$Ni_3O_4 + 8H^+ + 2e = 3Ni^{2+} + 4H_2O$	1.977	calcd	[9]
3 → 0	$Ni_2O_3 + 3H_2O + 6e = 2Ni + 6OH^-$	-0.450	calcd	[10], [11]
	$Ni_2O_3 + 4H^+ + 4e = 2Ni + \frac{1}{2}O_2 + 2H_2O$	0.012	calcd	[10], [11]
	$Ni_2O_3 + 6H^+ + 6e = 2Ni + 3H_2O$	0.418	calcd	[10], [11]
	$Ni(OH)_3 + 3H^+ + 3e = Ni + 3H_2O$	0.586	calcd	[10], [11]

3 → 2	$Ni_2O_3 + H_2O + 2e = 2HNiO_2^-$	-0.044	calcd	[10], [11]
	$Ni(OH)_3 + H^+ + e = Ni(OH)_2 + H_2O$	1.020	calcd	[9]
	$Ni_2O_3 + 2H^+ + 2e = 2NiO + H_2O$	1.032	calcd	[9]
	$Ni^{3+} = Ni^{2+} + e$	1.168 1.174	225°C; ref: Ni, H_2(atm) 250°C	[14] [14]
	$Ni_2O_3 + 6H^+ + 2e = 2Ni^{2+} + 3H_2O$	1.753	calcd	[9]
	$Ni(OH)_3 + 3H^+ + e = Ni^{2+} + 3H_2O$	2.08	calcd	[15]
	$Ni(OH)_3 + 3H^+ + e = Ni^{2+} + 3H_2O$	2.26	calcd	[12], [13]
3 → 2.67	$3NiO(OH) + H^+ + e = Ni_3O_2(OH)_4$	-0.478	calcd	[16]
	$3Ni_2O_3 + 2H^+ + 2e = 2Ni_3O_4 + H_2O$	1.305	calcd	[9]
4 → 0	$NiO_2 + 2H^+ + 2e = Ni + \frac{1}{2}O_2 + H_2O$	0.114	calcd	[10], [11]
	$NiO_2 + 4H^+ + 4e = Ni + 2H_2O$	1.678	calcd	[8]
4 → 2	$NiO_2 + 2H_2O + 2e = Ni(OH)_2 + 2OH^-$	0.49	exptl value	[4]
	$NiO_2(sol) + 2H_2O + 2e = Ni^{2+} + 4OH^-$	1.175	225°C; ref: Ni, H_2(atm)	[14]
	$NiO_2 + 4H^+ + 2e = Ni^{2+} + 2H_2O$	1.593	calcd	[9]
	$NiO_2 + 2H^+ + 2e = Ni^{2+} + 2OH^-$	1.75	calcd	[17]
4 → 3	$2NiO_2 + 2H^+ + 2e = Ni_2O_3 + H_2O$	1.434	calcd	[9]
	$NiO_2 + 2H_2O + e = Ni^{3+} + 4OH^-$	1.198 1.193	225°C; ref: Ni, H_2(atm) 250°C	[14] [14]
6 → 2	$NiO_4^{2-} + 8H^+ + 4e = Ni^{2+} + 4H_2O$	>1.8	calcd	[6]

TABLE 1.1.2. Standard Electrode Potentials Involving Hydroxy-Nickel Species

Half-cell	E (NHE) (V)	Remarks	Refs.
Ni, NiO(OH)/N $NiCl_2$	$E_{initial}$ = 1.17	Diminishes with time; pH = 5	[26]
Ni, NiO(OH)/N $Ni(ClO_4)_2$	1.19	pH = 5	[26]
Ni, NiO(OH)/N $NiBr_2$	$E_{initial}$ = 1.02	Diminishes with time; pH = 5	[26]
Ni, NiO(OH)/N $NiSO_4$	1.19	pH = 5	[26]
Ni, NiO(OH)/N NaOH	0.56	pH = 13.8	[12]
Ni, NiO(OH)/2.8 N KOH	$E_{initial}$ = 0.60	Diminishes with time	[26]
Ni, NiO(OH)/0.1 N $Na_2B_4O_7$	0.85	pH = 9.2	[12]
Ni, NiO(OH)/N Na_2CO_3	0.68	pH = 12	[12]
Pt/$Ni_3O_2(OH)_4$, NiO(OH)/0.1 N NaOH	0.15	N_2 atm	[27]
Pt/$Ni_3O_2(OH)_4$, NiO(OH)/N Na_2CO_3	0.24	-	[27]
Ni/$Ni(OH)_2$/0.1 N NaOH	-0.60	N_2 atm	[27]
Ni/$Ni(OH)_2$/25% KOH	-0.33	Ref: NCE	[28]
Ni/$Ni(OH)_2$/N Na_2CO_3	-0.51	-	[27]
Pt/$Ni(OH)_2 \cdot Ni_3O_2(OH)_4$/0.1 N NaOH	-0.35	N_2 atm	[15]
Pt/$Ni(OH)_2 \cdot Ni_3O_2(OH)_4$/N Na_2CO_3	-0.26	pH = 11.5	[15]

TABLE 1.1.3. Open Circuit Potentials of Galvanic Cells Incorporating Nickel Species

Galvanic cell	E (V)	Conditions	Refs.
NiO(OH)/25% KOH/H_2	1.305 (10°C)	E_{calcd} from $Ni(OH)_3 \rightarrow Ni(OH)_2$ at 10°C, 1.308	[20], [21]
NiO(OH)/25% KOH/H_2	1.266 (65°C)	—	[20], [21]
NiO(OH)/25% KOH + 20 g/l $Zn(OH)_2$/Zn(Hg)	1.75 (10-20°C)	10% Zn-90% Hg amalgam	[19]
NiO(OH)/N $NiSO_4$/Zn	1.630	—	[29]
NiO(OH)/satd $(NH_4)_2Ni(SO_4)_2$/Zn	1.643	—	[29]
NiO(OH)/N $NiSO_4$/Pb	1.139	—	[29]
NiO(OH)/satd $(NH_4)_2Ni(SO_4)_2$/Pb	1.138	—	[29]
NiO(OH)/N $NiSO_4$/Ni	1.086	—	[29]

(continued)

TABLE 1.1.3 (continued)

Galvanic cell	E (V)	Conditions	Refs.
NiO(OH)/satd $(NH_4)_2Ni(SO_4)_2$/Ni	1.075	—	[29]
NiO(OH)/N $NiSO_4$/Cu	0.551	—	[29]
NiO(OH)/satd $(NH_4)_2Ni(SO_4)_2$/Cu	0.560	—	[29]
NiO(OH), $Ni(OH)_2$/KOH/HgO, Hg	E ~ log C_{KOH} (up to 1 N)	0.01 N ≤ C_{KOH} ≤ 8 N	[30]
Ni/$NiO_{1.10}$/N KOH/HgO/Hg	0.423 ± 0.005	—	[13], [18]
Ni/$NiO_{1.20}$/N KOH/HgO/Hg	0.427	—	[13], [18]
Ni/$NiO_{1.25}$/N KOH/HgO/Hg	0.423	—	[13], [18]
Ni/$NiO_{1.25}$/0.01 N KOH/HgO/Hg	0.429	—	[13], [18]
Ni/$NiO_{1.25}$/0.1 N KOH/HgO/Hg	0.425	—	[13], [18]
Ni/$NiO_{1.25}$/1.01 N KOH/HgO/Hg	0.419	—	[13], [18]
Ni/$NiO_{1.25}$/7.3 N KOH/HgO/Hg	0.390	—	[13], [18]
Ni/$NiO_{1.25}$/14.6 N KOH/HgO/Hg	0.297	—	[13], [18]

TABLE 1.1.4. Standard Electrode Potentials of Nickel Sulfide Electrodes

Half-electrode	E (V)	Remarks	Refs.
Ni_3S_2/N $NiSO_4$	E' = -0.235	SCE	[31]
Ni_3S_2/200$NiSO_4$·7H_2O + 40Na_2SO_4 + 20H_3BO_3 + 3NaCl	E' = 0.07	NHE	[32]
Ni_3S_2/100 g/l H_2SO_4	E' = 0.202	SCE	[33]
NiS/satd NaCl	E' = -0.330	20°C; SCE	[34]
NiS(α) + 2e = Ni + S^{2-}	E° = -0.83	calcd	[6]
NiS(γ) + 2e = Ni + S^{2-}	E° = -1.04 E° = -1.07	calcd calcd	[6] [17]

TABLE 1.1.5. Equilibrium Constants of Selected Reactions Involving Nickel(II)

Reaction	K	Conditions	Refs.
$Ni(OH)_2 = Ni^{2+} + 2OH^-$	1.6×10^{-16}	-	[6]
	6.5×10^{-16}	-	[36]
	1.6×10^{-14}	-	[37]
	6.2×10^{-16}	-	[38]
$Ni^{2+} + H_2O = NiOH^+ + H^+$	5.0×10^{-9}	$NiSO_4$ var	[39], [40]
	3.2×10^{-7}	$NiCl_2$	[39], [40]
	4.0×10^{-10}	30°C, $\mu = 0.1$ (KCl)	[39], [41]
	3.2×10^{-10}	$NiSO_4$	[42]
	5.95×10^{-10}	$Ni(NO_3)_2$	[42]
	5.1×10^{-10}	$NiCl_2$	[42]
	5.02×10^{-10}	$Ni(ClO_4)_2$	[42]
	1.2×10^{-9}	$\mu = 0$(corr), 20°C	[39], [43]
	2.3×10^{-11}	$\mu = 0$(corr)	[39], [44]
	2.5×10^{-9}	dil $NiCl_2$	[39], [45]
	1.7×10^{-10}	$0.25 < \mu < 1$ ($NaClO_4$)	[46]
$NiOH^+ + H_2O = Ni(OH)_2 + H^+$	2.7×10^{-5} to 8.5×10^{-3}	calcd	[42]
$Ni^{2+} + H_2O = NiO + 2H^+$	-	$\log(Ni^{2+}) = 12.18 - 2pH$ (hydrated oxide)	[9]
	-	$\log(Ni^{2+}) = 12.41 - 2pH$ (unhydrated oxide)	[9]
$NiO + H_2O = HNiO_2^- + H^+$	-	$\log(HNiO_2^-) = -18.22 + pH$ (hydrated oxide)	[9]
	-	$\log(HNiO_2^-) = -17.99 + pH$ (unhydrated oxide)	[9]
$Ni(NH_3)_4^{2+} = Ni^{2+} + 4NH_3(aq)$	1×10^{-8}	-	[6]
$Ni(NH_3)_6^{2+} = Ni^{2+} + 6NH_3(aq)$	1.8×10^{-9}	-	[6]
$Ni(CN)_4^{2-} = Ni^{2+} + 4CN^-$	10^{-22}	-	[47]
$NiCO_3 = Ni^{2+} + CO_3^{2-}$	1.36×10^{-7}	-	[48]
$NiS(\alpha) = Ni^{2+} + S^{2-}$	3×10^{-21}	-	[49]
$NiS(\beta) = Ni^{2+} + S^{2-}$	1×10^{-26}	-	[49]
$NiS(\gamma) = Ni^{2+} + S^{2-}$	2×10^{-28}	-	[49]

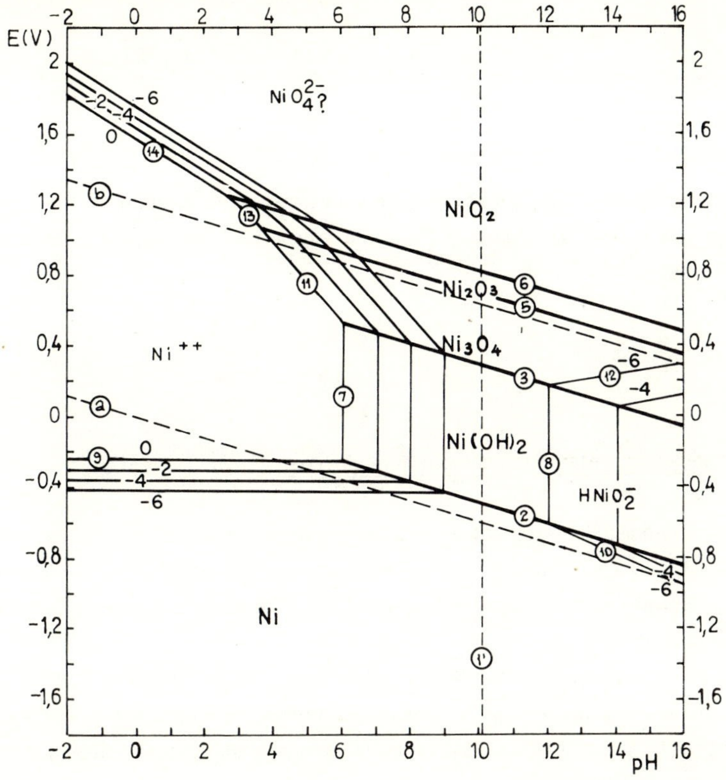

FIG. 1.1.1. Potential-pH equilibrium diagram for the system nickel-water at 25°C [9]. (2) $Ni(s) + H_2O(l) = NiO(s) + 2H^+ + 2e$; (3) $3NiO(s) + H_2O(l) = Ni_3O_4(s) + 2H^+ + 2e$; (5) $2Ni_3O_4(s) + H_2O(l) = 3Ni_2O_3(s) + 2H^+ + 2e$; (6) $Ni_2O_3(s) + H_2O(l) = 2NiO_2(s) + 2H^+ + 2e$; (7) $Ni^{2+} + H_2O(l) = NiO(s) + 2H^+$; (9) $Ni(s) = Ni^{2+} + 2e$; (10) $Ni(s) + 2H_2O(l) = HNiO_2^- + 3H^+ + 2e$; (11) $3Ni^{2+} + 4H_2O = Ni_3O_4(s) + 8H^+ + 2e$; (12) $3HNiO_2^- + H^+ = Ni_3O_4(s) + 2H_2O(l) + 2e$; (13) $2Ni^{2+} + 3H_2O(l) = Ni_2O_3(s) + 6H^+ + 2e$; (14) $Ni^{2+} + 2H_2O(l) = NiO_2(s) + 4H^+ + 2e$; (a) $H_2 = 2H^+ + 2e$; (b) $2H_2O = O_2 + 4H^+ + 4e$; (1') $Ni^{2+}/HNiO_2^-$, pH = 10.13. (By permission of Pergamon Press.)

1.1.2. Equilibrium Data

Equilibrium constants related to complex formation of interest in the electrochemistry of nickel are assembled in Table 1.1.5. The more reliable formal first hydrolysis constant of $Ni^{2+}(aq)$ is given, according to Arvía, Bolzan, and Jáuregui [46], by $pK'_a = 1692/T + 4.09$.

1.1.3. Potential/pH Diagrams

Potential/pH diagrams for nickel in aqueous solutions are calculated on the basis of the thermodynamic data shown in Tables 1.1.1 and 1.1.5, following the procedure indicated by Pourbaix [9]. The equilibrium diagram at 25°C for aqueous solutions is shown in Fig. 1.1.1.

Goret and Trémillon [50] established an approximate potential/p(H_2O) diagram for nickel in 53.1% NaOH + 46.9% KOH at 227°C.

1.2. NONAQUEOUS SOLUTION

The standard potential of the nickel electrode in nonaqueous systems has not been determined. Delgado, Posadas, and Arvía [51] measured the rest potential of a nickel electrode in HCl-DMSO solutions.

1.3. MOLTEN SALTS

1.3.1. Electrode Potentials and Electromotive Forces

The standard potential of Ni/Ni(II) was calculated from thermal data for the reaction: $Ni + X_2 = NiX_2$ (X = F, Cl, Br, I, O) occurring in a galvanic cell without liquid junction. The standard potentials of Ni/Ni(II) are therefore referred to the standard gas electrode, as shown in Table 1.3.1.

Grjotheim [56] studied concentration and Daniell-type cells with NiF_2 at 1082°K. Good agreement is obtained for the former between experimental and calculated electromotive forces. Results obtained from Daniell-type cells involving nickel and a different metal are shown in Table 1.3.2, where the first metal of each pair is the negative pole in the observation [56]. Jenkins, Mamantov, and Manning [57, 58] used the Ni/Ni(II) electrode as reference in molten $LiF-BeF_2-ZrF_4$ mixture at 500°C.

Formal potentials of nickel in molten chlorides and molten iodides were measured by different authors, as shown in Table 1.3.3. For chloride melts the following cells were employed:

$$Ni/NiCl_2(x)\text{-}MgCl_2\text{-}NaCl\text{-}KCl/\text{glass frit}/MgCl_2\text{-}NaCl\text{-}KCl/Cl_2, C$$

$$Ag/Ag(I)(x_1)\text{-}MgCl_2\text{-}KCl//Ni(II)(x)\text{-}MgCl_2\text{-}KCl/Ni$$

$$Ni/NiCl_2(x)\text{-}MgCl_2\text{-}KCl//MgCl_2\text{-}KCl/Cl_2, C$$

Formal electrode potentials measured in molten $MgCl_2$-KCl vs the Ag(I)/Ag(0) system as reference were converted to the standard Cl_2/Cl^- electrode, showing agreement within 1 to 6 mV.

Hamby and Scott [67] determined the thermodynamic properties, including activity coefficient, partial molar free energy, enthalpy, and entropy, of mixtures for the solute $NiCl_2$ (10^{-4} to 4×10^{-1} mole fraction) in solvents KCl, NaCl, LiCl and 1:1 NaCl-KCl at temperatures ranging from the melting point of the solvents up to 900°C, with the cell $Ni/NiCl_2(x)$, MCl/Cl_2, C, where M represented an alkali metal. Below 10^{-2} mole fraction, the potential at

TABLE 1.3.1. Potentials of Cells Involving Molten Salts of Nickel

Cell	Reaction	T (°C)	EMF (V)	Condition	Refs.
$Ni/NiF_2(\ell)/F_2, C$	$Ni + F_2 = NiF_2$	25	3.116	aq	[52]
$Ni/NiF_2(\ell)/F_2, C$	$Ni + F_2 = NiF_2$	25	3.226	s, uncertainty ± 0.02 V	[52]
$Ni/NiF_2(\ell)/F_2, C$	$Ni + F_2 = NiF_2$	100	3.169	s, uncertainty ± 0.02 V	[52]
$Ni/NiF_2(\ell)/F_2, C$	$Ni + F_2 = NiF_2$	200	3.096	s, uncertainty ± 0.02 V	[52]
$Ni/NiF_2(\ell)/F_2, C$	$Ni + F_2 = NiF_2$	300	3.026	s, uncertainty ± 0.02 V	[52]
$Ni/NiF_2(\ell)/F_2, C$	$Ni + F_2 = NiF_2$	400	2.958	s, uncertainty ± 0.02 V	[52]
$Ni/NiF_2(\ell)/F_2, C$	$Ni + F_2 = NiF_2$	500	2.890	s, uncertainty ± 0.02 V	[52]
$Ni/NiF_2(\ell)/F_2, C$	$Ni + F_2 = NiF_2$	600	2.825	s, uncertainty ± 0.02 V	[52]
$Ni/NiF_2(\ell)/F_2, C$	$Ni + F_2 = NiF_2$	800	2.697	s, uncertainty ± 0.02 V	[52]
$Ni/NiF_2(\ell)/F_2, C$	$Ni + F_2 = NiF_2$	1000	2.573	s, uncertainty ± 0.02 V	[52]
$Ni/NiF_2(\ell)/F_2, C$	$Ni + F_2 = NiF_2$	1500	2.338	s, uncertainty ± 0.02 V	[52]
$Ni/NiCl_2(\ell)/Cl_2, C$	$Ni + Cl_2 = NiCl_2$	25	1.610	aq	[53]
$Ni/NiCl_2(\ell)/Cl_2, C$	$Ni + Cl_2 = NiCl_2$	25	1.412	s, uncertainty ± 0.04 V	[53]
$Ni/NiCl_2(\ell)/Cl_2, C$	$Ni + Cl_2 = NiCl_2$	100	1.355	s, uncertainty ± 0.04 V	[53]
$Ni/NiCl_2(\ell)/Cl_2, C$	$Ni + Cl_2 = NiCl_2$	200	1.282	s, uncertainty ± 0.04 V	[53]
$Ni/NiCl_2(\ell)/Cl_2, C$	$Ni + Cl_2 = NiCl_2$	300	1.210	s, uncertainty ± 0.04 V	[53]
$Ni/NiCl_2(\ell)/Cl_2, C$	$Ni + Cl_2 = NiCl_2$	400	1.139	s, uncertainty ± 0.04 V	[53]
$Ni/NiCl_2(\ell)/Cl_2, C$	$Ni + Cl_2 = NiCl_2$	500	1.070	s, uncertainty ± 0.04 V	[53]
$Ni/NiCl_2(\ell)/Cl_2, C$	$Ni + Cl_2 = NiCl_2$	600	1.003	s, uncertainty ± 0.04 V	[53]
$Ni/NiCl_2(\ell)/Cl_2, C$	$Ni + Cl_2 = NiCl_2$	800	0.875	s, uncertainty ± 0.04 V	[53]
$Ni/NiCl_2(\ell)/Cl_2, C$	$Ni + Cl_2 = NiCl_2$	987	0.763	s, uncertainty ± 0.04 V	[53]
$Ni/NiBr_2(\ell)/Br_2, C$	$Ni + Br_2 = NiBr_2$	25	1.331	aq	[52]
$Ni/NiBr_2(\ell)/Br_2, C$	$Ni + Br_2 = NiBr_2$	25	1.106	s, uncertainty ± 0.04 V	[52]
$Ni/NiBr_2(\ell)/Br_2, C$	$Ni + Br_2 = NiBr_2$	100	1.049	s, uncertainty ± 0.04 V	[52]
$Ni/NiBr_2(\ell)/Br_2, C$	$Ni + Br_2 = NiBr_2$	200	0.977	s, uncertainty ± 0.04 V	[52]
$Ni/NiBr_2(\ell)/Br_2, C$	$Ni + Br_2 = NiBr_2$	300	0.907	s, uncertainty ± 0.04 V	[52]
$Ni/NiBr_2(\ell)/Br_2, C$	$Ni + Br_2 = NiBr_2$	400	0.838	s, uncertainty ± 0.04 V	[52]
$Ni/NiBr_2(\ell)/Br_2, C$	$Ni + Br_2 = NiBr_2$	500	0.771	s, uncertainty ± 0.04 V	[52]
$Ni/NiBr_2(\ell)/Br_2, C$	$Ni + Br_2 = NiBr_2$	600	0.705	s, uncertainty ± 0.04 V	[52]
$Ni/NiBr_2(\ell)/Br_2, C$	$Ni + Br_2 = NiBr_2$	800	0.576	s, uncertainty ± 0.04 V	[52]
$Ni/NiBr_2(\ell)/Br_2, C$	$Ni + Br_2 = NiBr_2$	1000	0.527	927°C vap	[52]
$Ni/NiI_2(\ell)/I_2, C$	$Ni + I_2 = NiI_2$	25	0.886	aq	[52]
$Ni/NiI_2(\ell)/I_2, C$	$Ni + I_2 = NiI_2$	25	0.542	s, uncertainty ± 0.02 V	[52]
$Ni/NiI_2(\ell)/I_2, C$	$Ni + I_2 = NiI_2$	100	0.487	s, uncertainty ± 0.02 V	[52]
$Ni/NiI_2(\ell)/I_2, C$	$Ni + I_2 = NiI_2$	200	0.415	s, uncertainty ± 0.02 V	[52]

(continued)

TABLE 1.3.1 (continued)

Cell	Reaction	T (°C)	EMF (V)	Condition	Refs.
$Ni/NiI_2(\ell)/I_2, C$	$Ni + I_2 = NiI_2$	300	0.346	s, uncertainty ± 0.02 V	[52]
$Ni/NiI_2(\ell)/I_2, C$	$Ni + I_2 = NiI_2$	400	0.287	s, uncertainty ± 0.02 V	[52]
$Ni/NiI_2(\ell)/I_2, C$	$Ni + I_2 = NiI_2$	500	0.212	s, uncertainty ± 0.02 V	[52]
$Ni/NiI_2(\ell)/I_2, C$	$Ni + I_2 = NiI_2$	600	0.147	s, uncertainty ± 0.02 V	[52]
$Ni/NiI_2(\ell)/I_2, C$	$Ni + I_2 = NiI_2$	800	0.053	747°C sub	[52]
$Ni/NiO/O_2$(1 atm)	$Ni + \frac{1}{2}O_2 = NiO$	25	1.119	aq, hydroxide	[54]
$Ni/NiO/O_2$(1 atm)	$Ni + \frac{1}{2}O_2 = NiO$	25	1.121	s, uncertainty ± 0.007 V	[54]
$Ni/NiO/O_2$(1 atm)	$Ni + \frac{1}{2}O_2 = NiO$	100	1.084	s, uncertainty ± 0.007 V	[54]
$Ni/NiO/O_2$(1 atm)	$Ni + \frac{1}{2}O_2 = NiO$	200	1.037	s, uncertainty ± 0.007 V	[54]
$Ni/NiO/O_2$(1 atm)	$Ni + \frac{1}{2}O_2 = NiO$	300	0.990	s, uncertainty ± 0.007 V	[54]
$Ni/NiO/O_2$(1 atm)	$Ni + \frac{1}{2}O_2 = NiO$	400	0.943	s, uncertainty ± 0.007 V	[54]
$Ni/NiO/O_2$(1 atm)	$Ni + \frac{1}{2}O_2 = NiO$	500	0.897	s, uncertainty ± 0.007 V	[54]
$Ni/NiO/O_2$(1 atm)	$Ni + \frac{1}{2}O_2 = NiO$	600	0.852	s, uncertainty ± 0.007 V	[54]
$Ni/NiO/O_2$(1 atm)	$Ni + \frac{1}{2}O_2 = NiO$	800	0.763	s, uncertainty ± 0.007 V	[54]
$Ni/NiO/O_2$(1 atm)	$Ni + \frac{1}{2}O_2 = NiO$	1000	0.677	s, uncertainty ± 0.007 V	[54]
$Ni/NiO/O_2$(1 atm)	$Ni + \frac{1}{2}O_2 = NiO$	1500	0.476	s, uncertainty ± 0.007 V	[54]
$Ni/NiO/O_2$(1 atm)	$Ni + \frac{1}{2}O_2 = NiO$	2000	0.253	s, uncertainty ± 0.007 V	[54]
$Ni/NiCO_3/CO_2(\frac{2}{3}$ atm), O_2 ($\frac{1}{3}$ atm)	$Ni + CO_2 + \frac{1}{2}O_2 = NiCO_3$	600	0.340	—	[55]

TABLE 1.3.2. Potentials of Various Metal-Nickel Cells with Molten NiF_2 (1082°K) Electrolyte

Metals	E_{obsd} (V)	T (°K)
Cr(III)-Ni(II)	0.70	1121
Mn(II)-Ni(II)	1.04	1118
Al(III)-Ni(II)	1.50	1118
Ni(II)-Ag(I)	0.64	1119
Fe(III)-Ni(II)	0.12	1121
Co(II)-Ni(II)	0.07	1120
Ni(II)-Cu(I)	0.48	1122

TABLE 1.3.3. Formal Potentials of $Ni/NiCl_2$ or Ni/NiI_2 in Several Melts

Reaction	$E^{\circ\prime}$ (V)	Condition	Refs.
$Ni + Cl_2 = NiCl_2$	-1.011	LiCl-KCl, 450°C	[59]
$Ni + Cl_2 = NiCl_2$	-0.868	$NaCl-KCl-MgCl_2$, 475°C	[60-62]
$Ni^{2+} + 2Ag = Ni + 2Ag^+$	-0.1673	$KCl-MgCl_2$, 475°C	[63]
$Ni + Cl_2 = NiCl_2$	-0.9761	$KCl-MgCl_2$, 475°C	[63], [64]
$Ni + I_2 = NiI_2$	0.44	5% mole NiI_2 in molten NaI, 600°C	[65]
$Ni + I_2 = NiI_2$	0.36	700°C	[65]
$Ni + I_2 = NiI_2$	0.29	800°C, (dE/dT) = 0.0008 V/°C	[65]
$Ni + I_2 = NiI_2$	0.20	NaI melt, CO_2 atm, 380°C	[66]
$Ni + I_2 = NiI_2$	0.38	$NaI-AlI_3$, 5% mole NiI_2, 400°C	[66]
$Ni + I_2 = NiI_2$	0.34	500°C	[66]
$Ni + I_2 = NiI_2$	0.30	600°C	[66]
$Ni + I_2 = NiI_2$	0.29	700°C, (dE/dT) = 0.0004 V/°C	[66]

TABLE 1.3.4. Values of the Constant a

	a (V)		
Solvent	700°C	800°C	900°C
LiCl	0.798	0.759	-
NaCl	-	0.864	0.824
KCl	-	0.982	-
1:1 NaCl-KCl	0.973	0.930	0.946

constant temperature may be expressed by $E = a - (2.3RT/2F) \ln x$, the average deviation at 700°C being 2 mV, and at 900°C 6 to 8 mV. Values for the constant a are given in Table 1.3.4.

The Ni/NiO electrode ($NiO + 2e = Ni + O^{2-}$) was investigated by Laitinen and Bathia [68] in fused LiCl-KCl eutectic for temperatures from 400 to 500°C. The nickel concentration calculated from the potentials is 3.3×10^{-4} M. The concentration found from a polarogram of a saturated solution of NiO in the melt was 3.2×10^{-4} M. An E vs $\log(O^{2-})$ plot indicated irreversibility of the system (experimental slope is 53 mV and the theoretical slope for a reversible half-reaction is 71.7 mV). The calculated standard potential is more positive than the potential expected from the Ni/Ni(II) couple and the solubility of NiO estimated polarographically. The explanation for this discrepancy may be attributed either to the formation of Ni_2O_3 or to the irreversibility of the Ni/Ni(II) system in the presence of an excess of oxide ion. Then the measured potential may be a mixed potential due to NiO and Ni_2O_3. The formation of the latter is promoted in the presence of Li_2O. Evidence for the existence of Ni_2O_3 at a temperature of 660°C and higher in molten Li_2SO_4-K_2SO_4 eutectic has also been given by Hill, Porter, and Gillespie [69]. Galvanic cells involving the Ni/NiO electrode in melts are shown in Table 1.3.5.

TABLE 1.3.5. Open Circuit Potentials of Various High-Temperature Cells

Galvanic cell	T (°C)	E (V)	Remarks	Refs.
Ni/NiO/58 mole % LiCl + 42 mole % KCl/Mg	380-500	1.70	Sintered Ni electrode	[70], [71]
Ni/NiO, Li_2O(CaO) in 59 mole % LiCl-41 mole % KCl/Pt(II)/Pt	450	1.23	irr 1 M Pt ref. electrode, $-\log(O^{2-}) = 0$	[68]
Ni/NiO, Li_2O(CaO) in 59 mole % LiCl-41 mole % KCl/Pt(II)/Pt	450	1.18	= 1	[68]
Ni/NiO, Li_2O(CaO) in 59 mole % LiCl-41 mole % KCl/Pt(II)/Pt	450	1.12	= 2	[68]
Ni/NiO, Li_2O(CaO) in 59 mole % LiCl-41 mole % KCl/Pt(II)/Pt	450	1.07	= 3	[68]
Ni, NiO/Pyrex/LiCl + KCl + kaolin/Mg	450	0.80	-	[72]
Ni, NiO/65 mole % $Na_2B_4O_7$ + 35 mole % $K_2B_4O_7/O_2$	780	0.71	-	[73]
Ni, NiO/borax/porcelain/borax, CuO/Cu	850-1000	0.7	-	[74]
Ni/NiO/borax(glass)/porcelain/Ag, O_2	980-1193	0.855	1000°C	[75]
Ni, NiO/CaO in Li_2SO_4 + K_2SO_4/O_2, Pt	658	0.788 ± 0.005	$E_{theor} = 0.792$; CaO concn 0.1-1%	[69]
Ni, NiO/CaO in Li_2SO_4 + K_2SO_4/O_2, Pt	682	0.884	Steady emf after turning off the gas	[75-77]
Ni, NiO/CaO in Li_2SO_4 + K_2SO_4/O_2, Pt	692	0.932	Steady emf after turning off the gas black oxide	
Ni, NiO/CaO in Li_2SO_4 + K_2SO_4/O_2, Pt	708	1.032	Steady emf after turning off the gas NiO	
Ni/NiO in $NaPO_3$/porcelain/Cu in $NaPO_3$/Cu	720	0.15	-	[78]
Ni/NiO in $Na_4P_2O_7$/porcelain/CuO in $Na_4P_2O_7$/Cu	1000	0.12	NiO concn 5 mole %	[78]

Kolotii [79] found that nickel acts as an oxygen electrode in $KCl-PbCl_2$ melts. When Ni(II) ions are present, the metal acquires its characteristic potential. The potential of nickel electrodes at 700°C in chloride eutectics, containing K_2CrO_4, was measured against a Ag/ LiCl-(or NaCl)-KCl-AgCl (0.1 N) reference electrode by Selis and McGinnis [80]. E_{Ni} (in $LiCl-KCl-K_2CrO_4$) = 0.24 V and E_{Ni} (in $NaCl-KCl-K_2CrO_4$) = 0.37 V. These authors also measured the cell $Mg/LiCl-KCl-K_2CrO_4/Ni$ at 440°C [71]. The average electromotive force with SiO_2 is 1.81 V and without SiO_2 1.77 V.

Nickel immersed in $Li_2CO_3-Na_2CO_3-K_2CO_3$ eutectic in an argon and CO_2 atmosphere exhibits an initial potential of -1.020 V which is stable for 11 to 15 hr after immersion. According to Penyagina, Manukhina, Ozeryanaya, and Smirnov [81], the potential corresponds to the following electrode cation exchange reaction: $(x + 1)NiO_z(s) = MO_{(z + x)}(s) + xNi^{n+}$ (melt) + xne, and is given by $E = const + (RT/nF) \ln (a_{M^{n+}}/a_{M(oxide\ phase)})$. The temperature dependence of the stationary potential is given by $E = -1.210 + 4.03 \times 10^{-4}T$. One concludes that nickel in alkali carbonate melts acts as an oxide electrode.

A Ni-Co spinel ($NiO-Co_2O_3$) oxygen electrode in molten carbonate electrolyte was reported by Dmitrenko, Ezerkii, Rezhikov, and Zyrin [82]. Rempel and Ozeryanaya [83] measured the electromotive force of the cell M/(1:1)NaCl-KCl + 0.1 N MCl_2/Cl_2, where M is nickel or its sulfide. At 690°C the potential of nickel is 1.1789 V. The measured potential for a galvanic cell involving Ni_3S_2 as the electrode, such as: Ni_3S_2/0.1 N $NiCl_2$ + (1:1)NaCl-KCl/Cl_2 at 690°C, is 1.2959 V. Chermak [84] studied the concentration cell: Ni-sulfide melt (C)/$NiCl_2-BaCl_2$-melt/Ni-sulfide melt (C_0 = 80.5 at % Ni) at 1200°C and evaluated the activity coefficients for nickel and sulfur. The cell 88 mole % Ni_3S_2, 12 mole % Ni/glass melt + Na_2S/Cu_2S + Ni_3S_2 at 1180°C attains a potential of 0.197 V according to Sryvalin and Esin [85].

Open circuit potentials of Ni-Pt couples in eutectic borates under air at 780°C were measured by Ilschner-Gensch [73] with the cell Ni/NiO/$Na_2B_4O_7$ + $K_2B_4O_7/O_2$, Pt(Au, Ag); E_{Ni-Pt} = 0.73 V; E_{Ni-Au} = 0.74 V; E_{Ni-Ag} = 0.49 V.

Ni/Pyrex and NiO/Pyrex electrodes, used by Panzer [72], suffered from the prime disadvantage of multilevel potentials owing to the multivalent character of the system.

Kornilov, Ilyushchenko, Rossokhin, and Belyaeva [86] determined the activities of nickel and beryllium as well as the integral thermodynamic values (ΔG, ΔS, and ΔH) of the BeNi system with the cell Be(s)NaCl-KCl + 10% wt $BeCl_2/BeNi_{1-x}(s)$ at 963 and 1113°K, x = 3-50 at.% Be. The extension to alloy compositions from 50 to 96 at.% Be was made by Ilyushchenko, Kornilov, and Rossokhin [87].

The potential of nickel in fused dimethylammonium chloride at 180°C was measured by Kisza in the mole fraction scale [88]. The value is -0.3723 V.

1.3.2. Equilibrium Diagrams

Ingram and Janz [55] calculated the potential/pCO_2 diagram in ternary alkaline carbonate eutectic at 600°C (shown in Fig. 1.3.1). The boundary between Ni^{2+} and NiO areas is

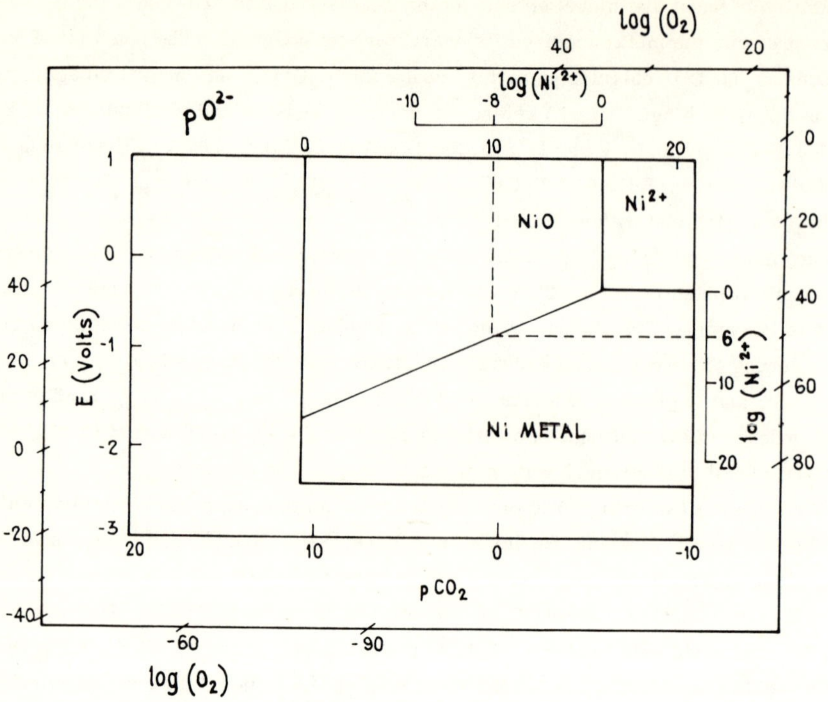

FIG. 1.3.1. Potential/pCO_2 diagram for the ternary alkaline carbonate eutectic at 600°C [55]. (By permission of Pergamon Press.)

the value of pCO_2 when NiO precipitates from pure $NiCO_3$ melt. For the reaction $NiCO_3$ = $NiO + CO_2$, the equilibrium constant is $10^{5.24}$. The dissociation pressure of NiO is $10^{-19.16}$ atm. Then the line on the diagram for $\log(O^{2-}) = -19.16$ is the boundary between NiO and Ni domains. If nickel is immersed in the carbonate melt under 0.9 atm CO_2 and 0.1 atm O_2, either the metal corrodes completely to oxide or the NiO will form a coherent film on the metal surface yielding passivation. The oxide formation is inevitable unless the CO_2 pressure is reduced to 10^{-19} atm. Reduction of the potential to -1.8 V should remove any oxide and render the metal immune.

Rahmel [89] calculated the potential/pSO_3 diagrams for nickel in a ternary alkaline sulfate eutectic with and without formation of nickel sulfide at 600°C, as illustrated in Figs. 1.3.2 and 1.3.3.

Potential/pO^{2-} diagrams for nickel-potassium chloride and nickel-lithium chloride were obtained by Littlewood [90]; the potential referred to the standard Cl_2/Cl^- electrode (see Figs. 1.3.4 and 1.3.5). The solubility products of NiO in these melts at 800°C are $\log(S)_{LiCl} = -7.8$ and $\log(S)_{KCl} = -19.7$.

Marchiano and Arvía [91, 92] calculated the potential/pO^{2-} diagrams for nickel in molten alkaline nitrites and for nickel in molten alkaline nitrates at 600°K. The corresponding diagrams are shown in Figs. 1.3.6 and 1.3.7.

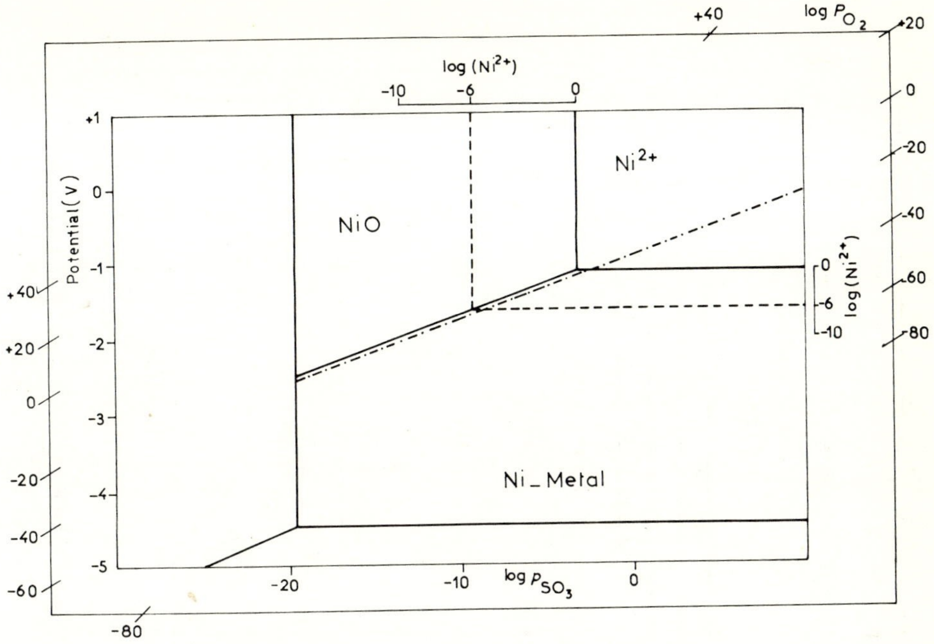

FIG. 1.3.2. Potential/pSO$_3$ diagram of nickel in tertiary alkaline sulfate eutectic at 600°C without consideration of nickel sulfide formation [89]. (By permission of Pergamon Press.)

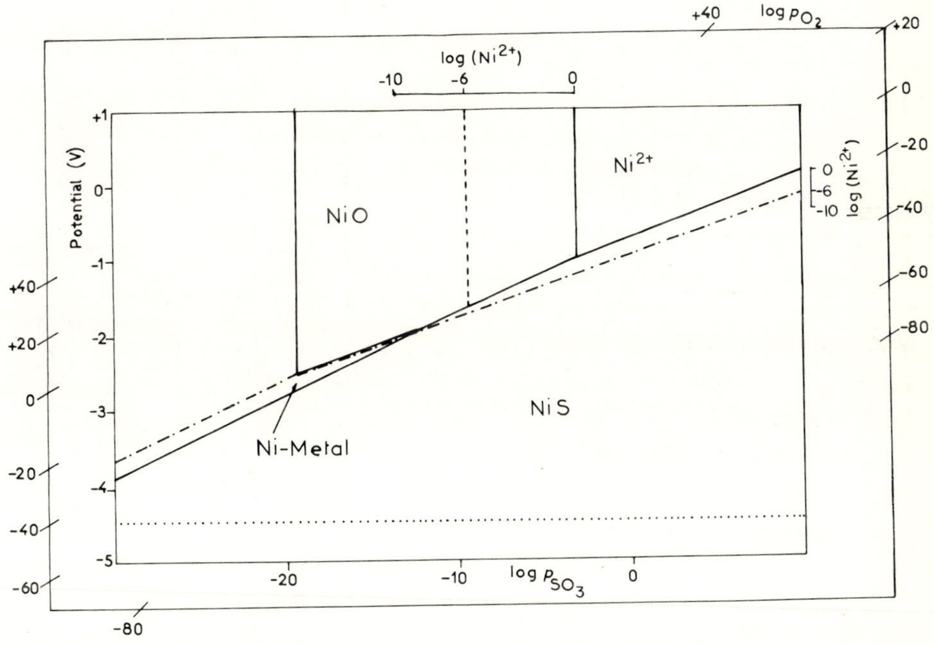

FIG. 1.3.3. Potential/pSO$_3$ diagram of nickel in tertiary alkaline sulfate eutectic at 600°C with consideration of nickel sulfide formation [89]. (By permission of Pergamon Press.)

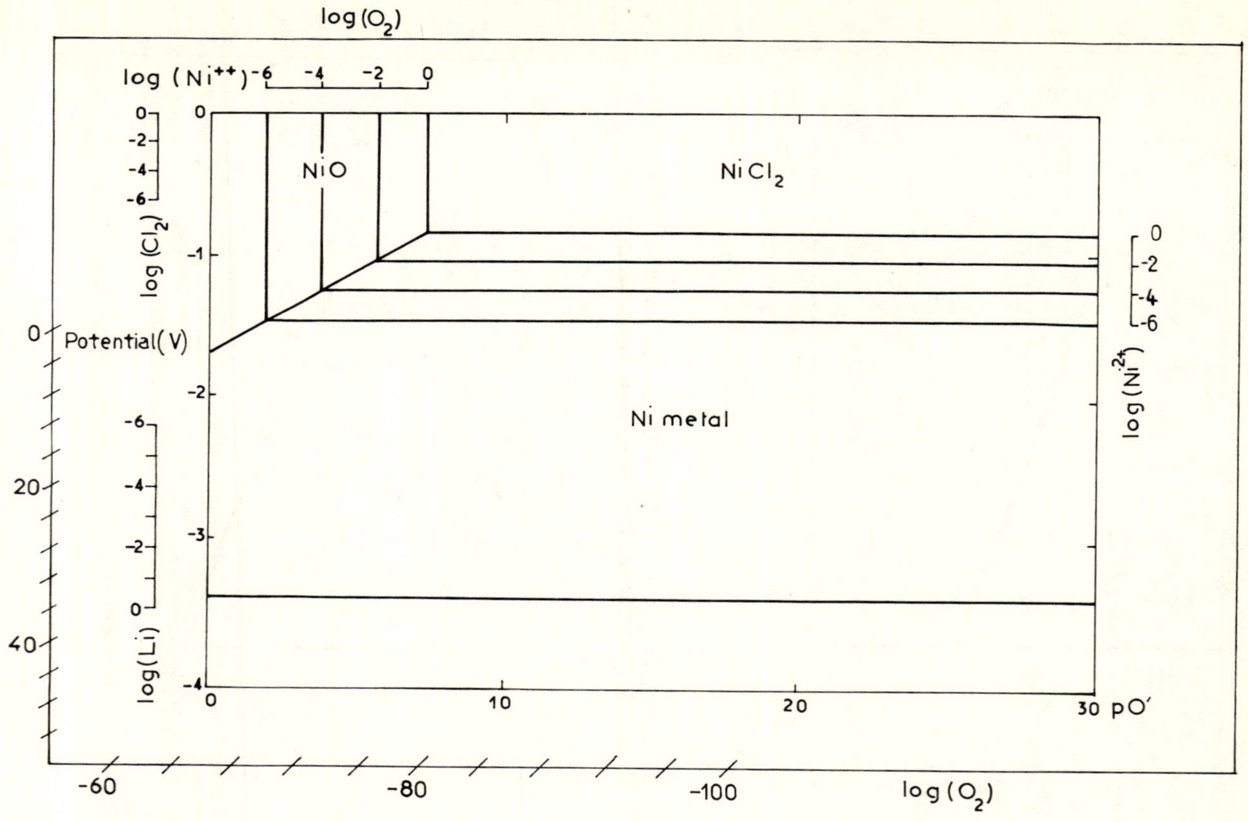

FIG. 1.3.4. Potential/pO^{2-} diagram for the system nickel/lithium chloride at 600°C (see Ref. 90). (By permission of the Electrochemical Society.)

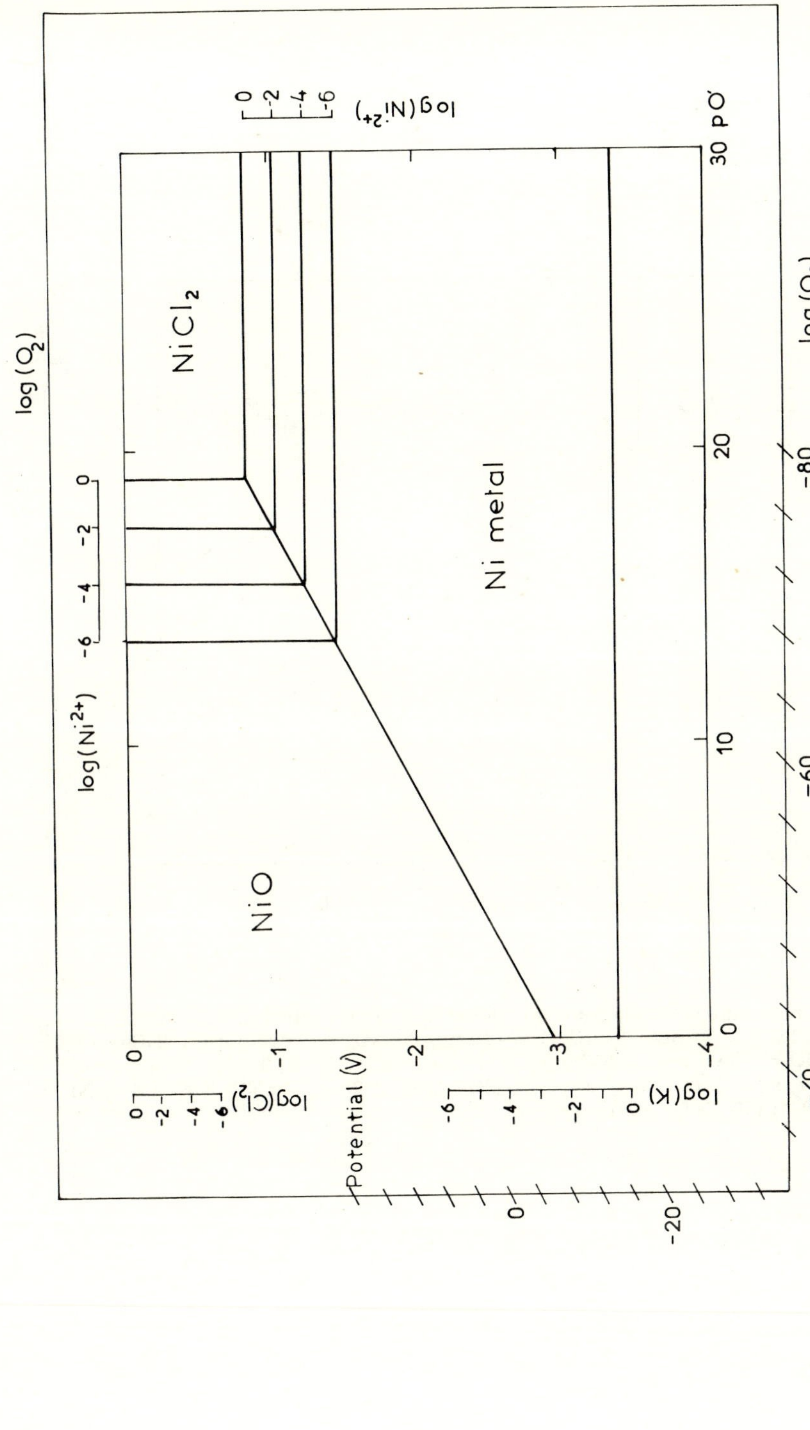

FIG. 1.3.5. Potential/pO^{2-} diagram for the system nickel/potassium chloride (see Ref. 90). (By permission of the Electrochemical Society.)

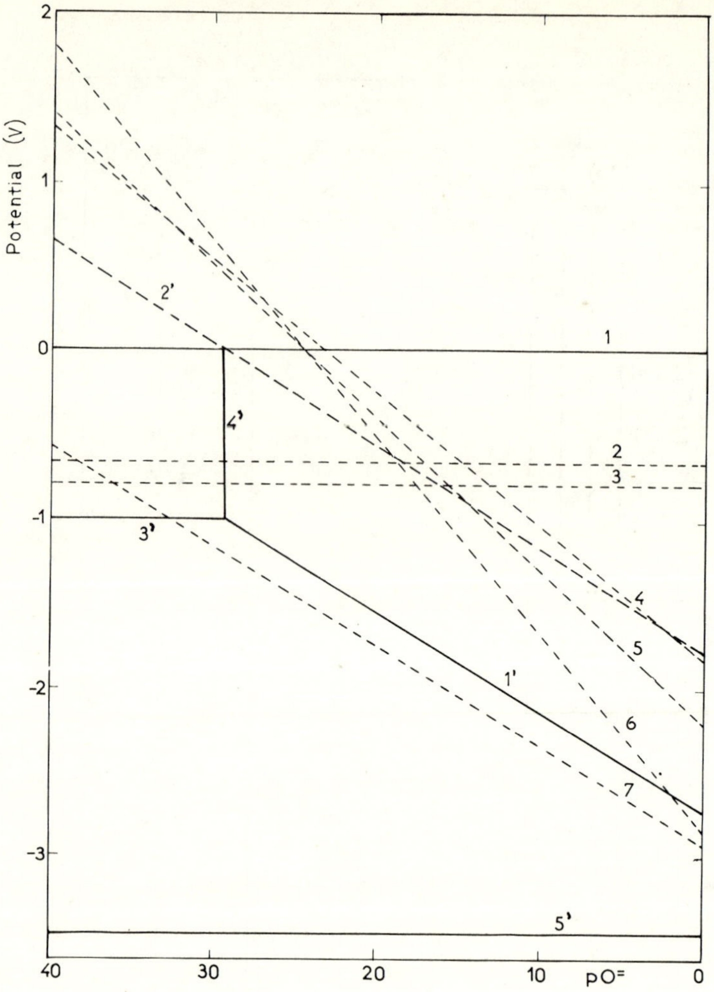

FIG. 1.3.6. Potential/pO^{2-} diagram for nickel in molten sodium nitrate at 600°K. (1) $2NO_2 + O_2 + 2e = 2NO_3^-$; (2) $NO + 1/2\ O_2 + e = NO_2^-$; (3) $NO_2 + e = NO_2^-$; (4) $2NO_2^- + 6e = N_2 + 4O^{2-}$; (5) $2NO_2^- + 4e = N_2O + 3O^{2-}$; (6) $NO_2^- + e = NO + O^{2-}$; (7) $NO_2 + NO_2^- + e = 2NO + 1/2\ O_2 + O^{2-}$; (1') $NiO + 2e = Ni + O^{2-}$; (2') $1/2\ O_2 + 2e = O^{2-}$; (3') $Ni^{2+} + 2e = Ni$; (4') $NiO = Ni^2 + O^{2-}$; (5') $Na^+ + e = Na$.

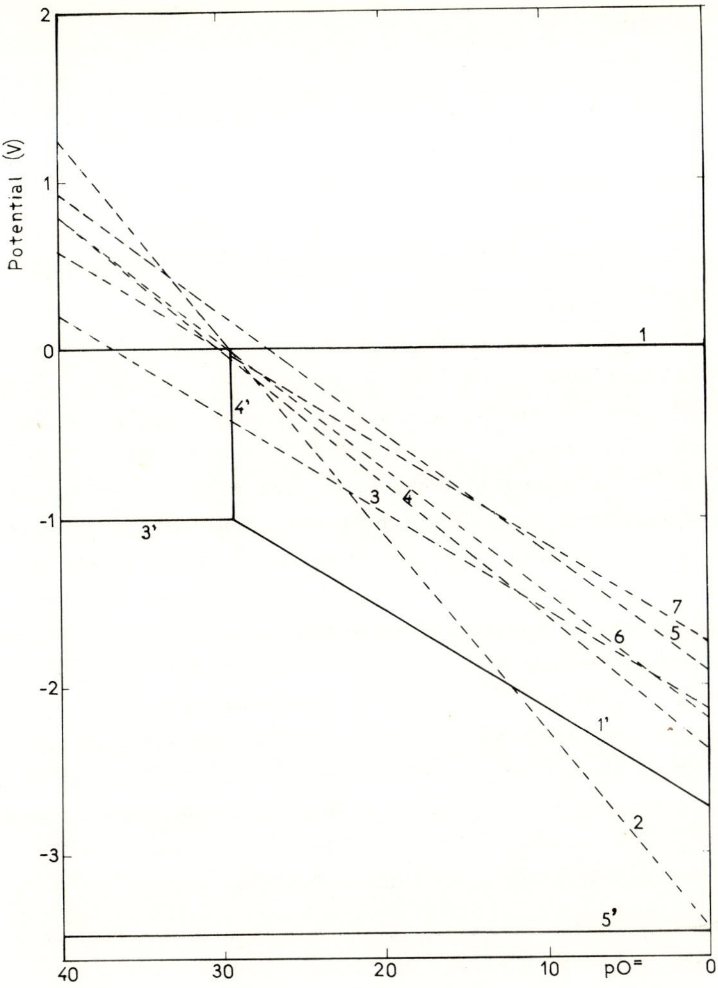

FIG. 1.3.7. Potential/pO^{2-} diagram for nickel in molten sodium nitrate at 600°C. (1) $2NO_2 + O_2 + 2e = 2NO_3^-$; (2) $NO_3^- + e = NO_2 + O^{2-}$; (3) $NO_3^- + 2e = NO_2^- + O^{2-}$; (4) $NO_3^- + 3e = NO + 2O^{2-}$; (5) $NO_3^- + 5e = 1/2\ N_2 + 3O^{2-}$; (6) $2NO_3^- + 8e = N_2O + 5O^{2-}$; (7) $1/2\ O_2 + 2e = O^{2-}$; (1') $NiO + 2e = Ni + O^{2-}$; (3') $Ni^{2+} + 2e = Ni$; (4') $NiO = Ni^{2+} + O^{2-}$; (5) $Na^+ + e = Na$.

1.4. SOLID ELECTROLYTES

NiO is the only nickel compound which is stable at 1 atm oxygen pressure and high temperature, behaving as a metal deficient p-type semiconductor. Mixtures of Ni and NiO were used at high temperature in solid galvanic cells as electrolyte for oxygen transfer (concentration cells) at constant oxygen activity, because the system behaves as a buffer to maintain constant composition. Electromotive forces of galvanic cells involving solid electrolytes and NiO are assembled in Table 1.4.1.

Kiukkola and Wagner [93] measured the free energy change for the reaction Ni(s) + $\frac{1}{2}O_2$(g) = NiO(s). Data from electromotive force measurements and calculation by Coughlin [99] from reduction of NiO by means of carbon monoxide were coincident. Pizzini and Morlotti [94] investigated the electrical conductivity and electromotive force on solid couples of nearly stoichiometric and nonstoichiometric cells. The Ni/NiO solid reference electrode was also employed in galvanic cells for gas equilibration techniques by Markin, Bones, and Wheeler [95]. Cells such as Ni, NiO/ZrO_2 + 15 mole % CaO/$UO_{2.07}$ and Ni, NiO/ZrO_2 + 8 mole % Y_2O_3/$UO_{2.07}$ were measured from 625 to 1025°C [95]. Horsley [100] used similar cells to determine oxygen potentials of nonstoichiometric oxides of uranium and plutonium and metal/metal oxide couples. The same reference electrode was employed by Cameron and Unger [101] and Goto and Matsushita [102] to evaluate thermodynamic properties of NiO-MnO solid solutions. Moebius [103] studied the same oxide mixtures with a galvanic cell for gas potentiometry and classified it as a second class reversible electrode. The Ni/NiO solid electrode was used by Rickert and Wagner [104] to determine the oxygen activity in molten copper, and by Patterson, Bogren, and Rapp [105] to investigate the mixed conduction in various electrolytes containing zirconium and thorium.

Mehandjiev [106] calculated the change of the thermodynamic potential for the reaction Ni + $(x/2)O_2$ = NiO_x, where x = [O]/[Ni] varied from 1.0 to 1.5.

The electromotive force of the cell Ni/NiO/ThO_2-$YO_{1.5}$/MnO/Mn was investigated by Alcock and Zador [96] at 923 to 1273°C. The standard free energy of the reaction NiO + Mn = MnO + Ni was evaluated. Pizzini, Morlotti, and Wagner [98] investigated the thermodynamics of solid solutions of Li_2O in NiO between 570 and 800°C. The mixture is ideal up to 1.5 at.% lithium. Deviations from ideality at higher concentrations can be explained with purely configurational energy terms.

A closed solid electrolyte electrochemical cell [cell reaction 2Ni + O_2(1 atm) = 2NiO] was used by Charette and Flengas [97] to investigate thermodynamic properties of nickel oxides at different temperatures (see Table 1.4.1). Lacy and Pask [107-109] used a solid electrolyte Pt/Ni, $NiO_{a=1}$ reference electrode to measure the activity of metal oxides in sodium dicilicate glasses, with the cell Pt/Ni, $NiO_{a=1}$/ZrO_2 + 7.5% CaO/X:$X_a < 1$, where X represents nickel, iron, or cobalt.

TABLE 1.4.1. Open Circuit Potentials of Solid Electrolyte Cells

Galvanic cell	T (°C)	E (V)	Remarks	Refs.
Ni, NiO/0.15CaO + 0.85ZrO_2/wustite, Fe	750	0.261 ± 0.002	ΔG° = -35.20 kcal/mole	[93]
Ni, NiO/0.15CaO + 0.85ZrO_2/wustite, Fe	800	0.266 ± 0.001	= -34.23	[93]
Ni, NiO/0.15CaO + 0.85ZrO_2/wustite, Fe	850	0.271 ± 0.001	= -33.29	[93]
Ni, NiO/0.15CaO + 0.85ZrO_2/wustite, Fe	900	0.276 ± 0.001	= -32.15	[93]
Ni, NiO/0.15CaO + 0.85ZrO_2/wustite, Fe	950	0.281 ± 0.001	= -31.19	[93]
Ni, NiO/0.15CaO + 0.85ZrO_2/wustite, Fe	1000	0.286 ± 0.002	= -30.15	[93]
Ni, NiO/0.15CaO + 0.85ZrO_2/wustite, Fe	1050	0.291 ± 0.002	= -29.12	[93]
Ni, NiO/0.15CaO + 0.85ZrO_2/wustite, Fe	1100	0.296 ± 0.002	= -28.11	[93]
Ni, NiO/0.15CaO + 0.85ZrO_2/wustite, Fe	1140	0.300 ± 0.001	= -27.22	[93]
Pt/NiO(p_1)/ZrO_2 + 18%CaO/NiO(p_2)/Pt	800	0.266 ± 0.002	-	[94]
Pt/NiO(p_1)/ZrO_2 + 18%CaO/NiO(p_2)/Pt	900	0.275	-	[94]
Pt/NiO(p_1)/ZrO_2 + 18%CaO/NiO(p_2)/Pt	1000	0.286	-	[94]
Pt/NiO(p_1)/ZrO_2 + 18%CaO/NiO(p_2)/Pt	1100	0.296	-	[94]
Ni, NiO/ZrO_2 + 15 mole % CaO/Cu_2O, Cu	350	0.304	2Ni + O_2(1 atm) = 2NiO	[95]

(continued)

TABLE 1.4.1 (continued)

Galvanic cell	T (°C)	E (V)	Remarks	Refs.
Ni, NiO/ZrO_2 + 15 mole % CaO/Cu_2O, Cu	450	0.297	2Ni + O_2(1 atm) = 2NiO	[95]
Ni, NiO/ZrO_2 + 15 mole % CaO/Cu_2O, Cu	550	0.289	2Ni + O_2(1 atm) = 2NiO	[95]
Ni, NiO/ZrO_2 + 15 mole % CaO/Cu_2O, Cu	650	0.282	2Ni + O_2(1 atm) = 2NiO	[95]
Ni, NiO/ZrO_2 + 15 mole % CaO/Cu_2O, Cu	750	0.275	2Ni + O_2(1 atm) = 2NiO	[95]
Ni, NiO/ZrO_2 + 15 mole % CaO/Cu_2O, Cu	850	0.267	2Ni + O_2(1 atm) = 2NiO	[95]
Ni, NiO/$ThO_2YO_{1.5}$/MnO/Mn	973	0.844	-	[96]
Ni, NiO/$ThO_2YO_{1.5}$/MnO/Mn	1273	0.858	-	[96]
Pt, O_2(1 atm)/electrolyte/Ni, NiO, Pt	700	0.767	Interpolated	[97]
Pt, O_2(1 atm)/electrolyte/Ni, NiO, Pt	800	0.725	Interpolated	[97]
Pt, O_2(1 atm)/electrolyte/Ni, NiO, Pt	900	0.684	Interpolated	[97]
Pt, O_2(1 atm)/electrolyte/Ni, NiO, Pt	1000	0.643	Interpolated	[97]
Pt, O_2(1 atm)/electrolyte/Ni, NiO, Pt	1110	0.601	Interpolated	[97]
Pt/NiO, $Li_2O(x_0)$//α-Li_2SO_4//NiO, $Li_2O(x_1)$/Pt	570 to 800	$(-9.8 - 0.0358T) \times 10^{-3}$	$(x_1/x_0) = 0.235/0.155$	[98]
Pt/NiO, $Li_2O(x_0)$//α-Li_2SO_4//NiO, $Li_2O(x_1)$/Pt	570 to 800	$(-55.8 - 0.260T) \times 10^{-3}$	= 1.965/0.155	[98]
Pt/NiO, $Li_2O(x_0)$//α-Li_2SO_4//NiO, $Li_2O(x_1)$/Pt	570 to 800	$(-89.6 - 0.373T) \times 10^{-3}$	= 5.265/0.155	[98]

2. VOLTAMMETRIC CHARACTERISTICS

2.1. POLAROGRAPHIC CHARACTERISTICS

2.1.1. Aqueous Solution

The polarographic reduction of aqueous Ni(II) ion from noncomplexing base electrolytes such as alkali chlorides, nitrates, and perchlorates containing gelatin as a maximum suppressor is characterized by an irreversible wave with a half-wave potential of approximately -1.1 V (SCE), as reported by Kolthoff and Lingane [110]. Details of the usual analytical methods can be found in standard texts [110-112]. Table 2.1.1 gives a number of supporting electrolytes for the polarographic determination of nickel. The influence of ionic strength on the half-wave potential, according to Vanderborgh and Sellers [142], is shown in Table 2.1.2.

Takahashi and Shirai [117] studied the reduction waves of Ni(II) in the presence of supporting electrolytes of different anions, such as nitrate, sulfate, chloride, and nitrite ions. They observed two reduction waves and studied their dependence on the concentration of supporting electrolyte. According to Zutshi [118] there is only one reduction wave in 0.1 M sodium formate, sodium sulfate, or sodium iodide as supporting electrolytes. As the concentration of the supporting electrolyte is increased from 0.5 to 1.5 M, two reduction waves are obtained. The total limiting current was diffusion controlled, thus providing a polarographic method for the determination of Ni(II). The ratio $(i_d)_{1st}/(i_d)_{2nd}$ diminishes as the supporting electrolyte concentration increases. The half-wave potential of the first wave corresponds to the reduction of aqueous nickel ion complex. The shift of the second half-wave potential indicates the coordination of the anion of the supporting electrolytes with nickel ion. In ethanol, propanol, and ethylene glycol-aqueous mixtures a single wave tends to form. Mechanistic interpretation of polarographic waves are discussed in Section 4.1.1.

2.1.2. Nonaqueous Solution

Table 2.1.3 contains the polarographic characteristics of nickel in nonaqueous solutions. There are few works on the polarography of Ni(II) in nonaqueous solvents, most of them carried out in organic nitriles and amides.

Kolthoff and Coetzee [156, 163] studied acetonitrile solutions and reported that the half-wave potential becomes more negative and the wave more irreversible as the amount of water increases, as shown in Table 2.1.4. This effect is explained as acid-base competition between water and acetonitrile. These experiments were repeated by Popov and Geske [157], except for the use of a Ag/AgCl/acetonitrile reference electrode. They observed two rather irreproducible polarographic waves and a linear relationship between the wave height and the

TABLE 2.1.1. Polarographic Data for Ni in Various Supporting Electrolytes[a]

Substance	Product	Conditions	$E_{1/2}$ (SCE) (V)	I	$E_{1/4} - E_{3/4}$ (mV)	Remarks	Refs.
Ni(II)	0	0.5 F NaF; 4.3 < pH < 6.4	-1.12	2.29	-	0.01% gel	[113], [114]
Ni(II)	0	1 M NaF	-1.12	2.29	-	0.01% gel	[114]
Ni(II)	0	0.1 M $NaClO_4$	-1.013	3.3	69	w; max supp	[111]
Ni(II)	0	1 M $NaClO_4$	-1.010	1.5	62	w; max supp	[111]
Ni(II)	0	8 M $NaClO_4$	-0.877	1.2	73	w; max supp	[111]
Ni(II)	0	$HClO_4$; pH = 0.2	-1.1	-	irr	-	[115]
Ni(II)	0	0.2 M KNO_3	-1.016	-	-	-	[116]
Ni(II)	0	0.5 M KNO_3; 30°C	-1.08	-	-	0.005% gel	[117]
Ni(II)	0	0.3 M K_2SO_4; 30°C	-1.12	-	-	0.005% gel	[117]
Ni(II)	0	0.5 M Na_2SO_4	-1.09 -1.52	-	53	temp coeff 2.85%; 0.001% TX-100	[118]
Ni(II)	0	7.3 M H_3PO_4	-1.18	-	-	i	[111]
Ni(II)	0	0.1 M NaH_2PO_2; 2 M HCl	-0.67	-	-	Indication of wave	[111]
Ni(II)	0	0.1 M $Al(NO_3)_3$ + 10^{-3} M Ni(II)	-1.10	-	-	0.0002% TX-100	[119]
Ni(II)	0	0.1 M $Sr(NO_3)_2$ + 10^{-3} M Ni(II)	-0.96	-	-	0.0002% TX-100	[119]
Ni(II)	0	0.1 M KNO_3 + 10^{-3} M Ni(II)	-0.99	-	-	0.0002% TX-100	[119]

Ni(II)	0	0.1 M HNO_3 + 10^{-3} M Ni(II)	-1.03	-	-	0.0002% TX-100	[119]
Ni(II)	0	0.1 M KCl + 10^{-3} M Ni(II)	-1.04	-	-	0.0002% TX-100	[119]
Ni(II)	0	0.1 M KSCN + 10^{-3} M Ni(II)	-0.70	-	-	0.0002% TX-100	[119]
Ni(II)	0	0.5 M KNO_2; 30°C	-0.85 -1.09	-	-	-	[117]
Ni(II)	0	2 M K_2CO_3; pH ~ 11	-1.2	-	-	i	[111]
Ni(II)	0	0.1 M KCl	-1.1	3.38	irr	w; max supp	[111]
Ni(II)	0	1 M KCl	-1.1	-	-	-	[120]
Ni(II)	0	1 M KCl; 30°C	-1.14	-	-	-	[121]
Ni(II)	0	6 M $CaCl_2$	-0.462	i_L/C = 2.31 μA/mM	38	-	[122]
Ni(II)	0	8 M HCl	-0.67	-	irr	-	[111]
Ni(II)	0	12 M HCl	-0.80	-	-	-	[111]
Ni(II)	0	0.5 M KBr; 30°C	-1.07	-	-	0.005% gel	[117]
Ni(II)	0	0.5 M NaI	-1.04 to -1.40	-	45	0.001% TX-100	[118]
$Ni(SCN)_x^{(2-x)+}$	0	0.5 M KSCN; 30°C	-0.75	-	-	-	[121]
$Ni(SCN)_x^{(2-x)+}$	0	1 M KSCN	-0.7	-	-	-	[123]
$Ni(SCN)_x^{(2-x)+}$	0	1 M KSCN	-0.685	3.59	34	fw	[111]
$Ni(SCN)_x^{(2-x)+}$	0	1 M KCl + 0.005 M KSCN	-0.725 to -0.91	-	-	0.005% gel	[124]

(continued)

TABLE 2.1.1 (continued)

Substance	Product	Conditions	$E_{1/2}$ (SCE) (V)	I	$E_{1/4} - E_{3/4}$ (mV)	Remarks	Refs.
$Ni(SCN)_x^{(2-x)+}$	0	0.01 M KSCN	-0.725 to -0.91	-	-	0.005% gel	[124]
$Ni(SCN)_x^{(2-x)+}$	0	0.04 M KSCN	-0.704	-	-	0.005% gel	[124]
$Ni(SCN)_x^{(2-x)+}$	0	0.1 M KSCN	-0.69	-	-	0.005% gel	[124]
$Ni(SCN)_x^{(2-x)+}$	0	0.5 M KSCN	-0.69	-	-	0.005% gel	[124]
$Ni(SCN)_x^{(2-x)+}$	0	2 M KSCN	-0.72	-	-	0.005% gel	[124]
$Ni(SCN)_x^{(2-x)+}$	0	0.01 M KSCN + KNO_3; $\mu = 2$	-0.764	-	-	0.005% gel	[124]
$Ni(SCN)_x^{(2-x)+}$	0	0.1 M KSCN	-0.713	-	-	-	[124]
$Ni(SCN)_x^{(2-x)+}$	0	2 M KSCN	-0.745	-	-	-	[124]
$[Ni(CN)_3]^{2-}$	$[Ni(CN)_4]^{2-}$	1 M KCN	(-0.8) anode	-	-	-	[125]
$[Ni(CN)_4]^{2-}$	0	0.1 M KCl + 0.01 M KCN	-1.47	-	-	-	[125]
$[Ni(CN)_4]^{2-}$	0	0.1 M KCl + 0.1 M KCN	-1.42	-	-	-	[125]
$[Ni(CN)_4]^{2-}$	0	0.1 M KCl + 1 M KCN	-1.36	-	-	-	[125]
$[Ni(NH_3)_6]^{2+}$	0	1 M NH_3	-1.05	-	-	-	[120]
$[Ni(NH_3)_6]^{2+}$	0	1 M NH_3 + 0.2 M NH_4Cl	-1.06	3.54	-	0.005% gel	[126]
$[Ni(NH_3)_6]^{2+}$	0	1 M NH_3 + 1 M NH_4Cl	-1.09	3.56	irr	0.005% gel; w	[126]
$[Ni(NH_3)_6]^{2+}$	0	1 M NH_3 + 3 M NH_4Cl	-1.12	3.33	-	-	[126]
$[Ni(NH_3)_6]^{2+}$	0	1 M NH_3 + 0.1 M NH_4Cl	-1.13	-	-	-	[127]

$[Ni(NH_3)_6]^{2+}$	0	2 M NH_3 + 0.1 M NH_4Cl	-1.18	-	-	-	[127]
$[Ni(NH_3)_6]^{2+}$	0	4 M NH_3 + 0.1 M NH_4Cl	-1.22	-	-	-	[127]
$[Ni(N_2O_3)_3]^{4-}$	0	10^{-3} M $Na_2N_2O_3$ + 0.01 M KCl	-0.900 to -1.130	-	-	-	[128]
$[Ni(N_2O_3)_3]^{4-}$	0	10^{-3} M $Na_2N_2O_3$ + 0.1 M KCl	-0.975 to -1.120	-	-	-	[128]
Ni(II) complex	0	1 M N_2H_4 + 1 M NH_3 + 1 M NH_4Cl	-1.079	-	31	fw, max	[111]
Ni(II) complex	0	0.1 M KOH + 0.03 M TEA	-1.35	-	-	-	[129]
Ni(II) complex	0	0.1 M KOH + 0.3 M TEA	-1.40	-	-	Height decreases on heating	[129]
Ni(II)	0	0.5 M NaHCOO + 10^{-3} M Ni(II)	-1.09 to -1.35	-	irr	0.001% TX-100; temp coeff 3%	[118]
Ni(II)	0	1.5 M NaHCOO + 10^{-3} M Ni(II)	-1.15 to -1.6	-	irr	0.001% TX-100	[118]
Ni(II)	0	2 F HOAc + 2F NH_4OAc; pH = 5	-1.1	-	-	0.01% gel; i	[111]
Ni(II)	0	0.8 M NaOAc; pH ~ 12	pptn.	-	-	-	[111]
Ni(II)(ox)	0	0.01 M $(NH_4)_2$ox; pH = 10 (ox = $C_2O_4^{2-}$)	-1.19	-	-	-	[127]
Ni(II)(ox)	0	0.1 M $(NH_4)_2$ox; pH = 10	-1.36	-	-	-	[127]
Ni(II)(ox)	0	0.3 M $(NH_4)_2$ox; pH = 10	-1.38	-	-	-	[127]
Ni(II)(ox)	0	0.05 M $(NH_4)_2$ox; pH = 5.1	-0.95	-	-	-	[127]
Ni(II)(ox)	0	0.05 M $(NH_4)_2$ox; pH = 6.0	-1.02	-	-	-	[127]

(continued)

TABLE 2.1.1 (continued)

Substance	Product	Conditions	$E_{1/2}$ (SCE) (V)	I	$E_{1/4}$ - $E_{3/4}$ (mV)	Remarks	Refs.
Ni(II)(ox)	0	0.05 M $(NH_4)_2$ox; pH = 8.0	-1.18	-	-	-	[127]
Ni(II)(ox)	0	0.05 M $(NH_4)_2$ox; pH = 10.0	-1.33	-	-	-	[127]
Ni(II)(ox)	0	0.1 M $(NH_4)_2$ox + 0.1 M NH_3	-1.15	-	-	-	[127]
Ni(II)(ox)	0	0.1 M $(NH_4)_2$ox + 0.1 M NH_3	-1.073	-	97	fw	[113]
Ni(II) complex	0	satd $MALH_2$	-0.97	-	irr	i	[111]
Ni(II) complex	0	0.1 M $NaClO_4$ + 9 X 10^{-4} Ni(II) + 0.052 M $MALH_2$; 2.1 < pH < 3.8	-1.042	1.21	72	0.004% gel	[130]
Ni(II) complex	0	0.5 M NaHMAL + 0.5 M Na_2MAL; pH ~ 4.8	-1.13	-	-	i	[111]
Ni(II) complex	0	0.5 M NaHMAL + 0.5 M Na_2MAL; pH ~ 4.8	-1.28	-	-	i	[111]
Ni(II) complex	0	0.5 M NaHMAL + 0.5 M Na_2MAL; pH ~ 4.8	-1.50	-	-	i	[111]
Ni(II) complex	0	0.1 M NH_3 + 0.1 M $(NH_4)_2$MAL	-0.983	-	75	w; max supp	[111]
Ni(II) complex	0	1 M Na_2MAL + Carbonate buffer; pH = 10	-1.14	-	-	i	[111]
Ni(II) complex	0	1 M Na_2MAL + Carbonate buffer; pH = 10	-1.36	-	-	fi	[111]

Ni(II) complex	0	0.1 M $NaClO_4$ + 0.05 M H_2SUCC + 9 X 10^{-4} M Ni(II); pH = 3.5	-1.072	1.68	58	0.004% gel	[130]
Ni(II) complex	0	0.1 M KCl + 2.5 X 10^{-2} M AcAcOH; pH = 4.2	-1.0	-	-	0.005% CMC; 1st wave	[131]
Ni(II) complex	0	0.1 M KCl + 2.5 X 10^{-2} M AcAcOH; pH = 6.3	-1.0	-	-	1st wave	[131]
Ni(II) complex	0	0.1 M KCl + 2.5 X 10^{-2} M AcAcOH; pH = 6.3	-1.5	-	-	2nd wave	[131]
Ni(II) complex	0	0.1 M KCl + 2.5 X 10^{-2} M AcAcOH; pH = 7.8	-1.5	-	-	2nd wave	[131]
Ni(II)(?)	0	0.1 M KHFt	-1.14	-	irr	-	[132]
Ni(II)(tart)	0	2% K_2tart; 7 < pH < 12	-1.0	-	-	-	[133]
Ni(II)(tart)	0	satd H_2tart	-1.05	-	-	fi	[111]
Ni(II)(tart)	0	0.25 M tart; 3 < pH < 4	-1.13	-	irr	i	[111]
Ni(II)(tart)	0	0.1 M NH_3 + 0.1 M $(NH_4)_2$tart	-0.960	3.16	52	w; max supp	[111]
Ni(II)(citr)	0	satd H_3citr	-0.98	-	70	i	[111]
Ni(II)(citr)	0	0.1 M NH_3 + 0.1 M $(NH_4)_3$citr; pH = 8.5	-1.09	-	82	i	[111]
Ni(II)(citr)	0	0.1 M NH_3 + 0.1 M $(NH_4)_3$citr; pH = 8.5	-1.395	-	10	fw	[111]
Ni(II)(glut)	0	0.05 M Hglut 0.2 M $NaClO_4$ + borate buffer; pH = 9.5	-1.24	-	-	-	[111]

(continued)

TABLE 2.1.1 (continued)

Substance	Product	Conditions	$E_{1/2}$ (SCE) (V)	I	$E_{1/4}$ - $E_{3/4}$ (mV)	Remarks	Refs.
Ni(II)(glut)	0	0.05 M Hglut 0.2 M $NaClO_4$ + borate buffer; pH = 9.5	-1.55	-	-	-	[111]
Ni(II) complex	0	0.1 M en + 0.1-1 M KCl or KNO_3	-1.60	-	irr	fw	[111]
Ni(II) complex		0.5 M en + 0.5 M K_3PO_4	-1.52	-	-	-	[111]
$[Ni_2(TEDA)_3]^{4+}$	0	0.01 M TEDA	-1.24	-	29.5	0.0006% TX-100	[134]
$[Ni(TEDA)(H_2O)_2]^{2+}$	0	10^{-6} M TEDA	-1.08	-	44.3	0.0006% TX-100	[134]
Ni(II)(TETA)	0	4×10^{-3} M TETA	-1.20	-	irr	0.008% TX-100	[135]
Ni(II) complex	0	0.5 M NaOH + 0.5 M bis-(2-hydroxybutyl)-2-hydroxyethylamine	-1.50	-	-	-	[136]
Ni(II)(4-CNO)	0	8.9 M KOH + 1.2×10^{-3} 4-CNO + 4.3×10^{-4} M $NiCl_2$	-1.51	-	-	-	[137]
Ni(IV)(DMG)	Ni(II)(DMG)	1.5 M NaOH + 0.1 M DMG + 5×10^{-3} M Ni(II)	-0.297	2.4 ± 0.1	30	-	[138]
$[Ni(Py)_6]^{2+}$	0	0.1 M Py + 0.1 M PyHCl	-0.751	2.6	29	w	[111]
$[Ni(Py)_6]^{2+}$	0	0.1 M Py + 0.1 M PyHCl	-1.16	2.6	irr	w; $i_2 = 0.14\ i_1$	[111]
$[Ni(Py)_6]^{2+}$	0	1 M KCl + 0.5 M Py	-0.78	-	-	0.01% gel	[126]
Ni(II) complex	0	0.2 M $NaClO_4$ + 5×10^{-4} M Ni(II) + 10^{-4} M Py	-1.025	-	-	3rd wave only	[139]

Ni(II) complex	0	0.2 M $NaClO_4$ + 5 X 10^{-4} M Ni(II) + 10^{-3} M Py	-0.76	-	-	-	[139]
Ni(II) complex	0	0.2 M $NaClO_4$ + 5 X 10^{-4} M Ni(II) + 10^{-3} M Py	-0.86	-	-	-	[139]
Ni(II) complex	0	0.2 M $NaClO_4$ + 5 X 10^{-4} M Ni(II) + 10^{-3} M Py	-1.035	-	-	-	[139]
Ni(II) complex	0	0.2 M $NaClO_4$ + 5 X 10^{-4} M Ni(II) + 2 X 10^{-3} M Py	-0.73	-	-	-	[139]
Ni(II) complex	0	0.2 M $NaClO_4$ + 5 X 10^{-4} M Ni(II) + 2 X 10^{-3} M Py	-0.81	-	-	-	[139]
Ni(II) complex	0	0.2 M $NaClO_4$ + 5 X 10^{-4} M Ni(II) + 2 X 10^{-3} M Py	-1.046	-	-	-	[139]
Ni(II) complex	0	0.2 M $NaClO_4$ + 5 X 10^{-4} M Ni(II) + 10^{-2} M Py	-0.73	-	-	1st and 3rd waves	[139]
Ni(II) complex	0	0.2 M $NaClO_4$ + 5 X 10^{-4} M Ni(II) + 10^{-2} M Py	-1.065	-	-	1st and 3rd waves	[139]
Ni(II) complex	0	0.2 M $NaClO_4$ + 5 X 10^{-4} M Ni(II) + 0.1 M Py	-0.735	-	-	1st wave only	[139]
Ni(II) complex	0	0.1 M KCl + 10^{-3} M Arg + 10^{-3} M Ni(II)	-0.87	-	-	-	[140]
Ni(II) complex	0	0.1 M KCl + 10^{-3} M Arg + 10^{-3} M Ni(II)	-0.94	-	-	-	[140]
Ni(II) complex	0	0.1 M KCl + 10^{-3} M Arg + 10^{-3} M Ni(II)	-1.09	-	-	-	[140]
Ni(II) complex	0	0.1 M NH_3 + 0.1 M NH_4Cl + 10^{-3} M Asp + 4 X 10^{-3} M Ni(II)	-0.92	-	-	-	[140]

(continued)

TABLE 2.1.1 (continued)

Substance	Product	Conditions	$E_{1/2}$ (SCE) (V)	I	$E_{1/4} - E_{3/4}$ (mV)	Remarks	Refs
Ni(II) complex	0	0.1 M NH_3 + 0.1 M NH_4Cl + 10^{-3} M Asp + 4 × 10^{-3} M Ni(II)	-1.16	-	-	-	[140]
Ni(II) complex	0	10^{-3} M HistHCl	-1.044	-	-	-	[140]
Ni(II) complex	0	10^{-3} M HistHCl	-1.37	-	-	-	[140]
Ni(II) complex	0	0.1 M KCl + 10^{-3} M HistHCl + 10^{-3} M Ni(II)	-0.96	-	-	-	[140]
Ni(II) complex	0	0.1 M KCl + 10^{-3} M HistHCl + 10^{-3} M Ni(II)	-1.11	-	-	-	[140]
Ni(II) complex	0	0.1 M KCl + 10^{-3} M Met + 2 X 10^{-3} M Ni(II)	-0.86	-	-	-	[140]
Ni(II) complex	0	0.1 M KCl + 10^{-3} M Met + 2 X 10^{-3} M Ni(II)	-1.03	-	-	-	[140]
Ni(II) complex	0	0.1 M KCl + 0.1 M gly; pH = 5	-0.98	-	-	-	[111]
Ni(II) complex	0	0.25 M NH_3 + 0.25 M NH_4Cl + 0.05 M Tiron	-1.48	-	-	-	[125]
Nickellocene	Nickello-cenium	50% H_2O + 50% ETOH + 0.5 M $HClO_4$	-0.08	-	-	Anode	[141]
Nickellocenium	Nickellocene	50% H_2O + 50% ETOH + 0.5 M $HClO_4$	-0.21	-	-	Cathode	[141]

[a] AcO = acetate; AcAcO = acetylacetate; Arg = Arginine; Asp = asparigine; CMC = carboxymethylcellulose; CNO = 4-carboxynioxime; citr = citrate; DMG = dimethylglyoxime; EtOH = ethanol; EN = ethylenediamine; Ft = phthalate; gel = gelatin; GLUT = glutamate; Gly = glycine; Hist = histidine; i = ill-defined wave; mal = malonate; max supp = maximum suppressor; Met = methionine; ox = oxalate; Py = pyridine; SUCC = succinate; tart = tartrate; TEA = triethanolaurine; TETA = triethylenetetramine; TEDA = triethylenediamine; TX-100 = Triton X-100; w = well-defined wave.

TABLE 2.1.2. Effect of Supporting Electrolyte Concentration on $E_{1/2}$ of Ni(II) [142] Solutions Containing 10^{-3} M $Ni(NO_3)_2$, 0.005% Gelatin, and Various Amounts of $LiNO_3$

Product	$LiNO_3$ concn (M)	$E_{1/2}$
(0)	0.5	-1.029
(0)	1.0	-1.018
(0)	1.5	-1.015
(0)	2.0	-1.009
(0)	3.0	-0.988
(0)	4.0	-0.975
(0)	5.0	-0.958
(0)	6.0	-0.936
(0)	7.0	-0.912

TABLE 2.1.3. Polarographic Data for Ni in Various Nonaqueous Solvents for Ni(II) → 0[a]

Solvent	Conditions	$E_{1/2}$ (V)	I	$E_{1/4} - E_{3/4}$ (mV)	Remarks	Refs.
Acetic acid	0.1 M $LiClO_4$	-0.645	-	127	liq junction potential correction	[143]
Acetic acid	0.25 M NH_4AcO	-0.975	-	86	liq junction potential correction	[143]
Acetic acid	0.5 M NH_4AcO	-0.825	-	80	0.005% TX-100	[144]
Ammonia	0.1 M KNO_3; -36°C	-0.79	4.69	42	-	[111]

(continued)

TABLE 2.1.3 (continued)

Solvent	Conditions	$E_{1/2}$ (V)	I	$E_{1/4} - E_{3/4}$ (mV)	Remarks	Refs.
Methanol + benzene (1:1)	0.3 M NaAcO + 2 X 10^{-3} M Ni octoate	-1.36	3.5	-	-	[145]
Methanol + benzene (1:1)	0.3 M NaAcO + 2 X 10^{-3} M Ni octoate	-1.70	1.85	-	-	[145]
Hydrazine	-	-1.102	3.12	-	-	[146]
Formamide	-	-0.98	0.99	83	-	[147]
Dimethylformamide	0.1 M $NaClO_4$	-1.33	-	-	-	[148]
Dimethylformamide	0.1 M $NaClO_4$	-1.06	2.59	105	-	[149]
Dimethylformamide	0.1 M $NaClO_4$ + less than 3 X 10^{-3} M H_2O	-0.53	-	64	Ag/AgCl ref electrode	[150]
Dimethylformamide	0.1 M Et_4NClO_4	-0.8	-	-	-	[151]
Dimethylformamide	0.1 M Et_4NClO_4	-2.09	-	-	-	[151]
Dimethylformamide	0.1 M Et_4NCl	-1.4	-	-	-	[151]
Dimethylformamide	0.1 M Et_4NCl	-2.09	-	-	-	[151]
Dimethylformamide	0.1 M Et_4NClO_4 + Et_4NBr	-1.0	-	-	-	[151]

Dimethylformamide	0.1 M Et_4NClO_4 + Et_4NBr	-2.15	-	-	-	[151]
Dimethylformamide	0.1 M Et_4NClO_4	-0.69	2.02	$\alpha n = 0.40$	"DMF" ref electrode	[152]
Dimethylformamide	0.1 M Et_4NClO_4	-0.92	2.80	60	-	[153]
Acetonitrile	0.1 M Et_4NClO_4 + 0.02 M H_2O + 8 X 10^{-5} M CDMBCl	-0.43	-	irr	-	[154]
Acetonitrile	0.1 M Et_4NClO_4 + 0.02 M H_2O + 8 X 10^{-5} M CDMBCl	-0.9	-	irr	-	[154]
Acetonitrile	0.1 M Et_4NClO_4 + 0.02 M H_2O + 10^{-3} M CDMBCl	-0.55	-	irr	-	[154]
Acetonitrile	0.1 M Et_4NClO_4 + 0.02 M H_2O + 10^{-3} M CDMBCl	-0.9	-	irr	-	[154]
Acetonitrile	0.1 M $NaClO_4$	-0.29	-	rev	-	[155]
Acetonitrile	0.1 M $NaClO_4$	-1.37	-	irr	-	[155]
Acetonitrile	0.1 M Et_4NClO_4	-0.33	4.04	31	-	[156]
Acetonitrile	0.1 M $(n\text{-}Bu)_4NClO_4$	-0.69	-	-	Irreproducible	[157]
Acetonitrile	0.1 M $(n\text{-}Bu)_4NClO_4$	-1.27	-	-	Irreproducible	[157]
Propionitrile	0.1 M Et_4NClO_4	-0.36	-	irr	-	[155]
Propionitrile	0.1 M Et_4NClO_4	-1.58	-	irr	-	[155]
Propionitrile	0.1 M Et_4NClO_4	-0.36	-	-	Irregular pol wave	[158]

(continued)

TABLE 2.1.3 (continued)

Solvent	Conditions	$E_{1/2}$ (V)	I	$E_{1/4} - E_{3/4}$ (mV)	Remarks	Refs.
Isobutyronitrile	0.05 M Et_4NClO_4	-0.33	1.83	200	-	[159]
Isobutyronitrile	0.05 M Et_4NClO_4	-1.55	-	-	-	[159]
Acrylonitrile	0.1 M Et_4NClO_4	-0.33	-	-	-	[158]
Benzonitrile	0.1 M Et_4NClO_4	-0.29	-	-	-	[155]
Benzonitrile	0.1 M Et_4NClO_4	-1.59	-	-	-	[155]
Benzonitrile	0.1 M Et_4NClO_4	-0.29	-	59	-	[158], [160]
Dimethylsulfoxide	0.1 M Et_4NClO_4	-1.08	-	-	max; Zn ref electrode	[161]
Dimethylsulfoxide	0.1 M $NaNO_3$	+0.032	-	69	-	[148]
Dimethylsulfoxide	0.1 M Et_4NClO_4	-0.95	1.71	$\alpha n = 0.40$	"DMSO" ref electrode	[152]
Dimethylsulfoxide	0.1 M Et_4NClO_4	-1.07	1.75	75	-	[153]
Sulfolane	0.1 M $NaClO_4$; 40°C	0.32	-	120	$Ag/AgCl/Et_4NCl$(sat)	[162]
Ethylene sulfite	0.1 M Et_4NClO_4	-0.60	-	irr	-	[159]

[a]CDMBCl = cetyldimethylbenzylammonium chloride; DMF = dimethylformamide; DMSO = dimethylsulfoxide; Et = ethyl; n-Bu = n-butyl.

TABLE 2.1.4. Effect of Water Content on $E_{1/2}$ of Ni(II) in Acetonitrile Solutions

$\%H_2O$	i_d (μA)	$E_{1/2}$ (V)	Remarks
0	14.0	-0.33	Reversible slope 0.031
0.4	13.8	-0.36	Slope 0.035
1.0	13.6	-0.70	Wave becomes drawn out
4.0	13.4	-1.2	Highly irr wave

nickel ion concentration in the solution. Larson and Iwamoto [154] studied, by polarography and spectrophotometry, the influence of chloride ion, dissolved as cetyldimethylbenzylammonium chloride, on the behavior of Ni(II) dissolved in acetonitrile containing ~ 0.02 M water. Two waves were observed, the first one probably related to the reduction of the complexes $NiCl_4^{2-}$, $NiCl_3^-$, $NiCl_2$, and $NiCl^+$, and the second wave related to the discharge of the hydrated nickel ion. The 0.005% methyl red used as a maximum suppressor probably acts to catalyze the reactions and eliminate the second wave. Ikeda and Itabashi [164] studied the influence of the cation of the supporting electrolyte on the reduction of Ni(II)-acetate dissolved in acetonitrile containing acetic acid and acetic anhydride. Depending on the composition of the solution either one or two steps were obtained, which were characterized by half-wave potentials, shown in Table 2.1.5 (see Section 4.1.8).

TABLE 2.1.5. Effect of Cation of the Supporting Electrolyte on the Reduction Waves of Ni(II) Acetate Dissolved in Acetonitrile

Cation (M)	Ni(II) concentration (mM)	$E_{1/2}$ (SCE) (V)	Slope X 10^3 (V)
Li^+	0.5	-0.47	82
Na^+	0.5	-0.58	80
$[(CH_3)_4N]^+$	1.0	-0.63	78
$[(CH_3)_4N]^+$	1.0	-1.05	135
$[(C_2H_5)_4N]^+$	1.0	-0.60	68
$[(C_2H_5)_4N]^+$	1.0	-1.1	max
$[(n\text{-}C_4H_9)_4N]^+$	4.0	-0.56	70
$[(n\text{-}C_4H_9)_4N]^+$	2.0	-1.2	max

Coetzee and MacGuire [155] and Larson and Iwamoto [158, 160] obtained Ni(II) polarograms in benzonitrile which indicate the presence of two species. The second wave showed a maximum which was suppressed by adding 0.001% of either methyl red or the sodium form of methyl red. On increasing the methyl red concentration, the second anodic wave shifted toward the first one until only one wave was obtained which was characterized by the half-wave potential of the first wave. The addition of water provoked a distortion of the second wave, which became more irreversible in character. Comparison of the half-wave potentials in benzonitrile, acrylonitrile, propionitrile, and acetonitrile, shows that Ni(II) reduction occurs easier in these solvents in the order written. This effect is related to the presence of conjugate double bonds in the solvents [158, 160]. Coetzee and Hedrick [159] investigated the behavior of Ni(II) in isobutyronitrile and compared the half-wave potentials with those known for other nitriles and water.

Brown and Al-Urfali [149] investigated the behavior of Ni(II) dissolved in dimethylformamide. The difference of the diffusion constants obtained in water, dimethylformamide, and acetonitrile is explained in terms of viscosity and solvodynamic radii of solvated ionic species. The same system in the presence of chloride, bromide, and iodide ions was studied by Ciana and Furlani [151]. The first wave increases with the chloride ion concentration and eventually the second wave tends to disappear. Only the half-wave potential of the first wave depends on the $[Cl^-]/[Ni(II)]$ ratio. In the presence of bromide ion the second wave is completely developed while the first one disappears. The half-wave potential of the second wave exhibits a slight dependence on bromide ion concentration. The system Ni(II)-methylformamide was studied by Brown and Hsiung [165].

Coulder and Iwamoto [143] investigated the polarographic behavior of Ni(II) in acetic acid in 1.0 F $LiClO_4$ and 0.25 F NH_4-acetate. The waves exhibit an irreversible character. The half-wave potentials mentioned in Table 2.1.3 have been corrected for the liquid junction potential difference between $LiClO_4$ solution and aqueous solution of the saturated calomel electrode. Ulery [144] studied the polarographic reduction of Ni(II) in anhydrous acetic acid (100%), in mixtures of acetic acid and water, and in 100% water. The half-wave potential increases from -0.825 V (SCE) for 100% acetic acid up to a maximum at about -1.200 V for a 50% mole water. On increasing the water content the potential shifts to -1.015 V for pure water solutions. The $\Delta E/\Delta \log[i/(i_d - i)]$ slopes obtained for different concentrations of water change from 0.080 to 0.115 V/decade. The limiting currents are ill-defined and the waves are irreversible. These results are different from those reported by Bachman and Astle [166], who found the half-wave potential in acetic acid (0.25 F NH_4-acetate) to be 0.1 to 0.2 V more negative than in water, but are in agreement with those reported by Migal and Agasleva [147]. The half-wave potential for acetic acid (0.01 M NH_4NO_3) lies 0.162 V more positive than in water.

The ratio $(i_d)_{H_2O}/(i_d)_{HOAc}$, as determined by Ulery [144], is 0.83, and the diffusion constant after correction for viscosity exhibits a very pronounced minimum at 13.7 M H_2O, which corresponds to the formation of $HOAc.H_2O$. On the other hand, the half-wave potential changes only a few millivolts with the NH_4-acetate concentration. The polarographic wave of Ni(II) in formic acid is obscured by the hydrogen wave over a large range of potentials [144].

Kolthoff and Reddy [161] studied Ni(II) reduction in dimethylsulfoxide solutions by different methods. In very dilute solutions the polarograms obtained with 0.1 M $NaClO_4$ exhibit a very pronounced maximum current. No surface-active substance tested was efficient for its complete suppression. However, well-defined waves are obtained with a rotating mercury pool electrode at 120 rpm.

Kumar and Pantony determined the half-wave potential for Ni(II) polarographic reduction in different solvents, using perchlorate supporting electrolytes [152] (see Table 2.1.6). This comparison, however, is not strictly valid because of the irreversible nature of the polarograms. Other variables such as ionic strength, αn, Goldschmidt radius, r_G, and Einstein-Stokes radius, r_{ES} were also compared. This is shown in Table 2.1.7.

Godeau, Berthon, Camps, and Bernard [146] studied Ni(II) polarography in hydrazine and mixtures of this solvent with water. The Ilkovic equation was satisfactorily obeyed. For hydrazine-water mixtures the diffusion coefficient exhibits a minimum value at 50 mole % water. At this composition the system presents its maximum viscosity. The half-wave potentials are in good agreement with those calculated with the Born free energy equation.

There are also references to polarograms obtained using mixtures of solvents. Kuta [145] obtained two waves in a 1:1 benzene-methanol mixture with 0.1×10^{-3} to 0.2×10^{-3} M nickel octoate. This behavior is the same exhibited by $NiCl_2$ in the same solvent. When the Ni(II) concentration diminishes, the second wave tends to disappear.

Headridge [162] investigated the reaction in sulfolane; Hubucki and Stasiewicz [167] in liquid ammonium thiocyanate ammoniate; Gutmann and Schoeber [168] and Gutmann and Nedbalek [169] in morpholine; and Pinfold and Sebba [170] in formic acid. Polarographic characteristics of nickel complexes involving organic ligands are shown in Table 2.1.8. General information about these complexes is given in Headridge's book [177].

TABLE 2.1.6. Half-wave Potentials for Ni(II) Polarographic Reduction in Different Solvents

DMSO	DMF	$HCONH_2$	H_2O	Sulfolane	CH_3CN
-1.48 V	-1.15 V	-1.13 V	-1.26 V	-0.51 V	-0.43 V

TABLE 2.1.7. Goldschmidt's and Einstein-Stokes' Radii for Ni(II) on Different Solvents

	r_{ES} (Å)		
r_G (Å)	DMF	DMSO	H_2O
0.68	14.7	9.94	3.16

TABLE 2.1.8. Polarographic Characteristics of Ni Complexes Involving Organic Ligands

Substance	Product	Conditions	$E_{1/2}$ (V)	$E_{1/4} - E_{3/4}$ (mV)	Remarks	Refs.
$Ni(mnt)_2^-$	-2	0.042 M $[(n\text{-}Pr)_4N]ClO_4$; solvent ACN	0.23	-	-	[171]
$Ni(dtfa)(mnt)^-$	-2	0.042 M $[(n\text{-}Pr)_4N]ClO_4$; solvent ACN	0.06	-	-	[171]
$Ni(dtb)_2^-$	-2	0.1 M $(Pr)_4NPF_6$; solvent CH_2Cl_2	-0.83	-	-	[171]
0	-1	0.1 M $(Pr)_4NPF_6$; solvent CH_2Cl_2	-0.01	-	-	[171]
$Ni(dbt)(dtfa)^-$	-2	0.1 M $(Pr)_4NPF_6$; solvent CH_2Cl_2	-0.55	-	-	[171]
0	-1	0.1 M $(Pr)_4NPF_6$; solvent CH_2Cl_2	0.35	-	-	[171]
$Ni(dtb)(mnt)^-$	-2	0.1 M $(Pr)_4NPF_6$; solvent CH_2Cl_2	-0.38	-	-	[171]
0	-1	0.1 M $(Pr)_4NPF_6$; solvent CH_2Cl_2	0.43	-	-	[171]
$Ni(dtfa)_2^-$	-2	0.1 M $(Pr)_4NPF_6$; solvent CH_2Cl_2	-0.22	-	-	[171]
0	-1	0.1 M $(Pr)_4NPF_6$; solvent CH_2Cl_2	0.80	-	-	[171]
$Ni(dtfa)(mnt)^-$	-2	0.1 M $(Pr)_4NPF_6$; solvent CH_2Cl_2	-0.01	-	-	[171]
$Ni(mnt)_2^-$	-2	0.1 M $(Pr)_4NPF_6$; solvent CH_2Cl_2	0.12	-	-	[171]
$Ni(pdt)_2^{-2}$	-1	0.1 M $[(n\text{-}C_3H_7)_4N]ClO_4$; solvent DMF	-1.24	-	Ref: $Ag/AgClO_4$	[172]
$Ni(xdt)_2^{-2}$	-1	0.1 M $[(n\text{-}C_3H_7)_4N]ClO_4$; solvent DMF	-1.14	-	Ref: $Ag/AgClO_4$	[172]

$Ni(tdt)_2^{-2}$	-1	0.1 M $[(n\text{-}C_3H_7)_4N]ClO_4$; solvent DMF	-1.07	-	Ref: $Ag/AgClO_4$	[172]
$Ni(bdt)_2^{-2}$	-1	0.1 M $[(n\text{-}C_3H_7)_4N]ClO_4$; solvent DMF	-1.05	-	Ref: $Ag/AgClO_4$	[172]
$Ni(tcdt)_2^{-2}$	-1	0.1 M $[(n\text{-}C_3H_7)_4N]ClO_4$; solvent DMF	-0.532	-	Ref: $Ag/AgClO_4$	[172]
$Ni(mnt)_2^{-2}$	-1	0.1 M $[(n\text{-}C_3H_7)_4N]ClO_4$; solvent DMF	-0.218	-	Ref: $Ag/AgClO_4$	[172]
$Ni[S_2C_2(CN)_2]_2^{2-}$	-1	0.1 M $LiClO_4$; solvent DMF	-0.259	-	Ref: $Ag/AgClO_4$ in 0.1 M $LiClO_4$(aq)	[173]
-1	0	0.1 M $LiClO_4$; solvent DMF	1.049	-	Ref: $Ag/AgClO_4$ in 0.1 M $LiClO_4$(aq)	[173]
$Ni[S_2C_2(CF_3)_2]_2^{2-}$	-1	0.1 M $LiClO_4$; solvent DMF	-0.088	-	Ref: $Ag/AgClO_4$ in 0.1 M $LiClO_4$(aq)	[174]
-1	0	0.1 M $LiClO_4$; solvent DMF	1.030	-	Ref: $Ag/AgClO_4$ in 0.1 M $LiClO_4$(aq)	[174]
$Ni[S_2C_2(p\text{-}C_6H_4Cl)_2]_2^{2-}$	-1	0.1 M $LiClO_4$; solvent DMF	-0.757	-	Ref: $Ag/AgClO_4$ in 0.1 M $LiClO_4$(aq)	[173]
-1	0	0.1 M $LiClO_4$; solvent DMF	0.218	-	Ref: $Ag/AgClO_4$ in 0.1 M $LiClO_4$(aq)	[173]
$Ni[S_2C_2(C_6H_5)_2]_2^{2-}$	-1	0.1 M $LiClO_4$; solvent DMF	-0.881	-	Ref: $Ag/AgClO_4$ in 0.1 M $LiClO_4$(aq)	[173]
-1	0	0.1 M $LiClO_4$; solvent DMF	0.134	-	Ref: $Ag/AgClO_4$ in 0.1 M $LiClO_4$(aq)	[173]
$Ni[S_2C_2H(C_6H_5)]_2^{2-}$	-1	0.1 M $LiClO_4$; solvent DMF	-0.879	-	Ref: $Ag/AgClO_4$ in 0.1 M $LiClO_4$(aq)	[173]

(continued)

TABLE 2.1.8 (continued)

Substance	Product	Conditions	$E_{1/2}$ (V)	$E_{1/4} - E_{3/4}$ (mV)	Remarks	Refs.
-1	0	0.1 M $LiClO_4$; solvent DMF	0.115	-	Ref: Ag/$AgClO_4$ in 0.1 M $LiClO_4$(aq)	[173]
$Ni[S_2C_2H_2]_2^{2-}$	-1	0.1 M $LiClO_4$; solvent DMF	-0.921	-	Ref: Ag/$AgClO_4$ in 0.1 M $LiClO_4$(aq)	[173]
-1	0	0.1 M $LiClO_4$; solvent DMF	0.120	-	Ref: Ag/$AgClO_4$ in 0.1 M $LiClO_4$(aq)	[173]
$Ni[S_2C_2(p\text{-}C_6H_4CH_3)_2]_2^{2-}$	-1	0.1 M $LiClO_4$; solvent DMF	-0.960	-	Ref: Ag/$AgClO_4$ in 0.1 M $LiClO_4$(aq)	[173]
-1	0	0.1 M $LiClO_4$; solvent DMF	0.083	-	Ref: Ag/$AgClO_4$ in 0.1 M $LiClO_4$(aq)	[173]
$Ni[S_2C_2CH_3(C_6H_5)]_2^{2-}$	-1	0.1 M $LiClO_4$; solvent DMF	-0.988	-	Ref: Ag/$AgClO_4$ in 0.1 M $LiClO_4$(aq)	[173]
-1	0	0.1 M $LiClO_4$; solvent DMF	0.025	-	Ref: Ag/$AgClO_4$ in 0.1 M $LiClO_4$(aq)	[173]
$Ni[S_2C_2(p\text{-}C_6H_4OCH_3)_2]_2^{2-}$	-1	0.1 M $LiClO_4$; solvent DMF	-0.945	-	Ref: Ag/$AgClO_4$ in 0.1 M $LiClO_4$(aq)	[173]
-1	0	0.1 M $LiClO_4$; solvent DMF	0.035	-	Ref: Ag/$AgClO_4$ in 0.1 M $LiClO_4$(aq)	[173]
$Ni[S_2C_2(CH_3)_2]_2^{2-}$	-1	0.1 M $LiClO_4$; solvent DMF	-1.114	-	Ref: Ag/$AgClO_4$ in 0.1 M $LiClO_4$(aq)	[173]

-1	0	0.1 M $LiClO_4$; solvent DMF	-0.107	-	Ref: $Ag/AgClO_4$ in 0.1 M $LiClO_4$(aq)	[173]
$Ni[S_2C_2(C_2H_5)_2]_2^{2-}$	-1	0.1 M $LiClO_4$; solvent DMF	-1.138	-	Ref: $Ag/AgClO_4$ in 0.1 M $LiClO_4$(aq)	[173]
-1	0	0.1 M $LiClO_4$; solvent DMF	-0.119	-	Ref: $Ag/AgClO_4$ in 0.1 M $LiClO_4$(aq)	[173]
$Ni[S_2C_2(n\text{-}C_3H_7)_2]_2^{2-}$	-1	0.1 M $LiClO_4$; solvent DMF	-1.138	-	Ref: $Ag/AgClO_4$ in 0.1 M $LiClO_4$(aq)	[173]
-1	0	0.1 M $LiClO_4$; solvent DMF	-0.121	-	Ref: $Ag/AgClO_4$ in 0.1 M $LiClO_4$(aq)	[173]
$Ni[S_2C_2(i\text{-}C_3H_7)_2]_2^{2-}$	-1	0.1 M $LiClO_4$; solvent DMF	-1.204	-	Ref: $Ag/AgClO_4$ in 0.1 M $LiClO_4$(aq)	[173]
-1	0	0.1 M $LiClO_4$; solvent DMF	-0.121	-	Ref: $Ag/AgClO_4$ in 0.1 M $LiClO_4$(aq)	[173]
$Ni[C_6H_4SNH]_2^{2-}$	-1	0.1 M $[(n\text{-}C_3H_7)_4N]ClO_4$; solvent DMSO	-1.04	-	Ref: SCE(aq)	[175]
-1	0	0.1 M $[(n\text{-}C_3H_7)_4N]ClO_4$; solvent DMSO	-0.19	-	Ref: SCE(aq)	[175]
$Ni[C_6H_4OS]_2^{2-}$	-1	0.1 M $[(n\text{-}C_3H_7)_4N]ClO_4$; solvent DMSO	-0.42	-	Ref: SCE(aq)	[175]
-1	0	0.1 M $[(n\text{-}C_3H_7)_4N]ClO_4$; solvent DMSO	0.38	-	Ref: SCE(aq)	[175]
$Ni[C_6H_4O_2]_2^{-2}$	-1	0.1 M $[(n\text{-}C_3H_7)_4N]ClO_4$; solvent DMSO	-0.29	-	Ref: SCE(aq)	[175]

(continued)

TABLE 2.1.8 (continued)

Substance	Product	Conditions	$E_{1/2}$ (V)	$E_{1/4} - E_{3/4}$ (mV)	Remarks	Refs.
-1	0	0.1 M $[(n\text{-}C_3H_7)_4N]ClO_4$; solvent DMSO	0.46	-	Ref: SCE(aq)	[175]
$Ni[C_6H_4S_2]_2^{-2}$	-1	0.1 M $[(n\text{-}C_3H_7)_4N]ClO_4$; solvent DMSO	-0.51	-	Ref: SCE(aq)	[175]
-1	0	0.1 M $[(n\text{-}C_3H_7)_4N]ClO_4$; solvent DMSO	0.45	-	Ref: SCE(aq)	[175]
$Ni[C_6H_4(NH)_2]_2$	-1	0.05 M $[(n\text{-}Pr)_4N]ClO_4$; solvent DMSO	-0.88	63	-	[176]
-1	-2	0.05 M $[(n\text{-}Pr)_4N]ClO_4$; solvent DMSO	-1.59	61	-	[176]
+2	+1	0.05 M $[(n\text{-}Pr)_4N]ClO_4$; solvent $(Me)_2CO$	0.73	-	-	[176]
+1	0	0.05 M $[(n\text{-}Pr)_4N]ClO_4$; solvent $(Me)_2CO$	-0.14	59	-	[176]
0	-1	0.05 M $[(n\text{-}Pr)_4N]ClO_4$; solvent $(Me)_2CO$	-0.89	60	-	[176]
-1	-2	0.05 M $[(n\text{-}Pr)_4N]ClO_4$; solvent $(Me)_2CO$	-1.43	63	-	[176]
$Ni[4\text{-}i\text{-}PrC_6H_4(NH_2)_2]_2^{2+}$	+1	0.05 M $[(n\text{-}Pr)_4N]ClO_4$; solvent $(Me)_2CO$	0.81	87	-	[176]
+1	0	0.05 M $[(n\text{-}Pr)_4N]ClO_4$; solvent $(Me)_2CO$	0.09	53	-	[176]

0	-1	0.05 M [(n-Pr)$_4$N]ClO$_4$; solvent (Me)$_2$CO	0.98	66	-	[176]
-1	-2	0.05 M [(n-Pr)$_4$N]ClO$_4$; solvent (Me)$_2$CO	-1.51	67	-	[176]
+2	+1	0.1 M (n-C$_4$H$_9$)$_4$NPF$_6$; solvent CH$_2$Cl$_2$	1.32	-	-	[176]
+1	0	0.1 M (n-C$_4$H$_9$)$_4$NPF$_6$; solvent CH$_2$Cl$_2$	0.44	-	-	[176]
0	-1	0.1 M (n-C$_4$H$_9$)$_4$NPF$_6$; solvent CH$_2$Cl$_2$	-0.77	-	-	[176]
-1	-2	0.1 M (n-C$_4$H$_9$)$_4$NPF$_6$; solvent CH$_2$Cl$_2$	-1.4	-	-	[176]
$Ni[CH_3CN(C_6H_5)_2]_2I_2^{2+}$	0	0.05 M [(n-Pr)$_4$N]ClO$_4$; solvent ACN	-0.61	35	-	[176]
0	-1	0.05 M [(n-Pr)$_4$N]ClO$_4$; solvent ACN	-1.60	56	-	[176]
-1	-2	0.05 M [(n-Pr)$_4$N]ClO$_4$; solvent ACN	-1.80	56	-	[176]
$[(Et_4N)Ni(CH_3C_6H_3S_2)_2]^-$	-2	0.05 M [(n-Pr)$_4$N]ClO$_4$; solvent ACN	-0.58	60	-	[176]
-1	-2	0.05 M [(n-Pr)$_4$N]ClO$_4$; solvent DMSO	-0.52	56	-	[176]
-1	-2	0.1 M [(n-Bu)$_4$N]ClO$_4$; solvent CH$_2$Cl$_2$	-0.40	-	-	[176]

[a] acac = acetylacetonato; bdt = bencene-1, 2-dithiolato; dtb = 1, 2-diphenyl-ethylene-1, 2-dithiolato; dtfa = 1, 2-di(trifluoromethyl) ethylene-1, 2-dithialato; gma =glyoxal-bis-(2-mercaptoanilo); mnt = maleonitrile-dithiolato; pdt = prehnitene-4, 5-dithiolato; tcdt = 3, 4, 5, 6-tetrachlorobencene-1, 2-dithiolato; tdt = bis(toluene-3, 4-dithiolato); xdt = o-xylene-4, 5-dithiolato.

TABLE 2.1.9. Polarographic Characteristics of Ni in Solvents Containing a Mixture of Water and a Nonaqueous Solvent

Solvent	Reaction	Conditions	$E_{1/2}$ (V)	I	$E_{1/4} - E_{3/4}$ (mV)	Remarks	Refs.
H_2O + 5% Acetamide	Ni(II) → 0	1.0 M KSCN	-1.562	-	-	0.01% Gelatin	[178]
H_2O + 10% Acetamide	Ni(II) → 0	1.0 M KSCN	-1.587	-	-	0.01% Gelatin	[178]
H_2O + 15% Acetamide	Ni(II) → 0	1.0 M KSCN	-1.604	-	-	0.01% Gelatin	[178]
H_2O + 20% Acetamide	Ni(II) → 0	1.0 M KSCN	-1.631	-	-	0.01% Gelatin	[178]
H_2O + 25% Acetamide	Ni(II) → 0	1.0 M KSCN	-1.640	-	-	0.01% Gelatin	[178]
H_2O + 30% Acetamide	Ni(II) → 0	1.0 M KSCN	-1.638	-	-	0.01% Gelatin	[178]
H_2O + 0% vol CH_3OH	Ni(II) complex → 0	0.1 M $NaClO_4$ + 0.052 M $malH_2$ + 9 × 10^{-9} M Ni(II)	-1.04	1.13	82	0.004% Gelatin; Ref: SCE	[130]
H_2O + 5% vol CH_3OH	Ni(II) complex → 0	0.1 M $NaClO_4$ + 0.052 M $malH_2$ + 9 × 10^{-9} M Ni(II)	-1.07	1.08	92	0.004% Gelatin; Ref: SCE	[130]
H_2O + 10% vol CH_3OH	Ni(II) complex → 0	0.1 M $NaClO_4$ + 0.052 M $malH_2$ + 9 × 10^{-9} M Ni(II)	-1.068	0.98	87	0.004% Gelatin; Ref: SCE	[130]
H_2O + 20% vol CH_3OH	Ni(II) complex → 0	0.1 M $NaClO_4$ + 0.052 M $malH_2$ + 9 × 10^{-9} M Ni(II)	-1.065	1.00	87	0.004% Gelatin; Ref: SCE	[130]
H_2O + 30% vol CH_3OH	Ni(II) complex → 0	0.1 M $NaClO_4$ + 0.052 M $malH_2$ + 9 × 10^{-9} M Ni(II)	-1.065	0.91	87	0.004% Gelatin; Ref: SCE	[130]
H_2O + 40% vol CH_3OH	Ni(II) complex → 0	0.1 M $NaClO_4$ + 0.052 M $malH_2$ + 9 × 10^{-9} M Ni(II)	-1.060	0.67	87	0.004% Gelatin; Ref: SCE	[130]
H_2O + 50% vol CH_3OH	Ni(II) complex → 0	0.1 M $NaClO_4$ + 0.052 M $malH_2$ + 9 × 10^{-9} M Ni(II)	-1.070	0.80	87	0.004% Gelatin; Ref: SCE	[130]
H_2O + 60% vol CH_3OH	NI(II) complex → 0	0.1 M $NaClO_4$ + 0.052 M $malH_2$ + 9 × 10^{-9} M Ni(II)	-1.075	0.80	87	0.004% Gelatin; Ref: SCE	[130]

H_2O + 80% vol CH_3OH	Ni(II) complex → 0	0.1 M $NaClO_4$ + 0.052 M $malH_2$ + 9 X 10^{-9} M Ni(II)	-1.072	0.75	110	0.004% Gelatin; Ref: SCE	[130]
H_2O + 90% EtOH	Ni(III) → Ni(II)	0.1 M $NaClO_4$	-0.08	-	rev	Cathode	[179]
H_2O + 90% EtOH	Ni(II) → Ni(III)	0.1 M $NaClO_4$	-0.08	-	rev	Anode	[179]
H_2O + 0% vol PrOH	Ni(II) complex → 0	0.1 M $NaClO_4$ + 0.05 M $succH_2$ + 9 X 10^{-4} M Ni(II)	-1.01	1.42	62	Ref: SCE	[130]
H_2O + 5% vol PrOH	Ni(II) complex → 0	0.1 M $NaClO_4$ + 0.05 M $succH_2$ + 9 X 10^{-4} M Ni(II)	-1.062	1.30	58	-	[130]
H_2O + 10% vol PrOH	Ni(II) complex → 0	0.1 M $NaClO_4$ + 0.05 M $succH_2$ + 9 X 10^{-4} M Ni(II)	-1.102	1.21	64	-	[130]
H_2O + 20% vol PrOH	Ni(II) complex → 0	0.1 M $NaClO_4$ + 0.05 M $succH_2$ + 9 X 10^{-4} M Ni(II)	-1.135	1.08	64	-	[130]
H_2O + 30% vol PrOH	Ni(II) complex → 0	0.1 M $NaClO_4$ + 0.05 M $succH_2$ + 9 X 10^{-4} M Ni(II)	-1.140	0.9	64	-	[130]
H_2O + 40% vol PrOH	Ni(II) complex → 0	0.1 M $NaClO_4$ + 0.05 M $succH_2$ + 9 X 10^{-4} M Ni(II)	-1.137	0.9	80	-	[130]
H_2O + 50% vol PrOH	Ni(II) complex → 0	0.1 M $NaClO_4$ + 0.5 M $succH_2$ + 9 X 10^{-4} M Ni(II)	-1.132	0.85	80	-	[130]
H_2O + 60% vol PrOH	Ni(II) complex → 0	0.1 M $NaClO_4$ + 0.05 M $succH_2$ + 9 X 10^{-4} M Ni(II)	-1.125	0.78	83	-	[130]
H_2O + 70% vol PrOH	Ni(II) complex → 0	0.1 M $NaClO_4$ + 0.05 M $succH_2$ + 9 X 10^{-4} M Ni(II)	-1.1	0.68	80	-	[130]
H_2O + 0% vol DMF	Ni(II) complex → 0	0.1 M $NaClO_4$ + 0.05 M $malH_2$	-1.04	1.21	83	Ref: SCE	[130]
H_2O + 5% vol DMF	Ni(II) complex → 0	0.1 M $NaClO_4$ + 0.05 M $malH_2$	-1.025	1.15	90	-	[130]
H_2O + 10% vol DMF	Ni(II) complex → 0	0.1 M $NaClO_4$ + 0.05 M $malH_2$	-1.033	1.20	90	-	[130]

(continued)

TABLE 2.1.9 (continued)

Solvent	Reaction	Conditions	$E_{1/2}$ (V)	I	$E_{1/4} - E_{3/4}$ (mV)	Remarks	Refs.
H_2O + 15% vol DMF	Ni(II) complex → 0	0.1 M $NaClO_4$ + 0.05 M $malH_2$	-1.015	1.27	90	-	[130]
H_2O + 20% vol DMF	Ni(II) complex → 0	0.1 M $NaClO_4$ + 0.05 M $malH_2$	-1.005	1.21	92	-	[130]
H_2O + 30% vol DMF	Ni(II) complex → 0	0.1 M $NaClO_4$ + 0.05 M $malH_2$	-	-	-	Wave not well defined	[130]
N_2H_4 + 0% H_2O	Ni(II) → 0	0.02 M $CaCl_2$ + 8.3×10^{-5} M Ni(II)	-0.102	6.25	-	-	[167]
N_2H_4 + 0.95% H_2O	Ni(II) → 0	0.02 M $CaCl_2$ + 8.3×10^{-5} M Ni(II)	-0.100	6.20	-	-	[167]
N_2H_4 + 4.6% H_2O	Ni(II) → 0	0.02 M $CaCl_2$ + 8.3×10^{-5} M Ni(II)	-0.096	7.0	-	-	[167]
N_2H_4 + 9% H_2O	Ni(II) → 0	0.02 M $CaCl_2$ + 8.3×10^{-5} M Ni(II)	-0.085	7.25	-	-	[167]
N_2H_4 + 17% H_2O	Ni(II) → 0	0.02 M $CaCl_2$ + 8.3×10^{-5} M Ni(II)	-0.070	7.55	-	-	[167]
N_2H_4 + 33% H_2O	Ni(II) → 0	0.02 M $CaCl_2$ + 8.3×10^{-5} M Ni(II)	-0.020	7.9	-	-	[167]
N_2H_4 + 44% H_2O	Ni(II) → 0	0.02 M $CaCl_2$ + 8.3×10^{-5} M Ni(II)	0.010	7.75	-	-	[167]
N_2H_4 + 50% H_2O	Ni(II) → 0	0.02 M $CaCl_2$ + 8.3×10^{-5} M Ni(II)	0.030	7.55	-	-	[167]
N_2H_4 + 60% H_2O	Ni(II) → 0	0.02 M $CaCl_2$ + 8.3×10^{-5} M Ni(II)	0.060	6.7	-	-	[167]
N_2H_4 + 67% H_2O	Ni(II) → 0	0.02 M $CaCl_2$ + 8.3×10^{-5} M Ni(II)	0.060	6.6	-	-	[167]

[a]DMF = dimethylformamide; EtOH = ethyl alcohol; gel = gelatin; $malH_2$ = malonic acid; PrOH = propyl alcohol; $succH_2$ = succinic acid.

TABLE 2.1.10. Polarographic Characteristics of Ni in Melts

Melt	Reaction	Conditions	Electrode	Technique	$E_{1/2}$ (V)	$E_{1/4} - E_{3/4}$ (mV)	Remarks	Refs.
Chlorides	Ni(II) → 0	NaCl-KCl-$AlCl_3$; 146°C	DME	dc pol	-0.830	f rev	Ref: Al	[178]
Chlorides	Ni(II) → 0	LiCl-KCl (59%-49%); 450°C	Bi	dc pol	-0.08	86	Ref: Ag/AgCl	[179]
Formates	Ni(II) → 0	NH_4(HCOO); 125°C	DME	dc pol	-0.450	irrev	Ref: Hg pool	[180]
Nitrates	Ni(II) → 0	$LiNO_3$-$NaNO_3$-KNO_3 (17%-30%-53%); 160°C	DME	dc pol	-0.361	45	Ref: Hg	[181]
Nitrates	Ni(II) → 0	$NaNO_3$-KNO_3 + 9 X 10^{-3} <u>M</u> Ni(II); 340°C	-	-	-1.04	74	-	[182]
Nitrates	Ni(II) → 0	$LiNO_3$-$NaNO_3$-KNO_3 (mp 120°C); 140°C	DME; hng Hg drp; Ni amalgam	dc pol imp	-0.700	f rev	Ref: Ag/$AgSO_4$	[183]
Nitrates	Ni(II) → 0	$LiNO_3$-$NaNO_3$-KNO_3; 156°C	DME	sq wave pol	-0.37	-	Ref: Hg/Hg_2Cl_2	[184]
Nitrates	Ni(II) → 0	$LiNO_3$-$NaNO_3$-KNO_3; 159°C	DME	cathode ray pol	-0.277	-	Ref: Ag/AgCl(sat)	[185]
Nitrates	Ni(II) → 0	$LiNO_3$-$NaNO_3$-KNO_3 + 2.5 10^{-4} <u>M</u> Ni(II); 143°C	DME	dc pol	-0.370	43	-	[186]

Polarographic characteristics of nickel in solutions containing a mixture of water and nonaqueous solvent are assembled in Table 2.1.9.

2.1.3. Fused Salts

The polarographic characteristics of nickel were evaluated in different ionic melts using conventional as well as oscillographic techniques employing in a few cases liquid mercury or bismuth electrodes. Results are summarized in Table 2.1.10, where the melts are listed in alphabetical order. A discussion of these results is given in Section 2.2.3.

2.2. VOLTAMMETRIC CHARACTERISTICS

2.2.1. Aqueous Solution

Voltammetric data of nickel ion in aqueous solutions have been obtained by current step and polarographic techniques on mercury electrodes in noncomplexing media and in the presence of different complexing agents. Data are assembled in Table 2.2.1. In noncomplexing media the results correspond to an irreversible reaction, and their interpretation is considered in Section 4.

2.2.2. Nonaqueous Solution

Table 2.2.2 comprises voltammetric data reported for nickel ion in nonaqueous solvents on platinum and mercury electrodes. Further information about these systems is found in Headridge's book [177].

2.2.3. Fused Salts

Various published reviews on molten salt voltammetry include information about nickel voltammetry [186, 198, 199]. Table 2.2.3 contains voltammetric data on nickel in different melts obtained with rotating solid microelectrodes and wire electrodes. Diffusion coefficients, D, and experimental activation energies for diffusion, ΔH_D^*, are shown in Table 2.2.4 including both polarographic and voltammetric results.

Schmidt [208] investigated the Ni(II) reaction in molten LiCl-KCl at 450°C by oscillographic polarography with square current pulses of 50 Hz. Although the occurrence of double waves, apparently caused by the formation of metallic monolayers at the electrolyte/solid electrode interface was observed, the diffusion coefficient of the cation was calculated. Schwabe and Ross [179] obtained a slope $\Delta E/\Delta \log[i/(i_d - i)]$ equal to 0.086 V/decade on a bismuth dropping electrode using the same system.

TABLE 2.2.1. Voltammetric Data for Nickel

Substance	Product	Condition	Electrode	Technique	Potential (SCE)	Rev	Remarks	Refs.
Ni(II)$(H_2O)_n$	0	4 M $NaClO_4$ + 10^{-2} M Ni(II)	DME	i stp	(-1.1 to -1.7)	irr	$(j\tau^{1/2})_{j=0}$ = 4370; $\mu As^{1/2}\ cm^{-2}$	[187]
Ni(II)$(H_2O)_n$	0	0.2 M KCl	DME	ac pol	E_p = -1.095	irr	i_p/i_d = 2.96; n = 1	[188]
Ni(II)$(H_2O)_n$	0	0.2 M KCl	Hg	i stp, 0.7 mA	-	-	$i\tau^{1/2}$ = 3654 $\mu As^{1/2}$	[188]
Ni(II)(?)	0	2 M NH_4OH, 1 M NH_4Cl	DME	ac pol	E_p = -0.85	-	Ref: Hg anode	[189]
$[Ni(II)(SCN)_n]^{(n-2)-}$	0	0.2 M KSCN	DME	ac pol	E_p = -0.600	rev	i_p/i_d = 4.31; n = 2	[188]
$[Ni(II)(SCN)_n]^{(n-2)-}$	0	0.2 M KSCN, 10^{-3} M Ni(II)	Hg	i stp, 0.7 mA	-	-	$i\tau^{1/2}$ = 3371.5 $\mu As^{1/2}$	[188]
Ni(II)$(EtNH_2)_n$	0	0.2 M KSCN, 0.2 M $EtNH_2$	DME	ac pol	E_p = -0.820	rev	i_p/i_d = 4.30; n = 2	[188]
Ni(II)$(EtNH_2)_n$	0	0.2 M KSCN, 0.2 M $EtNH_2$	Hg	i stp; 0.7 mA	-	-	$i\tau^{1/2}$ = 3308 $\mu As^{1/2}$	[188]
Ni(II)$(Et_2NH)_n$	0	0.2 M KSCN, 0.2 M Et_2NH	DME	ac pol	E_p = -0.975	irr	i_p/i_d = 3.14; n = 2	[188]
Ni(II)$(Et_2NH)_n$	0	0.2 M KSCN, 0.2 M Et_2NH	Hg	i stp; 0.7 mA	-	-	$i\tau^{1/2}$ = 2571 $\mu As^{1/2}$	[188]

(continued)

TABLE 2.2.1 (continued)

Substance	Prod-uct	Condition	Elec-trode	Technique	Potential (SCE)	Rev	Remarks	Refs.
$Ni(II)(Et_3N)_n$	0	0.2 M KSCN, 0.2 M Et_3N	DME	ac pol	$E_p = -1.125$	irr	$i_p/i_d = 2.75$; n = 1	[188]
$Ni(II)(Et_3N)_n$	0	0.2 M KSCN, 0.2 M Et_3N	Hg	i stp; 0.7 mA	-	-	$i\tau^{1/2} = 3481\ \mu As^{1/2}$	[188]
$Ni(II)(en)_n$	0	0.2 M KSCN, 0.2 M en	DME	ac pol	$E_p = -1.265$	irr	$i_p/i_d = 2.67$; n = 1	[188]
$Ni(II)(en)_n$	0	0.2 M KSCN, 0.2 M en	Hg	i stp; 0.7 mA	-	-	$i\tau^{1/2} = 1685\ \mu As^{1/2}$	[188]
$Ni(II)(en)_x(tart)_y$	0	0.2 M NaOH, 0.2 M (en) tart	DME	ac pol	$E_p = -1.251$	-	-	[188]
$Ni(II)(en)_x(tart)_y$	0	0.2 M NaOH, 0.2 M (en) tart	Hg	i stp; 0.7 mA	-	-	$i\tau^{1/2} = 1949\ \mu As^{1/2}$	[188]
Ni(II)(?)	0	Py/HCl	DME	i swp cathode ray pol	$E_{1/2} = -0.9$	-	w	[190]
Ni(II) complex	0	0.1 M KSCN + 5 X 10^{-4} M Ni(II)	Hg	CV	$E_{pc} = -0.695$	$E_p - E_{p/2} = 51$	$E_{pa} - E_{pc} = 205$	[139], [191]
Ni(II) complex	0	0.1 M KSCN + 5 X 10^{-4} M Ni(II) + 10^{-3} M Py	Hg	CV	$E_{pc} = -0.680$		$E_{pa} - E_{pc} = 54$	[190]

Ni(II) complex	0	0.1 M KSCN + 5 X 10^{-4} M Ni(II) + 2 X 10^{-3} M Py	Hg	CV	$E_{pc} = -0.685$	-	$E_{pa} - E_{pc} = 52$	[190]
Ni(II) complex	0	0.1 M KSCN + 5 X 10^{-4} M Ni(II) + 4 X 10^{-3} M Py	Hg	CV	$E_{pc} = -0.685$	-	$E_{pa} - E_{pc} = 48$	[195]
Ni(II) complex	0	0.1 M KSCN + 5 X 10^{-4} M Ni(II) + 1.6 X 10^{-2} M Py	Hg	CV	$E_{pc} = -0.730$	-	$E_{pa} - E_{pc} = 12$	[260]
Ni(II) complex	0	0.1 M KSCN + 5 X 10^{-4} M Ni(II) + 10^{-3} M γ-pic	Hg	CV	$E_{pc} = -0.675$	-	$E_{pa} - E_{pc} = 57$	[190]
Ni(II) complex	0	0.1 M KSCN + 5 X 10^{-4} M Ni(II) + 2 X 10^{-3} M γ-pic	Hg	CV	$E_{pc} = -0.720$	-	$E_{pa} - E_{pc} = 29$	[235]
Ni(II) complex	0	0.1 M KSCN + 5 X 10^{-4} M Ni(II) + 4 X 10^{-3} M γ-pic	Hg	CV	$E_{pc} = -0.770$	-	$E_{pa} - E_{pc} = 18$	[375]
Ni-TSC	0	Ammonia buffer, pH = 9.75; 10^{-3} M TSC, 10^{-4} M Ni(II)	drp Hg	OPPRV 50 Hz	$E_p = -1.56$	-	-	[192]
Ni-TSC	0	Ammonia buffer, pH = 9.75; 10^{-3} M Acetone-TSC	drp Hg	OPPRV	$E_p = -1.50$	-	-	[192]
Ni-cyst	0	Ammonia buffer, pH = 8.26; 10^{-3} M cyst	drp Hg	OPPRV	$E_p = -1.74$	-	-	[192]
Ni-8-MQ	0	Ammonia buffer, pH = 9.75; 10^{-3} M 8-MQ + 10% dioxane	drp Hg	OPPRV	$E_p = -1.72$	-	-	[192]

[a]cyst = cysteine; en = ethylendiamine; Et = ethanolamine; MQ = mercaptoquinoline; OPPRV = oscillographic pulse polarography with rectangular formed voltage; pic = picoline; py = pyridine; tart = tartrate; TSC = thiosemicarbazide.

TABLE 2.2.2. Voltammetric Data for Nickel in Nonaqueous Solvents

Substance	Product	Condition	Electrode	Technique	Potential	Rev	Remarks	Refs.
Ni(II)$(DMSO)_6$	0	0.1 M $KClO_4$; solvent DMSO	hng Hg drp	E swp	$E_p = -1.11$	-	-	[193]
Ni(II)$(DMSO)_6$	0	0.1 M $[(Et)_4N]ClO_4$; solvent DMSO	DME	ac pol	$E_p = -1.90$	-	-	[153]
Ni(II)$(DMSO)_6$	0	0.1 M $[(n\text{-}Pr)_4N]ClO_4$; 10^{-3} M Ni(II) complex; solvent DMSO	DME	ac pol	$E_{1/2} = -0.74$	-	-	[194]
Ni(II)	0	0.1 M $[(n\text{-}PR)_4N]ClO_4$; solvent DMF	DME	ac pol	$E_{1/2} = -1.309$	-	Ref: $Ag/AgClO_4$ 0.1 M $[(n\text{-}Pr)_4N]ClO_4$	[195]
$[Ni(o\text{-}C_6H_4NH)_2]^{2-}$	-1	0.05 M $[(n\text{-}Pr)_4N]ClO_4$; solvent DMSO	Pt	rot dsk	$E_{1/2} = -1.04$	-	-	[196]
-1	0	0.05 M $[(n\text{-}Pr)_4N]ClO_4$; solvent DMSO	Pt	rot dsk	$E_{1/2} = -0.19$	-	-	[196]
$[Ni(o\text{-}C_6H_4NH)_2]^{2-}$	-1	0.5 M $[(n\text{-}Bu)_4N]PF_6$; solvent CH_2Cl_2	Pt	rot dsk	$E_{1/2} = -0.93$	-	Ref: $Ag/AgClO_4$ 0.05 M $[(n\text{-}Bu)_4N]I$	[196]
-1	0	0.5 M $[(n\text{-}Bu)_4N]PF_6$; solvent CH_2Cl_2	Pt	rot dsk	$E_{1/2} = -0.03$	-	Ref: $Ag/AgClO_4$ 0.05 M $[(n\text{-}Bu)_4N]I$	[196]
0	+1	0.5 M $[(n\text{-}Bu)_4N]PF_6$; solvent CH_2Cl_2	Pt	rot dsk	$E_{1/2} = 1.05$	-	Ref: $Ag/AgClO_4$ 0.05 M $[(n\text{-}Bu)_4N]I$	[196]
$Ni(dbh)^{2-}$	0	0.05 M $[(n\text{-}Pr)_4N]ClO_4$; solvent DMSO	Pt	rot dsk	$E_{1/2} = -1.67$	-	-	[196]

-1	0	0.05 M [(n-Pr)$_4$N]ClO$_4$; solvent DMSO	Pt	rot dsk	$E_{1/2} = -0.92$	-	-	[196]
Ni(dbh)	0	0.05 M [(n-Pr)$_4$N]ClO$_4$; solvent ACN	Pt	rot dsk	$E_{1/2} = -1.00$	-	-	[196]
Ni(dbh)	0	0.5 M [(n-Bu)$_4$N]PF$_6$; solvent CH_2Cl_2	Pt	rot dsk	$E_{1/2} = -0.86$	-	Ref: Ag/AgClO$_4$ 0.05 M [(n-Bu)$_4$N]I	[196]
Ni(gma)$^{2-}$	-1	0.05 M [(n-Pr)$_4$N]ClO$_4$; solvent DMSO	Pt	rot dsk	$E_{1/2} = -1.05$	-	-	[196]
-1	0	0.05 M [(n-Pr)$_4$N]ClO$_4$; solvent DMSO	Pt	rot dsk	$E_{1/2} = -0.30$	-	-	[196]
Ni(dtbh)$^{2-}$	-1	0.05 M [(n-Pr)$_4$N]ClO$_4$; solvent DMSO	Pt	rot dsk	$E_{1/2} = -1.24$	-	-	[196]
-1	0	0.05 M [(n-Pr)$_4$N]ClO$_4$; solvent DMSO	Pt	rot dsk	$E_{1/2} = -0.53$	-	-	[196]
Ni(dtbh)$^{2-}$	-1	0.05 M [(n-Pr)$_4$N]ClO$_4$; solvent ACN	Pt	rot dsk	$E_{1/2} = -1.32$	-	-	[196]
-1	0	0.05 M [(n-Pr)$_4$N]ClO$_4$; solvent ACN	Pt	rot dsk	$E_{1/2} = -0.55$	-	-	[196]
Ni(dtbh)$^{2-}$	-1	0.5 M [(n-Bu)$_4$N]PF$_6$; solvent CH_2Cl_2	Pt	rot dsk	$E_{1/2} = -1.53$	-	Ref: Ag/AgClO$_4$ 0.05 M [(n-Bu)$_4$N]I	[196]
-1	0	0.5 M [(n-Bu)$_4$N]PF$_6$; solvent CH_2Cl_2	Pt	rot dsk	$E_{1/2} = -0.48$	-	-	[196]

(continued)

TABLE 2.2.2 (continued)

Substance	Product	Condition	Electrode	Technique	Potential	Rev	Remarks	Refs.
$Ni[SC(C_6H_5)NNH]_2^{2-}$	-1	0.05 M $[(n\text{-}Pr)_4N]ClO_4$; solvent DMSO	Pt	rot dsk	$E_{1/2} = -1.13$	-	-	[196]
-1	0	0.05 M $[(n\text{-}Pr)_4N]ClO_4$; solvent DMSO	Pt	rot dsk	$E_{1/2} = -0.14$	-	-	[196]
$Ni(dtb)_2^-$	-2	0.042 M $[(n\text{-}Pr)_4N]ClO_4$; solvent ACN	Pt	rot dsk	$E_{1/2} = -0.82$	-	-	[171]
0	-1	0.042 M $[(n\text{-}Pr)_4N]ClO_4$; solvent ACN	Pt	rot dsk	$E_{1/2} = 0.12$	-	-	[196]
$Ni(dtb)(dtfa)^-$	-2	0.042 M $[(n\text{-}Pr)_4N]ClO_4$; solvent ACN	Pt	rot dsk	$E_{1/2} = -0.55$	-	-	[171]
0	-1	0.042 M $[(n\text{-}Pr)_4N]ClO_4$; solvent ACN	Pt	rot dsk	$E_{1/2} = 0.43$	-	-	[171]
$Ni(dtb)(mnt)^-$	-2	0.042 M $[(n\text{-}Pr)_4N]ClO_4$; solvent ACN	Pt	rot dsk	$E_{1/2} = -0.35$	-	-	[171]
0	-1	0.042 M $[(n\text{-}Pr)_4N]ClO_4$; solvent ACN	Pt	rot dsk	$E_{1/2} = 0.57$	-	-	[171]
$Ni(dtfa)_2^-$	-2	0.042 M $[(n\text{-}Pr)_4N]ClO_4$; solvent ACN	Pt	rot dsk	$E_{1/2} = -0.12$	-	-	[171]
0	-1	0.042 M $[(n\text{-}Pr)_4N]ClO_4$; solvent ACN	Pt	rot dsk	$E_{1/2} = 0.92$	-	-	[171]

$Ni(tdt)^{2-}$	-1	0.1 M $[(n\text{-}Pr)_4N]ClO_4$; 10^{-3} M complex in DMF	Hg	ac pol	$E_{1/2} = -1.07$	-	Ref: $Ag/AgClO_4$ 0.1 M $[(n\text{-}Pr)_4N]ClO_4$	[195]
$Ni(mnt)^{2-}$	-1	0.1 M $[(n\text{-}Pr)_4N]ClO_4$; 10^{-3} M complex in DMF	Hg	ac pol	$E_{1/2} = -0.218$	-	Ref: $Ag/AgClO_4$ 0.1 M $[(n\text{-}Pr)_4N]ClO_4$	[195]
$Ni(tdt)_2^{2-}$	-1	0.1 M $[(n\text{-}Pr)_4N]ClO_4$; solvent DMF	drp Hg	ac pol	$E_{1/2} = -1.068$	-	Ref: $Ag/AgClO_4$	[197]
$Ni(mnt)_2^{2-}$	-1	0.1 M $[(n\text{-}Pr)_4N]ClO_4$; solvent DMF	drp Hg	ac pol	$E_{1/2} = -0.218$	-	Ref: $Ag/AgClO_4$	[197]
$Ni(gma)^{2-}$	-1	0.1 M $[(n\text{-}Pr)_4N]ClO_4$; solvent DMF	drp Hg	ac pol	$E_{1/2} = -1.605$	-	Ref: $Ag/AgClO_4$	[197]
-1	0	0.1 M $[(n\text{-}Pr)_4N]ClO_4$; solvent DMF	drp Hg	ac pol	$E_{1/2} = -0.823$	-	Ref: $Ag/AgClO_4$	[197]
$Ni[C_6H_4(NH)_2]_2^{2-}$	-1	0.1 M $[(n\text{-}Pr)_4N]ClO_4$; solvent DMF	drp Hg	ac pol	$E_{1/2} = -2.075$	-	Ref: $Ag/AgClO_4$	[197]
-1	0	0.1 M $[(n\text{-}Pr)_4N]ClO_4$; solvent DMF	drp Hg	ac pol	$E_{1/2} = -1.404$	-	Ref: $Ag/AgClO_4$	[197]
$Ni[C_6H_4SNH]_2^{2-}$	-1	0.1 M $[(n\text{-}Pr)_4N]ClO_4$; solvent DMF	drp Hg	ac pol	$E_{1/2} = -1.573$	-	Ref: $Ag/AgClO_4$	[197]
-1	0	0.1 M $[(n\text{-}Pr)_4N]ClO_4$; solvent DMF	drp Hg	ac pol	$E_{1/2} = -0.720$	-	Ref: $Ag/AgClO_4$	[197]

[a] ACN = acetonitrile; dbh = diacetylbis(benzolhydrazone); dtb = 1, 2-diphenylethylene-1, 2-dithiolato; DMSO = dimethylsulfoxide; DMF = dimethylformamide; dtbh = diacetylbis(thiobenzolhydrazone); dtfa = 1, 2-di(trifluoromethyl)ethylene-1, 2-dithiolato; gma = glyoxalbis(2-mercaptoanylo); mnt = maleonitrile-dithiolato; tdt = bis(toluene-3, 4-dithiolato).

TABLE 2.2.3. Voltammetric Data for Nickel in Different Melts

Melt	Reaction	Conditions	Electrode	Technique	$E_{1/2}$ (V)	$E_{1/4}$ - $E_{3/4}$ (mV)	Remarks	Refs.
Borates	Ni(II) → 0	$Na_2B_4O_7$; 820°C	rot Pt wr	-	-0.818	113	Ref: Pt	[200], [201]
Chlorides	Ni(II) → 0	LiCl-NaCl eutectic + 1.19 X 10^{-3} M Ni(II); 413°C	Pt	-	-0.7	70	Ref: Pt	[202]
Chlorides	Ni(II) → 0	NaCl-KCl + 3 X 10^{-4} M Ni(II); 660°C	Pt	-	-0.540	70	Ref: Pt	[203]
Chlorides	Ni(II) → 0	NaCl-KCl + (3 X 10^{-4} to 1.2 X 10^{-3}) M Ni(II); 710°C	-	-	-0.56	170	-	[204]
Chlorides	Ni(II) → 0	NaCl-KCl(1:1); 735°C	W	-	0.1	-	Ref: Ag/AgCl	[179]
Chlorides	Ni(II) → 0	LiCl-KCl (59-49%); 450°C	W	-	-0.11	-	Ref: Ag/AgCl	[179]
Fluorides	Ni(II) → 0	LiF-NaF-KF (46.5-11.5-42%) + 2.7X 10^{-3} M Ni(II); 500-600°C	Graphite	E swp	E_p = -0.2	E_p-$E_{p/2}$ = 45	Ref: Pt	[205]
Sulfamates	Ni(II) → 0	$NH_4(NH_2SO_3)$; 182°C	-	-	-0.306	-	Ref: Pb/Pb^{2+}	[206]
Sulfates	Ni(II) → 0	Li_2SO_4-Na_2SO_4-K_2SO_4 (78-8.5-13.3%); 550°C	Pt	-	-0.4	-	Ref: Ag/Ag^+	[207]

TABLE 2.2.4. Diffusion Coefficients for Ni(II) in Different Melts. Experimental Activation Energy for Diffusion.

Melt	Composition	T (°C)	D (cm^2/sec)	ΔH_D^* (kcal/mole)	Refs.
Chlorides	LiCl-KCl	407	-	5.6 (steady)	[202]
Chlorides	LiCl-KCl	407	-	4.3 (rot electrode)	[202]
Chlorides	LiCl-KCl + (0.05×10^{-4} to 0.5×10^{-4}) M Ni(II)	450	2.4×10^{-5}	-	[208]
Chlorides	NaCl-KCl + 0.065×10^{-4} M Ni(II)	710	1.60×10^{-5}	-	[209]
Chlorides	NaCl-KCl + 0.343×10^{-4} M Ni(II)	710	1.50×10^{-5}	-	[209]
Chlorides	NaCl-KCl + 0.660×10^{-4} M Ni(II)	710	1.43×10^{-5}	-	[209]
Chlorides	NaCl-KCl + 0.967×10^{-4} M Ni(II)	710	2.57×10^{-5}	-	[209]
Fluorides	LiF-NaF-KF	500	1.0×10^{-6}	18	[205]
Fluorides	LiF-NaF-KF	570	2.7×10^{-6}	-	[205]
Fluorides	LiF-NaF-KF	600	4.5×10^{-6}	-	[205]
Nitrates	$LiNO_3$-$NaNO_3$-KNO_3 + 1.06×10^{-4} M Ni(II)	160	0.68×10^{-6}	10	[181]
Nitrates	$LiNO_3$-$NaNO_3$-KNO_3 + 5.01×10^{-4} M Ni(II)	160	1.70×10^{-6}	-	[181]
Nitrates	$LiNO_3$-$NaNO_3$-KNO_3 + 8.43×10^{-4} M Ni(II)	160	1.59×10^{-6}	-	[181]
Nitrates	$LiNO_3$-$NaNO_3$-KNO_3 + 11.3×10^{-4} M Ni(II)	160	1.71×10^{-6}	-	[181]
Nitrates	$LiNO_3$-$NaNO_3$-KNO_3 + 13.3×10^{-4} M Ni(II)	160	1.48×10^{-6}	-	[181]
Nitrates	$LiNO_3$-$NaNO_3$-KNO_3 + 17.5×10^{-4} M Ni(II)	160	1.42×10^{-6}	-	[181]
Nitrates	$LiNO_3$-$NaNO_3$-KNO_3	159	0.6×10^{-6}	-	[185]
Nitrates	$LiNO_3$-$NaNO_3$-KNO_3 + (0.24×10^{-4} to 3.2×10^{-4} M Ni(II)	149 ± 5	$(0.35 \pm 0.19) \times 10^{-6}$	-	[186]
Sulfamate	$NH_4(NH_2SO_3)$	435	0.35×10^{-6}	-	[206]

Black and DeVries [202], using either stationary or rotating platinum disk electrodes in NaCl-KCl at 407°C, found a linear relationship between the limiting current and the square root of the rotation speed, which did not intercept the origin. The data also exhibited a linear relationship between the limiting current and the Ni(II) molar fraction. The $E_{1/2}$ was governed by the equation $E_{1/2} = E^{\circ}_{M} + 2.3(RT/nF)\log(f_s k') + 2.3(RT/nF)\log(N/2)$, where k' is a proportionality constant between concentration and mole fraction, N, and f_s is the activity coefficient of the metal ion near the electrode surface. Introducing the experimental results into the equation, a slope equal to 0.064 V/decade was obtained, which compares reasonably well with the theoretical one of 0.067 V/decade. The half-wave potential becomes more positive as the concentration of Ni(II) increases. The plot E vs log $(i_d - i)$ yields a straight line with a slope of 0.068 V/decade at 413°C. Delimarskii and Kuzimovich [203] studied the reduction of Ni(II) (from $NiCl_2$) as a function of concentration in molten NaCl-KCl at 660°C at a platinum microelectrode. The half-wave potential is nearly constant within the concentration range investigated. The $\Delta E/\Delta \log[(i_d - i)/i]$ values lie between 0.07 and 0.11 V/decade, while the theoretical slope for n = 2 is 0.097 V/decade. The same features are also observed at 710°C in the concentration range from 3×10^{-4} to 12×10^{-4} M. According to Maricle and Hume [210], the Ilkovic equation is obeyed within 8% in NaCl-KCl at 740°C, the resulting n value being 1.8. The experimental $\Delta E/\Delta \log[i/(i_d - i)]$ slope at 740°C is 0.147 V/decade. On the other hand, Gaur and Behl [211] observed a poorly defined Ni(II) reduction current plateau in molten NaCl-KCl-$MgCl_2$.

Molten nitrates are among the solvents more widely studied. Christie and Osteryoung [212] employed a dropping mercury electrode to determine the apparent equilibrium constants of nickel-chloride complexes in fused $LiNO_3$-KNO_3. The values at 180°C, obtained from the shift of the half-wave potentials with chloride concentration, are 26, 2, and 10 for $[NiCl_3]^-$, $[NiCl_4]^{2-}$, and $[NiCl_2]$ complexes, respectively. Chronopotentiometry was used by Narayan and Inman [213] to study Ni(II) reduction in molten $NaNO_3$-KNO_3 at 300°C. At 2.9×10^{-4} M concentration they observed that the product $i\tau^{1/2}$ was constant, $57.3 \pm 0.8\ \mu As^{1/2}$, and independent of the current between 0.057 and 0.300 mA.

Steinberg and Nachtrieb [181] investigated the reduction of Ni(II) in molten $LiNO_3$-$NaNO_3$-KNO_3 at 160°C at a dropping mercury electrode. They proved the constancy of the diffusion constant and calculated the diffusion coefficient as a function of Ni(II) concentration. The experimental slope $\Delta E/\Delta \log[i/(i_d - i)]$ was 0.045 V/decade, in good agreement with the theoretical one, assuming n = 2, of 0.043 V/decade. Inman, Lovering, and Narayan [186] studied the same system putting special emphasis in eliminating chloride ions, which transfer to the cathodic zone from the Ag/AgCl reference electrode compartment. The slopes of E vs $\log[i/(i_d - i)]$ curves were the same for the pure melt and for the melt containing chloride ion, but the half-wave potentials were -0.370 V for the former and -0.399 V for the latter.

Chovnik and Fomichev [214] described the polarograms of Ni(II) discharge on a platinum microelectrode in the Na_2SO_4-KCl eutectic at 555°C by the Ilkovic equation.

Delimarskii, Bojko, and Schilina [201] considered the electrochemical behavior of NiO dissolved in molten borax at 820°C on platinum electrodes. The well-developed limiting currents showed a linear dependence on the concentration of dissolved NiO. The average slope $\Delta E/\Delta \log[i/(i_d - i)]$ is 0.138 V/decade for n = 2, suggesting a possible irreversible behavior of the cathodic process.

2.3. ANODIC STRIPPING VOLTAMMETRY

The theory of anodic stripping on amalgams has been developed by Mamantov, Papoff, and Delahay [215] and on solid electrodes by Nicholson [216]. For the anodic stripping of nickel, the deposition must be performed near the maximum possible charge and the plating potential more closely controlled for nickel than for easily dischargeable metals. Sensitivity can be increased either by increasing the temperature or by stirring the solution or controlling both simultaneously during the plating step.

As yet there has been no agreement on the possibility of obtaining currents of the anodic oxidation of nickel amalgams [217-220]. In cases where it was possible to obtain anodic current before the oxidation of mercury, there has been no agreement on the nature of that current. Some authors [221, 222] consider that such a current is due to the oxidation of the nickel present in the liquid phase of the amalgam. Others [223, 224] ascribe the current to the oxidation of nickel which, under certain electrolysis conditions, is deposited on the surface of mercury in the form of a solid phase.

The highly irreversible nature of the Ni/Ni(II) aqueous couple requires the use of a complexing agent to decrease the potential for the dissolution of electrodeposited nickel. Porter and Cooke [225] used thiocyanate as complexing agent, and even in such medium the metal can not be anodically oxidized without oxidizing mercury.

Krasnova and Zebreva [226] obtained a nickel current peak on the polarogram when the metal is oxidized from its amalgam before the beginning of mercury oxidation. The shape of the peak depends on the nickel ion concentration in the solution and indicates the existence of two contributions, one depending on ion concentration and the other, which appears at high concentrations, on the anodic current. The nickel ions discharged on the surface of the mercury drop diffuse inward. However, depending on the ratio between the diffusion rate of nickel into mercury and its discharge rate, part of the nickel may form a solid deposit on the surface of the drop. Also, the nickel which had diffused into the depth of the drop also forms a solid phase after the solubility limit is reached, but in this case the solid phase is formed within the amalgam. As a result, the peak for the anodic oxidation of nickel from the first drop may reflect several anodic processes: 1) the anodic oxidation of nickel on the nickel surface; 2) the anodic oxidation of nickel directly on the mercury surface [239], and 3) the anodic oxidation of nickel in the liquid phase of the amalgam.

The calculated diffusion coefficient of nickel atoms in mercury, according to the Sevcik equation [227] derived for oscillographic polarography, is 1.9×10^{-5} cm^2/sec. The solubility of nickel in mercury is $(0.1 \pm 0.7) \times 10^{-4}$ g/l. The results of Krasnova and Zebreva [226] show that nickel diffuses into mercury in the case of its cathodic reduction on a hanging mercury drop, contrary to the assertion of Mindowicz [228]. The peaks corresponding to the anodic oxidation of nickel at the first drop may be due not only to the oxidation of the solid nickel phase from the surface but also to the oxidation of nickel from the liquid phase of the amalgams. With increasing nickel concentrations in the solution the amount of nickel reduced increases and, since the diffusion rate is constant, increasing amounts of nickel remain on the surface. At a certain concentration of nickel in the solution (8×10^{-5} M) the nickel deposited on the surface is subject to crystallization and the height of the peak decreases and then the peak disappears completely. This explains the negative results obtained by a number of investigators [218-220] who studied nickel amalgams by the same method.

Jangg and Jedlicka [229] reported that nickel-amalgam undergoes easy dissolution by the action of oxidizing agents without simultaneous attack on the mercury. Likewise, anodic dissolution at appreciable current density is possible without mercury dissolution. The dissolution process is preceded by a dewetting step induced by initially formed mercury ions. Data obtained by these authors is given in Table 2.3.1.

TABLE 2.3.1. Effect of Oxidizing Agents on Nickel-Amalgam Dissolution [229]

Composition	Amalgam	Remarks	Redox potential (NHE) (V)
10 g/l Fe(III) in HCl 1:1	0.55% Ni	Rapid reaction	0.107
10 g/l Fe(III) in 0.5 N HCl	0.55% Ni	Rapid reaction and calomel formation	0.100-0.107
10 g/l Cu(II) in HCl 1:1	0.55% Ni	Rapid reaction and calomel formation	0.102

Ni(Hg) in 0.5 N H_2SO_4: Potential for anodic decomposition 0.655V; rest potential (Hg, calomel satd soln) 0.641 V

Ni(Hg) in HCl 1:1: Potential for anodic decomposition 0.115 V; rest potential (Hg, calomel satd soln) 0.105 V

Experiments have also been carried out with solid electrodes. Nicholson [230] used platinum and gold in 0.1 M KSCN. When the electrodeposited layer is less than a monolayer, current peaks on both electrodes are observed. For larger layer thicknesses only one peak is obtained, and it coincides with the previous one at -0.4 V vs SCE. The shape of the dissolution curves are suitable for integration and analytical purposes. Table 2.3.2 shows the analytical data obtained with electrodes by an anodic scan run after only 2 min of plating. Electrodes cleaned with 1:1 nitric acid gave coincident results.

Interference with the anodic stripping method may be expected from substances which form deposits on the electrode, electrolytically or by adsorption, within the operating potential range or which oxidize the nickel by direct chemical reaction during the anodic scan.

TABLE 2.3.2. Anodic Stripping Data for Nickel at Solid Electrodes [230]

C°_{Ni}(II) (M)	t (min)	E_{max} (vs Ag/AgSCN)	Q × 10^{-6} (C)	(Q/C° t) × 10^{-6}	Δ(Q/C° t) × 10^{-6} deviation from av
5 × 10^{-8}	100	-0.12	2.76	0.55	+0.01
1 × 10^{-7}	60	-0.14	3.49	0.58	+0.04
2 × 10^{-7}	58	-0.18	5.54	0.48	-0.06
2 × 10^{-7}	62	-0.16	6.80	0.55	+0.01
2 × 10^{-7}	70	-0.15	6.49	0.47	-0.07
3 × 10^{-7}	35	-0.19	5.09	0.48	-0.06
3 × 10^{-7}	60	-0.23	8.46	0.47	-0.07
3 × 10^{-7}	60	-0.23	8.19	0.46	-0.07
5 × 10^{-7}	20	-0.21	8.74	0.62	+0.08
5 × 10^{-7}	30	-0.21	8.94	0.60	+0.06
7 × 10^{-7}	15	-0.20	5.58	0.53	-0.01
1 × 10^{-6}	30	-0.29	18.2	0.61	+0.07
1 × 10^{-5}	10	-0.23	66.1	0.66	+0.12
				Av 0.54	+0.06

3. KINETIC PARAMETERS AND DOUBLE-LAYER PROPERTIES

3.1. KINETIC PARAMETERS

3.1.1. Amalgam Electrodes

Table 3.1.1 shows the kinetic parameters of the reported investigations to date. They depend strongly on the potential applied to the electrode and on the solution composition, including the presence of complexing agents. Mechanistic interpretations and the dependence of the rate constants of these reactions on the double-layer properties are discussed in Section 4.

3.1.2. Solid Metal Electrodes

Table 3.1.2 contains kinetic parameters for the reduction of Ni(II) on solid nickel electrodes. The rate constant at equilibrium depends on the pH of the solution, and the transfer coefficients indicate the irreversible character of the electrode process. The kinetic parameters of the anodic dissolution of nickel are discussed together with the mechanism of the metal dissolution and corrosion in Section 4.

3.2. DOUBLE-LAYER PROPERTIES

3.2.1. Zero Charge Potential

The potential of zero charge, E_Z, for nickel was measured in aqueous solutions of different composition and with different experimental techniques as shown in Table 3.2.1 at 25°C. (See also Campanella's review [252].) Frumkin [253] estimated a potential of zero charge of -0.40 V from the nickel work function and in the absence of adsorbed species on the electrode surface. The recommended value at 25°C is -0.30 V, as given by Bockris, Argade, and Gileadi [251].

The potential of zero charge of nickel in molten electrolytes is given in Table 3.2.2. The electrical double-layer capacitance of nickel in fused alkali halides increased with temperature in going from chloride to iodide solutions, although no appreciable change of the zero charge potential was observed by Delimarskii and Kikhno [257].

The point of zero charge was also measured by Volkov, Ponomarev, and Yuriev [256] for electrodeposited nickel-zinc alloys containing 0 to 13 at.% zinc in 0.01 N K_2SO_4 solution. The increase of zinc concentration in the alloy shifts the zero charge from -0.33 V (for pure nickel) to -0.63 V (for pure zinc). The transition from one limiting value to the other occurs at 4.3 at.% zinc.

TABLE 3.1.1. Kinetic Parameters for Reaction on the Electrodes[a]

Reaction	Conditions	Electrode	Technique	Rate constant (cm/sec)	αn	Remarks	Refs.
Ni(II) → Ni(Hg)	0.1 M $NaClO_4$ + 9.44 × 10^{-4} M Ni(II)	DME	dc pol	3.3×10^{-9}	0.65	0.01% polyacrylamide	[231]
Ni(II) → Ni(Hg)	0.50 M $NaClO_4$ + 0.0254 M Ni(II)	DME	dc pol	5.14×10^{-9}	0.50	E = -0.75 to -0.86 (SCE)	[232]
Ni(II) → Ni(Hg)	0.50 M $NaClO_4$ + 0.0254 M Ni(II)	DME	dc pol	7.6×10^{-10}	0.73	E = -0.86 to -0.99 (SCE)	[232]
Ni(II) → Ni(Hg)	0.50 M $NaClO_4$ + 0.0254 M Ni(II) + 2.0 × 10^{-3} M n-octyl alcohol	DME	dc pol	6.06×10^{-16}	1.15	E = -0.98 to -1.06 (SCE)	[232]
Ni(II) → Ni(Hg)	0.10 M $NaClO_4$ + 0.0254 M Ni(II)	DME	dc pol	1.22×10^{-9}	0.63	E = -0.75 to -0.86 (SCE)	[232]
Ni(II) → Ni(Hg)	0.10 M $NaClO_4$ + 0.0254 M Ni(II)	DME	dc pol	3.02×10^{-9}	0.77	E = -0.86 to -0.98 (SCE)	[232]
Ni(II) → Ni(Hg)	0.10 M $NaClO_4$ + satd β-naphtol + 0.0254 M Ni(II)	DME	dc pol	7.17×10^{-12}	0.67	E = -0.97 to -1.01 (SCE)	[232]
Ni(II) → Ni(Hg)	0.10 M $NaClO_4$ + satd β-naphtol + 0.0254 M Ni(II)	DME	dc pol	2.92×10^{-16}	1.18	E = -1.01 to -1.09 (SCE)	[232]
Ni(II) → Ni(Hg)	0.10 M $NaClO_4$ + satd camphor + 0.0254 M Ni(II)	DME	dc pol	1.09×10^{-16}	0.70	E = -1.26 to -1.30 (SCE)	[232]

(continued)

TABLE 3.1.1 (continued)

Reaction	Conditions	Electrode	Technique	Rate constant (cm/sec)	αn	Remarks	Refs.
Ni(II) → Ni(Hg)	0.10 M $NaClO_4$ + satd camphor + 0.0254 M Ni(II)	DME	dc pol	1.18×10^{-23}	1.20	E = -1.31 to -1.35 (SCE)	[232]
Ni(II) → Ni(Hg)	0.0254 M $Ni(ClO_4)_2$	DME	dc pol	3.18×10^{-9}	0.57	E = -0.73 to -0.86 (SCE)	[232]
Ni(II) → Ni(Hg)	0.0254 M $Ni(ClO_4)_2$	DME	dc pol	3.44×10^{-10}	0.73	E = -0.86 to -0.98 (SCE)	[232]
Ni(II) → Ni(Hg)	1.00 M $NaClO_4$ + 0.0254 M Ni(II)	DME	dc pol	8.05×10^{-9}	0.46	E = -0.74 to -0.86 (SCE)	[232]
Ni(II) → Ni(Hg)	1.00 M $NaClO_4$ + 0.0254 M Ni(II)	DME	dc pol	2.4×10^{-10}	0.70	E = -0.86 to -1.00 (SCE)	[232]
Ni(II) → Ni(Hg)	2.00 M $NaClO_4$ + 0.0254 M Ni(II)	DME	dc pol	3.53×10^{-8}	0.39	E = -0.74 to -0.86 (SCE)	[232]
Ni(II) → Ni(Hg)	2.00 M $NaClO_4$ + 0.0254 M Ni(II)	DME	dc pol	1.03×10^{-9}	0.64	E = -0.86 to -0.99 (SCE)	[232]
Ni(II) → Ni(Hg)	3.00 M $NaClO_4$ + 0.0254 M Ni(II)	DME	dc pol	1.69×10^{-8}	0.47	E = -0.71 to -0.86 (SCE)	[232]
Ni(II) → Ni(Hg)	3.00 M $NaClO_4$ + 0.0254 M Ni(II)	DME	dc pol	1.56×10^{-9}	0.64	E = -0.86 to -0.98 (SCE)	[232]
Ni(II) → Ni(Hg)	0.1 M KNO_3 + 0.6×10^{-3} M Ni(II)	DME	dc pol	$\log k_c^{\circ} = -13.7$ (NHE) $\log k_s = (-10)$	$\alpha = 0.4$	–	[233]
Ni(II) → Ni(Hg)	0.1 M KNO_3 + 0.6×10^{-3} M Ni(II)	DME	dc pol	$\log k_c^{\circ} = -13.7$ (NHE)	$\alpha = 0.40$	0.01% Gelatin	[234], [235]

Ni(II) → Ni(Hg)	0.1 M KNO_3 + 4 × 10^{-6} M leo	DME	dc pol	log k_c^o = -13.7 (NHE)	0.85	-	[236]
Ni(II) → Ni(Hg)	0.5 M KNO_3 + 4 × 10^{-6} M leo	DME	dc pol	log k_c^o = -13.4 (NHE)	0.78	-	[236]
Ni(II) → Ni(Hg)	0.2 M KNO_3	DME	dc pol	1.24 × 10^{-10}	0.40	-	[116]
Ni(II) → Ni(Hg)	Nitrate melt, T = 140°C	D(NiHg)E	fi	(k_s = 4.2 × 10^{-3})	-	-	[183]
Ni(II) → Ni(Hg)	Nitrate melt, T = 140°C	DME	fi	(k_s = 6.2 × 10^{-3})	α = 0.41	-	[183]
Ni(II) → Ni(Hg)	0.12 M NaCl	DME	dc pol	k_c^o = 1 × 10^{-13} (NHE)	0.74	0.005% MC	[237]
Ni(II) → Ni(Hg)	0.18 M NaCl	DME	dc pol	k_c^o = 5 × 10^{-13} (NHE)	0.69	0.005% MC	[237]
Ni(II) → Ni(Hg)	0.24 M NaCl	DME	dc pol	k_c^o = 2.7 × 10^{-12} (NHE)	0.63	0.005% MC	[237]
Ni(II) → Ni(Hg)	0.1 M KCl + 2 × 10^{-6} M leo	DME	dc pol	log k_c^o = -13.5 (NHE)	0.84	-	[236]
Ni(II) → Ni(Hg)	1.0 M KCl, T = 30°C	DME	dc pol	2.95 × 10^{-13}	0.70	-	[121]
Ni(II) → Ni(Hg)	0.12 M NaBr	DME	dc pol	k_c^o = 1 × 10^{-13} (NHE)	0.75	0.005% MC	[237]
Ni(II) → Ni(Hg)	0.18 M NaBr	DME	dc pol	k_c^o = 4 × 10^{-13} (NHE)	0.69	0.005% MC	[237]
Ni(II) → Ni(Hg)	0.24 M NaBr	DME	dc pol	k_c^o = 2.7 × 10^{-12} (NHE)	0.63	0.005% MC	[237]
Ni(II) → Ni(Hg)	0.18 M NaI	DME	dc pol	k_c^o = 7 × 10^{-13} (NHE)	0.69	0.005% MC	[237]

(continued)

TABLE 3.1.1 (continued)

Reaction	Conditions	Electrode	Technique	Rate constant (cm/sec)	αn	Remarks	Refs.
$Ni(II) \rightarrow Ni(Hg)$	0.24 M NaI	DME	dc pol	$k_c^o = 8 \times 10^{-12}$ (NHE)	0.61	0.005% MC	[237]
$Ni(II) \rightarrow Ni(Hg)$	0.1 M KI + 4×10^{-6} M leo	DME	dc pol	($\log k_c^o = -14.3$) (NHE)	(0.87)	-	[236]
$Ni(II) \rightarrow Ni(Hg)$	0.4 M KI + 4×10^{-6} M leo	DME	dc pol	($\log k_c^o = -13.1$) (NHE)	(0.76)	-	[236]
$[Ni(SCN)_x]^{(2-x)+} \rightarrow Ni(Hg)$	0.2 M KSCN + 0.18 M $NaClO_4$	DME	dc pol and CV	0.25×10^{-4}	0.49	-	[191]
$[Ni(SCN)_x]^{(2-x)+} \rightarrow Ni(Hg)$	0.04 M KSCN + 0.16 M $NaClO_4$	DME	dc pol and CV	0.63×10^{-4}	0.50	-	[191]
$[Ni(SCN)_x]^{(2-x)+} \rightarrow Ni(Hg)$	0.07 M KSCN + 0.13 M $NaClO_4$	DME	dc pol and CV	2.3×10^{-4}	0.51	-	[191]
$[Ni(SCN)_x]^{(2-x)+} \rightarrow Ni(Hg)$	0.13 M KSCN + 0.07 M $NaClO_4$	DME	dc pol and CV	2.7×10^{-4}	0.53	-	[191]
$[Ni(SCN)_x]^{(2-x)+} \rightarrow Ni(Hg)$	0.2 M KSCN	DME	dc pol and CV	2.9×10^{-4}	0.60	-	[191]
$[Ni(SCN)_x]^{(2-x)+} \rightarrow Ni(Hg)$	0.6 M KSCN	DME	dc pol and CV	3.4×10^{-4}	0.66	-	[191]
$[Ni(SCN)_x]^{(2-x)+} \rightarrow Ni(Hg)$	1.8 M KSCN	DME	dc pol and CV	3.1×10^{-4}	0.73	-	[191]

$[Ni(SCN)_x]^{(2-x)+} \rightarrow Ni(Hg)$	5.5 M KSCN	DME	dc pol and CV	2.5×10^{-4}	0.75	-	[191]
$[Ni(SCN)_x]^{(2-x)+} \rightarrow Ni(Hg)$	0.5 M KSCN, T = 30°C	DME	dc pol	3.89×10^{-4}	1.33	-	[121]
$[Ni(SCN)_x]^{(2-x)+} \rightarrow Ni(Hg)$	1 M KSCN + 5% acetamide	DME	dc pol	2.20×10^{-20}	0.52	0.01% Gelatin	[238]
$[Ni(SCN)_x]^{(2-x)+} \rightarrow Ni(Hg)$	1 M KSCN + 10% acetamide	DME	dc pol	4.92×10^{-19}	0.49	0.01% Gelatin	
$[Ni(SCN)_x]^{(2-x)+} \rightarrow Ni(Hg)$	1 M KSCN + 15% acetamide	DME	dc pol	1.78×10^{-19}	0.47	0.01% Gelatin	
$[Ni(SCN)_x]^{(2-x)+} \rightarrow Ni(Hg)$	1 M KSCN + 20% acetamide	DME	dc pol	1.91×10^{-18}	0.42	0.01% Gelatin	
$[Ni(SCN)_x]^{(2-x)+} \rightarrow Ni(Hg)$	1 M KSCN + 25% acetamide	DME	dc pol	1.29×10^{-18}	0.41	0.01% Gelatin	
$[Ni(SCN)_x]^{(2-x)+} \rightarrow Ni(Hg)$	1 M KSCN + 30% acetamide	DME	dc pol	8.70×10^{-17}	0.40	0.01% Gelatin	
$[Ni(NH_3)_6]^{2+} \rightarrow Ni(Hg)$	0.1 N KNO_3 + 1 N NH_3	DME	dc pol	$\log k_c^\circ = -16.8$ (NHE) $\log k_s = (-7.0)$	$\alpha = 0.7$	-	[233]
$[Ni(NH_3)_6{}^{2+} \rightarrow Ni(Hg)$	0.1 M KNO_3 + 0.2 to 3 M NH_3	DME	dc pol	$\log k_c^\circ = -16.8$ (NHE)	$\alpha = 0.71$	0.01% Gelatin	[234], [235]
[Ni(II)(en) → Ni(Hg)	0.1 N KNO_3 + 1 M en	DME	dc pol	$\log k_c^\circ = -17.6$ (NHE) $\log k_s = (-6.5)$	$\alpha = 0.5$	0.02% Gelatin	[233]

[a] en = ethylenediamine; fi = faradaic impedance; gel = gelatin; leo = polyoxyethylenelaurylether; MC = methylcellulose.

TABLE 3.1.2. Kinetic Parameters for Reactions on Solid Nickel Electrodes

Reaction	Conditions	Electrode	Technique[a]	Rate constant i_o (in A/cm^2) k_s (in cm/sec)	Transfer coefficient	Remarks	Refs.
Ni(II) → 0	1 M $NiSO_4$; room temp	Ni	-	$\log i_0 = -8.7$	$\alpha = 0.5$	-	[239]
Ni(II) → 0	2.0 N H_2SO_4 + 0.01 N $NiSO_4$	Ni	cp	$i_0 = 8.3 \times 10^{-10}$	$\alpha = 0.35$–0.40	pH = 0.00	[240]
Ni(II) → 0	0.2 N H_2SO_4 + 0.01 N $NiSO_4$	Ni	cp	$i_0 = 2.4 \times 10^{-9}$	$\alpha = 0.35$–0.40	pH = 0.83	[240]
Ni(II) → 0	2.0 N Na_2SO_4 + 0.01 N $NiSO_4$	Ni	cp	$i_0 = 4.2 \times 10^{-8}$	$\Delta E/\Delta \log i_c = 84$ mV	pH = 5.86	[240]
Ni(II) → 0	AcO buffer 0.5 M $NiSO_4$, 20°C	Ni	-	$i_0 = 10^{-6}$	$\alpha n = 0.52$	pH = 2.3	[241], [242]
Ni(II) → 0	AcO buffer 0.5 M $NiSO_4$, 20°C	Ni	-	$i_0 = 10^{-6}$	$\alpha n = 0.72$	pH = 4.4	[241], [242]
Ni(II) → 0	AcO buffer 0.5 M $NiSO_4$, 20°C	Ni	-	$i_0 = 10^{-7}$	$\alpha n = 0.89$	pH = 5.5	[241], [242]
Ni(II) → 0	AcO buffer 0.5 M $NiSO_4$, 20°C	Ni	-	$i_0 = 10^{-5}$	$\alpha n = 0.56$	pH = 6.7	[241], [242]
Ni(II) → 0	AcO buffer 0.5 M $NiSO_4$, 20°C	Ni	-	$i_0 = 10-5$	$\alpha n = 0.62$	pH = 7.9	[241], [242]
Ni(II) → 0	2.0 N HCl + 1.0 N $NiCl_2$	Ni	cp	$i_0 = 1.1 \times 10^{-8}$	$\alpha = 0.35$–0.40	pH = 0.28	[240]
Ni(II) → 0	1.0 N HCl + 1.0 N KCl + 0.01 M $NiCl_2$	Ni	cp	$i_0 = 6.9 \times 10^{-9}$	$\alpha = 0.35$–0.40	pH = 0.01	[240]
Ni(II) → 0	0.2 N HCl + 1.5 N KCl + 0.01 M $NiCl_2$	Ni	cp	$i_0 = 1.7 \times 10^{-9}$	$\alpha = 0.35$–0.40	pH = 0.60	[240]
Ni(II) → 0	0.2 N HCl + 0.01 M $NiCl_2$	Ni	cp	$i_0 = 3.9 \times 10^{-9}$	$\alpha = 0.35$–0.40	pH = 0.67	[240]
Ni(II) → 0	2 N KCl + 0.01 M $NiCl_2$	Ni	cp	$i_0 = 1.0 \times 10^{-8}$	$\alpha = 0.35$–0.40	pH = 5.96	[240]
Ni(II) → 0	Fused LiCl-KCl + Ni(II); 450°C	Ni	dp	$k_s = 0.1$ $i_0 = 110 \pm 20$	$\alpha = 0.25 \pm 0.06$	Ref: NHE	[243], [244]

[a] cp = current/potential measurement. dp = double pulse technique.

TABLE 3.2.1. Potential of Zero Charge for Nickel under Various Conditions

Composition of solution	E_Z (NHE)	Method	Refs.
0.001 M HCl	-0.06	Salt effect on hydrogen evolution	[245]
0.0015 M HCl + 0.0015 M HBr	-0.28	Hydrogen overvoltage	[246]
$H_2SO_4 + Na_2SO_4 + NaOH$, $\mu = 0.02$	-0.21 (pH = 1)	Capacitance minimum	[247]
$H_2SO_4 + Na_2SO_4 + NaOH$, $\mu = 0.02$	-0.33 (pH = 2)	Capacitance minimum	[247]
$H_2SO_4 + Na_2SO_4 + NaOH$, $\mu = 0.02$	-0.37 (pH = 5)	Capacitance minimum	[247]
0.01 M LiCl in anhydrous methanol	0.088	Immersion	[248]
0.01 M LiCl in anhydrous ethanol	-0.020	Immersion	[248]
0.01 M KCl aq	-0.193	Immersion	[249]
0.01 M LiCl in isopropanol	-0.100	Immersion	[250]
0.01 M, LiCl in 2-butanol	-0.105	Immersion	[250]
x M $NaClO_4$, 5×10^{-5} M naphtalene, pH = 13; $0.01 \le x \le 1.0$	-0.32 ± 0.03	Organic adsorption	[251]
0.1 M $NaClO_4$, pH = 12.5	-0.30	Friction method	[251]
10^{-4} M NaOH	-0.26 ± 0.025	Capacitance	[251]
5×10^{-4} M $KClO_4$; pH 10.3	-0.24	Capacitance	[251]

TABLE 3.2.2. Potential of Zero Charge for Nickel in Molten Electrolytes

Composition	E_Z (V)	Method	Refs.
NaCl-KCl (1:1); Ref: (Ag/AgCl 0.05 M); 700°C	-0.72	Capacitance	[254], [255]
0.01 N K_2SO_4; 700°C	-0.330	Capacitance	[256]

TABLE 3.2.3. Scraping Potential (E_{sp}) in 1 N K_2SO_4 or 1 N $NaClO_4$ [258]

pH	E_{sp} (NHE) (V)
1	-0.10
7	-0.34
11	-0.46

TABLE 3.2.4. Scraping Potential (E_{sp}) in Solution with Different Solvents

Composition	E_{sp} (NHE) (V)
0.1 N $KClO_4$ (H_2O)	-0.35
0.1 N $KClO_4$ (DMSO)	-0.50
Satd soln $KClO_4$ (C_2H_5OH)	-0.34

Andersen, Anderson, and Eyring [258] measured the "scraping potential" of nickel, E_{sp}, when the metal surface is renewed up to 400 times/min. In the absence of any reaction the scraping potential would correspond to the zero charge potential. Although this is not the case for nickel in the whole pH range in aqueous media (Table 3.2.3), it would be valid in other solvents (Table 3.2.4).

3.2.2. Electrode Capacitance

Most of the electrode capacitance data obtained for nickel are referred to the geometrical area, which in turn depends on the pretreatment of the electrode surface. Table 3.2.5 shows the results obtained by different authors.

3.2.2.1. Galvanostatic Charging Method. Brodd and Hackerman [264] obtained the capacitances of electrodeposited nickel electrodes in 1 M Na_2SO_4 aqueous solution at 30°C and -0.4 V (SCE) on the assumption that the latter was roughly the zero charge potential of the metal. The actual electrode area was measured by the BET adsorption method. The reported value is $28.8 \pm 0.5\ \mu F/cm^2$. Because there is no trend in polarization capacity with the total area of the electrodes, the authors suggest it can be used to estimate the actual surface area of nickel electrodes. The roughness factor of the plated nickel electrodes varied from 80 to 175.

TABLE 3.2.5. Double-Layer Capacitance of Nickel

Pretreatment	Media	C (μF/ cm^2)	E (SCE) (V)	Method	Refs.
-	1 N H_2SO_4	39	-	-	[25]
-	1 N NaOH	22 to 27	-	-	[259]
-	0.01 N HCl	28	-	-	[260]
-	0.1 N HCl	37	-	-	[260]
-	1.0 N HCl	41	-	-	[260]
-	0.006 N NaOH	22	-	-	[260]
-	0.12 N NaOH	27	-	-	[260]
Polished emery paper (Ni 99%)	0.5 M $NiCl_2$	35	0 to 0.15	Galvan charge	[261]
Polished emery paper (Ni 99.5%)	1 N H_2SO_4	31 ± 2	Open circuit	Galvan charge	[262]
Polished emery paper (Ni 99.97%)	1 N H_2SO_4	23 ± 1	Open circuit	Galvan charge	[262]
Polished emery paper	1 N H_2SO_4 + 1.2% NaCl	25	Open circuit	Galvan charge	[263]

3.2.2.2. Bridge Methods. Capacity measurements at solid electrodes by the ac impedance bridge method show frequency dispersion. Several authors have obtained data for nickel at a fixed frequency. Batrakov, Gamilikhana, Mikhailova, and Iofa [265] measured the capacity and resistance of electrodes at 800 Hz in 1 N H_2SO_4 aqueous solutions at different potentials. Within the potential range +0.2 to -0.6 V, the capacity values are between 20 and 40 μF/cm^2, and no definite dependence was found. Yoshida, Nomoto, and Okada [266] reported a hyperbolic capacitance/potential relationship for the same system at 1 KHz.

3.2.2.3. Potentiostatic Voltage Step Method. Berndt [267] obtained the differential capacitance of the electrical double layer and its potential dependence for sintered plates and smooth nickel electrodes in 1 N NaOH, as shown in Figs. 3.2.1 and 3.2.2. The electrochemical active surface area given by this method coincides with the BET surface of the sintered plates.

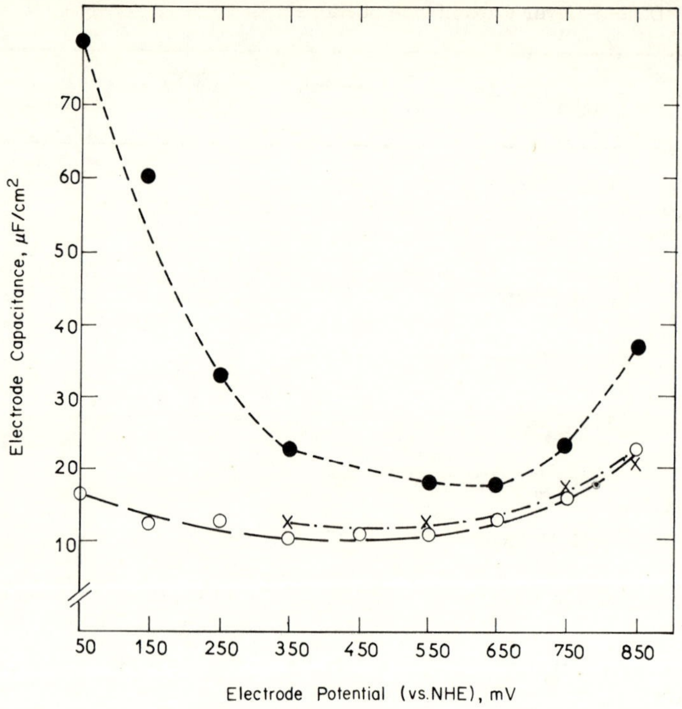

FIG. 3.2.1. Capacitance vs potential diagram for polished nickel plate electrodes in 1 N KOH. (●) and (X) correspond to a 12 cm^2 electrode area; (O) to 24 cm^2. Six and 15 mV ac voltage is superimposed [267]. (By permission of Pergamon Press.)

3.2.3. Adsorption of Ions

Lukovtsev, Levina, and Frumkin [245], Legran and Levina [246], and Mikhailova and Iofa [268] investigated the influence of specific adsorption of halide ions on nickel for the hydrogen evolution reaction. The latter employed a 1 N H_2SO_4 solution measuring the electrode capacitance by the impedance method at 800 Hz. Schwabe and Schwenke [269] showed the dependence between concentration of adsorbed bromide ion and potential of a nickel electrode with solutions containing Na_2SO_4 as supporting electrolyte. The results are seen in Fig. 3.2.3.

Grahame [270] studied the role of Ni(II) in the electrical double layer by measuring the capacity of a dropping mercury electrode in 0.1 N $NiCl_2$ solution at 25°C. Results are shown in Fig. 3.2.4.

Partridge, Tansley, and Porter [271] compared the effect of the adsorption of coumarin on the "deposition potential" of nickel with that at the dropping mercury electrode. Ni(II) ion is specifically adsorbed on its parent metal. The electrocapillary curves in 1 M $NiSO_4$ solution and 1 M $NiSO_4$ + 1 M Na_2SO_4 are shown in Fig. 3.2.5.

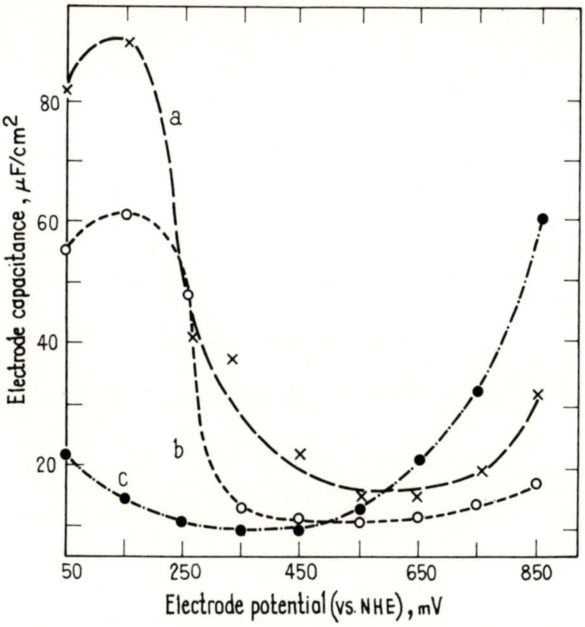

FIG. 3.2.2. Capacitance vs potential diagram for nickel plate electrodes in 1 N KOH after different treatments. Electrode area, 12 cm². (X) Measured after previous cathodic polarization; (●) measured after previous anodic polarization; (O) measured after a previous cathodic polarization at 450 mV. Six and 15 mV ac voltage is superimposed [267]. (By permission of Pergamon Press.)

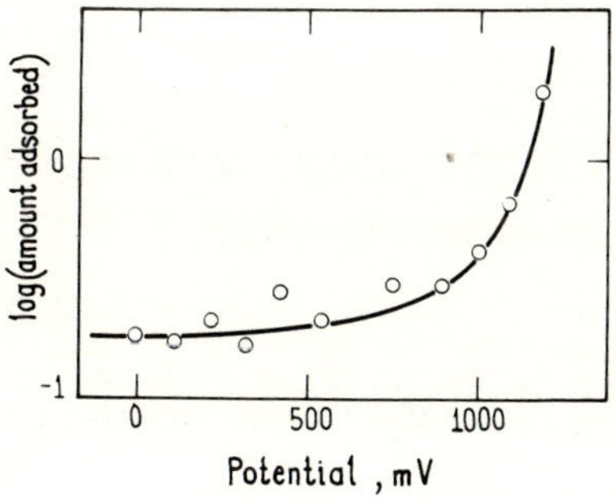

FIG. 3.2.3. Semilogarithmic plot of the potential dependence of Br^- ion adsorption on nickel electrode. Solution composition: 8.7 × 10^{-5} M Br^- + 0.1 M Na_2SO_4 [269]. (By permission of Pergamon Press.)

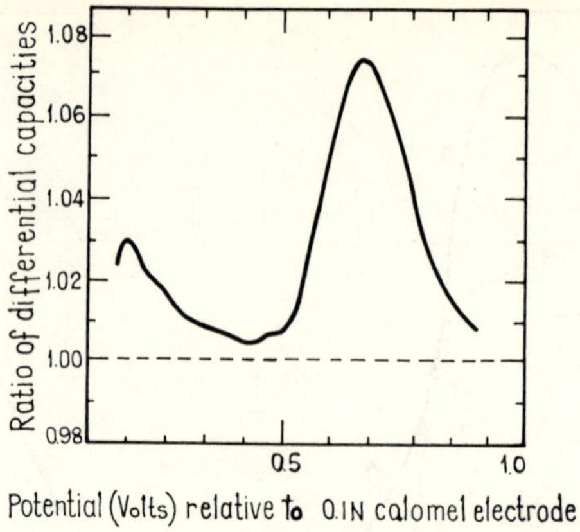

FIG. 3.2.4. Ratio of differential capacity of mercury-solution interface in 0.1 N $NiCl_2$ to differential capacity in KCl [270]. (By permission of the Electrochemical Society.)

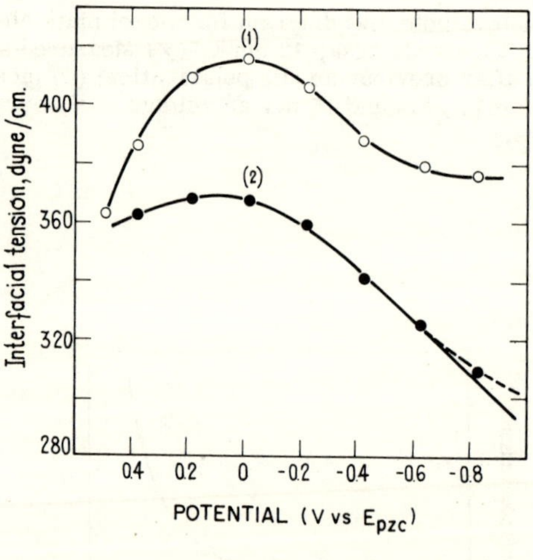

FIG. 3.2.5. Electrocapillary curves for sulfate solutions, pH 4.6, 20°C. (1) 1 M sodium sulfate + 1 M nickel sulfate; (2) Solution (1) saturated with coumarin [271]. (By permission of Pergamon Press.)

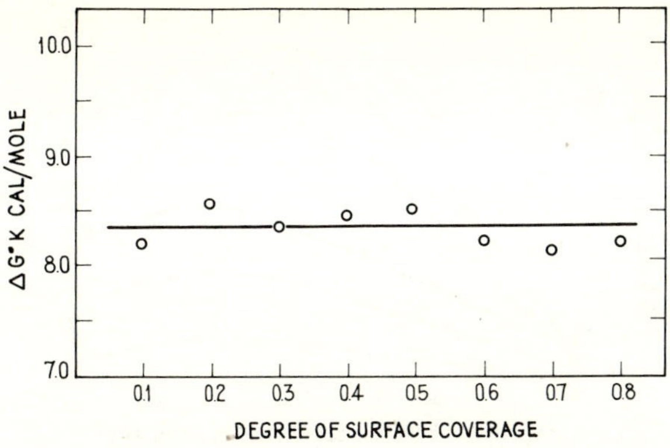

FIG. 3.2.6. Apparent standard free energy of adsorption of acridine on nickel at 22°C as a function of fractional coverage [272]. (By permission of Elsevier.)

3.2.4. Adsorption of Organic Compounds

Studies on the adsorption of organic compounds on nickel electrodes by direct methods are scarce. Most of them refer either to corrosion inhibition or brightening or leveling agents but are not specifically orientated to physical chemistry (see Section 4.3.3).

Barradas and Conway [272] measured the equilibrium adsorption of quinoline, pyridine, 2,3-benzoacridine, and N-methylacridinium ion on nickel. The electrode was kept near the reversible potential and the change of adsorbate concentration when equilibrium was reached was determined by UV spectrophotometry. Thus, the change of the standard free energy of adsorption, ΔG°_{ads}, as a function of the degree of surface coverage was evaluated as shown in Fig. 3.2.6. They also found evidence for the adsorption of the ionic form of the N-methyl-quaternized salt of the organic bases.

Bockris and Swinkels [273], and Bockris, Green, and Swinkels [274], using a radiotracer method, determined the amount of ^{14}C labeled naphthalene and n-decylamine adsorbed per unit geometric area as a function of both potential and concentration of adsorbate, as shown in Figs. 3.2.7 and 3.2.8. The scatter of data varies from 10 to 25%. The surface coverage was evaluated in each case using the roughness factor as determined by BET; quantities relevant to adsorption are given in Table 3.2.6. The results are discussed in terms of a "competition with water" model of the electrode solution interface according to Bockris, Devanathan, and Mueller [275]. There is evidence for π-bonding between the cyclic organic molecules and the metal.

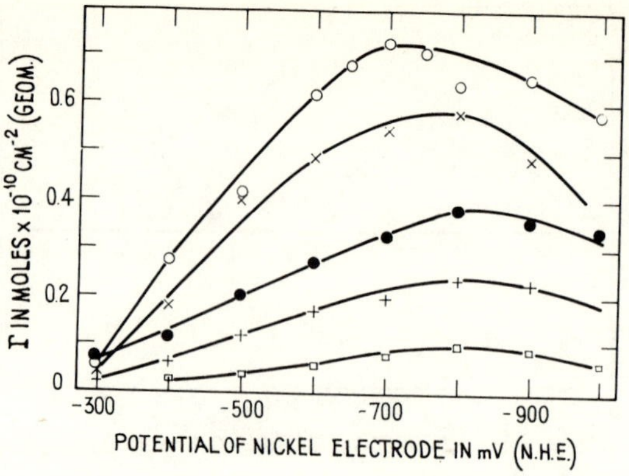

FIG. 3.2.7. Adsorption of naphthalene on nickel from 1 M $NaClO_4$, pH 12. The naphthalene values are 7.5×10^{-5} (O), 5.6×10^{-5} (X), 3.8×10^{-5}, (●), 1.9×10^{-5}, (+), and 0.8×10^{-5} M (□) [273]. (By permission of the Electrochemical Society.)

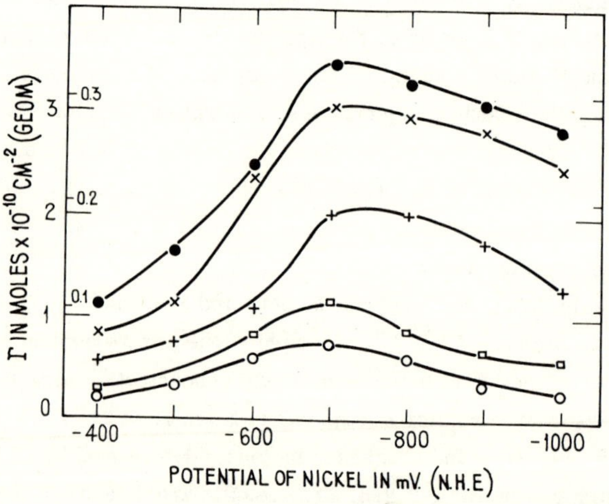

FIG. 3.2.8. Adsorption of n-decylamine on nickel from 0.9 N $NaClO_4$, pH 12. The n-decylamine values are 7.5×10^{-5} (●), 5×10^{-5} (X), 2.5×10^{-5} (+), 1×10^{-5} (□), and 0.5×10^{-5} M (O) [273]. (By permission of the Electrochemical Society.)

TABLE 3.2.6. Data Related to the Adsorption of Organic Compounds on Nickel Electrodes [273, 274]

Roughness factor	Solution composition	E_{max} ($\theta \to 0$) (V)	ΔG°_{ads} at E_{max} ($\theta \to 0$) (kcal/mole)
1.32	0.9 M $NaClO_4$ + 0.1 M NaOH + (0.5 to 7.5) $\times 10^{-5}$ M n-decylamine	-0.7	-6.8
1.32	1 M $NaClO_4$ + (0.8 to 7.5) $\times 10^{-5}$ M naphthalene; pH = 12	-0.8	-6.0

4. ELECTROCHEMICAL STUDIES

4.1. MECHANISMS OF Ni(II) REDUCTION AT THE DROPPING MERCURY ELECTRODE

4.1.1. Aqueous Solution

The kinetic parameters of Ni(II) reduction at the dropping mercury electrode indicate the occurrence of an irreversible electrochemical process. Strassner and Delahay [276] studied the influence of gelatin on αn and $\Delta G^{\ddagger}$ in 1 M KNO_3 aqueous solutions. The results are shown in Table 4.1.1. The polarographic behavior of Ni(II) in concentrated $NaClO_4$ solutions offers a good example of double-layer effects on a process with charge transfer preceded by a chemical reaction [187, 277, 278]. Ni(II) polarograms are considerably distorted at high $NaClO_4$ concentrations, as seen in Fig. 4.1.1. This distortion had been previously interpreted as evidence for a two-step reduction of Ni(II) (see Table 2.1.1), but this interpretation was disapproved by Gierst [278]. The distortion in 2 M chloride solution of the alkali halides increases from lithium to cesium ions.

Gierst found that the diffusion current was a function of the rational potential, ϕ, and ionic strength, μ. Plots of ψ_0, the potential of the diffusional double layer at the Helmholtz planc, vs ϕ at a constant rate of reaction v*, and of v* vs ψ_0, allow the author to conclude that after an initial increase, the total rate is independent of $\phi - \psi_0$. The dependence of v* on ψ_0 is given by

$$v^* = v_0 \exp\left[-\frac{nF}{RT}\psi_0\right] \tag{4.1.1}$$

TABLE 4.1.1. Influence of Gelatin on αn and ΔG^* [276]

Gelatin (%)	αn	ΔG^* (kcal/mole)
0.001	0.56	24.1
0.005	0.45	21.8
0.01	0.32	19.5

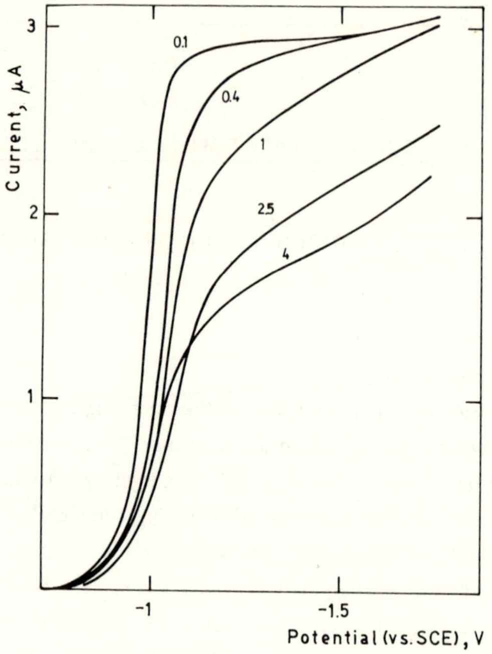

FIG. 4.1.1. Polarographic current/potential curves for Ni(II) ion in sodium perchlorate medium at five different molarities (indicated on the curves). Nickel ion concentration, 10^{-2} M [278]. (By permission of the Electrochemical Society.)

v_0 is the reaction rate after correction for the double-layer effect (ψ-static effect). The result is n = 2 and $v_0 = 10^{-4}$ cm/sec. The fact that v* depends on ψ_0 and is independent of $\phi - \psi_0$ suggests that the rate-determining step is a preceding relatively slow dehydration step. There is no dependence of v* on the different anions present in the solution. Quantitative data involving ψ_0 variation are given by these authors, but calculation of ψ_0 from Gouy-Chapman theory is very uncertain at such high concentrations as pointed out by Delahay [277]. For a 10^{-3} M Ni(II) concentration in 0.25 M $NaClO_4$, the i/i_d ratio obtained at

TABLE 4.1.2. i/i_d Ratio at Different Potentials for Ni(II) Reduction at the DME [278]

E (NHE) (V)	i/i_d
-0.672	0.1
-0.708	0.2
-0.730	0.3
-0.746	0.4
-0.761	0.5
-0.777	0.7

different potentials is shown in Table 4.1.2, as given by Gierst [278]. Kinetic data were calculated by applying the equation

$$\frac{i_t}{i_d} = \frac{1 - \exp(nF\eta/RT)\varphi(x)}{1 + (D_0/D_R)^{1/2}\exp(nF\eta/RT)} \qquad (4.1.2)$$

to a purely cathodic wave [279]. Equation (4.1.2) is derived for the reaction $0 + ne = R$; i_t is the instantaneous current at the end of the drop-time, t_d; $\eta = E - E^0$; D_0 and D_R are the diffusion coefficients of the reaction species; and $\varphi(x)$ is a function tabulated by Koutecky [280]. Taking $E^0 = -0.245$ V (NHE), the following parameters are obtained: $k_s = 4.0 \times 10^{-9}$ cm/sec and $\alpha n = 0.62$. With a simplified version of Eq. (4.1.2) the calculated values are: $k_s = 2.4 \times 10^{-9}$ cm/sec and $\alpha n = 0.66$. These figures are comparable to those obtained by Delahay and Mattax [121] from chronopotentiometry. The activation energy obtained by Dandoy and Gierst [187] is 11.6 kcal/mole.

Hush and Scarrot [281], determined the activation energies of both diffusional process ($\Delta H_D^* \approx 4.5 \pm 0.4$ kcal/mole) and of the previous equilibrium ($\Delta H_1^* - \frac{1}{2}\Delta H_2^* = 7.6 \pm 1.6$ kcal/mole), the latter derived from the temperature dependence of the product of the rate constants, $k_1 k_2^{-1/2}$, following Koutecky procedure for a transfer reaction preceded by a chemical step [280]. They postulated a model where the $Ni(H_2O)_6^{2+}$ undergoes a S_N2-type reaction to yield an unstable heptahydrate with a C_{2v} symmetry with two ligands lying on a square pyramid. Then the activation energy of the process related to k_2 would be $\Delta H_2^* \approx \Delta H_D^*$, yielding, therefore, $\Delta H_1^* = 9.9 \pm 2$ kcal/mole. This figure is comparable to the value of 11.6 kcal/mole obtained by Swift and Connick [282] for the homogeneous water exchange reaction. This model is consistent with ligand field theory which establishes slow ligand exchange reactions for ions with a $t_{2g}3$ and $t_{2g}6\ e_g2$ configuration such as hydrated Ni(II) ion.

The model of Hush and Scarrot [281] was modified by Verdier and Pernicelli [283] assuming a different structure of the complex at the dropping mercury electrode interface. They postulated the loss of a water molecule to form the activated complex as follows:

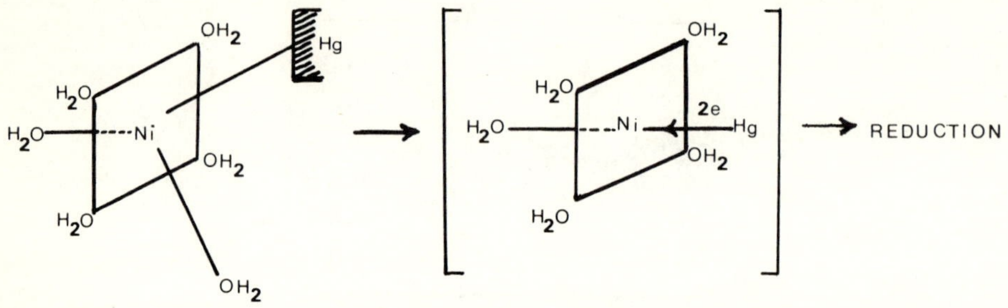

The activated complex finally discharges and leads to the electrochemical destruction of the complex. This model would agree with the positive entropy change found by Swift and Connick (0.6 u.e.) [282] for a one molecule homogeneous water exchange reaction.

The reduction of $[Ni(CN)_4]^{2-}$ ion is one of the examples of nickel complex ions where the reduction reaction is fairly well established both from the kinetic and mechanistic standpoints (see Heyrovsky and Kuta [284]). $[Ni(CN)_4]^{2-}$ ion is very stable, and it is already formed although the amount of cyanide ion in solution is lower than the stoichiometric one (see Refs. 125, 285, 286, and Table 2.1.1).

During a study of the two-electron reduction of this complex, Vlcek [287, 288] found either a reversibly reoxidizable product containing zero-valent nickel or an irreversibly reoxidizable complex of monovalent nickel. The form of the cathodic wave corresponded neither to a reversible two-electron nor to a reversible one-electron reduction. The waveheight pointed to the consumption of two electrons. The steepness of the wave and the halfwave potential depended on drop-time and on the composition of the solution, but not, however, on the depolarizer concentration. Assuming that chemical reactions are involved in the reduction of this complex, they should have a monomolecular character. To interpret these facts, it was assumed that the reduction of Ni(II) in cyanide medium proceeds via Ni(I). Vlcek [287, 288] suggests the following mechanism to explain the polarographic behavior of $[Ni(CN)_4]^{2-}$:

$$\begin{array}{ccccc} Ni(II) & \rightleftarrows & Ni(I) & \rightleftarrows & Ni(0) \\ & & k_1 \downarrow & & \downarrow k_2 \\ & & y_1 & & y_2 \end{array}$$

y_1 and y_2 are inactive products of the chemical reactions. The solution of the reaction scheme can be expressed by the equation given by Vlcek [287, 288]:

$$\frac{\bar{i}}{i_d} = \frac{2 + \sqrt{(K'P')}}{2[1 + \sqrt{(K'P')} + P']} \qquad (4.1.3)$$

where K' = K/a. K is the constant for Ni(I) formation and a is a parameter characterizing the reaction mechanism. For the case

$$K' = 0.81K \times \frac{k_1}{k_2^{1/2}} t^{1/2}$$

and $a = 0.81\ k_2^{1/2} t_d^{1/2}$, t_d is the drop-time. The quantity P' is a function of the potential:

$$P' = \exp\left[\frac{2F(E - E_{1/2})}{RT}\right] = \frac{1}{a} \exp\left[\frac{2F(E - E^0)}{RT}\right] \qquad (4.1.4)$$

where E^0 denotes the standard oxidation-reduction potential for the totally reduced system.

Torsi and Papoff [289] studied the effect of the supporting electrolyte on the reduction of $[Ni(CN)_4]^{2-}$, and thus could explain their results by the theory of Gierst and by the intramolecular arrangement suggested by Vlcek [287, 288].

Galus and Jeftic [191] studied the behavior of Ni(II) ions in thiocyanate and thiocyanate-pyridine media (see Table 2.1.1) by applying both polarography and cyclic voltammetry. For high SCN^- ion concentration, the rate constants varied between 2×10^{-4} and 3×10^{-4} cm/sec. The three waves observed at low SCN^- ion concentration were explained by the reduction of $[Ni(H_2O)_3(SCN)_3]^-$, $[Ni(H_2O)_5SCN]^+$, and hydrated Ni(II). The following mechanism for the first kinetic wave was suggested:

$$[Ni(SCN)]^+(\text{bulk}) \xrightarrow{\text{diff}} [Ni(SCN)]^+(\text{surface}) \underset{\text{II}}{\overset{SCN^-}{\rightleftharpoons}} [Ni(SCN)_2](\text{surface})$$

$$[Ni(SCN)_2](\text{surface}) \underset{\text{III}}{\overset{SCN^-}{\rightleftharpoons}} [Ni(SCN)_3]^- \rightarrow \text{reduction product}$$

Turiyan and Malyavinskaya [290] proposed that the slower step must be Step II and not Step III as earlier suggested. Similar mechanisms were postulated for the reduction of $[Ni(H_2O)_5SCN]^+$ and Ni(II) ions [291, 292].

Turiyan and Serova [293] observed a kinetic polarographic wave for Ni(II) in the presence of pyridine (Py) which was assigned to a chemical reaction on the electrode, and a reduction of the Ni(II)-py complex. Mark and Reilley [294] reported a polarographic catalytic prewave of Ni(II) in the presence of pyridine. They suggested that the reacting species, $[Ni(Py)_x(H_2O)_y]^{2+}$, follows the catalytic reaction path:

$$[Ni(Py)_x(H_2O)_y]^{2+} + e \xrightarrow{\text{electrochem}} Ni(Hg) + xPy + yH_2O$$

$$xPy + [Ni(H_2O)_6]^{2+} \xrightarrow{\text{chem}} [Ni(Py)_x(H_2O)_y]^{2+} + (6 - y)H_2O$$

As the reacting species would have a lower half-wave potential than the hydrated Ni(II) ion, it would be discharged first, thus originating the prewave. Analytical applications of this type of prewave have been considered by the authors [295].

Kémula, Jeftic, and Galus [139] obtained either two or three waves for the discharge of Ni(II) in the presence of pyridine at high and low [Py]/[Ni(II)] concentration ratios, respectively. When the three waves appear, the two first would be related to the reduction of the $[NiPy_2]^{2+}$ ion and the $[NiPy]^{2+}$ ion, whereas the latter would be due to the hydrated Ni(II) ion. These conclusions would be supported by chronovoltammetric data and differential capacity measurements.

Turiyan and Malyaniskaya [296] studied the effect of concentration and nature of supporting electrolyte, pH, temperature, and adsorption of pyridine on the Ni(II) catalytic polarographic prewave in the presence of pyridine. The following reaction scheme was proposed:

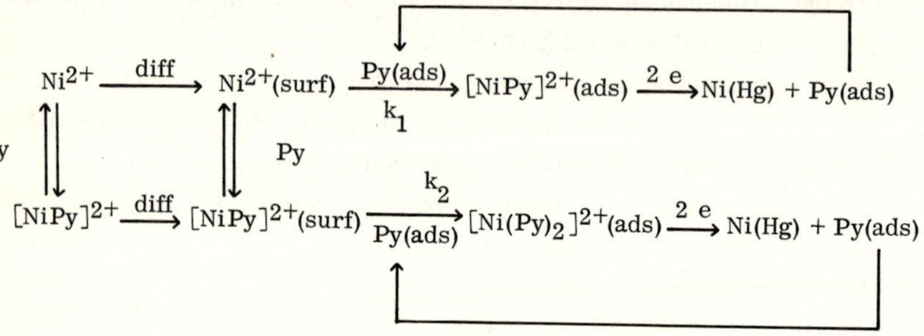

The instability constants of the complexes, as well as the double-layer-corrected kinetic equations, that interpreted the data were also given by the authors [296]. Marianovskii, Gulimyai, and Lisitsina [297] found that Ni(II) ions, in the presence of pyridine, counteract markedly the inhibition of persulfate reduction, whereas pyridine does not affect the reduction kinetics. This indicates the great degree of adsorption of Ni(II)-Py complexes on the electrode surface at their discharge potentials.

According to Vlcek [298], there is evidence that for dipyridyl, ethylendiamine, pyridine, ammonia, and chloride complexes of Ni(II) (in 5 M $CaCl_2$), the proper reduction of the complex occurs, the intermediate being rapidly decomposed either into the metal or oxidized to a higher degree. The half-life of the zero valent complexes is < 0.01 sec.

The reduction of Ni(II)-ethylenediamine complex at pH 7 exhibits one prewave which was assigned to the complex reduction. At pH 5 a postwave has been observed, which Mark and Schwartz [299] related to catalytic hydrogen discharge, associated in some way with ethylenediamine. Both are not reducible at the expected potentials on the dropping mercury electrode. The effect of Ni(II) to increase the catalytic current is similar to that reported for the hydrogen catalytic wave in the presence of amino acids and cysteine with Co(II), Co(III), and Ni(II) species [110, 300, 301]. The presence of Ni(II) only changes the peak current. It is thought that $[Ni(en)_3]^{2+}$ reduction occurs at a potential slightly more positive than that of $[Ni(H_2O)_6]^{+2}$.

Mark [302] reported a catalytic polarographic current in the reduction of Ni(II)-o-phenylenediamine complex. The half-wave potential of the prewave is ~0.30 V more positive than that of $[Ni(H_2O)_6]^{2+}$. The mechanism to explain the prewave involves a chemical reaction preceding the electron transfer step. An adsorption of o-phenylenediamine (o-ph) occurs, and the limiting current is a function of the degree of surface coverage. This suggests that an adsorption equilibrium exists which controls the effective reaction area. The adsorbed o-phenylenediamine reacts with $[Ni(H_2O)_6]^{2+}$ to yield a surface complex prior to the occurrence of the electron-transfer step. The proposed reaction mechanism is

$$[Ni(\text{o-ph})_y(H_2O)_{4-2y}]^{2+}(\text{bulk}) \rightleftharpoons [Ni(H_2O)_6]^{2+}(\text{bulk}) + \text{o-ph(bulk)} \rightleftharpoons \text{o-ph(surf)}$$

$$y'\,\text{o-ph(surf)} + [Ni(H_2O)_6]^{2+}(\text{surf}) \xrightarrow[\text{slow}]{k_{chem}} [Ni(\text{o-ph})_{y'}(H_2O)_{4-2y'}]^{2+}(\text{surf}) + 2y'H_2O$$

$$[Ni(\text{o-ph})_{y'}(H_2O)_{4-2y'}]^{2+}(\text{surf}) + 2e \xrightarrow[\text{fast}]{k_{el}} Ni(Hg) + y'\,\text{o-ph(surf)} + (4-2y')H_2O$$

Thiourea adsorption onto mercury has a pronounced effect on the αn value for Ni(II) reduction [303]. Turiyan and Ruvinskii [304] studied the polarographic reduction of Ni(II) in aqueous solutions in the presence of thiourea (Tu) as a function of the nature and concentration of the supporting electrolyte, thiourea, and Ni(II) concentrations, temperature, pH, and gelatin addition. Two catalytic waves were found. They were explained by the formation and successive reduction of the $[NiTu]^{2+}$ complex and thiourea regeneration on the electrode according to

$$Ni^{2+} \xrightarrow{\text{diff}} Ni^{2+} + Tu \xrightarrow{k_a} [NiTu]^{2+} \xrightarrow{2e} Ni(Hg) + Tu$$

The first wave corresponds to complex formation with adsorbed thiourea taking part; the second wave is the result of the contribution of the formation of a complex (or eventually an association) in the adsorption layer and in the thicker reaction layer. They also considered [305] electrical double-layer effects and adsorption of thiourea on the catalytic prewave [$E_{1/2} = -0.58$ V (SCE)], caused by the formation and subsequent reduction of Ni(II)-monothiourea complex. The previous mechanism [304] was slightly modified to account for the interaction between $[Ni(H_2O)_6]^{2+}$ ion and thiourea adsorbed at the mercury surface. Its rate constant at $\psi_1 = 0$ is $k_a = (2\text{-}4) \times 10^{-2}$ l/(mole)(sec). The same authors also studied Ni(II)-thiosemicarbazide complexes.

4.1.2. Nonaqueous Solution

Ikeda and Itabashi [164] (see Section 2.1.8) studied the Ni(II)-acetate system in acetonitrile containing acetic acid and acetic anhydride to establish the reaction mechanism for Ni(II) polarographic reduction. They considered the effect of acetic acid, acetic anhydride, water, and $LiClO_4$ concentrations on the two cathodic waves. When there is only one wave, the electrochemical process is controlled by diffusion and kinetics. The second wave is a kinetic wave. The probable mechanism of the reaction is represented by

$$Ni(OAc)_2(HOAc)_x(H_2O)_{4-x} + 2MClO_4 + nCH_3CN$$

$$\underset{\longleftarrow}{\overset{K_{eq}}{\longrightarrow}} [Ni(CH_3CN)_n]^{2+} + 2ClO_4^- + 2MOAc + xHOAc + (4 - x)H_2O$$

and

$$[Ni(CH_3CN)_n]^{2+} + 2e = Ni(Hg) + nCH_3CN$$

The same authors [306] considered the effect of halide ions on Ni(II) reduction in the same solvent. It appears that the presence of halides promote a catalytic prewave which depends both on Ni(II) and halide concentrations. The catalytic activity of halide ions in $LiClO_4$-supporting electrolyte diminishes in the order $I^- > Br^- > Cl^-$, but in solutions containing quaternary ammonium perchlorates the order is reversed. The proposed mechanism to explain these results comprises two steps. One of them is an adsorption equilibrium between the bulk and adsorbed halide ions and the cations of the supporting electrolyte. The other implies a complexation reaction of Ni(II) and adsorbed halide ions. This mechanism also explains the selective effect of cations on the catalytic current.

4.2. MECHANISM OF NICKEL DISSOLUTION AND DEPOSITION AT SOLID ELECTRODES

4.2.1. Aqueous Solutions

4.2.1.1. Active Anodic Dissolution and Cathodic Deposition. The dissolution reaction in aqueous media, either spontaneous or under an applied potential, is related to corrosion and passivation processes, and the deposition reaction usually occurs with the simultaneous formation of hydrogen. The rates of these reactions depend upon the electrode potential, pH, concentration of metal ions, nature of the anions present in the solution, and the state of the electrode surface. The polarization state often varies with time, resulting in a hysteresis of the polarization curve, as first noted by Foerster and Krueger [307]. Advances made in understanding the kinetics of the electrode processes through application of the single pulse polarization technique, led to a reasonable, although not yet complete, interpretation of the electrode reaction in aqueous solutions.

TABLE 4.2.1. Kinetic Parameters for the Anodic Dissolution of Nickel [261]

Composition	pH	T (°C)	j_0 (mA/cm^2)	$\partial E/\partial \log j_a$ (V)	j_a range (mA/cm^2)
1.0 M KCl	2.6	25	1.4×10^{-7}	0.084	0.25-13.0
1.0 M KCl	2.6	35	1.8×10^{-7}	0.082	0.25-11.0
1.0 M KCl	2.6	45	2.8×10^{-7}	0.080	0.22-12.0
0.1 M KCl	2.5	45	1.8×10^{-7}	0.084	0.05-4.0
0.01 M KCl	2.5	45	1.0×10^{-7}	0.084	0.05-1.2
1.0 M $NaClO_4$	0.9	27	2.3×10^{-7}	0.125	0.5-9.0
1.0 M $NaClO_4$	0.9	44	5.0×10^{-7}	0.13	0.6-9.0
0.25 M $NaClO_4$	2.5	45	3.0×10^{-7}	0.11	0.55-3.0
1.0 M $NaClO_4$	2.5	45	3.0×10^{-7}	0.11	0.55-3.0
0.5 M $NiCl_2$	2.5	25	1.35×10^{-7}	0.090	0.08-7.0
0.5 M $NiCl_2$	2.5	35	1.27×10^{-7}	0.085	0.07-7.0
0.5 M $NiCl_2$	2.5	45	3.1×10^{-7}	0.090	0.08-9.0
0.5 M $NiSO_4$	1.0	45	3.6×10^{-7}	0.115	0.07-4.0

Kronenberg, Banter, Yeager, and Hovorka [261] studied the anodic dissolution of 99% purity nickel in acidified solutions containing chloride, sulfate, or perchlorate ions under a variety of nonpassivating conditions. The Tafel slopes at 45°C, as shown in Table 4.2.1, were between 0.085 and 0.120 V/decade. The nickel anode potential was independent of Ni(II) concentration under conditions for which the backward reaction is negligible and no pH dependence has been found for $1 < pH < 2.5$. The experimental activation energy in solutions containing chloride ions is 15 kcal/mole. In solutions containing sulfate and perchlorate ions, the Tafel slope involved an αn value of 0.52 and provided evidence favoring a charge transfer step as rate determining. The lowest Tafel slopes and reaction rates were obtained in chloride solutions and were attributed to its specific adsorption.

Sato and Okamoto [308] investigated the anodic dissolution of 99.6% nickel in H_2SO_4 at 0.5 M constant sulfate ion concentration ($0 < pH < 3$) and 45°C. As seen in Table 4.2.2, the

TABLE 4.2.2. Effect of pH on the Anodic Tafel Slope for Nickel [308]

pH	$\partial E/\partial \log j_a$ (V)
0.45	0.098
0.45	0.091
0.45	0.100
0.45	0.109
1.10	0.092
1.10	0.099
1.73	0.096
1.73	0.110
2.75	0.130
2.75	0.120

Tafel slope obtained by a rapid polarization technique depended on pH, as did the rest potential ($\partial E_r/\partial pH = 0.105$ V). The steady-state anodic curve yielded a Tafel slope close to 0.030 V/decade in the range $0.45 \leq pH \leq 2.75$. When the pH of the solution was increased, the anodic polarization curve shifted in the direction of less noble potential, and therefore the dissolution rate at constant potential increased with the pH. In the case of both rapid and steady-state polarization, $(\partial \ln j_a/\partial pH)_E = 1$. The following consecutive reaction scheme for the anodic dissolution reaction has been suggested:

$$Ni + OH^- \rightleftharpoons NiOH_{ad} + e \qquad (4.2.1a)$$

$$NiOH_{ad} \rightleftharpoons NiOH^+ + e \qquad (4.2.1b)$$

$$NiOH^+ \rightleftharpoons Ni^{2+} + OH^- \qquad (4.2.1c)$$

The rate-determining step changes from Step (4.2.1a) to (4.2.1c) as the $NiOH^+$ ion concentration increases in the vicinity of the surface. The rate of anodic dissolution is also controlled by the amount of hydroxyl ion involved in a reaction cycle consisting of the three steps.

Schwabe and Voigt [309] studied the influence of concentrated neutral salt solutions on nickel corrosion, using 99.9% Ni. No straight Tafel line was found for stationary measurements of the active anodic dissolution, but S-shaped curves resulted. The calculated reaction order for hydrogen ion lies between -0.5 and -0.6, either through an increase of the hydrogen ion activity by an increase of the acid concentration or through an increase of the activity coefficient by the addition of neutral salt.

Matulis and Slizys [310] studied the potential of Ni(II) discharge at room temperature in a hydrogen atmosphere. The changes in the potential of both the nonpolarized and polarized electrodes are due to gradual changes in the acidity of the solution near the cathode. One minute after nickel immersion the ΔpH in the vicinity of the electrode surface rises from 0.0 to 2.5, depending on the pH of the solution. Upon cathodic polarization, the pH rises to 7 to 8. Within the pH range of 3.5 to 4.5, an arrest in its increase is observed. The changes in pH in the vicinity of the electrode surface are accounted for by the formation of basic nickel compounds and their adsorption on the electrode surface. Slizys [311] measured the change of pH in the diffusion layer during the electrodeposition of nickel. At 700 to 800 μm distance from the cathode surface, the pH increases from 4 to 7. At 10 μm its value is between 7.5 and 8. Consequently, the metal ion concentration in the solution layers nearest the cathode should be significantly less than in the bulk of the solution. Both hydrogen atom adsorption and $NiOH^+$ ion play an important role in these processes.

Heusler and Gaiser [312] studied the kinetics of the anodic dissolution and cathodic deposition of nickel in acid solutions at 65°C at different hydrogen and nickel ion activities in the presence of either 1 M $NaClO_4$ or 1 M $Ba(ClO_4)_2$ (μ = 1.5) with polycrystalline electrodes of spectroscopic quality. Stationary and nonstationary current/potential curves were traced. The evaluated kinetic parameters are shown in Table 4.2.3. These kinetic parameters fit the following mechanism, which involves either water or hydroxyl ion participation in the metal dissolution process:

$$\mathrm{Ni}\left[\mathrm{Ni(OH)_{ad}}, \mathrm{Ni(OHNi)_{ad}}\right] + \mathrm{OH^-} = \mathrm{NiOH^+} + \mathrm{Cat} + 2e \qquad (4.2.2a)$$

$$\mathrm{NiOH^+} + \mathrm{H^+} = \mathrm{Ni^{2+}(aq)} \qquad (4.2.2b)$$

Reaction (4.2.2a) comprises three elementary steps:

$$\mathrm{Ni} + \mathrm{H_2O} \underset{k_{-1}}{\overset{k_1}{\rightleftharpoons}} \mathrm{Ni(OH)_{ad}} + \mathrm{H^+} + e \qquad (1)$$

$$\mathrm{Ni} + \mathrm{Ni^{2+}(aq)} + e \underset{k_{-2}}{\overset{k_2}{\rightleftharpoons}} \mathrm{Ni(OHNi)_{ad}} + \mathrm{H^+} \qquad (2)$$

$$\mathrm{Ni(OH)_{ad}} + \mathrm{Ni^{2+}(aq)} + 2e \underset{k_{-3}}{\overset{k_3}{\rightleftharpoons}} \mathrm{Ni(OHNi)_{ad}} \qquad (3)$$

where the surface complexes $Ni(OH)_{ad}$ and $Ni(OHNi)_{ad}$ are referred to as catalysts (Cat) in Reaction (4.2.2a). Assuming that Step (3) is rate determining, the rate of the overall reaction is given by

$$j = 2Fk_3 a_{\mathrm{Ni(OHNi)_{ad}}} a_{\mathrm{OH^-}} \exp\left[\frac{2\alpha FE}{RT}\right] \qquad (4.2.3)$$

Steps (1) and (2) can be regarded as in quasi-equilibrium:

$$a_{\mathrm{(NiOH)_{ad}}} = \frac{k_1}{k_{-1}} a_{\mathrm{Ni}} a_{\mathrm{OH^-}} \exp\left[\frac{FE}{RT}\right] \qquad (4.2.4)$$

TABLE 4.2.3. Kinetic Parameters for Nickel Electrodissolution after Heusler and Gaiser [312]

Stationary E/j curves	Nonstationary E/j curves
Tafel slope for H_2 evolution: 130 mV	$(\partial E/\partial \log j_c) = 67 \pm 8$ mV
Tafel slope for Ni dissolution: 37 ± 4 mV	$(\partial E/\partial \log j_a) = 66 \pm 7$ mV
Reaction order (OH^-): 1.75 ± 0.3	Capacitance: $20 \pm 2\ \mu F/cm^2$ (20 μsec)
Tafel slope for Ni deposition: 30 ± 5 mV	
Reaction order (Ni^{2+}): 1.9	

$$a_{Ni(OHNi)_{ad}} = \frac{k_2}{k_{-2}} a_{Ni} a_{(NiOH)_{ad}} \exp\left[-\frac{FE}{RT}\right] \tag{4.2.5}$$

Substituting Eqs. (4.2.4) and (4.2.5) into Eq. (4.2.3) yields

$$j = 2F \frac{k_1 k_2 k_3}{k_{-1} k_{-2} k_{-3}} a_{Ni}^{2} a_{OH^-}^{2} \exp\left[\frac{(1+2\alpha)FE}{RT}\right] \tag{4.2.6}$$

The slopes of Tafel equation for the stationary and nonstationary conditions can be calculated from Eqs. (4.2.6) and (4.2.3), respectively, assuming that the symmetry factor α of the rate-determining step is 0.5. For the stationary condition, $(\partial \log j_a/\partial E) = 0.0276$ V/decade and for nonstationary condition $(\partial \log j_a/\partial E) = 0.0552$ V/decade; in this case $a_{Ni(OHNi)ad}$ in Eq. (4.2.3) was regarded as constant. Kinetic parameters derived from this mechanism agree well with the experimental values (see Table 4.2.3).

Sayano and Nobe [262] made continuous galvanostatic and potentiostatic measurements and applied pulse polarization techniques to study nickel dissolution in 1 N H_2SO_4. The anodic Tafel slopes were 0.048 V/decade for 99.5% Ni and 0.050 V/decade for 99.97% Ni. The corresponding apparent electrode capacitances are seen in Table 3.2.5.

The dissolution characteristics of high purity polycrystalline nickel in 1 N H_2SO_4 (H_2 saturated) at 25°C were studied by Condit [313] using atomic absorption analysis. Potentiostatic dissolution was a logarithmic function of time, the exponent of time being 0.7. According to the author, the dissolution involved three electrons per atom. The Tafel slope is closer to a value of 0.040 V/decade and would suggest a mechanism similar to that given by Kelly [314, 315] for the iron electrode.

Piatti, Arvía, and Podestá [316, 317] investigated the electrochemical behavior of nickel in acid aqueous solution saturated with hydrogen by means of stationary and nonstationary techniques at 25, 40, and 60°C using Ni (99.8%, with traces of Fe, Mn, Si, and Mg).

The nickel ion concentration was changed from 0.004 to 0.5 M in either 2 M NaCl or 2 M $NaClO_4$ at $0.5 \leq pH \leq 6.5$. The galvanostatic anodic yield for nickel dissolution was practically 100% at about 100 μA. The kinetic parameters obtained from the anodic curves were the anodic Tafel slope, an extrapolated current density $(j_{0, Ni})_a$ at the nickel reversible potential, and an extrapolated current density $j_{0, r}$ at the rest potential. The anodic Tafel slope at 25°C was close to 0.060 V/decade. $j_{0, Ni}$ and $j_{0, r}$ were pH-dependent and their values lay between 10^{-8} and 10^{-10} A/cm² and between 10^{-5} and 10^{-6} A/cm², respectively. At lower pH's the average Tafel slope for hydrogen evolution was about 0.120 V/decade at 25°C; the values of $j_{0, H}$ were of the order of 10^{-7} A/cm². At higher pH a convective diffusion limiting current density, $j_{d, H}$, was observed. When nickel ions were present in the solution at pH > 4 and $E < E^{\circ}_{Ni}$, a cathodic Tafel region was observed, corresponding to the simultaneous electrodeposition of nickel and hydrogen. The experimental Tafel slope was about 0.060 V/decade at 25°C and the Tafel line for nickel deposition, after correction for the hydrogen evolution reaction in perchlorate solutions, yielded a slope close to 0.120 V/decade in the range of 10^{-5} to 10^{-3} A/cm² at pH = 4.5. The extrapolated exchange current density $(j_{0, Ni})_c$ at the reversible potential of the nickel electrode was 2×10^{-6} A/cm². In chloride solutions the cathodic Tafel line for nickel electrodeposition was not so well defined as for perchlorate solutions. The figure for $(j_{0, Ni})_c$ was not coincident with that resulting from the extrapolation of the anodic Tafel line at the reversible nickel electrode potential. The cathodic Tafel line for nickel electrodeposition could only be evaluated in the pH range from ~4 to 6. An increase of temperature caused an appreciable decrease of polarization both in the anodic and cathodic regions. From anodic experiments, the average activation energy for chloride solutions was $\Delta H^* = 13 \pm 3$ kcal/mole, and for perchlorate solutions $\Delta H^* = 18 \pm 3$ kcal/mole.

The rest potential of nickel in acid oxygen-free solutions was independent of the nickelous ion activity, but changed with the logarithm of the hydrogen-ion activity. Except at very low pH, the initial potential was more positive than the corresponding hydrogen and nickel electrode potentials. Therefore, as pH is increased, at least in the acid region, the nickel electrode showed an increasing passivity. While the metal exhibited a positive potential with respect to the hydrogen electrode at 25°C, this became negative when the temperature was increased either to 40 or 60°C. In most of the experiments at lower pH the initial potential was close to the potential of the hydrogen electrode.

The rest potentials depended on the anion present in the system. Thus, at constant pH and temperature, the nickel electrode was more active in chloride solutions than in perchlorate solutions. The apparent electrode capacitance of nickel electrodes in the presence of chloride ions obtained from the overvoltage decay was about 35 μF/cm². For perchlorate ion solution the capacitance decreased from about 25 μF/cm² at pH 1.0 to about 6 μF/cm² at pH 4. At the rest potential it was between 46 and 96 μF/cm² in chloride ion solutions and between 11 and 37 μF/cm² in perchlorate ion solutions. No definite pH dependence was established in any case. The buildup curves exhibited a superpolarization which was assigned to the pH change at the interface. The nonstationary current/potential curves involved a Tafel slope equal to RT/F. Results indicated it is unlikely that in acid aqueous solutions a clean nickel metal solution interface occurs, and the interface structure should depend upon the anion in solution. The different kinetic parameters were explained in terms of the following reaction mechanisms:

Anodic reaction:

$$Ni + H_2O \rightleftharpoons (NiOH)_{ad} + H^+ + e \qquad (4.2.7a)$$

$$(NiOH)_{ad} \xrightarrow{rds^*} NiOH^+ + e \qquad (4.2.7b)$$

$$NiOH^+ \rightleftharpoons Ni^{2+} + OH^- \qquad (4.2.7c)$$

*rds = rate determining step

Cathodic reaction:

$$Ni^{2+} + H_2O \rightleftharpoons NiOH^+ + H^+ \qquad (4.2.7d)$$

$$NiOH^+ + e \xrightarrow{rds} (NiOH)_{ad} \qquad (4.2.7e)$$

$$(NiOH)_{ad} + H^+ + e \rightleftharpoons Ni + H_2O \qquad (4.2.7f)$$

and

Anodic reaction:

$$Ni + OH^- \xrightarrow{rds} (NiOH)^+_{ad} + 2e \qquad (4.2.8a)$$

$$(NiOH)^+_{ad} \rightleftharpoons Ni^{2+} + OH^- \qquad (4.2.8b)$$

Cathodic reaction:

$$Ni^{2+} + H_2O \rightleftharpoons (NiOH)^+_{ad} + H^+ \qquad (4.2.8c)$$

$$(NiOH)^+_{ad} + 2e \xrightarrow{rds} Ni + OH^- \qquad (4.2.8d)$$

Theoretical kinetic parameters deduced from Mechanisms (4.2.7) and (4.2.8) are assembled in Table 4.2.4; the symmetry factor of the rate determining step was assumed to be 0.5. In principle, Mechanism (4.2.7) explains formally most of the experimental results, particularly at low pH. However, the anodic Tafel slope is larger than the predicted one and coincides with the theoretical anodic Tafel slope derived from Mechanism (4.2.8). A slope larger than 2RT/3F may result if the electron transfer process takes place on a surface covered by a film of the oxide type. Therefore, the kinetic parameters favors Mechanism (4.2.7) over Mechanism (4.2.8).

The effect of chloride ions was explained through its specific adsorption on the metal, thus decreasing the potential of the inner plane of the Helmholtz double layer and

TABLE 4.2.4. Theoretical and Experimental Parameters Deduced from Mechanisms (4.2.7) and (4.2.8) [316, 317]

	Mechanism (4.2.7)	Mechanism (4.2.8)	Experimental (25° C)
$\left(\frac{\partial E}{\partial \ln j_a}\right)_{a_{OH^-}, a_{Ni^{2+}}}$	$\frac{2RT}{3F}$	$\frac{RT}{F}$	0.055 ± 0.005 V
$\left(\frac{\partial E}{\partial \ln j_c}\right)_{a_{OH^-}, a_{Ni^{2+}}}$	$-\frac{2RT}{F}$	$-\frac{RT}{F}$	-0.120 ± 0.015 V
$\left(\frac{\partial \ln j}{\partial \ln a_{Ni^{2+}}}\right)_{a_{OH^-}, \eta}$	1	1	1.0 ± 0.3
$\left(\frac{\partial \ln j_{0, Ni}}{\partial \ln a_{OH^-}}\right)_{a_{Ni^{2+}}}$	1	1	1.0 ± 0.2
$\left(\frac{\partial E}{\partial \ln a_{OH^-}}\right)_{a_{Ni^{2+}}, j}$	$-\frac{2RT}{3F}$	$-\frac{RT}{F}$	-0.040 ± 0.005 V
$\left(\frac{\partial \ln j_{0, Ni}}{\partial \ln a_{a_{Ni^{2+}}}}\right)_{a_{OH^-}}$	0.75	0.5	0.8 ± 0.3
$\left(\frac{\partial E_r}{\partial \ln a_{OH^-}}\right)_{a_{Ni^{2+}}}$	$-\frac{RT}{F}$	$-\frac{4RT}{3F}$	-0.035 ± 0.010 -0.065 ± 0.010
$\left(\frac{\partial \ln j_{0, r}}{\partial \ln a_{OH^-}}\right)_{a_{Ni^{2+}}}$	-0.5	-0.33	-0.3 ± 0.1

consequently the anodic overvoltage. Assuming that the nickel surface is not completely free of oxygen-containing species, it is likely that chloride-ion interaction probably occurs by overlapping of the chloride-ion orbitals, which are distorted due to the high local field strength at the electrical double layer with part of the orbitals of nickel. The participation of these orbitals requires a kind of distorted ligand. The outer electronic structure of solid nickel [318, 319] results from a statistical average of at least three states: $(3d)^{10}(4s)^0$, $(3d)^9(4s)^1$, and $(3d)^8(4s)^2$. The average value may be written as $(3d)^{9.4}(4s)^{0.6}$, indicating that the d-orbital contribution in the solid is more relevant than in the free gaseous nickel atom in the normal state. Taking into account that aqueous nickel ion has a $(3d)^8(4s)^2$ configuration, it is evident that an electronic energy redistribution must take place for the dissolution as well as for the electrodeposition process to occur. As far as the electronic structure and the participation of chloride ions is concerned, the effect of the latter, through its adsorption on the metal, may promote d-electrons, making the dissolution process easier.

Vagramyan and Uvarov [320, 321], Vagramyan, Zhamagortsyan, and Uvarov [322], and Vagramyan, Zhamagortsyan, Uvarov, and Yavich [323-325] investigated the potential of a nickel electrode and the kinetic parameters of the processes of deposition and dissolution of Ni (99.99%) in 1 N aqueous $NiSO_4$ at pH 1.5 and 1 N aqueous $NiCl_2$ at pH 1.5 in the temperature range of 25 to 275°C. Temperature coefficients of the overvoltage for nickel deposition and dissolution are given in Table 4.2.5. The influence of temperature on the kinetic parameters are given in Table 4.2.6 for 1 N $NiCl_2$ solution at pH 1.5. According to these authors the electrochemical behavior of nickel in aqueous solutions is due to chemisorption of foreign particles on the electrode surface exerting an influence both on the stationary potential and on the overvoltage in the deposition and dissolution of the metal. In the temperature range of 175 to 200°C the inhibiting effect of foreign particles is eliminated. This fact allowed for the first time the determination of the true kinetic parameters of the electrochemical reaction for the discharge-ionization in the processes of nickel deposition and dissolution. The effect of temperature on the reaction rates was tentatively assigned to an increase in the activity of the discharging ions and a decrease in the inhibiting effect of foreign particles.

TABLE 4.2.5. Overvoltage Coefficients for Nickel Deposition and Dissolution [320-325]

Solution	$\partial E_c/\partial T$ (mV/°C)		$\partial E_a/\partial T$ (mV/°C)	
	25-75°C	175-200°C	25-75°C	175-200°C
$NiCl_2$	2.4	0.2	3.5	0.4
$NiSO_4$	3.3	0.4	4.8	0.7

TABLE 4.2.6. Influence of Temperature on the Electrochemical Kinetic Parameters for Nickel [320-325]

T (°C)	$\partial E_c/\partial \log j_c$ (V)	$j_{0,c}$ (A/cm²)	$\partial E_a/\partial \log j_a$ (V)	$j_{0,a}$ (A/cm²)
25	0.123	1.3×10^{-4}	0.098	3.5×10^{-6}
50	0.115	3.2×10^{-4}	0.080	1.0×10^{-5}
75	0.110	1.0×10^{-3}	0.072	4.6×10^{-5}
100	0.107	2.4×10^{-3}	0.064	2.0×10^{-4}
125	0.101	4.6×10^{-3}	0.055	7.2×10^{-4}
150	0.100	9.1×10^{-3}	0.054	1.6×10^{-3}
175	(0.098)	1.4×10^{-2}	(0.048)	3.5×10^{-3}
200	-	2.2×10^{-2}	-	6.7×10^{-3}

Ovari and Rotinyan [326] observed the overvoltage related to the simultaneous discharge of nickel(II) and hydrogen ions on nickel cathodes from $NiSO_4 + H_2SO_4$ solutions ($0.55 < pH < 3$ and $0.25\ \underline{M} < C_{Ni^{2+}} < 2.0\ \underline{M}$) at 25°C. Tafel plots corrected for the overvoltage due to hydrogen evolution were obtained. The Tafel slopes varied between 0.124 and 0.130 V/decade and were independent of nickelous ion concentration. For solutions at pH 1.65, a first-order dependence on hydrogen ion concentration was observed. At higher pH's the reaction order was <1.

The kinetics of electrode processes of nickel electrodes in $NiCl_2$ aqueous solutions at 55°C was investigated by Ovari and Rotinyan [327], covering a pH range of 0.75 to 3.0. Cathodic polarization was studied on an electrode with a 3 to 5 μ thick electrolytic coating. The rates of the cathodic and anodic processes were independent of pH, so the authors concluded that in a chloride electrolyte no hydroxo-complexes of metal ions, but aquated (or halide) complexes, participated in the cathodic process. The rate of the cathodic process is first order with respect to nickelous ion activity, and the rate of the anodic reaction is independent of it. They proposed a mechanism, following Fedotiev and Dmitreshova [328], which consists of

$$Ni^{2+} + e \underset{}{\overset{rds}{\rightleftarrows}} Ni^{+} \quad (4.2.9a)$$

$$Ni^{+} + e \rightleftarrows Ni \quad (4.2.9b)$$

where the first reaction is rate determining. A comparison of the experimental and theoretical parameters is given in Table 4.2.7.

Oshe and Lobachev [329] investigated the kinetics of the anodic oxidation of Ni (99.9%) by potentiostatic chronoamperometry with the following solutions: $0.8\ \underline{N}\ K_2SO_4 + 0.2\ \underline{N}\ H_2SO_4$, pH 1.6; 1 $\underline{N}$ $HClO_4$, pH 0.1; and 7 $\underline{N}$ NH_4OH, pH 12. A spontaneous increase in the current with time, arising in the region of potentials of active dissolution, was detected. This current was termed autocatalytic and cannot be explained either by a change of the pH of the solution in the layer near the electrode or by a change in the concentration of any component of the solution. The autocatalytic current is almost entirely due to the reaction

$$Ni + H_2O = NiO + 2H^{+} + 2e \quad (4.2.10)$$

Considering the fact that in the active dissolution region the electrode is already coated with an oxide film even in weakly acid solutions [330, 331], as well as the fact that stirring the solution does not affect the shape of the current/time curves, the dependence obtained should be explained by the fact that the oxidation of nickel is controlled by the rate of delivery (or removal) of some species through the oxide layer. Using a photoelectric polarization method, it was demonstrated that in the active dissolution region the nickel oxide layer represents a phase of variable composition with p-type conductivity. The concentration variation of 3d holes was correlated with the variation of the anodic dissolution current. Therefore it was

TABLE 4.2.7. A Comparison of Experimental and Theoretical Kinetic Parameters for the Ni(II)/Ni Reaction in Aqueous $NiCl_2$ Solutions [327]

Conditions	Value of derivative	Numerical values	
		Theoretical	Experimental
Ni deposition	$-\partial E_c/\partial \log j_c$	0.130	0.096-0.102
Ni deposition	$\partial E_c/\partial \log[Ni^{2+}]$	0.130	0.115-0.132
Ni deposition	$\partial E_c/\partial (pH)_s$	0.00	0.0-0.01
Ni dissolution	$\partial E_a/\partial \log j_a$	0.043	0.041
Ni dissolution	$\partial E_a/\partial \log[Ni^{2+}]$	0.00	0.0-0.01
Ni dissolution	$\partial E_a/\partial pH_s$	0.00	0.0-0.01
H_2 evolution	$-\partial E_c/\partial \log j_{H_2}$	0.130	0.115-0.130
H_2 evolution	$\partial E_c/\partial \log[Ni^{2+}]$	0.00	0.00
H_2 evolution	$-\partial E_c/\partial pH_s$	0.130	0.130
Equilibrium potential	$\partial E_{eq}/\partial \log[Ni^{2+}]$	0.032	0.032
Equilibrium potential	$\partial E_{eq}/\partial pH_s$	0.00	0.0-0.01
Exchange current	$\partial \log j_0/\partial \log[Ni^{2+}]$	0.75	0.74
Exchange current	$\partial \log j_0/\partial pH_s$	0.00	0.0-0.01
Corrosion characteristics	$\partial \log j_{corr}/\partial pH_s$	0.75	0.74
Corrosion characteristics	$-\partial E_{corr}/\partial pH_s$	0.032	0.034

proposed that the current controlling polarization in the oxide layer was associated with polarization along the 3d holes. The occurrence of the processes:

$$3Ni + 4H_2O = Ni_3O_4 + 8H^+ + 8e \qquad (4.2.11)$$

and

$$2Ni + 3H_2O = Ni_2O_3 + 6H^+ + 6e \qquad (4.2.12)$$

in the region of potentials where the catalytic current is observed, can lead to the formation of solid solutions of Ni_2O_3 in NiO. This means an increase in Ni(III) ion concentration (3d holes) in the oxide. If the nickel oxidation rate is determined by the 3d hole concentration in the reaction zone, it is clear that an increase in the latter during the formation of the solid solution should cause an acceleration in the oxidation.

In studying the mechanism of the nickel electrode in HCl solutions, Penov, Kosakovskaya, Botneva, Andreeva, and Zhuk [332] observed a potential region in which the corrosion rate was independent of the potential. In this region the experimental activation energy was about 15 kcal/mole, a value approximately 1.5 times that for the self-dissolution process.

Epelboin [333] considered the effect of double-layer structure on the electrodeposition of nickel in Watts solutions. Results obtained at varying pH or nickelous ion concentration prove that the adsorbed species $(NiOH)_{ad}$ does not play a catalytic role but is involved as an intermediate in the electron transfer, being responsible for the inductive component of the impedance. The surface concentration of $(NiOH)_{ad}$ is related to the coverage through the Langmuir isotherm. In the presence of acetylenic alcohol, as an inhibitor, one merely observes a decrease in the double-layer capacity, an increase of the inductive component of the impedance and a change in the reaction rates.

The chemical dissolution of nickel in 8 N HNO_3 at 73 to 93°C has been studied by Dorofeeva and Kalinichenko [334]. It is a first-order reaction with respect to acid concentration and the rate constants are 2.8×10^{-3}, 3.75×10^{-3}, and 5.75×10^{-3} sec^{-1} at 73, 83, and 93°C, respectively. The experimental activation energy is 12.1 kcal/mole.

The corrosion rate in 1 N H_2SO_4 acid under the influence of 50 Hz ac was investigated by Yamuna and Subramanyan [335]. The rates at 5 and 25 mA/cm^2 are 3 to 10 times higher than those of normal corrosion.

Valeev and Ushanova [336] studied photochemical effects in the anodic dissolution of electrolytic nickel in 11.5 M H_2SO_4 at 40°C and potentials between 0.14 and 2.0 V. A relationship between the structural changes of the surface and the amount and character of the photoelectrochemical effect (change of electrode potential by the action of light) was established. The structural changes occur in the presence of an oxide film on the electrode, and the saturation diffusion layer changes with the oxide composition.

4.2.1.2. Passivity. The passivation of nickel has been investigated in a systematic way for a long time [337]; studies include hydrogen-ion and chloride-ion influence [307, 308], the participation of anions [339-341], the formation of a porous solid salt layer [342-347], passivation times in relation to stirring [348, 349], the oxidation state, and the influence of metal purity [350, 351]. Piontelli and Serravalle [352] determined the conditions leading to corrosion without passivation, to passivation with attack, and to complete passivity at 25 and 65°C for aqueous solutions containing different anions at various pH's. The main peculiarities of the potential/time curves, when passivation occurs, are the existence of maxima, especially at the start of the current flow, the occurrence of intermediate steps preceding the attainment of steady or quasi-steady conditions, oscillatory phenomena, and hysteresis effects. Periodic phenomena at a nickel electrode in H_2SO_4 solutions were also reported by Osterwald and Feller [353].

Hickling and Spice [25] studied anodic passivation by means of the potential change produced after a definite amount of charge had been passed prior to oxygen evolution. The ratio O/Ni, related to the formation of the passive film, was equal to 3.9/2. MacGillavry, Singer, Rosenbaum, and Swenson [354, 355] studied phase boundary potentials in aqueous solutions containing hydroxyl, carbonate, and phosphate ions at different pH values at 30°C. These results, as well as those reported by Lopez-Lopez [5], which referred to the influence of surface structure and pH on electrode potentials, can be interpreted in terms of potential/pH diagrams.

After the interruption of an anodic polarization of nickel in deaerated solutions, Turner [356] observed that the potential decreased to a value which was more active than the original open circuit potential. Subsequently, the potential slowly increased to the original value. The more active potential observed immediately after cutoff of anodic polarization was attributed to desorption of oxygen and other impurities from the electrode surface during polarization. After a longer period of time subsequent to cutoff, readsorption of impurities on the electrode occurred, bringing the open circuit potential back to its original value. Turner explained this behavior by assuming that during the initial phase of the anodic polarization of nickel in sulfuric acid solutions the reaction was $Ni = Ni^{2+} + 2e$, causing the precipitation of $NiSO_4$ in the vicinity of the electrode surface. At sufficiently high current density, the rate of formation of $NiSO_4$ was greater than its rate of dissolution, causing the electrode surface to be covered with a film. As a result, the ohmic resistance increased, restricting the passage of the current and giving rise to the increase in potential just before passivation of nickel. The passivation was due to a change in mechanism from nickel dissolution to the formation of nickel oxide with the oxide film replacing the nickel sulfate film. However, other authors maintain that a high pH at the interface must be reached for the onset of passivation [348, 349]. Although passivation may start with adsorption or precipitation of compounds on the electrode surface, such salt films may not produce complete passivity [352-355, 357, 358].

When the potential of the passivated metal is increased, oxygen evolution occurs. The potential at open circuit shows a very rapid decrease from 2.2 to 1.6 V. Then a second step occurs at 0.26 V and finally a rest potential at 0.02 V is attained. The step at 1.6 V would correspond to the decomposition of nickel oxide mixtures [356, 359]. The step at 0.26 V was attributed to a nickel oxide film which does not have passivating properties and finally the rest potential, which corresponds to equal rates of nickel salt precipitation and dissolution.

Arnold and Vetter [360] found that the Flade potential, E_F, at $0 \leq pH \leq 3$, is given by: $E_F = E_F^\circ - 0.059\ pH$ at 25°C, where $E_F^\circ = 0.355$ V (NHE). The passive layer corresponds to NiO, in agreement with results of Pfisterer, Politycky, and Fuchs [361], and Okamoto, Takaishi, and Sato [362]. Passivating oxide films about 40 Å thick were measured [360] and it was assumed that at the Flade potential a nonporous monolayer was formed which grew to that thickness only at more anodic potentials. Raub and Disam [363] obtained various Flade potentials (E_F^o) from anodic polarization curves of differently electrodeposited nickel in 0.1 N H_2SO_4. The values fall into distinct groups: 0.11 V; 0.30 to 0.60 V; 0.41 and 0.67 V. Three different Flade potentials at 0.14, 0.24, and 0.42 V were reported by Greene [364] in 1 N H_2SO_4. According to Huq, Rosenberg, and Makrides [365], the Flade potential of nickel in H_2SO_4 was equal to 0.30 V, while in alkaline solutions it was 0.14 V. These results suggest that Ni_3O_4 and NiO are the species involved at the Flade potential. Weininger and Breiter [366], after several passivation-activation cycles using single nickel crystals in 0.2 N KOH, attained a Flade potential (E_F^o) value of 0.32 to 0.34 V. In certain cases two E_F^o values were found (0.32 and 0.40 V). An E_F^o value equal to 0.30 to 0.40 V in alkaline $NaClO_4$ solutions has also been found [367]. Capacitance measurements indicated $Ni(OH)_2$ formation commenced at 0.106 V and the passivation potential at 0.3 V was due to completion of surface coverage by $Ni(OH)_2$.

Davies and Barker [368] observed two potential arrests during galvanostatic cathodic polarization measurements in NaOH containing sodium carbonate and sodium borate. The upper one, not well defined, could correspond either to the equilibrium NiO_2/Ni_3O_4 or to NiO_2/Ni_2O_3. The lower one would relate to Ni/NiO equilibrium over the whole pH range studied.

De Gromoboy and Shreir [369] investigated the behavior of spectroscopic nickel in H_2SO_4 and acetic acid-acetate buffers at various pH values. Depending on the nature of the solution, passivation during potentiostatic polarization can occur at four different potentials, each of which corresponds to a reversible Ni_xO_y/Ni potential. These authors conclude that different oxides, not necessarily NiO or $Ni(OH)_2$, are formed on the metal surface by direct metal oxidation and passivation in these solutions. The most frequent E_F^o value is 0.36 V, probably related to the Ni_3O_4/Ni system.

Okamoto and Sato [357] observed that the potential decay of a nickel electrode after anodic polarization showed two distinct potential arrests. In H_2SO_4 solutions passivation occurs at 0.15 V [358], but some corrosion still exists at 1.35 V. At 1.5 V a basic nickel sulfate film has been detected [344]. In 5 N H_2SO_4 the film is $NiSO_4 \cdot 7H_2O$ and it is more stable as the acid concentration is increased. The film undergoes transformation to oxide (Ni_2O_3) above 2 V and its stability would be related to the passivity of the metal [370-373]. Keddam and Epelboin [374] observed the formation of $NiSO_4 \cdot 6H_2O$ on nickel in concentrated sulfuric acid solutions in both the active and transpassive regions. Results dealing with the dependence of the Flade potential on pH, given by different authors, are shown in Fig. 4.2.1.

Schwabe and Dietz [372] investigated the passivity of nickel in weak acid aqueous solutions in the absence of oxygen. The metal potential was independent of nickelous activity but increased with pH. The potential/pH diagram exhibited a slope which depended on the anion present, as seen in Table 4.2.7a. These results were explained by considering that the electrode potential is a mixed potential and that formation of a chemisorbed oxygen layer on the electrode occurs. According to Schwabe [375], a porous film does not constitute a necessary requirement for the passivation of nickel. The extent of the resulting passivation depends on

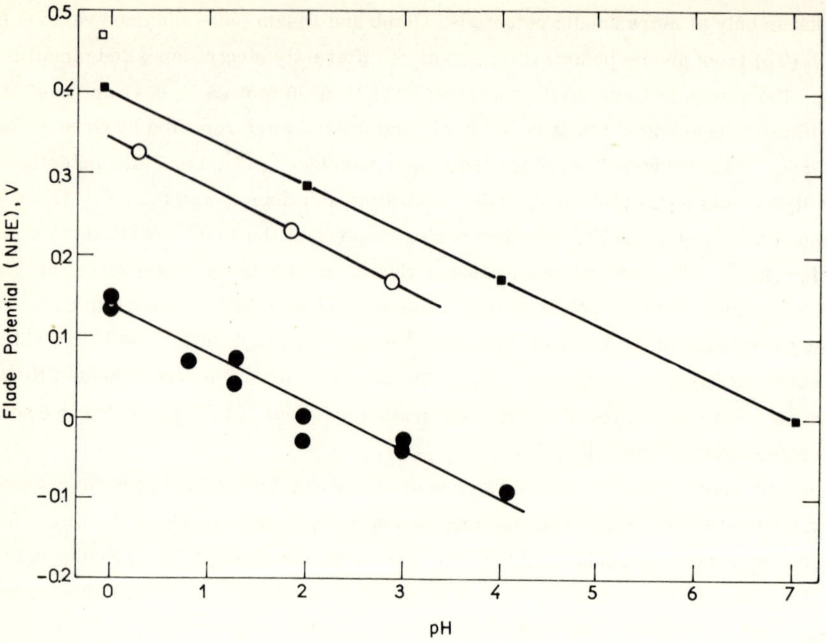

FIG. 4.2.1. Dependence of the Flade potential of nickel on pH at 25°C: (O) Arnold and Vetter [360, 378], (●) Osterwald and Feller [353], (□) Okamoto, Takaishi, and Sato [362], and (■) Markovic [11].

TABLE 4.2.7a. Influence of the Anion on the Slope of Potential/pH Diagrams [372]

Solution	ΔE/ΔpH (2 < pH < 6) (mV/pH)
N $NiCl_2$	21
N $NiSO_4$	33
N $Ni(ClO_4)$	42
Pure acetate	52

the nature of the anions present in the solution, since these anions will tend to displace oxygen with variable rapidity.

Trueb, Truempler, and Ibl [373] considered the behavior of the phase boundary impedance of anodically polarized nickel rod electrodes in 0.2 to 10 N H_2SO_4 by means of ac measurements in the audio range. The behavior of the capacitive element was interpreted as being due to adsorption and reciprocal displacement of sulfate ions and oxygen. The ohmic element increases little at passivation. It was suggested the initiation of passivation was due to the chemisorption of oxygen.

The activation potential of nickel previously passivated at 0.1 V was studied by Kandler, Romer, and Hausler [376] in 1 N KOH and 8 N H_2SO_4 by means of capacitance measurements at about 0.15 V. The addition of alcohols started one type of anodic reaction at 0.10 to 0.15 V, and others at 0.32 to 0.42 V and 0.42 to 0.45 V, depending on the alcohol added. Results were explained either on the basis of oxygen adsorption or by the presence of the various oxides of nickel.

The pH dependence of passive nickel corrosion as given by Vetter and Arnold [360, 377, 378] is similar to that found by Okamoto and Sato [357], Osterwald and Feller [353], and Uhlig and Feller [379]. Within a certain potential range where a Tafel line is valid, the corrosion current density seems to be independent of the pH value because the change of oxide composition due to the potential and the change of the potential difference at the phase boundary oxide/electrolyte due to the pH value of the electrolyte compensate each other. The states and compositions of the oxides are always equal for any pH value at the same potential vs the hydrogen electrode for the same corresponding pH value. The corrosion current densities were compared at different pH values with the same oxide at the same degree of oxidation as shown in Fig. 4.2.2. For the equivalent state of the oxide, the corrosion current density depends on the pH value or on the potential difference between oxide and electrolyte, respectively, according to a Tafel relation (Fig. 4.2.3). The straight line corresponds to a transfer reaction of the nickelous ion, with a transfer coefficient equal to 0.23. The change of the corrosion current density by the potential is related to the corresponding change in the degree of oxidation.

Sato and Okamoto [359] studied the anodic passivation of electroplated (0.88% Co, <0.04 Al, <0.015 Mg, and traces of Ca, Ag, Cu, and Fe) and rolled (0.28% Co, <0.04 Al, <0.03 Mn, and traces of Ca, Zn, Cu, and Fe) nickel samples in sulfuric acid solutions (N_2-saturated) of pH ranging from 0.4 to 1.4 at 25 and 40°C. The arrest potentials observed in the decay curve correspond to the equilibrium potentials for the NiO/Ni, Ni_3O_4/NiO, and Ni_2O_3/Ni_3O_4 electrodes, respectively. The second potential is the Flade potential (0.435 ± 0.02 V vs SCE). The formation of the passive film is postulated in terms of

$$Ni + OH^- \rightleftharpoons NiOH^+ + 2e \tag{4.2.13a}$$

$$3NiOH^+ + OH^- \rightleftharpoons Ni_3O_4 + 4H^+ + 2e \tag{4.2.13b}$$

The rate of nickel dissolution in the passive potential region, j_a, is independent of the electrode potential. It depends on the pH in the acid region according to $\log j_a = K - 0.46$ pH, where K = 1.9 $\mu A/cm^2$ at 25°C and 10.9 $\mu A/cm^2$ at 40°C. No effect of the concentration of either sulfate ion or nickel ion on the dissolution current was observed in acid solutions. When the concentration of nickel ion is small, the rate of the backward reaction is negligibly small compared with that of the forward reaction. The formation and the

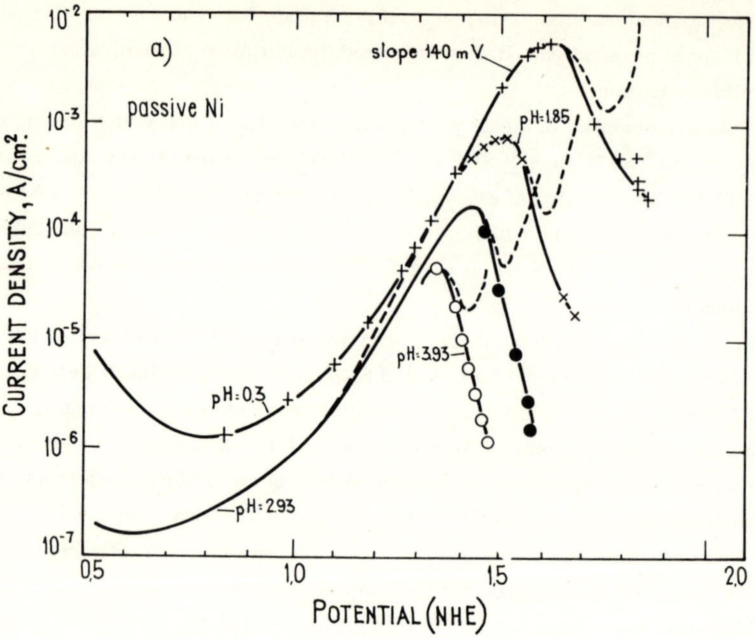

FIG. 4.2.2. Dependence of the corrosion current density of nickel at 25°C on potential at different pH values. Solution: x$\underline{M}$ H_2SO_4 + (0.5 - x)$\underline{M}$ K_2SO_4. The full line corresponds to the corrosion current density; the dotted line corresponds to the total current density [378]. (By permission of the Bunsengesellschaft für Physikalische Chemie.)

dissolution of the passive film proceed simultaneously, and the rates of those reactions are exactly the same at a steady state of polarization so as to keep the thickness of the film constant. The dissolution reaction may be expressed by

$$(Ni^{2+}, O^{2-}) + H^+(aq) \rightleftarrows Ni^{2+}(aq) + OH^-(aq) \tag{4.2.14}$$

The rate of this reaction is not affected by the electrode potential of nickel, but controlled only by the Galvani-potential difference across the oxide/solution interface. Reaction (4.2.14) is a coupled reaction of two processes:

$$O^{2-}(\text{oxide}) + H^+(aq) \rightleftarrows OH^-(aq) \qquad z = -2 \tag{4.2.15}$$

$$Ni^{2+}(\text{oxide}) \rightleftarrows Ni^{2+}(aq) \qquad z = +2 \tag{4.2.16}$$

where z is the charge of the ion passing across the interphase in the anodic direction. If the nickel ion is dissolved as the complex ion $NiOH^+$, Reaction (4.2.16) could be expressed as follows:

$$Ni^{2+}(\text{oxide}) + OH^-(aq) \rightleftarrows NiOH^+(aq) \qquad z = +2 \tag{4.2.17}$$

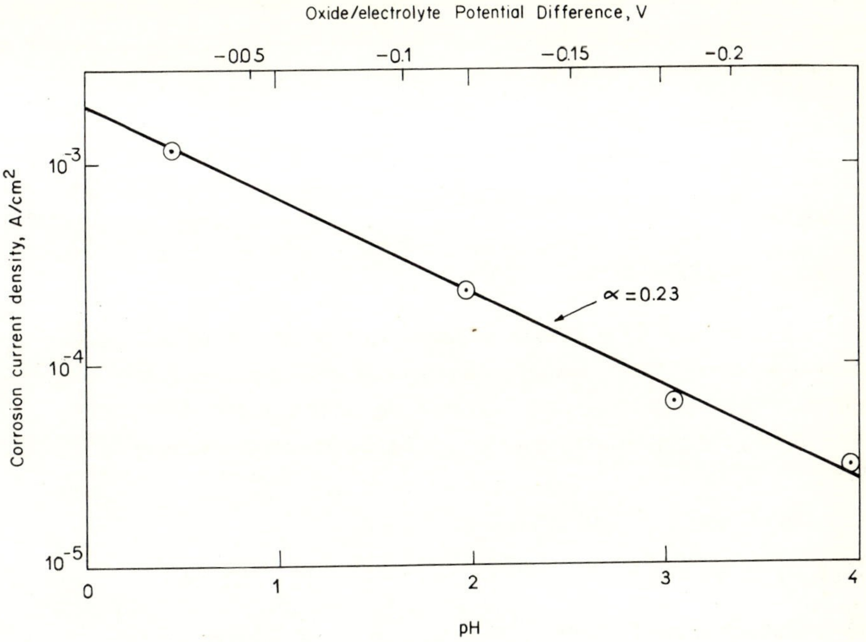

FIG. 4.2.3. Dependence of the corrosion current density of nickel on pH at a fixed potential (1.5 V) against the NHE at the same pH value [378]. (By permission of the Bunsengesellschaft für Physikalische Chemie.)

In the same way, Reactions (4.2.15) to (4.2.17) can be divided, respectively, into the following steps:

$$H^+(aq) \rightleftarrows H^+(s) \qquad z = -1 \qquad (4.2.15a)$$

$$O^{2-}(oxide) + H^+(s) \rightleftarrows OH^-(s) \qquad z = 0 \qquad (4.2.15b)$$

$$OH^-(s) \rightleftarrows OH^-(aq) \qquad z = -1 \qquad (4.2.15c)$$

$$Ni^{2+}(oxide) \rightleftarrows Ni^{2+}(s) \qquad z = 0 \qquad (4.2.16a)$$

$$Ni^{2+}(s) \rightleftarrows Ni^{2+}(aq) \qquad z = +2 \qquad (4.2.16b)$$

$$Ni^{2+}(oxide) + OH^-(a) \rightleftarrows NiOH^+(s) \qquad z = +1 \qquad (4.2.17a)$$

$$NiOH^+(s) \rightleftarrows NiOH^+(aq) \qquad z = +1 \qquad (4.2.17b)$$

$$NiOH^+(aq) \rightleftarrows Ni^{2+}(aq) + OH^-(aq) \qquad z = 0 \qquad (4.2.17c)$$

where (s) denotes an ion located in the surface layer of the passive oxide. Since Processes (4.2.15) are in equilibrium, the Galvani-potential difference, E_g, between the passive oxide and the solution can be expressed as

$$
\begin{aligned}
E_g &= E_{g0} + (RT/F) \ln [H^+(aq)/H^+(s)] \\
&= E_{g0'} + (RT/F) \ln [H^+(aq)/H^+(s)] \qquad (4.2.18) \\
&= E_{g0''} + (RT/F) \ln [O^{2-}(oxide)/OH^-(aq)]
\end{aligned}
$$

From the pH dependence of the dissolution rate ($\partial \log j_a / \partial \text{pH} = -0.46$) was observed for the sulfuric acid solutions, it was concluded that the dissolution of passive film proceeds through Process (4.2.17) in which Step (4.2.17b) is rate controlling.

The following mechanism was suggested for the dissolution of passive nickel in alkaline solutions:

$$Ni^{2+}(oxide) + 2OH^-(s) \rightarrow NiO_2H^-(s) + H^+ \qquad z = 0 \qquad (4.2.19a)$$

$$NiO_2H^-(s) \rightarrow NiO_2H^-(aq) \qquad z = 1 \qquad (4.2.19b)$$

The pH-dependence of the dissolution rate is +0.5 for the rate-determining step (4.2.19b). Here, the sign is reversed compared to that for the acid solution. The experimental results obtained in the alkaline region fit the mechanistic prediction although the quantitative agreement was not perfect.

The steady-state anodic polarization curves of nickel for the sulfuric acid solutions at various pH were measured in the potential region extending from the passive to the transpassive region. The anodic current observed in the transpassive region may be regarded as the sum of the dissolution current of passive film, i_A, and the dissolution current of nickel, i_B, through the active patches in the passive film. A Tafel relationship was observed between the potential and the current i_B, and it was scarcely affected by the hydrogen ion concentration in solution. The results for the estimation of the Tafel constant are shown in Table 4.2.8. The mean value of the Tafel slope is 0.120 V/decade, which is close to that observed by Vetter and Arnold [360, 378]. The concentrations of sulfate ion and nickel ion, as well as that of hydrogen ion, do not affect the dissolution current i_B. However, the maximum value of i_B observed at the critical potential, E_{cp}, decreased with the increase in pH, according to the equation $\log j_B^{max} = -2.13 - 0.46$ pH, (A/cm^2), 25°C. The relation between the critical potential and pH can be expressed by the equation $E_{cp} \approx 1.68 - 0.08$ pH, (V), which is not affected by the temperature between 25 and 40°C. The value of this potential and its pH-dependence are in good agreement with those of the third arrest potential observed in the potential decay curve of anodized nickel electrode. It is concluded that the dissolution reaction in the transpassive region proceeds through the consecutive reaction scheme:

TABLE 4.2.8. Tafel Constant for the Dissolution of Overpassive Nickel in 0.5 M SO_4^{2-} Solutions at Different pH Values

pH	Tafel slope	(RT/F) X (∂ ln i/∂E)
0.45	0.123	0.480
1.55	0.129	0.458
1.80	0.122	0.484
2.10	0.125	0.472
2.80	0.131	0.451
3.10	0.131	0.451
4.01	0.125	0.472
4.10	0.121	0.488

$$Ni + H_2O(aq) \rightleftarrows NiOH(ads) + H^+ + e \tag{4.2.20a}$$

$$NiOH(ads) \rightleftarrows NiOH^+ + e \tag{4.2.20b}$$

in which Step (4.2.20a) is rate controlling.

The effect of temperature on the steady anodic current of passive nickel was investigated by Ishikawa and Okamoto [380] in 5% H_2SO_4 solutions. The Arrhenius plot gave 21.2 kcal/mole at +0.6 V (N_2-saturation) by steadily raising the temperature at a constant rate of 12.7°C/min. However, the effect of temperature on the dissolution of passive metal appears to vary with a change of the rate of temperature rise when the change in the surface condition induced by the latter is appreciable.

DiBari and Petrocelli [319] studied the effect of composition (presence of C, Si, S, Se, Te, and P) and structure on the electrochemical reactivity of nickel in terms of the parameters of the anodic polarization curve as determined potentiostatically. It was concluded that small amounts of those elements affect the reactivity of nickel to a greater extent than large changes in structure. They increase reactivity and inhibit passivity significantly. The cold working of Ni (electronickel) interfered with its tendency to become passive. Cold work would be expected to increase the number of lattice imperfections which interferes with the readiness of passivating film formations. Imperfections will influence the initiation and growth of the film. Additives especially affect the value of the primary passivating current. At equal atomic concentration the order of effectiveness is S > Se > Te > P > C > Si. The mechanism of passivity involves the formation of a good conducting film as suggested previously [381, 382] rather than the presence of either adsorbed or absorbed oxygen on the metal [383]. The idea of an oxygen monolayer related to the beginning of passivity has been postulated by Trusov, Gochalieva, and Novakovskii [384].

One important contribution to the mechanism of passivity of nickel was made by Bockris, Reddy, and Rao [330] who investigated the reaction by ellipsometry to solve the basic cause of nickel passivity, ascertaining whether a tridimensional oxide film or a two-dimensional chemisorbed oxygen film is involved. Experiments were performed at 20°C in 0.01 N H_2SO_4 + 0.5 M K_2SO_4 at pH 3.15 under N_2 with 99.5% Ni. At -0.250 V (vs NHE) nickel is film-free since optical constants calculated from the relative phase retardation Δ and relative amplitude diminution χ coincide with those found in the literature for the pure metal. Within the range -0.250 to -0.025 V, current/potential curves indicate anodic metal dissolution without film formation. A sudden large change in χ and Δ within a narrow potential range between -0.025 and 0.000 V is the ellipsometric evidence of film formation. Ellipsometry indicates the film thickness is not less than 45 Å. Coulometric data yields a film thickness of about 60 Å. The onset of film formation is not accompanied by any decrease in the dissolution current. The inescapable conclusion is that the film formed at 0.000 V does not cause passivity per se. It can only be a precursor or prepassive film. Nickel dissolves anodically to form a soluble intermediate, $NiOH^+$, which accumulates near the electrode until there is precipitation of $Ni(OH)_2$. The estimation of the potential at which film formation takes places gives +0.048 V, a figure which is close, within experimental error, to the observed value of 0.000 V.

The refractive index n and the absorption coefficient κ of the film show two marked qualitative changes, the first of which is associated with film formation. The second of these two changes occurs at the passivation potential. The decreasing trend of κ, which began at the film formation potential, is reversed and κ begins to increase. Thus the changes in κ_F are a direct measure of the potential variation of electrical conductivity at the optical frequencies. It decreases to the passivation potential and immediately afterward an increase in the conduction is observed. The final values of the electrical conductivity are of the order of those observed with semiconductors at optical frequency. This indicates that the passivity process is associated with the formation of a semiconducting film.

The changes of n from an initial precursor film of $Ni(OH)_2$ to the final semiconducting passive film probably arise from the anodic oxidation of $Ni(OH)_2$. Some information concerning the nature of the passive film can be obtained by comparing the n calculated theoretically for various types of oxides with those observed experimentally from ellipsometry. It was found that the refractive indices of the compounds $NiO_{1.5}$-$NiO_{1.7}$ varied from 3.3 to 3.7. The experimental n is 3.5. Therefore the probable composition of the passive layer is $NiO_{1.5-1.7}$.

The conversion of $Ni(OH)_2$ to nonstoichiometric oxides has been studied by Briggs and Wynne-Jones [385] by means of X-ray diffraction and electrochemical measurements; they suggested that the final product is $NiO_{1.7}$, in agreement with ellipsometric determinations. The existence of a semiconducting passive film does not preclude the possibility of oxygen chemisorption on its surface, which can occur at ~1.5 V. During the galvanostatic reduction, the charge used indicates that only chemisorbed oxygen on top of the passive oxide is reduced before the potential falls to that corresponding to hydrogen evolution; the passive film is not reduced under such conditions. After the galvanostatic reduction, the ellipsometer still reveals a film-covered surface in the presence of hydrogen evolution. Therefore, according to the authors [330], the essential cause of passivity is the conversion in the multilayer from ionic to electronic conduction at the passivation potential, probably by the introduction of nonstoichiometry.

Ebersbach, Schwabe, and Ritter [386-388] investigated nickel passivation in sulfuric acid at various concentrations and temperatures. The kinetics of the passivation process in aqueous solutions can be regarded as a competition between anodic oxide formation with water and the reactions that remove the oxide. Due to the low exchange current density for active dissolution, this competition is favored. Experimental current/potential curves for the passivation process can be matched with calculated curves when correct parameters are chosen. The computations show, moreover, that no normal passivation can occur at negative potentials when the oxygen passive layer is rapidly removed. In this case salt layers formed directly on the metal can impede the anodic current to a considerable extent. The following reaction schemes were suggested for active dissolution, passivation, and reactivation processes:

$$Ni + H_2O \rightleftharpoons NiOH + H^+ + e \qquad (4.2.21a)$$

$$Ni + H_2O \longrightarrow NiOH^+ + H^+ + 2e \qquad (4.2.21b)$$

$$NiOH^+ + H^+ \rightleftharpoons Ni^{2+} + H_2O \qquad (4.2.21c)$$

$$Ni + H_2O \leftrightarrows NiOH + H^+ + e \qquad (4.2.22a)$$

$$NiOH + H_2O \longrightarrow Ni(OH)_2 + H^+ + e \qquad (4.2.22b)$$

$$Ni(OH)_2 \rightleftharpoons NiO + H_2O \qquad (4.2.22c)$$

$$NiO + 2H^+ \rightleftharpoons Ni^{2+} + H_2O \qquad (4.2.23a)$$

$$H_2O + NiO + 2A^- \longrightarrow NiA_2 + 2OH^- \qquad (4.2.23b)$$

where A^- represents an anion in solution. Ebersbach, Schwabe, and Koenig [389] extended the previous theory to study the anodic current/time curves at constant potential using a rotating nickel disk electrode in 0.5 M H_2SO_4 and 1 M $HClO_4$. No dependence of the rate of passivation upon the rate of stirring or solubility of the nickel salts was found. It was dependent upon the primary formation of a salt layer during passivation. At constant potential the current density fell exponentially with time in the passivation region and the slopes of the (log j)/t straight line increased exponentially with potential. Both dependences can be explained in terms of the occurrence of active dissolution simultaneously with the passivation reaction, the latter comprising a simultaneous two electron transfer step.

Kesten and Feller [390] used 1 N H_2SO_4 (N_2-, air- and H_2-saturated) to study the adsorption effect of sulfate ions on the anodic behavior of Ni (99.998%) at 1 V/hr. At -0.10 to +0.15 V (NHE) a current density decrease was observed, but at $E > 0.15$ V a considerable increase occurred. Responsible for the activation on the surface in this region were the sulfide ions formed on the surface by direct reaction of the chemisorbed sulfate ions with the metal surface atoms. The formation of NiS has been demonstrated. Depending on the concentration of the ions on the surface, passivation is delayed until 0.5 V. Polarization of the {110} and {111} single crystal planes shows that acid sulfate and hydrogen sulfide ion are adsorbed differently on these planes.

The nickel passivity caused by layers of salts in sulfuric acid was considered by Gilli, Borea, Zucchi, and Trabanelli [391] in 1 to 18 M H_2SO_4 by means of anodic polarization curves. A progressive increase in the passive current was observed in solutions of H_2SO_4 from 1.0 up to 10 M. At this concentration the anodic dissolution becomes independent of the potential and is determined by diffusion across a saturated layer of nickel sulfate. At concentrations higher than 10 M the gradual decrease of solubility of the salt results in its precipitation as a layer on the electrode. This film becomes more compact and less conducting as the concentration of acid rises, causing pseudo-passivation by a layer of salts. This film has been identified by X-ray diffraction as monoclinic β-$NiSO_4 \cdot 6H_2O$.

The anodic behavior of passive nickel in phosphate solutions at different pH, at different phosphate concentrations at constant pH, and in phosphoric acid from 1 to 14.5 M was investigated by Zucchi and Trabanelli [392]. The passivation occurs by successive steps and the potential decay curves show two arrests which are linearly related to pH with a slope of 60 mV/pH. An oxide of higher oxygen content, probably Ni_2O_3, was assumed in the stable potential region.

The effect of halide ions on nickel passivation has been investigated by different authors. Truempler and Keller [393] have shown that chloride ions increased the passive and critical current densities. In addition, they observed that chloride ions displaced the primary passivation potential toward more noble values. Evans [394] postulated that the chloride ions will be strongly adsorbed on the metal. The adsorbed chloride ions permit the metallic cations to pass into the solution with little activation energy. If a film is formed in the presence of chloride ions, it will be less protective than in the absence of chloride ions. This theory explains the acceleration of nickel corrosion by chloride ions. Piron and Nobe [395] investigated the effect of chloride ions in 1 N H_2SO_4. Passivation could not be achieved above 0.5% NaCl concentration. Below that concentration, the passive current density increased with increasing chloride concentration, as did the primary passivation potential and critical current density. The corrosion potential and the open circuit differential capacitance of nickel decreased and increased, respectively, with an increase in chloride ion, in reasonable agreement with Truempler and Keller [393].

The nickel passivity in H_2SO_4, HNO_3, and H_3PO_4 solutions and mixtures of the three acids with HCL was studied by Tousek [396]. An oxide film is formed depending on the electrode potential. At 1 V (SCE) in sulfate and chloride ion solutions the dissolution provokes a punctual attack which is not observed in the presence of either phosphate or nitrate

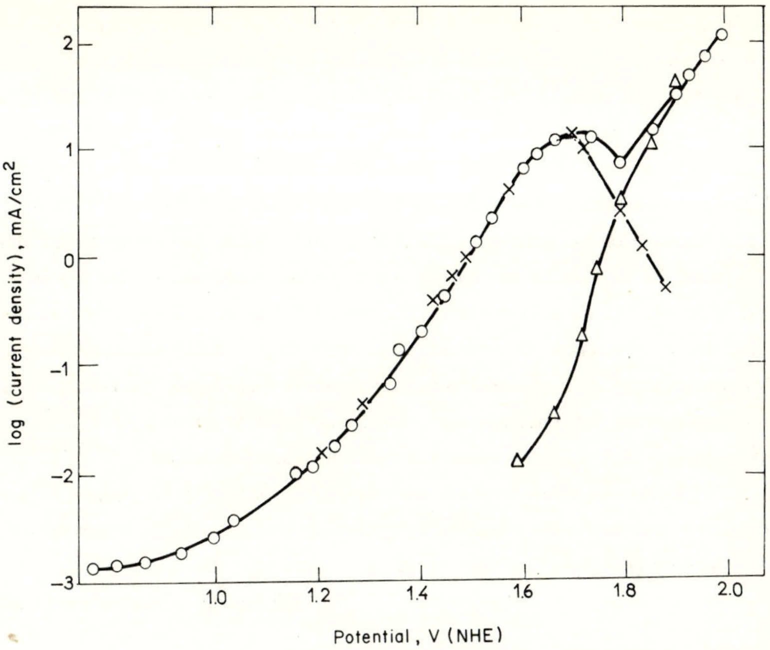

FIG. 4.2.4. Semilogarithmic plot of a stationary current density/potential curve: (O) total, (X) dissolution, and (Δ) oxygen evolution [398]. (By permission of Pergamon Press.)

ions. The pitting corrosion region is related to the secondary passivity portion of the current/potential curve where oxygen adsorption on the electrode surface probably takes place.

Ammar and Darwish [397-399] studied the passivity of Ni (containing traces of Si and Cu, 0.007% Fe, 0.01% C, and <0.04% Mg) under potentiostatic conditions in 1.0 N H_2SO_4 at 25°C in the potential range 0.76 to 1.20 V (NHE). Results show a potential maximum at 1.720 V (Fig. 4.2.4). Evidence is given for a pore-free passivating oxide which undergoes oxidation from NiO to NiO_2 with the rise in potential. Nickelous ion has no effect on the kinetics. Boric acid causes a slight decrease of the stationary current with increase of concentration; oxygen and $KMnO_4$ shift the position of the maximum. A critical passivating concentration of 10^{-2} M $KMnO_4$ is observed. KNO_3 has no effect. The postulated mechanism for passive film dissolution is:

$$Ni^{2+}_{metal} \rightleftharpoons Ni^{2+}_{oxide,\ inner} \tag{4.2.24a}$$

$$Ni^{2+}_{oxide,\ inner} \longrightarrow Ni^{2+}_{oxide,\ outer} \tag{4.2.24b}$$

$$Ni^{2+}_{oxide,\ outer} \longrightarrow Ni^{2+}_{solution} \tag{4.2.24c}$$

$$H_2O_{solution} \rightleftharpoons O^{2-}_{oxide,\ outer} + 2H^{+}_{solution} \tag{4.2.24d}$$

The first reversible reaction is responsible for the potential at the metal/oxide interphase. The composition of the solid passive layer is estimated as 20% NiO_2 and 80% NiO.

Postlethwaite [400] investigated the breakdown of nickel passivity at 25 to 275°C in either nitrogen- or argon-saturated alkaline chloride solutions (0.001 to 5 N NaOH and NaCl up to saturation). The critical $[Cl^-]/[OH^-]$ ratio required for passivity breakdown increases with hydroxyl ion concentration. The induction time for passivity breakdown is between 0.1 and 2000 min and depends on the potential, the $[Cl^-]/[OH^-]$ ratio, and the hydroxyl ion concentration. Up to 175°C a temperature increase stimulates metal attack, but above this value the trend is reversed, the critical potential moves to higher values, and both the induction time and the critical chloride concentration are raised. Passivation breakdown is accompanied by the precipitation of nickelous hydroxide over the whole temperature range, but the attack changes from the characteristic pitting at 25°C to more general attack at elevated temperatures where metal is removed over an expanding area.

Gressman [414] found that the activation time for nickel passivity decreases with increasing chloride ion concentration, increasing potential, and increasing temperature, but decreases with increasing sulfate ion concentration. Up to pH 2 the activation time increases with increasing pH, but decreases from pH 2 to 7. The high energy of activation is related to

$$Ni\text{-}O_{ad} + 2H^+ + 2Cl^- \rightleftarrows Ni^{2+} + 2Cl^- + H_2O \qquad (4.2.25)$$

Postlethwaite and Freese [402] made a potentiokinetic study [dE/dt = 0.95 V/hr, potential sweep (vs NHE) -0.5 to +1.2 V] of the effects of sodium chloride, sodium bromide, and sodium iodide in the range 0.0001 to 0.1 M on the anodic behavior of nickel in deaerated 0.1 N H_2SO_4 at both active and passive regions. All the halides lowered the corrosion potential, raised the primary passivation potential, and increased the critical passivation current density.

The effect of bromide ion on nickel passivity was considered by Ammar, Darwish, and Riad [403] in 1 N H_2SO_4 solutions containing 10^{-2} to 6×10^{-2} N KBr with the anode rotated at 600 rpm. Activation begins at potentials below the discharge potential of the halide ion and is primarily initiated by ions. Discharge of molecular halogen leads to attack by direct chemical reaction with NiO and causes the appearance of a second maximum in the potential/log (current density) relationship.

The anodic passivation in chloride solutions at pH 4.65, containing small amounts of hydrogen sulfide, was investigated by Chatfield and Shreir [404]. Passivation is due to the formation of a NiS film, which has been confirmed by X-ray diffraction. At low hydrogen sulfide concentrations the HS^- is the dominant species in the formation of NiS, whereas at higher concentrations the S^{2-} becomes dominant:

$$HS^- + Ni^{2+} \rightleftarrows NiS + H^+ \qquad (4.2.26)$$

$$S^{2-} + Ni^{2+} \rightleftarrows NiS \qquad (4.2.27)$$

Many authors investigated the transpassivity of nickel under quasi-steady conditions, principally in sulfuric acid solutions [308, 378, 405, 406]. Results obtained under potentiostatic conditions are not easily comparable to those obtained at equilibrium because the former are strongly influenced by the potential sweep rate [407, 408]. The transpassivity zone of nickel is characterized by a current maximum, i_M, followed by a minimum, i_m, indicating secondary passivity. Petit [409] has studied the influence of temperature, pH, speed, and direction of polarization on the potential E_M and E_m, corresponding to i_M and i_m, with pure nickel and H_2SO_4 + Na_2SO_4 solutions at constant sulfate ion concentration and 20°C. Under a potentiostatic regime, E_M is related to pH by $E_M = E_0 - KpH$. E_0 and K depend on the polarization rate. With a potential scan experiment, the phenomena are irreversible; the variations of E_M, E_m, I_M, and I_m with polarization speed are linked with the direction in which the polarization is changed; with increasing potential the variations are continuous with polarization speed, whereas with decreasing potential a change of slope or a maximum appears at a polarization speed between 0.100 and 0.250 V/min. With increasing potential the activation energy is a continuous function of polarization speed, whereas with decreasing potential it shows a minimum at 0.250 V/min. Impedance measurements at nickel electrodes during a potential scan experiment showed that the reactions occur in several stages.

Jouanneau and Petit [410] investigated the same reaction in the potential range between 0.8 and 1.8 V (SCE). The potential scan curve shows a maximum and considerable hysteresis. The polarization parameters depend on the degree of surface coverage by the adsorbed intermediate species.

Desestrat, Epelboin, Froment, Keddam, and Morel [407] investigated, by means of potential scan current/potential curves, secondary passivity of nickel and stainless steels. Results reveal the preponderant influence of carbon on secondary passivity of differently prepared nickel samples whose carbon content varied between 0.002% and 0.12%. The mechanism proposed to explain the effect is based on the presence in the anode layer of carbon or silica favoring formation of materials that limit the diffusion of the reacting substances to sites of formation of an oxide of spinel structure.

4.2.1.3. Pitting Corrosion. Various authors dealt with pitting corrosion of pure nickel [359, 368, 383, 399, 400, 401, 411-416]. The three main factors to be considered for the initiation of pits are: 1) the ease of dissolution (reactivity) of the metal itself; 2) the inhibiting action of the passive film, and 3) the adsorbing or penetrating ability of the aggressive ions. Since the last factor is dependent on the nature of the passive film, being large if the film is thin or unstable and small if the film is thick or stable, only the first two factors are in fact relevant to pitting. Microscopic observations indicate that scratches and grain boundaries are the nucleation sites for pitting. The active sites for pitting are closely related to the active sites for active dissolution.

Ammar and Darwish [397-399] investigated the pitting corrosion of a passivated nickel electrode rotated at 600 rpm in solutions of 1.0 N H_2SO_4 + (10^{-3} to 6 X 10^{-2} N) HCl at 25°C. Activation occurs after an induction period that depends on chloride ion concentration and potential. A very small effect for 10^{-3} N HCl on the passive behavior of nickel over a period of 24 hr is observed, but the attack is vigorous above 6 X 10^{-2} N. An induction period is noticed in the range 1.0 to 1.6 V. The mechanism of pitting is based on the effect of both chloride ion and chlorine. Chlorine formation explains the occurrence of a second maximum in the current/potential relation.

Garz, Worch, and Schatt [411] and Schatt and Worch [417] investigated the nucleation sites of pits on nickel single crystals (99.4 and 99.93% purity) in 0.5 M $NiCl_2$. On polarizing nickel surfaces they found the pitting potentials, E_p, of specific crystal surface were such that $E_p\{100\} < E_p\{110\} < E_p\{111\}$. The effect was attributed to the relative adsorption ability of the chloride ion. The pits were formed mainly at subboundaries and this was correlated with the existence of dislocations at the subboundaries. The active sites will be formed preferentially at subboundary branchings and to a lesser degree within subgrains. The effect of impurities on the current/potential curves and the positions of the breakdown potentials were determined.

Tokuda and Ives [418] made microscopic observations in situ on the pitting corrosion of polycrystalline and single crystal nickel, at a passive potential, in 1 N H_2SO_4 solution containing chloride ions. Corrosion pits nucleate at scratches and grain boundaries. Comparison of anodic currents in the Tafel regions suggests the following order of reactivities, R_A, for active dissolution:

$$R_A[\text{mechanical polycrystalline}] > R_A[\text{mechanical}\{111\}] > R_A[\text{electropolished}\{110\}] >$$

$$R_A[\text{electropolished}\{100\}] > R_A[\text{electropolished}\{111\}]$$

The order of electropolished surfaces agrees with the commonly accepted concept that more densely-packed crystal planes are less reactive. The passivation potentials, E_F, are ordered as follows:

$$E_F[\text{mechanical or electropolished}\{111\}] \leq E_F[\text{electropolished}\{110\}] <$$

$$E_F[\text{electropolished}\{100\}] < E_F[\text{mechanical polycrystalline}]$$

The pitting susceptibilities, S_p, measured from the relative anodic current, are arranged as follows:

$$S_p[\text{mechanical polycrystalline}] > S_p[\text{mechanical}\{111\}] > S_p[\text{electropolished}\{100\}] >$$

$$S_p[\text{electropolished}\{110\}] > S_p[\text{electropolished}\{111\}]$$

4.2.2. Hydrogen Fluoride and Hydrogen Fluoride-Water Solution

4.2.2.1. Dissolution of Nickel. Nickel anodes have been used in electrochemical fluorination. Watanabe and Chang [419] investigated the kinetics of the anodic behavior of Ni (99.9%) in HF-H_2O system (1, 10, and 53 wt% HF) at 0°C using a hydrogen reference electrode. The rest potential change consisted of a fast change into a nonnoble potential and a subsequent slow change into a noble equilibrium potential after 30 min. Nickel metal placed in HF solution undergoes the following reaction:

$$Ni(s) + 2HF(l) = NiF_2(s) + H_2(g) \qquad (4.2.28)$$

($\Delta G^\circ = -17.8$ kcal/mole), which consists of two elementary processes as follows:

Anodic process:

$$Ni = Ni^{2+} + 2e^- \qquad (4.2.29a)$$

$$Ni^{2+} + 2F^- = NiF_2 \qquad (4.2.29b)$$

Cathodic process:

$$2H^+ + 2e^- = H_2 \qquad (4.2.29c)$$

Polarization curves measured by the potential sweep method at 0.125 V/min exhibit in the potential range from -0.2 to 1 V two anodic peaks which depend on the nickel immersion time and their current ratio depends on water content. Below 0.2 V, currents are mainly due to the dissolution of nickel according to Reaction (4.2.29a). Furthermore, polarization curves obtained by decreasing the anode potential showed that electrode processes became more reversible as the water concentration was increased. It is concluded that the nickel anode is passivated at the potential of the peak current located at a more positive value. Polarization curves for the dissolution process obtained by means of rapid and steady-state polarizations show Tafel slopes of 0.059 ± 0.003 and 0.056 ± 0.002 V/decade for the former in 90 and 47% solutions, respectively. The Tafel slope for stationary conditions approaches a slope which is nearly one half of that for the rapid polarization. The Tafel equation is not obeyed by nickel electrodes in 99% HF solution. Results indicate that either water or the hydroxyl ion in these solutions seem to play a part in nickel dissolution, according to the Heusler and Gaiser mechanism [312].

4.2.2.2. Passivity. Hackerman, Snavely, and Fiel [420] showed that a trace of water in liquid hydrogen fluoride results in the formation of a passive film on an anodically polarized nickel electrode, and that a porous NiF_2 film formed on the electrode surface strongly sorbs large amounts of fluorine at fluorine evolution potentials. Donohue and Zletz [421] have concluded from studies on the electrolysis of wet hydrogen fluoride with a Ni/NiF_2 anode that oxygen difluoride, a principal product, is formed by a reaction involving fluorine derived from the anode, and that water in hydrogen fluoride solution passivates the Ni/NiF_2 anode by making the NiF_2 conductive, probably by stabilizing the formation of higher valent nickel species. Nickel is anodically passivated in liquid HF containing 0.01 wt% H_2O with the formation of a thin, tightly adhering NiF_2 film [420]. In anhydrous acid the anode corrodes rapidly at high positive potentials with simultaneous evolution of fluorine and formation of a thick NiF_2 film. A Tafel slope of 0.600 V/decade in the potential range of fluorine evolution was determined. They considered that the transfer coefficient α (equal to 0.917) could be a composite of α's for several reactions occurring by a growth of film on the nickel anode, and thus reach unusually large values. The current efficiency for anodic dissolution of passive nickel in anhydrous HF + 0.01 M NaF at -20°C was 75.6% (as NiF_2) at a potential of 6.0 V against a Hg/Hg_2F_2 reference electrode. The open circuit loss was 0.05 mg/(cm^2)(hr) [422]. Similar studies were conducted by Vijh [423] both on nickel and Monel.

Watanabe and Chang [419] studied passivity in hydrogen fluoride solutions. In 99% HF solutions the current/potential curves exhibit two peaks suggesting the formation of at least two kinds of nickel oxide films, as proposed by Sato and Okamoto [359]. However, there is a difference in the current/potential curves with concentration in going from 47 to 99% HF. The passivated nickel was rapidly and reversibly reactivated in the 47% HF solution, but slowly and irreversibly in the 90% HF solution. A partial recovery of current by the reactivation of the passivated nickel suggests that the anode is covered not only by the oxide but also by the fluoride. A mixed film of fluoride and oxide is probably formed on the nickel anode surface at potentials below fluorine evolution.

Tafel lines with a slope equal to 0.50 V/decade for a new nickel anode and 0.31 V/decade for a nickel anode which had been put in the HF + 1% H_2O solution at 5 V for several hours were found. The transfer coefficients were 0.89 and 0.82, respectively. The anodic overvoltage decay showed an arrest at the reversible potential of fluorine evolution (2.86 V). The potential decays linearly with the logarithm of time between 0.1 and several milliseconds. The average decay slope is 0.140 V/decade. Linearity is no longer obeyed at >7 msec. The decay slope seems more reasonable for the charge transfer process than the Tafel slope from stationary measurements. The average electrode capacitance is 1.9 F/cm^2 and was assigned to the surface film. It corresponds to an average film thickness of 200 Å if the surface roughness factor is 3 and the dielectric constant of the film is taken as 15.

4.2.3. Nonaqueous Solution

Piontelli, Bertocci, and Sternheim [424] studied the behavior of nickel in various nonaqueous systems using a $Ag/AgNO_3$ 0.5 M (pyridine) electrode. The results are shown in Table 4.2.9. Addition of water lowers the current efficiency for nickel dissolution. The presence of chloride ions actually makes the anodic attack easier. Nickel dissolves

TABLE 4.2.9. Anodic Behavior of Nickel Electrode in Pyridine Solutions [424]

Solution	j (mA/cm^2)	E^a (mV)	Current efficiency (%)
$HClO_4$ 1 M, 4.0 g H_2O/100 g Py	1	1150	
	10	1315	3
	100	1530	
HNO_3 0.3 M, 5.0 g H_2O/100 g Py	1	1120	
	10	1280	11
	100	1995	
$LiNO_3$ 1 M, 1.07 g H_2O/100 g Py	1	1200	
	10	1610	21
	100	1800	
HCl 0.24 M, 0.6 g H_2O/100 g Py	1	-20	
	10	+1100	36
	100	+1300	
LiCl 1 M, 0.5 g H_2O/100 g Py	1	850	
	10	1010	0
	100	1200	

[a]Versus Ag/0.5 M $AgNO_3$ in Py (pyridine).

spontaneously in hydrogen chloride solutions at an appreciable rate. The formation of layers of sparingly soluble $NiCl_2$ occurs at the anode, and stirring is very effective in lowering the overvoltage. A concurrent electrode reaction, involving chloride and the solvent, occurs at potentials corresponding to nickel attack. Water addition enhances passivity in hydrogen chloride solutions. At increasing current density, oscillatory phenomena (overvoltage maximum of the order of 2 V) are observed and then destruction of the layers occur.

Heitz [425] used a rotating disk to investigate the kinetics of the diffusion-influenced corrosion on homogeneous Ni (99.8%, containing Fe, Si, C, and S) surfaces in ethanolic solutions of HCl and H_2SO_4 in the presence of 0.1 M LiCl and oxygen by measuring stationary current/potential curves. The corrosion rate depends on hydrogen ion concentration, oxygen partial pressure, and rotation speed. The results are shown in Table 4.2.10.

Schwabe and Schmidt [426] studied water influence on nickel passivity by recording current/potential curves of Ni (99.9%) in dimethylformamide and acetonitrile solutions containing

TABLE 4.2.10. Corrosion of Rotating Nickel Disk Electrodes in 0.1 M LiCl in Ethanol [425]

Chemical agents	Limiting cases	Reaction order with respect to $[H^+]$	$[O_2]$	$[Cl^-]$
$HCl + LiCl + O_2$	$[H^+] \geq 10^{-1}$ M	0	0.75	(1.0)
	$[O_2] \geq 10^{-2}$ M		(kinetic)	
	$[H^+] > [O_2]$	0	0.8-1.0 (kinetic + electro-chemical)	1.0 (electro-chemical)
	$[O_2] > [H^+]$	1.0 (kinetic + electro-chemical)	0	(1.0)

1 M/kg H_2SO_4 and different concentrations of water. Passivity occurs in acetonitrile above 3% water content and in dimethylformamide above 0.2% water content. The critical passivity potential becomes more negative as the water content increases. The critical passivation current depends on water content and goes through a minimum. This is explained by the activation reaction

$$NiO_{ads} + H_2SO_4 + nH_2O \rightleftarrows [Ni^{2+}(n+1)H_2O] + SO_4^{2-} \qquad (4.2.30)$$

since nickelous ion hydration is faster than its solvation with dimethylformamide. The oxide solubility in this solvent and in acetonitrile is largely increased when water concentration increases.

The kinetics and mechanism of the nickel electrode in solution of hydrogen chloride in dimethylsulfoxide was investigated by Delgado, Posadas, and Arvía [51]. The anodic Tafel line at 25°C is close to 0.060 V/decade.

4.2.4. Fused Salts

Littlewood and Argent [427] measured the potential of Spec-pure nickel wires in NaCl-KCl melt at 700°C under argon. The potential was 50 mV higher than the value calculated by Hamer, Malmberg, and Rubin [53], suggesting complex formation. NiO was identified on the electrode, the origin of which should be ascribed to ionic corrosion products. There is an effect on the rate of potential change due to the area of the specimen, and the kinetics is largely determined by diffusion. No sign of passivity was found in a nitrate-free melt in contact with air. Amosse, Bouteillon, and Barbier [428] applied the potentiostatic double pulse method to the kinetic study of the Ni/Ni(II) system in molten LiCl-KCl at 450°C. The exchange current density is 597 mA/cm^2.

Pizzini, Morlotti, and Roemer [429] investigated the behavior of nickel in molten fluorides eutectics (NiF_2-KF, LiF-NaF-KF, NaF-KF, and LiF-KF) from 450 to 800°C. In oxide-free melts the Tafel law is followed in the current density range from 10^{-3} to 10^{-1} A/cm^2, with a slope equal to RT/2F, in good agreement with the prediction of a diffusion-controlled anodic dissolution process. In oxide-containing melts, nickel behaves as a "covering-layer electrode."

Nickel corrodes to NiO in ternary alkali carbonates eutectics at 600°C [55]. Janz and Conte [430] investigated the reaction in molten ternary Li_2CO_3-Na_2CO_3-K_2CO_3 eutectic from 600 to 700°C using a Ag/Ag^+ reference electrode. At 600°C, the corrosion potential is -1.166 V, and the anodic potentiostatic current/potential curve demonstrates metal passivation. A primary passive potential is found at -1.015 ± 5 V, but the protection conferred appears far from complete. The phenomenon of carbon deposition at the cathode is reported and examined in the light of free energy calculations. Drossbach and Sauerman [431] studied the influence of nitrogen, carbon dioxide, oxygen, methane, carbon monoxide, and hydrogen on nickel corrosion in molten ternary alkaline carbonates.

The corrosion of nickel was studied by the weight change after short immersion of the samples in alkali nitrate melts at temperatures from 400 to 500°C [432-434]. According to Piontelli, Bertocci, and Sternheim [424], nickel electrodes in fused nitrate baths ($NaNO_3$ + KNO_3 50% weight) at temperatures from the melting point up to 360°C show the spontaneous formation of a compact and uniform yellow layer of cubic NiO. The rate of oxide film formation increases with temperature. The stable potential is 400 to 550 mV nobler than the initial one. At a polarized nickel electrode, gas evolution occurs, but it corresponds to only a small fraction of the current, the main part corresponding to attack of the metal. The current efficiency, as referred to the formation of Ni(II), is of the order of 80 to 90%. Under anodic polarization a black layer is formed which contains "active oxygen," and it corresponds to a formula $NiO_{(1+n)}$, where $0 < n < 1$. The anode attack occurs, however, also by dissolution of the nickel into the ionic form which is then precipitated in the bath as black flaky NiO. The potential after current interruption is nobler than that of the spontaneously oxidized electrode. If chloride ions are present (more than 25%), strong gas evolution suddenly occurs. At this intermediate stage, polishing of the electrode occurs, leaving a very bright surface. From the residual current/voltage curves for solid nickel microelectrode in the eutectic at 250°C, chemical attack and an erratic trace on anodic polarization was observed.

Arvía, Podestá, and Piatti [435] investigated the onset of nickel passivity in molten $NaNO_3$-KNO_3 at 230 to 320°C. Once the metal is immersed in the melt its potential lies at about -0.3 V with respect to the Ag/Ag^+ (nitrate melt) 0.03 M reference electrode and steadily becomes more positive until it finally attains, after 4 hr at 270°C, a potential of -0.175 V. If the electrode is then anodically polarized, with an overpotential of 0.025 V for a few minutes, and the current is interrupted, the potential returns to -0.175 V. But after anodic polarization at about 0.5 V above the immersion potential, the rest potential after current interruption is located at about 0.15 V. The difference between the first and the second rest potentials is independent of temperature. The average thickness of nickel oxide involved at the onset of passivity, according to galvanostatic transients [435], is of the order of 11 Å. The spontaneous passivation, as deduced from the potential/pO^{2-} diagram [109, 110], is related to the overall reaction

$$2(Na,K)NO_3 + Ni = NiO + 2NO_2 + (Na,K)_2O \tag{4.2.31}$$

Potentiostatic current/potential curves exhibit passivation phenomena on a freshly polished nickel anode. When the potential reaches about 0.4 V above the passivation potential, a depolarization of the electrode occurs and a Tafel line region is approached with a slope close to RT/F. By increasing the potential still further, a second Tafel line portion occurs involving a slope which is close to 2RT/F. By running the potentiostatic curve in reverse, only the Tafel line regions and the depolarization region are observed. The rest potential attained is then located about 0.35 to 0.40 V more positive than the passivation potential. If the experiment is continued by running a second polarization curve with the same electrode starting from the rest potential, the curve is reproduced according to the description already given for the returning curve. Within the temperature range from 230 to 287°C the current density extrapolated at the passivation potential increases monotonously with temperature and it fits reasonably well an Arrhenius plot which comprises an experimental activation energy equal to 14 ± 4 kcal/mole. At high anodic potentials other processes take place, provoking a dissolution of the metal either under the NiO or nickelous ion. Nitrogen oxides are also formed. It was concluded that the nickel oxide related to the rest potential has, on the average, the composition $NiO_{1.36}$.

The same authors also investigated the passivation of nickel in molten alkali nitrite [436]. As the metal is immersed into the molten salt it acquires a passivate state which is characterized by the spontaneous shift of the electrode potential at null current from -0.5 to -0.30 V at 252°C. Simultaneously, its metallic brightness disappears. The average thickness of the passive film is equal to 10 ± 2 Å. The rest potential after electrolysis at high current density lies higher than the immersion potential, a fact which suggests the existence of a nickel oxide of higher oxygen content than NiO. This effect was interpreted in terms of an accumulation of oxygen in the NiO lattice. The probable composition of the nonstoichiometric compound occurs after the high anodic polarization is $NiO_{1.15}$. When a fresh electrode is anodically polarized, the current increases and polarization at 0.1 V above the immersion potential shows a maximum of the apparent current density. The initial portion of the polarization curve may be assigned to the active electrochemical dissolution of the nickel anode. The active region, however, covers a small extension as temperature increases. A further potential increase of ~0.2 V above the immersion potential provokes passivation of the metal. On polarizing the anode still further, the current increases again and, at 0.3 V above the potential where the passivation effect has occurred, a Tafel region exists with a slope close to RT/F at any temperature. At higher anodic potentials another break in the potential/log(current density) plot is observed. This break corresponds to another oxidation, probably of nitrate or oxide formed during the previous anodic process.

Insoluble NiO is formed at low current density, soluble nickel species are found in the melt, and no appreciable amounts of nitrogen oxides are observed. The anodic nickel corrosion yield is 75% of the total charge passed. Most of the dissolved metal appears as insoluble nickel oxide, and the weight ratio NiO/Ni^{2+} is about 12. At high current density the nickel anode undergoes attack, and gases are simultaneously evolved. These consist predominantly of a mixture of NO_2, NO, and N_2O at room temperature. Analysis of the melt after electrolysis yielded a small content of soluble nickel and a precipitate of nickel oxide. The polarization curves were interpreted as the sum of two processes, i.e., the corrosion of the passive film and the nitrite ion anodic oxidation which also provokes the dissolution of the passivating film.

Arvía and Videla [437] measured the galvanic behavior of nickel in molten $KHSO_4$ at 260 to 280°C. The rest potential vs a Pt/H_2(1 atm)/$KHSO_4$ electrode is -0.330 ± 0.025 V.

Krueger, Rahmel, and Schwenk [438] measured the following rest potentials for nickel in NaOH 99.8% (0.025% H_2O) at 500°C against a $Pt(O_2, H_2O/NaOH)$ electrode: -0.148 V (O_2-atmosphere) and -0.182 V (N_2-atmosphere).

4.3. ELECTROCRYSTALLIZATION KINETICS

4.3.1. Effects of Face Orientation

Piontelli, Poli, and Serravalle [439] and Piontelli, Lazzari, and Rivolta [440] studied the electrochemical behavior of $\{111\}$, $\{110\}$, and $\{100\}$ oriented nickel surfaces and observed, for 0.5 M $NiCl_2$ + 0.5 M H_3BO_3 + HCl solutions at 30°C, a slight influence of the electrode orientation on the open circuit potentials. The influence of the surface orientation on the nickel electrode reaction is shown in Table 4.3.1. Nickel cathodes exposing oriented single crystal surfaces to 1.0 M $NiCl_2$ + 0.33 M H_3BO_3 and 1.0 M $Ni(NH_2SO_3)_2$ + 0.33 M H_3BO_3 at pH > 2 fit the Tafel law, though sometimes two branches occur with different slopes. At 25°C the $\{100\}$ plane exhibited bright deposits oriented parallel to the base in both sulfamate and chloride baths at 1.3 < pH < 5.0 and pH 3.1, respectively. The $\{111\}$ plane deposits were rather nonoriented and had a milky appearance. The $\{110\}$ plane apparently had an intermediate behavior.

Anodic attack, especially in the high pH range, often leads to localized attack, giving etch figures, sometimes coexistent with a distributed attack. Deposition and, to a much greater extent, anodic attack depend on the existence of inclusions and defects. For instance, the anodic attack on a $\{110\}$ face, derived from a polygonized region of the crystal,

TABLE 4.3.1. Effect of Surface Orientation on Several Reactions at a Nickel Electrode

Reaction	Electrolyte	Order	Refs.
Hydrogen overpotential	0.1 M $HClO_4$	{100} < {111} ≤ {110}	[441]
Hydrogen overpotential	1 N H_2SO_4	{100} < {111}, {110}	[442]
Hydrogen overpotential	0.1–0.15 M $HClO_4$	{100} < {111} < {110}	[443], [444]
Hydrogen overpotential	0.1 M HCl	{111} < {100} < {110}	[443], [444]
Hydrogen overpotential	0.1 N HNH_2SO_3	{111} < {100} < {110}	[443], [444]
Hydrogen overpotential	0.1 M NaOH	{111} < {100} < {110}	[443], [444]
Ni^{2+} deposition overpotential	1.0 M $NiCl_2$ + 0.33 M H_3BO_3	{100} < {110} < {111}	[439]
Anodic overpotential	1.0 M $NiCl_2$ + 0.33 M H_3BO_3	{100} ≈ {110} < {111}	[439]
Anodic reactivity	Strongly alkaline	{111} < {100} < {110}	[366]
Anodic reactivity	1 N H_2SO_4	{111} < {100} < {110}	[442]

is exclusively localized on the polygon boundaries. Also, the source of the etch figures appears to correspond to defects or inclusions. On the {111} face, especially, the presence of more readily attacked planes causes the attack to be exclusively localized there, with an etch resulting, while the surrounding surface may remain unaffected. Chloride ions are especially conducive to the exploitation of any structural weakness of the surface, and the sulfamate bath (as well as sulfate, perchlorate, and others) acts just to the contrary.

Mieluch [442] found, for anodic {111}, {110}, and {100} orientations in 1 N H_2SO_4, the following sequence of anodic kinetic parameters at a scan rate of 250 mV/min: $b_a^{111} > b_a^{100} > b_a^{110}$ and $a_a^{111} > a_a^{100} > a_a^{110}$, where b_a is the Tafel slope and $a_a = -b_a \log j_0$. Under the same conditions the resistive and capacitive components were measured for both the active and passive regions in anodic dissolution by superimposing a modulated square wave component on a direct current (see Fig. 4.3.1). In the active region the following orders were established: $C_p^{111} \simeq C_p^{110} < C_p^{100}$ and $R_p^{100} \simeq R_p^{111} < R_p^{110}$, and in the passive region: $C_p^{110} < C_p^{111} < C_p^{100}$.

Conway and Bockris [445] calculated the activation energy for the process Ni^{2+}(double layer)$\rightleftarrows$Ni(lattice) in the different cases. The results are shown in Table 4.3.2. The last column corresponds to the values of the heat contents of the states in the course of a transfer reaction to a surface site (see also Pangarov [446]).

Garz, Worch, and Schatt [411] studied the {100}, {110}, and {111} planes during anodic polarization in 0.5 M $NiCl_2$ + HCl at pH 3.5. Three stages of dissolution were distinguished:

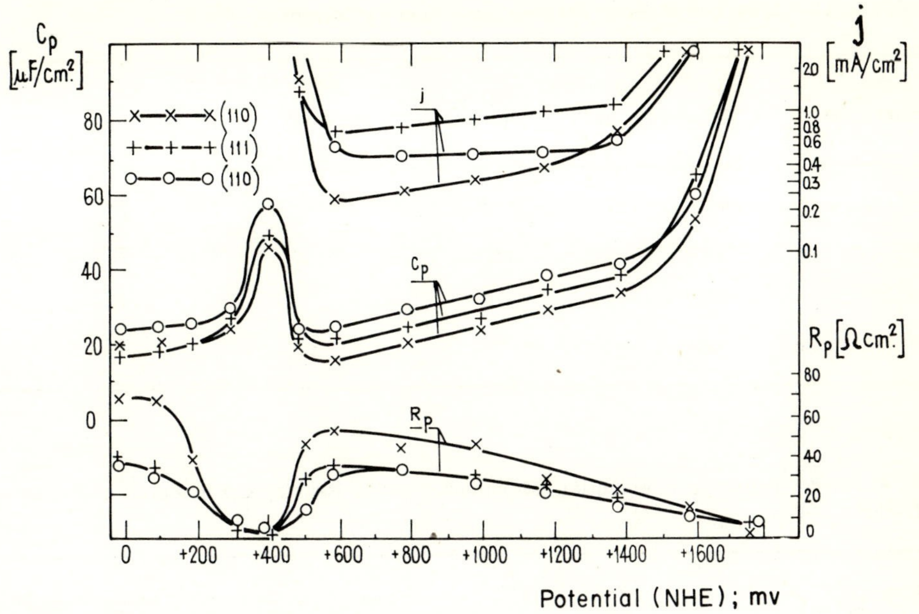

FIG. 4.3.1. Effect of the potential on the capacitance (C_p), resistance (R_p), and polarization current density (j) for three crystallographic orientations of nickel electrodes in 1 N H_2SO_4 at 20°C, obtained by the square wave technique [442]. (By permission of the Polish Academy of Sciences.)

TABLE 4.3.2. Activation Energies for Nickel Electrocrystallization [445]

Sites	ΔH* (kcal/mole)	ΔH_t (kcal/mole)
Planar surface	130	-15
Edge	190	-29
Kink	>190	-46
Superficial vacancy	≫190	-

1) uniform corrosion; 2) formation of a small but constant number of pits at the pitting potential E_p; and 3) a marked increase of the number of pits at the so-called "accumulation potential," E_{acc}. The three stages are characterized by typical current/potential relationships and the crystal structure is important in the nucleation of pits. Garz and Haefke [447] investigated the influence of metal structure on nickel passivation. The results are summarized in Tables 4.3.3 and 4.3.4.

TABLE 4.3.3. Influence of Metal Structure on Nickel Passivation [447]

Crystal face	E(i-decay) (mV)	E_p (SCE) (mV)	b_a (mV)	α	E_{acc} (SCE) (mV)
{100}	-220	-160	151	0.2	30
{110}	-160	-100	151	0.2	75
{111}	-130	-70	96	0.3	75
Polycrystalline	-160	-70	-	-	-

TABLE 4.3.4. Passivation Potential, Maximum Current Density and Potential of Passivity Breakdown [447]

Sample	E_p (SCE) (mV)	j_{max} ($\mu A/cm^2$)	E_{pb} (SCE) (mV)
Single crystal without deformation	-104 ± 4	203 ± 72	905 ± 10
Single crystal, 6% deformation	-125 ± 4	316 ± 39	880 ± 5
Single crystal, 19% deformation	-161 ± 4	520 ± 53	860 ± 5
Single crystal, 45% deformation	-170 ± 0	1567 ± 24	845 ± 5
Polycrystal, coarse	-120 ± 0	1280 ± 80	830 ± 10
Polycrystal, fine	-157 ± 0	1377 ± 240	750 ± 10

Vagopoliskaya [448] studied the effect of crystallographic orientation on the anodic oxidation of single crystals of nickel in 1 N H_2SO_4 at 25°C. The {111} plane exhibited the highest resistance to oxidation and the {110} the lowest. At 65°C the rate of oxidation is so rapid that the polarization curves are independent of the crystallographic orientation. The oxidizability of the planes decreased in the sequence: {110} > {100} > {111}, and can be correlated with the work functions of the same planes, 4.64, 4.84, and 5.22 eV, respectively. The adsorption of oxygen proceeds predominantly on the planes with the lowest energy state and consequently with the lowest work function.

Windfeldt [449] considered the effect of plastic flow on the anodic dissolution of Ni (99.9% purity) in acid $NiCl_2$ solutions. Plastic flow has a stimulating effect on the dissolution and increases with the ratio strain rate/dissolution rate. It disappears shortly after the deformation process is stopped. Results are in accordance with a model involving direct solutions at kinks in crystal steps on the surface.

4.3.2. Crystal Growth on Nickel Electrodeposits

Many types of textures of nickel electrodeposits have been reported. Glocker and Kaupp [450] found $\{100\}$ axes of preferred orientation; Finch and Sun [451] reported $\{110\}$ for thick deposits; Makarieva [452] detected $\{110\}$, $\{112\}$, $\{113\}$, and $\{100\}$ textures; Banerjee and Goswami [453, 454] claimed $\{100\}$, $\{110\}$, $\{112\}$, and $\{210\}$ and Schloetterer [455] obtained the $\{100\}$ and $\{111\}$ orientation. No substrate (single crystal, random-polycrystalline or fiber textured) influence was observed on the orientation during the final stage of growth [451].

Finch, Wilman, and Yang [456] studied the three stages of growth (epitaxial, medium, and final) for nickel deposition on different types of surfaces. The results are summarized in Tables 4.3.5 and 4.3.6. Wilman [457-459] suggested that the conditions of electrodeposition favor the preferential growth of those crystals which have two or more main growth-faces simultaneously perpendicular or nearly perpendicular to the substrate. The atom row at the common intersection of these faces becomes the orientation axis. The effect of pH and temperature on the orientation of nickel electrodeposits, according to Banarjee and Goswami [453, 454], are shown in Table 4.3.7. Above pH 5, deviations from the "free grow" trend occurred. A hexagonal orientation appears along with fcc $\{211\}$, and at higher temperatures a $\{100\}$ orientation is observed. Chloride ions generally counterbalance the outgrowth condition due to impurities and favor lateral growth. This was attributed to the desorption of the impurities which facilitate the electron transfer processes. The addition of hydrogen peroxide (1 g/l) to a sulfate-chloride bath changes the orientation from $\{210\}$ to $\{210\}$ + $\{100\}$, and the addition of nitrate ions (0.426 g/l) determines the shift from $\{210\}$ to $\{110\}$. Reddy and Rajagopalan [460, 461] attributed the change to nitrate ions being perhaps more easily reduced, thus removing atomic hydrogen through the reduction reaction. Banerjee and Walker [462] studied the orientation order under different experimental conditions, and the structure and topography of electrodeposits by electron microscopic techniques [463].

Banerjee [464] studied mild steel with nickel electrodeposits having $\{100\}$, $\{10\bar{1}0\}$ + $\{211\}$, $\{110\}$, and $\{210\}$ orientations. The $\{210\}$ orientation gave the best protection to iron in hydrogen chloride solution. On the other hand, in neutral sodium chloride solution no appreciable difference in the dissolution rate has been observed. These results have been discussed with reference to the precipitation of basic iron salts, the electrochemical anisotropy of different faces of a single crystal, and also the texture of the deposits.

Pangarov, Vitkova, and Uzunova [465] studied the influence of a polished platinum substrate on the degree of $\{110\}$ oriented nickel electrodeposits in relation to layer thickness and current density. Experimental data confirm the two-dimensional theory of preferred orientation, showing that the nucleus formation is the determining factor for the appearance

TABLE 4.3.5. Deposition on Single-Crystal Substrates

Substrate surface	Concentration of bath (g/l)		T (°C)	j (A/dm^2)	Time of deposit	Type of deposit crystals[a]
			Deposition on Cu Single Crystals			
{100}	$NiSO_4 \cdot 7H_2O$	60	50	0.2	30 sec	P, T, slight R
{100}	H_3BO_3	30	50	0.2	2.5 hr	P, faint K
			Deposition on Fe Single Crystals			
{110}	As above		50	0.2	Various	R always

[a]P = lattice parallel to substrate crystal lattice; T = twinning on octahedral planes; R = randomly disposed crystals; K = Kikuchi line pattern observed, showing smooth surface and perfect lattice.

TABLE 4.3.6. Orientation of Nickel Electrodeposits

Crystal system	Electrolytic solution (g/l)		T (°C)	j (A/dm^2)	Current efficiency (%)	Deposit thickness (10^3 Å)	Orientation of the deposit[a]
Thick deposits, plane of orientation perpendicular to the most densely-packed atom rows in the most densely-packed lattice plane.							
fcc	$NiSO_4 \cdot 7H_2O$	240	17	0.65	45	20-45	W {110}
	H_3BO_3	30	17	1.25	75	20-40	M {110}
Thick deposits whose plane of orientation is a less densely-packed lattice plane.							
fcc	$NiSO_4 \cdot 7H_2O$	240	55	1.25	80	50-120	S {100}
	H_3BO_3	30					
fcc	$NiCl_2 \cdot 6H_2O$	140	17	0.25	75	31-57	S {211} cubic
hcp	H_3BO_3	30	17	0.65	80	20-45	S {10$\bar{1}$0} hexagonal

[a]W = weak; M = medium, S = strong.

TABLE 4.3.7. Effect of pH and Temperature on Nickel Electrodeposits [453, 454]

pH	Temperature (°C)	Orientations
2	15	{210}
2	25	{210} + {100}
2	45	{100} + {110}
2	55-75	{110}
4	15	{210} + {100}
4	30	{100}
4	45	{100} + {110}
4	55-75	{110}

of a definite axis of crystal orientation in thick electrodeposits. Pangarov, Uvarov, and Vagramyan [466] obtained nickel electrodeposits from a 2.5 M $NiCl_2$ solution at pH 3 and 260°C which have a {111} orientation axis and the {111} plane parallel to the substrate. Rashkov and Pangarov [467] have studied the influence of pH (Tables 4.3.8 and 4.3.9), temperature, and overvoltage on the preferred orientation of electrodeposits from baths with 280 g/l $NiSO_4 \cdot 7H_2O$ with or without boric acid. The texture axes change in the sequence {110} → {110} + {100} → {100} → {100} + {111} as the temperature increases and the overvoltage decreases.

Pangarov [446], studying nickel deposits on copper substrates, found that crystal size and orientation gradually departs from that of the substrate until at deposit depths of 30,000 Å, or generally less, the deposit has a crystal size and orientation dependent on the deposition conditions. Electron microscopy observations of thin epitaxial deposits <1500 Å on {100} and {111} faces of copper single crystal substrates made by Thompson and Lawless [468] revealed that nickel films become continuous quite rapidly (<<50 Å), suggesting that the stresses in the thin epitaxial electrodeposited films result from the equilibrium misfit strain between the deposit and its substrate at the time of nuclei coalescence. Bozorth's theory [469] (see Pangarov [470]) suggests that the orientation is associated with the strain developed during deposition.

Evans [471] found that the {110} axis of orientation is associated with the higher values of internal stresses and hardness while {100} was associated with the lower values. Ives, Edington, and Rothwell [472] studied by transmission electron microscopy nickel films electrodeposited from a Watts bath at 1 mA/cm^2 onto vapor-deposited {100} copper films. The nickel films are continuous but have a nonuniform thickness, some regions being up to 5 times

TABLE 4.3.8. Influence of Temperature and Overvoltage on Preferred Orientations [467]

	1 A/dm^2		4 A/dm^2		6 A/dm^2	
Temp (°C)	Overvoltage (V)	Orientation axis	Overvoltage (V)	Orientation axis	Overvoltage (V)	Orientation axis
15	-	-	0.570	{110}	0.590	{110}
25	0.450	{110}	0.505	{110}	0.522	{110}
35	0.410	{110} + {100}	0.458	{110}	0.472	{110}
45	0.366	{100}	0.405	{100} + {110}	0.420	{110}
60	0.300	{100} + {111}	0.346	{100}	0.360	{100}
70	0.262	{100} + {111}	0.303	{100} + {111}	0.316	{100} + {111}

TABLE 4.3.9. Influence of pH on Preferred Orientations [467]

	pH 5.5			pH 4		
Temp (°C)	Current density (A/dm^2)	Overvoltage (V)	Orientation axis	Current density (A/dm^2)	Overvoltage (V)	Orientation axis
40	1	0.330	{100} + {111}	1	0.332	{100} + {111}
40	2	0.365	{110}	2	0.357	{100} + {110}
40	4	0.392	{110}	4	0.380	{110}
40	6	0.410	{110}	6	0.398	{110}
70	1	0.245	{100} + {111}	1	0.220	{100} + {110}
70	2	0.274	{100} + {111}	2	0.240	{100} + {110}
70	4	0.300	{100} + {111}	4	0.266	{100} + {111}
70	6	0.320	{100} + {111}	6	0.280	{100} + {111}

thicker than the average. Small pH changes during deposition result in the formation of triangular single crystals of $Ni(OH)_2$ with the orientation relationship: $\{01\bar{1}1\}$ $Ni(OH)_2$ parallel to {100} Ni and $\{\bar{1}\bar{1}23\}$ $Ni(OH)_2$ parallel to {310} Ni. Large increases in pH give rise to the formation of amorphous $Ni(OH)_2$.

Jones and Kenez [473] reported results on the initial stages of nickel deposition from the Watts solution, particularly those related to crystal growth phenomena. The primary sites

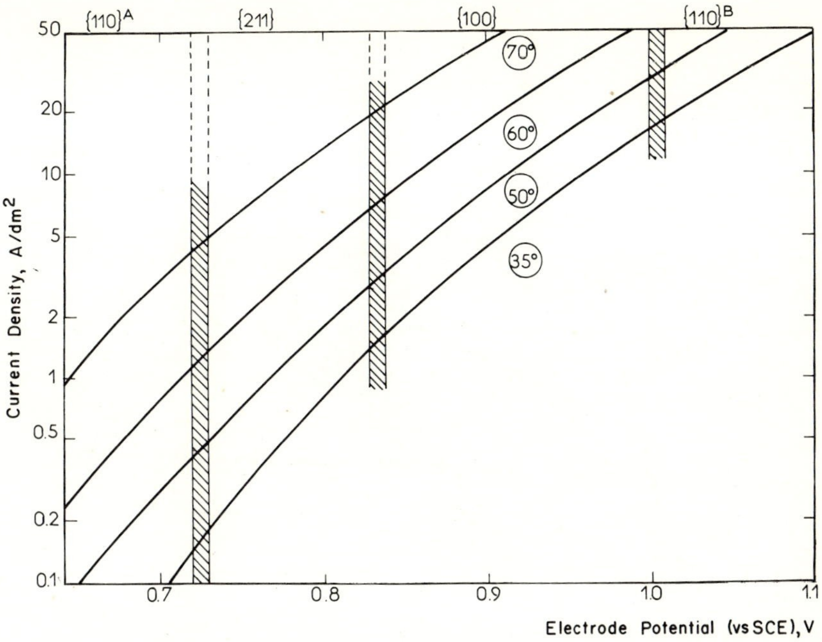

FIG. 4.3.2. Current/potential curves for the cathodic deposition of nickel. Shaded areas correspond to different textures [474]. (By permission of the American Electroplaters Society.)

of nucleation are grain boundaries and defect locations within the grains. Deposition occurred preferentially on grains with orientations close to the $\{111\}$ index plane. Growth from the nucleation phase was found to be mainly in the lateral direction.

Epelboin and Froment [474], Froment, Maurin, and Thévenin [475], and Froment and Maurin [476] carried out systematic studies on the axis of texture for nickel in terms of the various electrolyte parameters. The influence of temperature and current density is shown in Figs. 4.3.2 and 4.3.3 for a Watts-type bath at pH 4.5. By varying the cathodic potential and pH, they obtained a diagram of the Pourbaix type with five areas corresponding to different axis textures. Froment, Maurin, and Thévenin [477] and Epelboin and Froment [474] observed a change of texture by the addition of 2-butyne-1, 4-diol, but in some cases it produced a reduction of the size grains and an increase of the reflecting power of the deposits. The effect occurred only with deposits having a determined texture.

Froment, Maurin, and Thévenin [478] observed two phases during the epitaxial growth of the electrolytic nickel deposits: 1) epitaxial growth, for a monocrystal the initial deposit is monocrystalline and of the same orientation; 2) polycrystalline growth with a fiber texture where the axis depends only on the electrolysis conditions.

Reddy [479] and Reddy and Rajagopalan [460, 461] have developed a theory of preferred orientations on nickel, called "geometrical selection theory" by Pangarov [470]. The

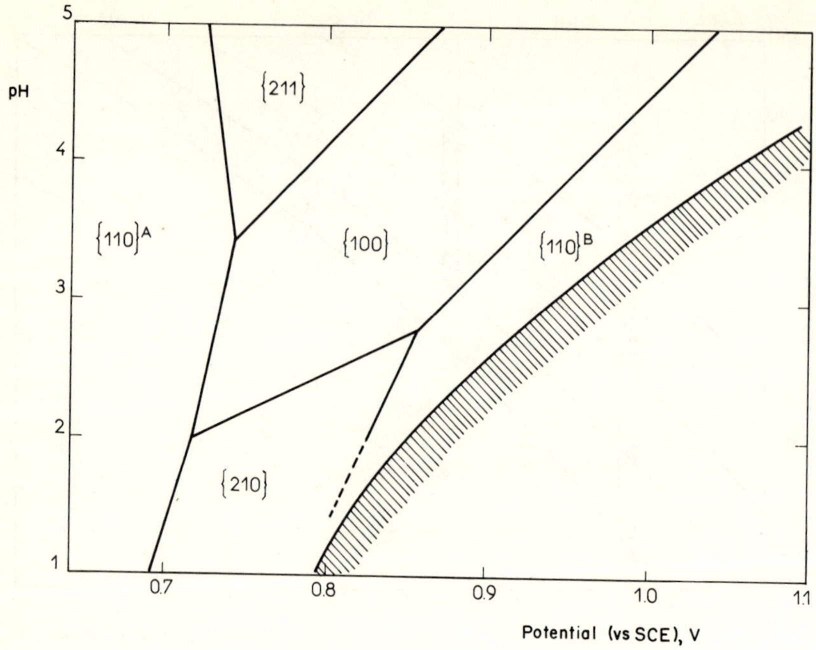

FIG. 4.3.3. Potential/pH curve for the cathodic deposition of nickel showing areas of different textures [474]. (By permission of the American Electroplaters Society.)

electrodeposition mechanism with unadulterated bath comprises: 1) the preferential formation of facets of a particular type, and 2) the alignment of these facets normal to the substrate. The result of these two growth processes is the development of a preferred orientation in thick electrodeposits. From data available, the following systematic changes in textures were pointed out [479]:

Change in deposition conditions	Trend of change of preferred orientations
Increase of temperature	Change toward free growth
Increase of current density	Slight change toward free growth
Addition of oxidizing agents	Change toward free growth

The variations in texture refer to those observed when a Watts-type bath has been used. However, similar trends were also obtained with a pure chloride bath. It has been suggested that an increase in bath temperature leads to a decrease in the average surface concentration of adsorbed hydrogen atoms, and thus to changes of textures toward free growth; i.e., $\{210\} \rightarrow \{100\} \rightarrow \{110\}$. This conclusion is in agreement with experiments conducted at pH values less than 5. The slight free growth trend observed with an increase of current density was tentatively ascribed to differences in the slopes of the free growth current/

potential curves for different single crystal faces of nickel. The addition of oxidizing agents reduces surface coverage and hence produces texture changes toward free growth.

The theory of texture development for nickel electrodeposits can be extended to include: 1) lateral growths, 2) deposits of other metals, and 3) deposits from adulterated baths. It has been suggested that when deposits adopt the lateral mode of growth, the slowest growing $\{hkl\}$ facets are formed — in the ideal case — parallel to the substrate. The normal to these facets becomes the texture axis. In practice, the $\{hkl\}$ facets may deviate from parallelism and the texture axis then deviates from its ideal position. Addition agents can influence the hydrogen evolution reaction, the nucleation rate, and the mode of growth. However, according to Pangarov [470] this theory does not answer the question as to what determines the mode of crystal growth.

Tajima and Ogata [480] found, during the electrocrystallization of nickel dendrites from chloride and sulfate solutions at 95°C, two forms, so-called "dendrites" and "linearly stretched." In the latter the crystal was stretching in the direction of convection. Dendrite growth is influenced by current density, concentration, temperature, voltage, and anion present. When the bath voltage is higher (20 to 40 V), the diameter of the dendrite becomes smaller and finally it becomes fibrous and woolly. The current density at the growth site is much higher than the limiting current density of the metal deposition in the stationary state (~2000 A/cm^2) (see Diggle [481] and Tajima and Ogata [482]).

4.3.3. Effect of Additives

Many substances may be added to electrolytic solutions and nickel plating baths either to inhibit corrosion or to produce leveling or brightening in the deposition of the metal. The phenomenon of leveling was first discussed by Gardam [483], who found that Watts' nickel plating bath containing small amounts of cadmium or zinc showed leveling.

Leidheiser [484], Foulkes and Kardos [485], Kardos [486], Thomas [487], and Watson and Edwards [488] studied nickel plating baths containing various additives and proposed different mechanisms of leveling. Raub [489] investigated metal microdistribution from various plating solutions and the transition from macrothrowing power to microthrowing power.

Beacom [490], Beacom and Riley [491], Watson [492], Edwards [493], Voronko and Kaikaris [494, 495], and Kudriavtsev and Kruglikov [496-499] tested previous explanations of leveling. Literature surveys on microthrowing and leveling are given by Leffler and Leidheiser [500], Vagramyan and Solovyova [501], and Kardos and Foulkes [502].

Fischer [503] studied the influence of inhibitors on stationary current/potential curves for the cathodic deposition and anodic dissolution of nickel. The results are shown in Table 4.3.10. The first two substances provoke an increase of the anodic and cathodic overvoltage. For the last two the cathodic overvoltage increases much more than the anodic one. Volk and Fischer [504] also investigated the dependence of the current density on both the

TABLE 4.3.10. Influence of Several Inhibitors on Current-Potential Curve Parameters [503]

Substance	C (M)	$(b_{Tafel})_a$ (V)	$(b_{Tafel})_c$ (V)	Current density range (A/cm^2)
Without inhibitor	-	0.060	0.060	-
Acridine	0.1×10^{-3}	0.030	-	-
Acridine	0.3×10^{-3}	-	0.040	10^{-3} to 5×10^{-2}
2-Butyne-1, 4-diol	0.3×10^{-3}	0.030	-	-
β-Naphthoquinoline	0.3×10^{-3}	-	0.035	-
Propargyl alcohol	10^{-4}	0.018	-	10^{-3} to 5×10^{-2}
Propargyl alcohol	10^{-3}	-	0.015	-
N-Ethylquinolinium iodide	10^{-4}	0.01	-	10^{-3} to 5×10^{-2}
N-Ethylquinolinium iodide	10^{-3}	-	0.01	-

electrode capacitance and degree of surface coverage using a nickel electrode in 1 M $NiCl_2$ solution with 20 g/l H_3BO_3 at $2 \leq pH \leq 5$. The results are shown in Table 4.3.11. The degree of surface coverage is indicated within parentheses. The inhibitors are grouped according to their effect on anodic overvoltage and hydrogen evolution overvoltage in various classes, as shown in Table 4.3.12. The sorption effects of these substances reflect the characteristics of the electrode double-layer capacity which clearly diminishes on increasing the additive concentration. The capacitance minimum either shows no shift or shifts slightly toward negative potentials. This is the case for 2-butyne-1, 4-diol and propargyl alcohols [505]. Various additives such as coumarin, alcohols, and thiourea have been studied more in detail from a fundamental viewpoint.

Rogers and Taylor [506, 507] studied the effect of coumarin in a Watts bath on a rotating disk electrode and found that coumarin reduces the cathodic efficiency for nickel deposition whereas the cathodic efficiency for hydrogen evolution is increased. The reaction of the additives may be either transport-controlled or electrochemically controlled, depending on conditions. The maxima in leveling power are a function of concentration resulting from the change of the rate determining factor of coumarin consumption, be it transport or electrochemical control. Coumarin reduces at the cathode mainly to melilotic acid [506]; an equimolecular mixture of o-hydroxyphenylpropanol and o-propylphenol is also formed in a secondary reaction. A decrease in pH favors reduction of primary alcoholic groups, compared with reduction of $>C=C<$ or $-C\equiv C-$, and supports the view that reduction of the alcoholic group proceeds as long as the molecule remains attached to the nickel surface. Reactions leading

TABLE 4.3.11. Effect of Inhibitor Surface Coverage on Current Density[a] [504]

Substance	C (M)	$j = 10^{-3}$ A/cm^2	$j = 5 \times 10^{-3}$ A/cm^2	$j = 10^{-2}$ A/cm^2
-	0	48 (0)	50 (0)	58 (0)
2-Butyne-1, 4-diol	(10^{-3})	23 (0.52)	33 (0.34)	45 (0.22)
2-Butyne-1, 4-diol	3×10^{-3}	19 (0.6)	25 (0.5)	34 (0.41)
2-Butyne-1, 4-diol	10^{-2}	15 (0.68)	17 (0.65)	22 (0.62)
Propargyl alcohol	10^{-4}	36 (0.25)	53 (0)	63 (0)
Propargyl alcohol	10^{-3}	16 (0.66)	30 (0.4)	50 (0.14)
Propargyl alcohol	3×10^{-3}	11 (0.76)	20 (0.4)	32 (0.45)
β-Naphthoquinoline	10^{-4}	36 (0.25)	32 (0.36)	35 (0.40)
β-Naphthoquinoline	3×10^{-4}	28 (0.41)	12 (0.75)	4 (0.93)
β-Naphthoquinoline	10^{-3}	23 (0.52)	3 (0.96)	4 (0.93)

[a]Surface coverage given in parentheses.

TABLE 4.3.12. Different Classes of Inhibitors [504]

Inhibition by film formation		β-Naphthoquinoline
		Acridine
Sorptive inhibition	Physical sorption	Tetraethylammonium ion
	Chemical sorption	2-Butyne-1, 4-diol
		Propargyl alcohol
		β-Naphthoquinoline
		Acridine
	Physical and chemical sorption	β-Naphthoquinolinium ion
		Acrydinium ion
		N-Ethylquinolinium ion

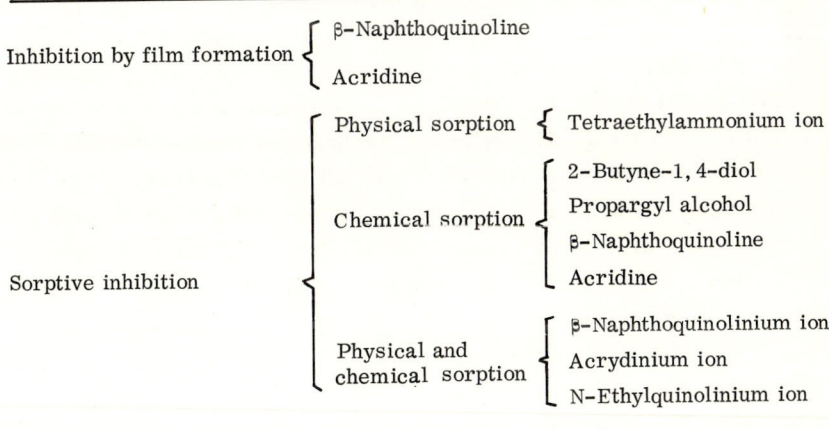

TABLE 4.3.13. Rate of Loss of Coumarin from Watts/Coumarin Plating Solution at 48.5°C

Coumarin concn (m$\underline{M}$)	Rate of coumarin loss [μmole/(l)(sec)]	Concentration/solubility
1.62-1.66	1	0.17
1.70-1.74	1	0.18
2.93-2.98	2	0.31
3.01	8	0.32
3.09	11	0.32
4.81	23	0.51
4.88	29	0.51
4.95	59	0.52

to more highly reduced products are likely to be more important in inhibiting nickel deposition.

The effect of coumarin on electrodeposition at higher potentials [508] is to inhibit the deposition and/or change the deposit nature from a columnar structure to a structureless one. The reactivity of the surface for coumarin increases with current density, suggesting an increase in surface density of active sites. The dependence of the rate of loss of coumarin in a Watts bath on concentration of additive above 2.9×10^{-3} $\underline{M}$ is summarized in Table 4.3.13 [509].

Partridge, Tansley, and Porter [271, 510] studied the adsorption of coumarin on mercury and its effects on nickel deposition. An increase in polarization due to the field formed by an array of oriented dipoles of the added agent was observed. Their results are shown in Figs. 4.3.4, 4.3.5, and 4.3.6.

Kruglikov, Kudriavtsev, and Sobolev [511] measured the double-layer capacitance of secondary brighteners (coumarin and thiourea) in a Watts-type plating solution on a nickel rotating disk electrode. These additives increase cathodic polarization and reduce differential capacitance. Secondary brighteners are more effective at greater rotation rates because their adsorption is diffusion controlled at the cathode during electrodeposition. Raub, Knoedler, Dissam, and Kawase [512] considered the inhibiting effects of alcohols, diols, phosphonium salts, and metal salts on electrodeposits of nickel. Relations between the inhibitor character and the structure of the organic additives were established. Vegys, Skuaene, and Boduevas [513] investigated cathode consumption, the polarizing action, and the inclusion of butynediol in electrodeposits as a function of additive concentration. The extent of carbon inclusion was also measured. Wiart [514] examined the effects of 1,4-butynediol (I), propargyl alcohol (II), and $PhSO_3Na$ on the electrocrystallization kinetics of nickel in a Watts

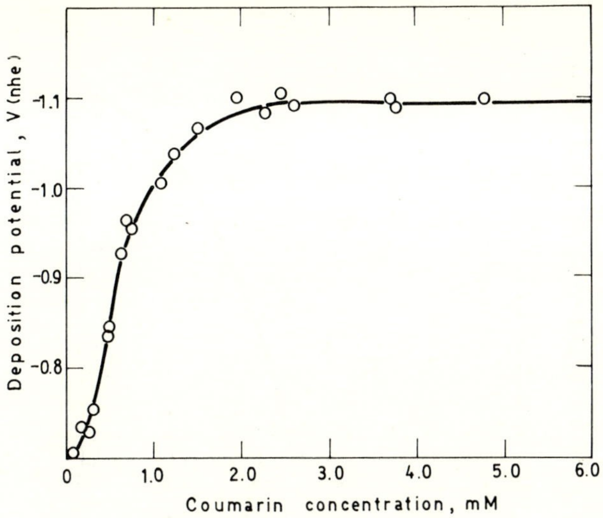

FIG. 4.3.4. The effect of adsorbed coumarin on the deposition potential of nickel. Nickel concentration: 50 m<u>M</u> in 1 <u>M</u> sodium sulfate; pH 4.6; 20°C [510]. (By permission of Pergamon Press.)

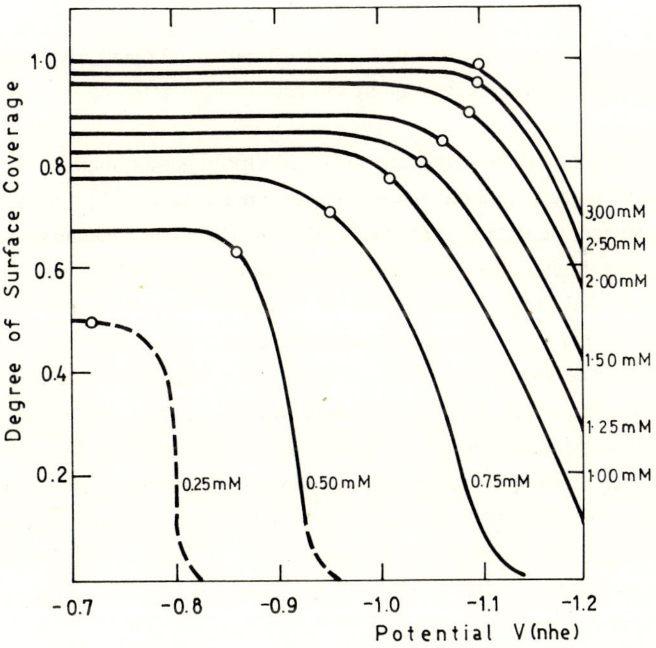

FIG. 4.3.5. Adsorption of coumarin on mercury from 1 <u>M</u> sodium sulfate as a function of coumarin concentration and potential; pH 4.6; 20°C [510]. (O) Deposition potential for nickel. (By permission of Pergamon Press.)

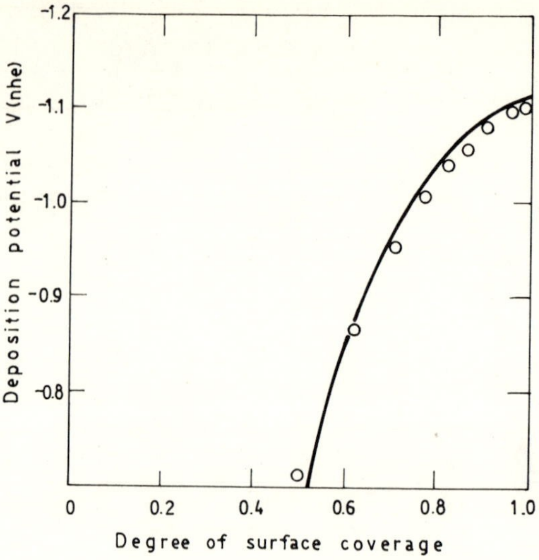

FIG. 4.3.6. Calculated and observed effect of coumarin on the deposition potential of nickel; pH 4.6; 20°C. (—) Calculated values; (O) experimental values [510]. (By permission of Pergamon Press.)

bath. The rate of transport of I and II is under diffusion and convective control but the inhibiting effect of $PhSO_3Na$ is controlled by the rate of adsorption on the electrode or by a chemical transformation.

Both 1-methylquinolinium iodide and succinodinitrile act as brighteners of second class in nickel electrodeposits, as indicated by Edwards and Levett [515]. Alaune and Lazauskiene [516] and Alaune and Talaikyte [517] studied the reactions and consumption of pyridine and quinoline additives during electrodeposition from Watts-type baths.

Kruglikov and Volkov [518] studied the interrelation between the rate of inclusion (r_i) and the rate of total consumption (r_c) of the 1-methyl-quinolinium-iodide as leveling agent, the extent of the electrode surface coverage, θ, and current density in a Watts-type bath, at pH 2.5 to 5.5, 20°C, and 1-7 A/dm^2, using a disk cathode rotating at 400 to 1510 rpm. The rate of inclusion increased with pH; the higher the concentration of the leveling agent, the greater the effect. Both r_i and r_c increased with rate of stirring. An increase in concentration of the leveling agent also increased its supply to the electrode and hence the degree of surface coverage. An interrelation between r_i, r_c, θ, and current density and their combined effect on the rate of inclusion and cathodic reduction of the leveling agent has been proved.

Knoedler [519] studied the effect of sulfur containing unsaturated compounds on the texture, the strength improvement and the decrease in the occurrence of fissures in the deposit.

Matulis [520] reported that thiourea accelerated hydrogen discharge, altered the cathodic processes because of the formation of a new phase near the cathode, and lowered the limiting current density of Ni(II) ions at a rotating disk electrode immersed in a Watts electrolyte during nickel electrodeposition.

Jugarav, Guruviah, and Parameswara Iyer [521] determined the influence of the addition of cystine as brightener. A noticeable shift in the cathode potential accompanies the production of a satisfactory bright deposit. The stress in the deposit becomes compressive and hardness increases. Hockstra and Trivich [522] observed that the uptake of radioactive ^{35}S from thiourea by single crystal spheres of nickel increased with its concentration, but showed no preference for any crystal face. Allyl-sulfonate labeled with ^{35}S was not taken up by nickel under several conditions of simple immersion, but was included under conditions of nickel electrodeposition. Dissolution of the electrodeposits in hydrochloric acid caused the included sulfur to be evolved as hydrogen sulfide, as shown by tracer experiments, quantitative analysis, and x-ray diffraction studies.

Zeimyte, Bodnevas, and Matulis [523] studied the cathodic polarization and sulfur inclusion of sodium benzene-sulfonate, sulfanilic acid, benzene-sulfonamide (I), p-amino-benzene-sulfonamide (II) or sodium thiosulfate together with urotropin (III) in nickel deposits obtained from Watts bath. Sulfur inclusion from I was considerably increased when low concentrations of II were present in the solution. The interaction of II with III was limited by the competitive adsorption on the surface.

Reklyte, Bubelis, and Matulis [524] studied the effect of inorganic salts such as $CdSO_4$, $ZnSO_4$, $FeSO_4$, $Fe_2(SO_4)_3$, $CuSO_4$, and $CoSO_4$, along with saccharin (I), benzene-sulfonamide (II), allyl alcohol, or propargyl alcohol on hydrogen adsorption of nickel coatings in Watts-type baths. Low concentrations of salts of metals with a hydrogen overpotential higher than that of nickel greatly increase the hydrogen adsorption, while $CoSO_4$ decreases it. When I or II is added to an electrolyte containing cobaltous or ferrous ions, hydrogen adsorption decreases. Addition of any or both alcohols to the electrolyte containing those ions increases hydrogen adsorption.

Bukaveckas and Matulis [525] observed that the double-layer capacity during nickel electrodeposition in the presence of saccharin and butynediol increases with current density and decreases with additive concentration. The adsorption rate of the former is one order of magnitude greater than the latter. Saccharin in Watts-type baths lowered both the limiting current density and the current efficiency for metal deposition but increased the current efficiency for hydrogen evolution. The rate of consumption of saccharin was not diffusion-controlled but was limited by the rate of renewal of the metal surface [526]. The effect of decomposition products of butynediol and saccharin, butynediol combined with saccharin and p-amino-benzene-sulfonamide in Watts-type baths was also considered [527]. The rate of adsorption of saccharin or o-toluene-sulfamide on the nickel cathode is markedly higher than that of benzamide. Saccharin is adsorbed more on the surface of a nickel cathode than benzamide and o-toluene-sulfamide both of which increase cathode polarization less than saccharin.

Kruglikov and Volkov [528, 529] considered the behavior of the p-toluene sulfonamide using a Watts-type bath and a nickel rotating disk electrode at 20°C and pH 4.3. At low concentrations ($\sim 10^{-7}$ $\underline{M}$) the rate of sulfur inclusion was diffusion controlled, and a diffusion limitation was observed even at 3×10^{-3} M. The sulfur content in the deposit increases with a decrease in current density. The consumption rate of additive shows that all the sulfur atoms included in the deposits are in the form of sulfide.

Kruglikov and Antipova [530] studied the mechanism of the decrease in the leveling action of nickel plating electrolytes at low pH values, with coumarin, adiponitrile, saccharin, and 8-quinoline-sulfonic acids in the concentration range 10^{-5} to 10^{-3} M. An increase in the acidity causes a decrease in the inhibiting and leveling actions of additives on sulfate and sulfamate baths. These effects were attributed to an increase in the amount of atomic hydrogen at the nickel electrode at lower pH.

Crossley, Brook, and Cuthbertson [531] investigated the effect of different additives on Watts bath by electron microscopy. The most effective brighteners produce a considerable reduction in size grain and develop a uniform polycrystalline surface. The mechanism of electroreduction of naphthalene-sulfonic-acid derivatives involves the splitting off of the sulfonic acid group in a one electron step. The reduction product at constant potential is naphthalene [532]. The effect of addition of 1-5-naphthalene-disulfonic acid (NDA) and NDA and chloral hydrate at different ratios in a Watts bath, on nickel electrodeposits was reported by Weil and Jacobus [533]. The two structures observed consisted of alternate bands of columnar growth and more impeded fine crystallite growth, and the crevices were very small crystallites or depressed grain boundaries. They are explained in terms of the preferential codeposition of foreign substances which would be able to impede growth.

Dubsky and Kozak [534] studied the cathodic reactions of sulfonates, sulfonamides, sulfinoles, and saccharin, added in amounts of <0.5 M in a Watts-type bath at pH 4.0 to 5.2 and 60°C. The aromatic additives form simple aromatic compounds, which are desorbed from the cathode surface and go into solution, during nickel plating. Benzene-sulfonamide, benzene-sulfinate, benzene-sulfonate, and benzene-disulfonate formed benzene at the cathode; p-toluene-sulfonate, p-toluene-sulfinate, p- and o-toluene-sulfonamide formed toluene; naphthalene-sulfonate forms naphthalene, etc. Every compound containing the $-SO_2NH_2$ group simultaneously formed ammonia, which was detected from the increase of the ammonium-ion concentration in solution. The hydrogenolytic decomposition of the C—S bond may represent a cathodic reaction of these additives.

Alaune and Lazauskiene [535] studied the desulfurizing reaction of β-naphthalene-sulfonic acid during nickel electrodeposition from a Watts bath. The naphthalene formed is an indication of the sulfur content of the plated metal. Less naphthalene is formed at lower pH.

Dye and Klingenmaier [536] studied the inclusion of fuchsin in bright nickel deposits. The amount of fuchsin recovered per gram of nickel was roughly linear with fuchsin concentration and decreased with increasing current density. The amount of fuchsin included and its brightening effect are correlated with a model which assumes preferential adsorption at grain corners and edges.

Attempts to correlate the textures with brightness [537, 538] were not very successful [539-541]. However, Weil and Paquin [542] and Weil and Cook [543] found that with nickel deposits from Watts baths containing several different additives, bright deposits always had a very fine-grained structure. There was a linear relationship between the fraction of surface area having a roughness less than 0.15 μ in depth and the logarithm of the reflected light, but no direct relationship between brightness and degree of preferred orientation. Cliffe and Farr [544] determined a restricted range of temperatures and current densities over which the surface topography of growing nickel reflects.

The influence of linear solution flow past a horizontal anode on its rate of dissolution under limiting current density conditions giving anodic brightening has been examined by Rothwell and Hoar [545] for polycrystalline nickel in aqueous 10 M H_2SO_4. The limiting current density increases linearly with the square root of flow rate, but an inflection at moderate flow rates was observed. This effect was discussed in terms of field-induced change of the compact solid film on the anode.

The experimental results, as far as leveling is concerned, can be summarized as follows: 1) "true" leveling, i.e., that additional to the "geometrical" leveling that takes place in uniform metal microdistribution, is caused by special additives (leveling agents); 2) all leveling agents increase cathode polarization; 3) leveling agents are incorporated in the electrodeposits or otherwise consumed at the cathodic surface, and incorporation is greater at the protrusions than in the hollows of a rough surface; and 4) the rate of consumption of some leveling agents, under certain conditions, is independent of current density but increases with agitation.

Two different hypotheses [485-488, 490, 491] were proposed to explain the mechanism of leveling, both based on the suggestion that the unusual metal microdistribution peculiar to leveling solutions is due to a considerable difference in the inhibiting action produced by the addition agent on peaks and in recesses. The different inhibiting action is ascribed to variations in leveling agent concentration on the cathode surface, the concentration at peaks being greater than that at recessed areas. To explain this, one hypothesis assumes preferential adsorption of the leveling agents onto the high points of an irregular surface [487, 490, 491], whereas the other [485, 486, 488] considers different diffusion rates of the leveling agent to peaks and into recesses to be responsible for surface concentration variations, thus implying that the inhibiting action produced by leveling agents is diffusion controlled. The hypothesis which assumes diffusion control of inhibition is in fair agreement with experimental results, as shown by Kruglikov, Kudriavtsev, Vorobiova, and Antonov [546]. Slowing down of the cathodic process seems to be controlled by the diffusion of the additive toward the electrode.

4.4. NICKEL OXIDE AND NICKEL HYDROXIDE ELECTRODES

4.4.1. Aqueous Solutions

4.4.1.1. Compounds of Electrochemical Interest. Nickel oxides and hydroxides play an important role in the passivity of nickel in aqueous solutions, in oxidizing ionic melts, and in the formation of electrodes related to nickel-iron and nickel-cadmium batteries.

TABLE 4.4.1. Hydrated Nickel Oxides

Compound	Reaction	Identification	Refs.
$Ni_2O_3.nH_2O$ ($1 < n < 4$)		X-ray	[558], [559]
$Ni_2O_3.H_2O$		X-ray	[560]
$Ni_2O_3.2H_2O$		X-ray	[560]
$Ni_3O_4.2H_2O$		X-ray	[560]
α-NiOOH	oxidn of $K_2Ni(CN)_4$ and $K_2S_2O_8$	X-ray; electronic micrography	[561-563]
β-NiOOH	oxidn Ni(II) with Br_2 or NaClO or $NaBrO_2$	X-ray; electronic micrography	[561-563]
γ-NiOOH	oxidn of Ni with Na_2O_2	X-ray; electronic micrography	[561-563]
$Ni_3O_2(OH)_4$		Hexagonal	[561-563]
γ_1-$NaNi_3O_6.2H_2O$		X-ray brucite type structure with additional Na/O intermediate layers	[564]

Formation conditions and the structure of oxides and hydroxides of nickel are given by Feitknecht, Studer, and Meyer [547] and Bédert [548]. Various authors supposed part of nickel may oxidize to Ni(IV) [549-556] because nonstoichiometric oxides with a O/Ni ratio larger than 1.5 could be obtained. Vogel [557] indicated the existence of a nickel oxide with a valency as high as 3.6 in hot KOH solutions (83 to 84 wt% KOH, 1 to 2 wt% K_2CO_3 and 14 to 16 wt% H_2O) at 225 to 250°C. Table 4.4.1 shows some of the hydrated nickel oxides which were structurally identified.

Aia [565] studied the crystal growth and phase changes in the system NiO-H_2O from 200 to 400°C and water pressures from 0.2 to 2.1 kbar. $Ni(OH)_2$ and NiO were the only solid phases found with brucite and rocksalt structures, respectively. At 285°C $Ni(OH)_2$ converted to NiO. The reverse reaction was not achieved. Hydrothermal treatment of $Ni_2O_3.H_2O$ gave a black material with an X-ray diffraction pattern containing the γ-NiOOH. Differential thermal analysis was combined with X-ray diffraction analysis to follow both the electrolyte development in KOH solutions and the thermal decomposition of the charged state of sintered nickel electrodes impregnated with $Ni(OH)_2$. The charged nickel oxide electrode contains active oxygen, insoluble potassium, water of constitution, and water of hydration. The electrochemical capacity is close to the electrochemical equivalent of the active oxygen formed on charging.

Bode, Dehmelt, and Witte [566] described the Ni(II) hydroxyhydrate including its structure. It transforms as follows:

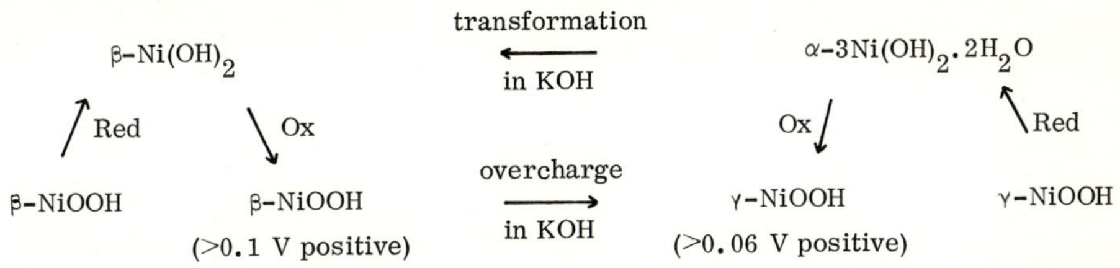
transformation
in KOH
β-Ni(OH)2
α-3Ni(OH)2.2H2O
Red
Ox
Ox
Red
overcharge
in KOH
β-NiOOH
β-NiOOH
(>0.1 V positive)
γ-NiOOH
(>0.06 V positive)
γ-NiOOH

For the reaction:

$$\alpha\text{-Ni(OH)}_2\cdot(2/3)\text{H}_2\text{O} \rightleftharpoons \beta\text{-Ni(OH)}_2 + 2/3\ \text{H}_2\text{O} \qquad (4.4.1)$$

$\Delta G°$ is equal to -3.4 kcal/mole, and the logarithms of the solubility products of active and inactive $Ni(OH)_2$ are, respectively, -14.7 and -17.2.

Dennstedt and Loeser [567] investigated the influence of the water content in $Ni(OH)_2$ on its electrochemical reactivity, which is different in α- and β-nickel hydroxides. Water of the β-$Ni(OH)_2$ exceeding stoichiometry is completely removed at 160°C. The corresponding reactions are:

β-$Ni(OH)_2$ (no water)

$$\text{Ni(OH)}_2 = \text{NiOOH} + \text{H}^+ + e \qquad (4.4.2)$$

β-$Ni(OH)_2$ (with water molecules in the external layer)

$$\text{Ni(OH)}_2\cdot\text{H}_2\text{O} = \text{Ni(OH)}_3 + \text{H}^+ + e \qquad (4.4.3)$$

$$\text{Ni(OH)}_3 = \text{NiOOH}\cdot\text{H}_2\text{O} \qquad (4.4.4)$$

α-$Ni(OH)_2$ (with water in the inner layer)

$$\text{Ni[(OH)}_2\text{]}\cdot\text{H}_2\text{O} = [\text{NiOOH}]\cdot(\text{H}_3\text{O})^+ + e \qquad (4.4.5)$$

$$[\text{NiOOH}]\cdot(\text{H}_3\text{O})^+ = [\text{NiOOH}]\cdot\text{H}_2\text{O} + \text{H}^+ \qquad (4.4.6)$$

Brownsword and Farr [568] studied the electrochemical influence of NiO as an impurity phase in nickel. Slow solidification, such as that corresponding to the condition of zone refining or crystal growing, may result in local concentration of an impurity by segregation.

4.4.1.2. Voltammetric Studies. Weininger and Breiter [366, 367] investigated the mechanism of anodization at surfaces of nickel single crystals when only a few layers of reaction product are formed under the influence of a linearly increasing potential with 99.9% nickel single crystals and 99.97% polycrystalline nickel samples in 0.2 N KOH solutions. Current/potential voltammograms are shown in Fig. 4.4.1, and results are given in Table 4.4.2. There is a change of the voltammogram shapes obtained for the 1st, 2nd, and 3rd sweeps

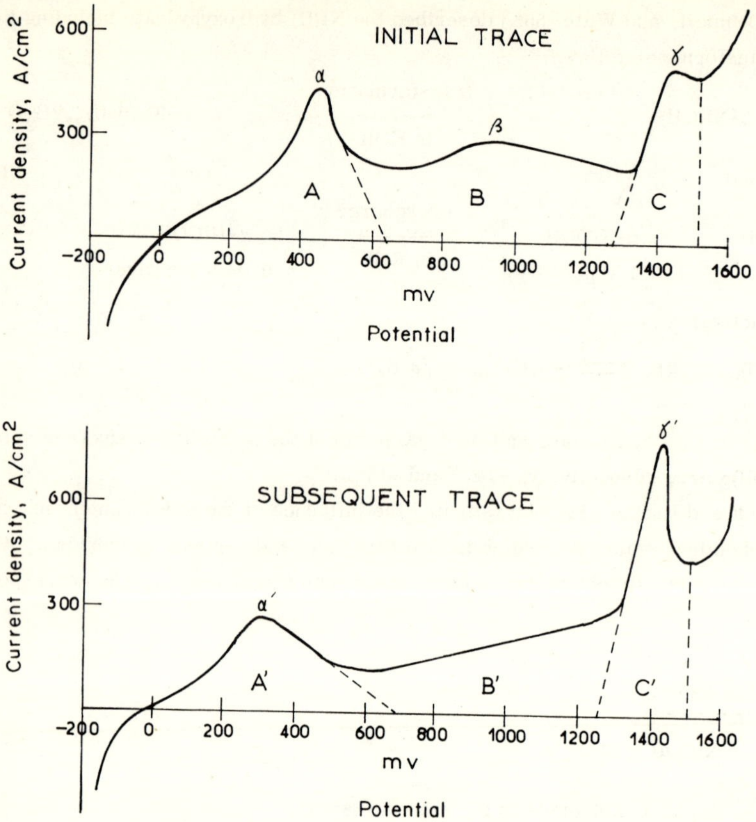

FIG. 4.4.1. Schematic current/potential curves of single crystal (initial trace) and disordered surfaces (subsequent trace) of nickel [366, 367]. (By permission of the Electrochemical Society.)

between -0.2 and 1.6 V at 0.11 V/sec. The rate of oxide formation increases with decreasing atomic packing: $\{111\} < \{100\} < \{110\}$. Cycling carried out under 0.51 V produces no surface disorder. During the first cycle three current peaks are distinguished. Between 0.2 and 0.6 V, $Ni(OH)_2$ formation occurs:

$$Ni + 2OH^- \rightleftarrows Ni(OH)_2 + 2e \qquad (4.4.7)$$

Above 1.4 V:

$$Ni(OH)_2 + OH^- \rightleftarrows \beta\text{-}NiOOH + e + H_2O \qquad (4.4.8)$$

and at sufficiently higher overvoltages:

$$Ni + 3OH^- \rightleftarrows \beta\text{-}NiOOH + H_2O + 3e \qquad (4.4.9)$$

Above 1.5 V, oxygen evolution occurs and probably Ni(IV) formation, the latter with a 1 to 2% yield [569]. The third peak disappears after the first cycle. A solid $Ni(OH)_2$—β-NiOOH

TABLE 4.4.2. Characteristic Features of Current-Voltage Curves for Anodization of Nickel

		{110}	{100}	{111}	Polycrystalline
Voltage measurements, mV	α	370 ± 20	450 ± 20	480 ± 20	370 ± 20
Voltage measurements, mV	β	880 ± 20	815 ± 20	870 ± 20	850 ± 20
Voltage measurements, mV	γ	1440 ± 20	1430 ± 20	1435 ± 20	1435 ± 20
Voltage measurements, mV	α'	325 ± 20	320 ± 20	340 ± 20	320 ± 20
Voltage measurements, mV	γ'	1435 ± 20	1445 ± 20	1425 ± 20	1450 ± 20
Charge measurements, mC/cm^2	A	1.4 ± 30%	1.6 ± 30%	1.3 ± 30%	1.3 ± 30%
Charge measurements, mC/cm^2	B	3.9 ± 30%	3.6 ± 30%	2.7 ± 30%	1.5 ± 30%
Charge measurements, mC/cm^2	C	0.8 ± 30%	0.7 ± 30%	1.1 ± 30%	1.0 ± 30%
Charge measurements, mC/cm^2	A'	1.2 ± 30%	1.9 ± 30%	1.1 ± 30%	1.1 ± 30%
Charge measurements, mC/cm^2	B'	2.9 ± 30%	2.2 ± 30%	2.7 ± 30%	1.4 ± 30%
Charge measurements, mC/cm^2	C'	2.0 ± 30%	1.6 ± 30%	1.2 ± 30%	1.0 ± 30%
Charge measurements, mC/cm^2	A + B + C	6.1 ± 10%	5.9 ± 10%	5.1 ± 10%	3.8 ± 10%
Charge measurements, mC/cm^2	A' + B' + C'	6.1 ± 10%	5.7 ± 10%	5.0 ± 10%	3.5 ± 10%

solution is probably formed. The cathodic half-cycle shows the reduction peaks of the two compounds.

Double-layer capacitance curves, at 100 Hz, as a function of potential, on polycrystalline nickel in 4 N KOH saturated with argon, exhibit a maximum at about 0.2 V — related to the formation of two layers of $Ni(OH)_2$ — followed by a minimum of about 20 $\mu F/cm^2$ when passivation initiates (Fig. 4.4.2). This minimum indicates a maximum electrode resistance due to $Ni(OH)_2$ [570]. As soon as β-NiOOH begins to be formed at about 0.7 V, the electrode capacitance increases. The successive apparent capacitance/potential curves differ from the initial one, especially in the region where $Ni(OH)_2$ is formed, in the region following the minimum and in the oxygen evolution region. The maximum apparent electrode capacitance decreases during the three cycles.

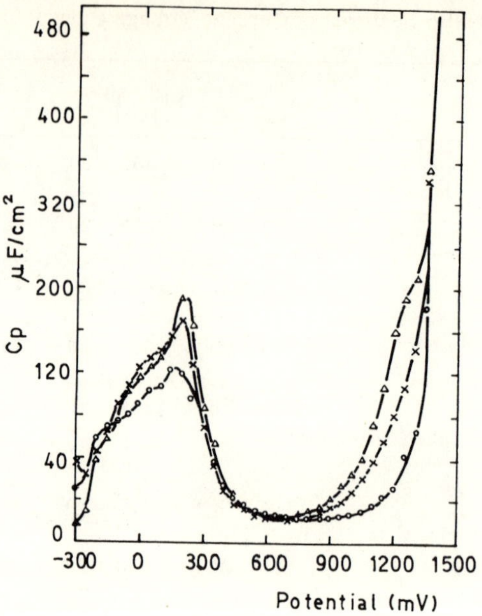

FIG. 4.4.2. Plot of capacity C_p as a function of potential for a polycrystalline Ni electrode in 4 N KOH saturated with Ar and measured at 100 Hz: (O) first sweep; (X) second sweep; and (Δ) third sweep [367]. (By permission of the Electrochemical Society.)

Recently, Chernykh and Yakuvleva [571] recorded potential/current curves for nickel electrodes immersed in 2 M KOH and 2 M LiOH at 25°C, within the potential range -0.2 to 1.6 V (alkaline hydrogen electrode). They observed four peaks which they assigned to the initial formation of $Ni(OH)_2$, growth of layer of $Ni(OH)_2$, formation of Ni_3O_4, and formation of Ni_2O_3 or NiO_2. Oxygen evolution occurred at ~1.5 V. The cathodic branch exhibited a single peak caused by the reduction of the higher oxide.

4.4.1.3. Anodic and Cathodic Charging Curves on Nickel Electrodes. The charging and discharging curves of nickel electrodes in alkaline solution have been considered in detail by different authors. Typical curves are shown in Fig. 4.4.3. When a fresh nickel surface is dipped into a 1 M KOH solution, a passivating $Ni(OH)_2$ film is formed on the electrode [366, 367] which reaches a thickness of about 15 atom units after 16 hr. A brown oxide film can be formed on a nickel anode in NaOH solution, while the potential increases slowly with time [572]. When current flows through a $Ni(OH)_2$ grain, it acquires a dark color on the anode side and gradually the color extends over the whole grain. The oxidized grain becomes an electrical conductor, a fact which was attributed to the presence of active oxygen [572, 573]. During the galvanostatic anodic polarization of an oxygen-free nickel surface, within the potential range of water decomposition, a current plateau was observed

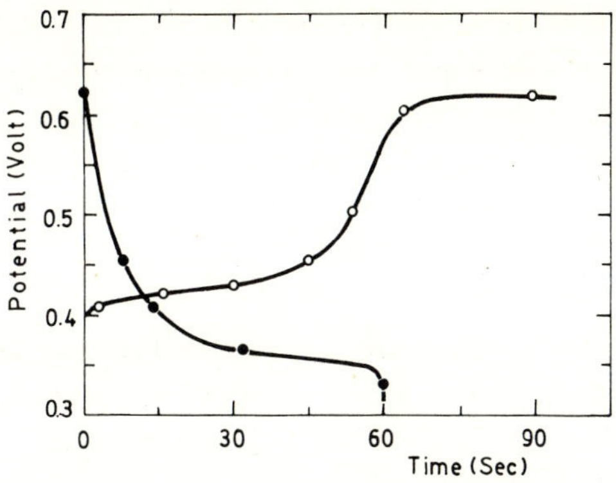

FIG. 4.4.3. Constant current charging curves on Ni electrodes in 1 N KOH solution at 2 mA/cm^2: (O) anodic and (●) cathodic [581]. (By permission of the Chemical Society.)

at -0.6 V (NHE) which corresponded to the oxidation of nickel to $Ni(OH)_2$ [27, 368]. Two $Ni(OH)_2$ layers were required to attain this potential. If a $Ni(OH)_2$-impregnated nickel electrode was used, the current/potential curve initiates at about 0.4 V, as reported by Briggs, Jones, and Wynne-Jones [574].

Studies made by different authors [575, 576] showed that $Ni(OH)_2$ is a poor electronic conductor but a good proton conductor and the rate of oxygen diffusion through it is insignificant [368, 577]. When the film is already formed, at constant current, the potential increases up to 0.4 V where the electron transfer process Ni(II) → Ni(III) takes place [368, 562, 574, 575, 577-579]:

$$Ni(OH)_2 \rightleftharpoons \beta\text{-}NiOOH + H^+ + e \quad (4.4.10)$$

However, some authors claim that the charging potential/time curve obtained at 5 $\mu A/cm^2$ exhibits a small step at 0.4 to 0.5 V [27, 549, 553, 580]; this has not been confirmed [368].

The process at potentials between 0.4 and 0.6 V probably involves a continuous series of solid solutions formed between both species [368, 385, 574, 575, 582-586]. The species $Ni_3O_2(OH)_4$ corresponds to the case where the ratio $Ni(OH)_2/\beta$-NiOOH in the film is equal to 1:2. The nucleation and growth of β-NiOOH has been considered by Briggs and Fleischmann [587, 588] who also studied the oxidation of solid α-$Ni(OH)_2$ layers to γ-NiOOH, their subsequent reduction, and the extent to which the processes are controlled by phase growth and diffusion mechanisms, respectively. The discharge of fully formed layers of NiOOH as well as the recharging process, although it follows a composite pattern, is normally controlled by diffusion of species through or within the layer. The diffusion part of the process coincides with the study made by McArthur [589]. During polarization, $Ni(OH)_2$ and β-NiOOH X-ray patterns were detected, and during the charging process the $Ni(OH)_2$ lines disappear. Therefore, it is likely that the composition of the anodic film in the potential region of the plateau of the charging curve is a solid solution of $Ni(OH)_2$ and β-NiOOH [590]. Although β-NiOOH is the species of highest oxidation degree observed by X-ray, amorphous NiO_2 cannot be definitely excluded [581, 591]. If a conversion of Ni(III) into Ni(IV) yielding non-stoichiometric nickel oxides in the β-NiOOH lattice takes place, it would be incorrect to consider the film as solid solutions of β-NiOOH and NiO_2 [561, 592]. Among the different theories proposed for the structure of the film, the single lattice theory of Wynne-Jones agrees best with the X-ray results.

At potentials higher than 0.6 V, steady oxygen evolution takes place. The plateau of the charging curves obtained with freshly prepared $Ni(OH)_2$ electrodes are less pronounced than those obtained with aged electrodes [385, 593]. Oxygen overvoltage is larger on aged $Ni(OH)_2$ electrodes.

X-ray analysis shows open laminar structures for fresh electrodes which are more disordered than the normal structure of aged $Ni(OH)_2$ electrodes. Fresh $Ni(OH)_2$ electrodes are both chemically and electrochemically more reactive than aged ones [582, 586]. As the number of hydroxyl groups is large at the film formed on the nickel electrode, it exhibits ionic exchange properties and absorption of H_2O, producing potential changes related to the KOH and H_2O activities [30, 385, 594].

The galvanostatic discharge curve of the charged nickel electrode in alkali solutions exhibits a plateau at about 0.45 V [25, 574], which corresponds to the reduction of β-NiOOH to $Ni(OH)_2$. Only under very special conditions, ~0.0 V, does a second small step appear in the curve [27, 595]; this is attributed to the reduction to $Ni_3O_2(OH)_4$. At potentials lower than -1.1 V [574], hydrogen evolution occurs.

The potential of the plateau for the charging process is nobler than the one for the discharge. This indicates that activation energies are involved in the formation and reduction of β-NiOOH. During the discharge of a completely charged nickel electrode, β-NiOOH [579] remains unaltered until the 0.45 V plateau is reached, yielding $Ni(OH)_2$ at the end of the plateau, according to X-ray analysis.

Kober [596, 597] recorded the IR spectrum of the sintered nickel oxide electrode and showed the discharged state to be $Ni(OH)_2$ having a hexagonal layer structure isomorphic with the space group D^3_{3d}. The hydroxyl ions are parallel to the c-axis of the crystal and are "free" in the sense that hydrogen bonding is absent. Associated with this structure is a relatively small amount of water which is trapped in the crystal lattice through the formation of coordinate-covalent bonds with the nickel ions. The charging process is accompanied by the formation of a hydrogen bond structure possessing a higher degree of crystal symmetry than that found in the discharged state. During discharge these hydrogen bonds are continuously broken and the structure transforms back to a free hydroxyl configuration. The octahedral Ni—O coordination in the charged state has been suggested by Tuomi [598]. Through the use of $Ni(OH)_2$ electrodes at various stages of charge, it has been possible to make definite bond assigments to the structures of the active material. The relative intensity of the bond due to hydrogen bonding in the charged state is shown to be a measure of the electrochemical capacity of the system. The charging reaction gives rise to the formation of active oxygen sites.

4.4.1.4. Proton Diffusion Mechanism. Electrochemical measurements by Kunchinskii and Ershler [573] on a large grain nickel oxide electrode in an alkaline electrolyte established the existence of a controlling solid-state diffusion mechanism which leads to the depletion of conductive species in the surface layer of the grain and, ultimately, to the formation of an insulating $Ni(OH)_2$ layer at the conductor/nickel oxide interphase, which prevents the reduction of the conductive, higher valence interior of the grain. Such a mechanism explains the lower coulombic efficiency of thicker nickel oxide layers if the specific surface area of nickel oxide does not remain constant as the layer thickness is increased. Appelt [599] measured the surface conductivities of nickel hydroxides and their dependences on lattice structure.

Extensive studies of cathodic polarization of nickel oxide were made by Wynne-Jones and Jones [578] and by Lukovtsev [600], who independently suggested proton diffusion in the hydroxide as a step in the mechanism of oxidation of $Ni(OH)_2$ to NiOOH. This was verified experimentally by Lukovtsev and Slaidin [575]. X-ray diffraction studies of the anodization of nickel in alkaline solution were performed by Briggs and Wynne-Jones [385, 579, 601], by Falk [591], and by Salkind and Bruins [590] who elucidated the electrode process in the nickel-cadmium cell. The electrochemical oxidation of $Ni(OH)_2$ yields solid solutions of (NiO)OH in $Ni(OH)_2$ [385, 579, 601]. The oxidation state goes further than $NiO_{1.7}$. The charging and discharging processes are represented by

$$Ni\begin{matrix} \diagup OH \\ \diagdown OH \end{matrix} \rightleftharpoons Ni\begin{matrix} \diagup O^- \\ \diagdown OH \end{matrix} + H^+ \qquad (4.4.11)$$

and

$$Ni^{2+} \rightleftharpoons Ni^{3+} + e \qquad (4.4.12)$$

Lukovtsev and Slaidin [575] investigated the proton diffusion process through nickel oxide in alkaline solutions and established that the reaction occurring at the electrode surface is

$$(Ni^{3+}.O^{2-}.OH^-)_S + H_2O + e = (Ni^{2+}.2OH^-)_S + OH^- \qquad (4.4.13a)$$

and later protons diffuse from the surface to the bulk of the oxide according to

$$(Ni^{2+}.2OH^-)_S + (Ni^{3+}.O^{2-}.OH^-)_B = (Ni^{3+}.O^{2-}.OH^-)_S + (Ni^{2+}.2OH^-)_B \qquad (4.4.13b)$$

with the corresponding electron transfer from Ni^{2+} to Ni^{3+}. The second step controls the diffusion rate of protons into the nickel oxide. The equilibrium potential of a Ni/Ni-oxide electrode depends on the concentration of proton defects in the superficial layers of the $Ni(OH)_2$ lattice. Later, Ewe and Kalberlah [576] studied the diffusion of protons in an oxidized sintered nickel powder electrode placed as a separating diaphragm. Nickel oxides were produced by successive anodic oxidation, causing a decrease of porosity, an increase in flow resistance, and a decrease of the electrical diaphragm resistance. They determined that the electric current is carried by ions of the electrolyte in the pores and the charge transported by proton diffusion on the oxidized nickel surface. The diffusion coefficient for protons was 3×10^{-5} cm^2/sec.

Aleshkevich, Golovchenko, Morozov, and Sagoyan [602] concluded from a theoretical study that the amount of energy required to detach a proton from β-NiOOH with no electric field being applied is smaller than the amount of energy expended in the case of $Ni(OH)_2$. This indicates that Ni(III) is directly converted into Ni(IV) at the current conductor/active mass/electrolyte interphase. Proton supply to the separator surface limits the rate of the process.

The anion transfer which is the basis of the oxidation of $Ni(OH)_2$ obtained in thin layers by cathodic electrodeposition was studied by Feuillade and Jacoud [603] by the isotopic method. The chemical properties related to stoichiometry variations were determined by the irreversibility of the ionic and local electronic transport during the anodic oxidation of the $Ni(OH)_2$.

4.4.1.5. Open Circuit Potential Decay of a Charged Ni/NiO Electrode. The oxide layers of a charged nickel electrode are unstable. At open circuit the potential decays first rapidly and later more slowly after the $Ni(OH)_2$ composition is attained [27]. Conway and Bourgault [604] observed that after charging the Ni/NiO electrode to a chemical state represented by the nonstoichiometric oxide NiO_x ($1.4 \le x \le 1.8$), loss of oxygen and decay of potential at open circuit occur. The self-discharge in aqueous KOH solutions was examined by: 1) the rate of decay of electromotive force at open circuit; 2) the rate of oxygen evolution at open circuit; 3) the electrochemical capacitance of the electrode and; 4) the charging curve of the electrode. During the decay, abrupt changes occur at certain electrode potentials indicating changes in the rate-determining mechanism of the self-discharge process. The results are shown in Tables 4.4.3 to 4.4.5. The decay slope b_j is equal to $\partial E/\partial \log(\text{time})$. The general reaction scheme involves a mixed Ni(III) and Ni(IV) nonstoichiometric hydrated oxide lattice [561, 579]:

$$(Ni^{4+}2O^{2-})_s + OH^- \rightleftarrows (Ni^{4+}2O^{2-})_s + (OH)_{ad} + (e)_{lattice} \quad (4.4.14a)$$

$$(Ni^{4+}2O^{2-})_s + (e)_{lattice} \underset{\text{transfer}}{\overset{\text{electron}}{\rightleftarrows}} (Ni^{3+}2O^{2-})_s \quad (4.4.14b)$$

$$(Ni^{3+}2O^{2-})_s + H_2O \underset{\text{transfer}}{\overset{\text{proton}}{\rightleftarrows}} (Ni^{3+}O^{2-}OH^-) + OH^- \quad (4.4.14c)$$

If there is electronic and protonic mobility in the lattice, $(Ni^{4+}2O^{2-})$ can be regenerated at the surface:

$$(Ni^{3+}O^{2-}OH^-)_s + (Ni^{4+}2O^{2-})_B \overset{e,\,H^+}{\rightleftarrows} (Ni^{4+}2O^{2-})_s + (Ni^{3+}O^{2-}OH^-)_B \quad (4.4.14d)$$

TABLE 4.4.3. Tafel Slopes Observed in the Decay of Electromotive Force of the Nickel Oxide Electrode

Concn of KOH (M)	Mean slope in high potential region (V)	Mean slope in low potential region (V)
14.6	0.028 ± 0.0005	-
9.4	0.032 ± 0.0005	-
9.0	0.032 ± 0.001	-
7.1	0.038 ± 0.0002	0.020 ± 0.0003
2.9	0.034	0.016
0.93	0.040 ± 0.0003	0.023 ± 0.0003
0.327	0.038	0.019
0.145	0.037 ± 0.0015	0.024 ± 0.0004
0.077	0.040	0.021
0.015	0.039 ± 0.002	0.020 ± 0.000
0.0015	0.040 ± 0.002	0.023 ± 0.002

TABLE 4.4.4. Rates of Self-Discharge (j_0) at the Reversible Oxygen Potential

Method	From emf decay (A/g)	From O_2 evolution (A/g)
Upper potential region	11.8×10^{-8}	7.0×10^{-8}
Lower potential region	3.5×10^{-10}	4.3×10^{-10}

TABLE 4.4.5. Electrode Capacity and Charge

Concn of KOH (M)	Active oxygen / Nickel	Capacitance (F/equiv $\times 10^{-5}$)			
		From degree of surface coverage	O_2/emf I	O_2/emf II	Dc charging
14.6	0.57	17.8	21	41	-
9.4	0.51	26	32	64.6	-
7.0	0.48	33.4	41	62.5	-
2.9	0.45	31.9	39.5	70.3	37
0.93	0.43	47.5	57	-	39
0.145	0.39	25.1	44	170	25
0.015	0.36	21	33	140	20

The first step can then continue until all Ni^{4+} is converted to Ni^{3+}. Similar self-discharge of the ($Ni^{3+}O^{2-}OH^-$) phase to ($Ni^{2+}2OH^-$) could also occur if the potential of the electrode remained positive to that of the reversible oxygen electrode. Oxygen evolution via the first process then proceeds because of the overall reaction

$$2(OH)_{ad} = H_2O + \frac{1}{2}O_2 \tag{4.4.15}$$

The slopes were interpreted mechanistically with a set of consecutive reactions for oxygen evolution.

Further evidence that the rate-determining step in self-discharge of the nickel oxide electrode is the anodic partial reaction of oxygen evolution [($MO + MOH \rightarrow MHO_2$) or ($MO + OH^- \rightarrow MHO_2 + e$)] is based on the comparison of: 1) The activation energy, ΔH^*, for open circuit oxygen evolution and for dc anodic polarization with oxygen evolution; 2) the current/potential for dc anodic polarization and for rates of oxygen evolution at open circuit as a function of potential; and 3) the H/D isotope effects on the open circuit and dc polarization behavior. In the latter case an unusual and characteristic inverse isotope effect is observed [605]. Tafel slopes (Table 4.4.6) calculated from the slopes, b_d, of the potential decay lines, taking into account the dependence of surface charge upon potential, were interpreted in terms of possible mechanisms of oxygen evolution, considering the dependence of ΔH^* on surface coverage by adsorbed intermediates. ΔH^* (7 M KOH) = 15.2 ± 1 kcal/mole (at E_{O_2} = 0.18 V); ΔH^* = 20.2 (±1.6) kcal/mole (at the reversible potential). The observed slope in the upper potential region could correspond to the final recombination step at appreciable coverages under "activated adsorption" conditions and the change of slope with decreasing potentials (and coverage) could correspond to a real change of mechanism. The probable oxide composition at the reversible potential is $NiO_{1.25}$.

Bourgault and Conway [18] also investigated the electrode potential of nickel oxide electrode as a function of water and solute activity in aqueous solutions of KOH. The electrode was charged to an average oxidation state corresponding to 50% Ni(II) and 50% Ni(III) and was examined after long periods of time by cathodic and anodic potential decay measurements after polarization in order to establish the quasi-reversible potentials. For the cell: Hg/HgO/KOH(a_{KOH}), $H_2O(a_{H_2O})$/NiO. OH*, $Ni(OH)_2^*$ (where the asterisk refers to unspecified and probably nonstoichiometric quantities of associated water and adsorbed KOH), the electromotive force is given by

$$E \text{ (in V)} = 0.420\ (\pm 0.002) - 0.004\ (\pm 0.001)\log(a_{KOH}) + 0.105\ (\pm 0.01)\log(aH_2O) \tag{4.4.16}$$

which corresponds to the formal reaction

$$[2NiO.OH](0.14KOH_{ad}) + Hg + 3.56H_2O = 2[Ni(OH)_2.1.28H_2O] + HgO + 0.14KOH_{aq} \tag{4.4.17}$$

This result is in qualitative agreement with that of Zedner [19-21] with respect to the dependence of the electromotive force on H_2O activity. The residual dependence on KOH activity is consistent with and in fair agreement with that deduced by Kornfeil [30] using KOH/KF

TABLE 4.4.6. True Tafel Slopes from Electromotive Force Decay

Conditions (potential region)	b_d (V)
0.15 M KOH (25°C) upper	0.045
0.15 M KOH (25°C) lower	0.022
7 M KOH (25°C) upper	0.045
7 M KOH (25°C) dc polarization (upper)	0.049
0.15 M KOH (60.5°C) upper	0.051
0.15 M KOH (60.5°C) lower	0.027
1.5 M KOH (60.5°C) upper	0.044
1.5 M KOH (60.5°C) lower	0.021

mixtures. Chemically, adsorption of KOH by the higher oxide is not inconsistent with the general properties of higher oxides, which are usually more acidic than the corresponding lower ones. For the single electrode process at the NiO electrode:

$$[2NiO.OH](0.14KOH_{ad}) + 4.56H_2O + 2e = 2[Ni(OH)_2.1.28H_2O] + 2OH^- + 0.14KOH_{aq} \quad (4.4.18)$$

Zedner [19-21] derived

$$2NiO.OH.H_2O + 4H_2O + 2e = 2Ni(OH)_2.2H_2O + 2OH^- \quad (4.4.19)$$

and Foerster [22, 23] derived

$$2NiO.OH.0.1H_2O + 1.8\ H_2O + 2e = 2Ni(OH)_2 + 2OH^- \quad (4.4.20)$$

These authors discussed their results in terms of the stoichiometry of the potential-determining reaction, making a distinction between mixed and reversible potentials.

Conway, Sattar, and Gilroy [606] reported that oxygen evolution at nickel and oxidized nickel exhibits some inhibition effects (self-passivation). The inhibiting species is the surface oxide which is directly involved in the overall reaction. Conway and Sattar [607] studied a anodically-formed thin-film oxide layers of nickel in alkaline solutions in relation to the behavior of bulk nickel oxide examined in previous works. The stoichiometry depended both on time and potential prior to polarization. The surface oxide formation or reduction charges were found to be logarithmic with potential up to and well beyond apparent monolayer coverage.

Jost and Rufenacht [608] made polarization studies on porous, sintered plate nickel oxide electrodes containing varied amounts of active material (nickel hydroxide) with a single pulse interruption technique. Double-layer capacitance measurements and long-time overvoltage decay suggest that the nickel hydroxide, deposited in the pores in sequential impregnations, is deposited as a sponge of discrete particles. The available surface area proved, therefore, to be proportional to the $Ni(OH)_2$ weight. The long decay times indicate solid-state diffusion or relaxation phenomena on top of liquid concentration overvoltage, which becomes increasingly important with heavily impregnated electrodes.

The effect of ultrasound on nickel oxide electrode, as well as the neutron irradiation of a nonstoichiometric nickel oxide electrode, has been investigated by Kukoz and Skalozubov [609]. The amount of adsorbed oxygen increases in the latter [610]. Irradiation produces positive holes in the oxide film and the electrons of the loosely adsorbed oxygen atoms are trapped in the acceptor levels, thus producing strongly bound oxygen.

4.4.1.6. Dissolution of NiO. Anodic dissolution in the presence of oxidizing agents was investigated by Riga, Greef, and Yeager [611]. The dissolution rate is increased by anodic polarization. The dissolution of NiO prepared by thermal decomposition was investigated by Nii [612]. The rate is affected remarkably by the oxygen pressure prevailing during the preparation of the sample, stirring period and atmosphere after preparation, and by the nature and concentration of ions in the solution. These effects are ascribed to the adsorption of reducing molecules or ions on the defect sites of the NiO surface. If a small amount of reductant, such as ferrous or iodide ions, is present in the solution, it will adsorb preferentially at the defect sites and thus hinder initiation of dissolution.

The role of lattice defects in the anodic dissolution of NiO was studied by Yoneyama and Tamura [613]. According to Toshima [614], the flat band potential of passive nickel resembles that of NiO. Independent of the mechanistic viewpoint of dissolution of passive nickel, this process is referred to Ni^{2+} in the film and not to Ni^{3+}. It has been observed that the activation energy for NiO electric conduction is lower than that corresponding to passive nickel dissolution at values of resistance up to 1000 Ω/cm.

4.4.2. Lithiated Nickel Oxide Electrodes in Aqueous Solution

Lithiated nickel oxide is one of the few transition metal oxides with well-characterized semiconducting properties. NiO has a very wide energy band gap [615] and therefore is an insulator in the pure state. When doped with lithium ions ($Li_xNi^{2+}_{1-2x}Ni^{3+}_xO$), it becomes a p-type semiconductor. Verwey, Haayman, and Romeyin [616] first suggested that this conduction arises because the addition of lithium generates an equal number of Ni(IV) ions. The mechanism of conduction is then thought to consist of the exchange of holes between Ni^{2+} and Ni^{3+} ions in the cation lattice (hole hopping mechanism). Exchange only takes place over an energy barrier, hence the drift mobility of the holes is very small and increases with temperature [617].

Adams, Bacon, and Watson [618] found that the corrosion resistance of biporous nickel electrodes could be improved significantly by soaking them in lithium hydroxide solution followed by heating at 700°C in air. This treatment produces a thin surface layer of lithiated nickel oxide, which has a very low concentration of nickel cation vacancies compared to pure nickel oxide and is therefore more corrosion-resistant. In addition, the incorporation of each lithium cation into the nickel-oxide lattice leads to the creation of a Ni^{3+} cation to preserve electroneutrality. This process increases the p-type semiconductivity by introducing acceptor levels into the nickel oxide band gap and produces a profound influence on its adsorptive properties, and consequently the electrocatalytic properties.

Rouse and Weininger [569] determined the distribution of charge and potential at single crystals of lithiated-Ni-oxide/electrolyte interface. The system nearly behaves as an ideally polarizable electrode over an appreciable potential range. Near zero potential (NHE) there is some evolution of hydrogen, and above 1 V there is evidence for faradaic processes. A space charge region due to the exhaustion of holes near the surface of the NiO is observed at weakly anodic potentials. The flat band potential for NiO is given by $E_{FB} = 0.94 - 0.050$ pH. At the flat band potential the potential drop in the Helmholtz double-layer region is controlled by an adsorbed layer on the oxide in equilibrium with protons in solution. The resistive component at the flat band potential is determined by the crystal resistance. The measured capacitance at potentials more anodic than the flat band potential is not controlled by the semiconductor. The behavior is complex and is probably related to the faradaic processes at the electrode surface.

Yohe, Riga, Greef, and Yeager [619] reported that linear sweep voltammetry of lithiated NiO electrodes in the form of mosaic crystals, in aqueous H_2SO_4, HCl, $HClO_4$, and NaOH solutions, reveals a reversible peak at 0.9 V (NHE) which closely corresponds in potential to that assigned to the Ni(II)/Ni(IV) couple. In H_2SO_4 solutions a second peak at 1.4 V closely corresponds to the potential for the Ni(III)/Ni(IV). Dissolution rates in HCl solution at elevated temperature are surprisingly small. A step-like increment in dissolution rate occurs at potentials corresponding to the first peak. Various redox couples (ferrocyanide/ferricyanide, oxygen reduction, etc.) on lithiated NiO electrodes impose a relative inhibition of cathodic processes with respect to the anodic ones, as would be expected generally for a p-type semiconductor.

Tseung, Hobbs and Tantran [620] found that the activity of NiO electrodes doped with various percentages of lithium and chromium exhibit a sixfold increase in performance at 220°C compared to those tested at 150°C, as referred to oxygen reduction in KOH. It was suggested that the rate-controlling process is the chemisorption of oxygen and that the enhanced performance at 220°C is due to oxygen being dissociatively chemisorbed above the Neel point of NiO (220 to 250°C), thus bypassing the peroxide intermediate which is usually formed at lower temperatures. At 220°C, NiO doped with 10 at% Li gave 300 mA/cm^2 at 0.9 V vs dynamic hydrogen electrode, compared with 160 mA/cm^2 at 0.9 V for a commercial platinum black electrode.

An X-ray diffraction study of the phases produced and their geometric distribution during formation of Edison-type tubular positive nickel electrodes was made by Tuomi [598] using a 4.3 N KOH-1.0 N LiOH solution. The initial charging in lithiated KOH electrolytes converts the $Ni(OH)_2$ to a β-phase having a solid solution range for Ni(II) and Ni(IV). The β-crystalline structure is related to $LiNiO_2$. This phase can be further oxidized to α-phase, a tetravalent nickelate having appreciable Ni(III) solid solubility. α-Phase is most readily formed during $Ni(OH)_2$ oxidation in concentrated KOH or NaOH electrolytes, but does not form in LiOH electrolytes. The addition of LiOH to the battery electrolyte contributes to α-phase discharge and limits α-phase re-formation. The contribution of the α-phase to electrochemical capacity is limited by a low effective exchange current as compared to the β-phase.

Tuomi and Crawford [621] observed that the anodic oxidation of $Ni(OH)_2$ in pure LiOH electrolytes involves lithium migration into the active material. The β-phase is characterized at room temperature by Seebeck coefficients near -100 μV/°K and electrical conductivities of 1 mho/cm, in contrast to the positive Seebeck coefficient for lithium nickelate.

Kober and Lublin [622] investigated the potassium content of sintered nickel electrodes impregnated with nickel hydroxide by means of electron probe analysis. The electrodes were cycled in 31% KOH each 3 hr. The charged state appears to have a higher potassium content than the discharged state. The higher potassium concentration is observed during the initial charging. It is associated within the $Ni(OH)_2$ structure in the discharge state. A residual amount always remains incorporated with the crystal structure. X-ray scans reveal localized areas of high potassium concentration. The addition of LiOH to the electrolyte of alkaline cells improves their performance [581]. The oxygen overvoltage increases from KOH to LiOH. Slaidin and Lukovtsev [623] concluded that alkali ions displace Ni^{2+} or Ni^{3+} from the oxide lattice and affect the proton transfer rate through the oxide film. Proton diffusion is highest in Li^+-containing electrodes.

4.4.3. NiO Electrode in Fused Electrolyte

Hill, Porter, and Gillespie [69] measured the potential developed between a nickel covered with NiO and a platinum wire over which air or oxygen was swept. The electrolyte was the Li_2SO_4-K_2SO_4 eutectic mixture containing dissolved CaO and operated between 550 and 750°C. As the metal oxide is very slightly soluble in this melt, the measured potential corresponds to the formation of the metal oxide starting from the elements. The formation of NiO occurs below 658°C and probably a solid solution of $LiNiO_2$ in NiO is formed at higher temperatures. Nickel electrodes dipped into oxo-anion-containing melts are covered with an oxide layer under certain circumstances (see Section 1.3.A and 4.2.D). The good adhesion of NiO often observed is due to the fact that the coefficients of thermal expansion of both oxide and metal are very similar [624, 625].

4.4.4. Solid Electrolytes

The specific conductivity, κ, of NiO as a solid electrolyte is a function of oxygen partial pressure according to

$$\kappa \sim p_{O_2}^{1/n} \tag{4.4.21}$$

where, according to Baumbach and Wagner [626], n is equal to 4, but according to Mitoff [627] is equal to 6. Pizzini and Morlotti [94] obtained the following expression for the specific conductivity in the temperature range 800 to 1050°C.

$$\kappa = \kappa^{\circ}\, p_{O_2}^{1/n} \exp\left[-\frac{\Delta H^*}{RT}\right] \tag{4.4.22}$$

At $p_{O_2} < 10^{-1}$ atm, n is about 6. The overall activation enthalpy, ΔH^*, varies from 38.9 to 18.7 kcal/mole depending on sample composition and preparation.

Sockel and Schmalzried [628] determined the existence of double ionized cationic vacancies in NiO as the existent punctual defects by coulometric titration according to Wagner [629]:

$$\begin{bmatrix}\text{neutral}\\ \text{cationic}\\ \text{vacancy}\end{bmatrix} = \begin{bmatrix}\text{single}\\ \text{ionized}\\ \text{cationic vacancy}\end{bmatrix} + \begin{bmatrix}\text{electron}\\ \text{in}\\ \text{defect}\end{bmatrix} \tag{4.4.23}$$

The net positive charges are responsible for the conduction mechanism. The conductivity will depend on the number of vacancies formed. The latter are generated if oxygen is introduced into the crystal.

Aiken and Jordan [630] investigated the electrical transport of single crystal NiO, grown by halide decomposition and by the flame fusion method. The dc conductivity is characterized by a single activation energy over a wide temperature range. At sufficiently low temperature the activation energy begins to decrease with decreasing temperature. The ac conductivity is frequency dependent, with the dispersion becoming more pronounced as the temperature is lowered. Experimental data are indicative of a mechanism involving a wide distribution of relaxation times. A model based on the hopping of a small polaron between a pair of randomly distributed acceptors provides a possible explanation of the frequency dependence. The experimental temperature dependence suggests that hopping of carriers between groups of three or more acceptors is also important. In addition to the interacceptor hopping conduction, a second mechanism involving intraacceptor hopping has been observed in samples grown by halide decomposition.

Neuimin, Paliguev, Strekalovskii, and Burov [631] studied the electrical conductivity in the system ZrO_2-CaO-NiO (85:25:2 to 5 mole % NiO). Over a wide range of temperatures, the electrical conductivity falls monotonously and remains practically purely ionic, as in the case of the ZrO_2-CaO mixture without additives, with an increase in the NiO content of the mixture up to 30 mole %.

4.5. DISSOLUTION AND PASSIVATION OF NICKEL ALLOYS

4.5.1. Binary Alloys

4.5.1.1. Nickel-Copper Alloys. Osterwald and Uhlig [405] obtained potentiostatic anodic polarization curves on nickel and nickel-copper alloys at 25 and 40°C in deaerated 1 N H_2SO_4. The passive film formed on copper-nickel alloys is probably similar in structure to that on pure nickel and continues to exist on alloys containing more than about 30% nickel. The critical current density for passivity and the minimum current density to maintain passivity move to higher values as copper is alloyed with nickel. Above about 70% copper, vestiges of passivity are lost and the alloy behaves as pure copper. The critical alloy composition is related to a filled d band electronic structure corresponding to a relatively short life of the adsorbed passive film above 70% copper. The anodic potentiostatic behavior of alloys of the nickel-copper system in 0.5 N $NaClO_4$ shows that alloys with less than 51 at.% Ni do not passivate. The passivation potential decreased with increasing nickel content. At 74 to 88% Ni it attains a constant value at 0.23 V, indicating a uniform surface state. The formation of NiO may be the cause of passivation [632, 633].

Rubin [634] recorded the individual rates of dissolution of the components of a binary alloy of copper and nickel by assembling two mercury pools, one before and the other after the alloy electrode, electrically balanced in such a way that the latter one is very sensitive to small quantities of copper dissolved from the alloy. Thus current densities for copper and nickel could be obtained and independently plotted as a function of potential.

4.5.1.2. Nickel-Chromium Alloys. Bond and Uhlig [635] measured the corrosion characteristics in sulfuric and nitric acid of nickel-chromium alloys containing up to 29% chromium. Specific alloying proportions of passive compositions were evaluated in relation to electron configuration of the alloy system by critical current density measurements.

Horvath and Uhlig [636] investigated the critical pitting potentials for binary chromium-nickel alloys in 0.1 N NaCl. They became more resistant to pitting with increasing chromium content, particularly in the 10 to 20% chromium region.

Tikkanen and Hyavarinen [637] studied the passivity of nickel-chromium sintered alloys in deaerated 1 N H_2SO_4. Sintered alloys were less active than ordinary cast alloys. A critical composition of 14 to 15% Cr divides alloys into active and passive in deaerated solutions. The critical point occurs at ~ 8 to 9% Cr in highly aerated solutions.

4.5.1.3. Nickel-Iron Alloys. Dahms and Croll [638] studied the electrodeposition of nickel-iron alloys on a rotating disk electrode and observed the suppression of nickel discharge only when the surface pH is high enough to cause hydroxide formation. The adsorption of $Fe(OH)_2$ suppresses the deposition of nickel but permits a high rate of iron discharge. The effect of temperature and other experimental parameters on the electrodeposition of nickel-iron were also considered and satisfactorily compared to theoretical rate equations derived from reaction schemes [639, 640].

Nickel-iron alloys were studied by Economy, Speiser, Beck, and Fontana [641] using potentiostatic and amperostatic methods to find the effect of the composition and pH on the electrochemical properties and corrosion resistance.

4.5.1.4. Nickel-Molybdenum Alloys. Uhlig, Bond, and Feller [642] studied the corrosion rates of 3 to 25% molybdenum-nickel alloys in hydrogen chloride solutions including the corrosion potentials, the critical current density for passivity, and they measured the anodic polarization curves in the active region. Nickel-molybdenum alloys corrode under an anodic control; their corrosion potentials are more active than the Flade potentials which correspond to the onset of passivity.

Nickel containing 3.2% molybdenum shows a lower critical potential. Higher percent molybdenum-nickel alloys appear to fall within the transpassive region corroding anodically as MoO_4^{2-} plus Ni^{2+} without pitting [635].

According to Borggraefe and Feller [643], the corrosion rate of nickel-molybdenum alloys is not lowered continuously with increasing molybdenum content. The current/potential curves for 14% nickel-molybdenum alloy shows a break near the Curie temperature. The effect was explained by the difference in the electronic configuration of the alloys in the ferromagnetic and in the paramagnetic state, and the additional influence of adsorbed ions.

4.5.1.5. Nickel-Tin Alloys. The electrochemical behavior of nickel-tin alloys in $NiCl_2$, $SnCl_4$, $SnCl_2 + NiCl_2$, $FeCl_3$, H_2SO_4, $HNO_3 + HCl$, and H_3PO_4 solutions, at different concentrations, pH, and temperatures, was investigated by various authors [644, 645]. The alloy immersed in a solution of $NiCl_2$ or of $SnCl_4$ separately acquires a passive state. Passivity was absent with mixtures of those substances.

4.5.2. Ternary Alloys

4.5.2.1. Copper-Nickel-Zinc Alloys. According to Mansfeld and Uhlig [646], potentiostatic anodic polarization data show that a transition from active to passive state is observed for nickel-copper-zinc alloys when the 3d band of energy levels in the alloy is unfilled. The calculated critical composition comes close to the observed value, assuming that alloyed copper donates one electron and alloyed zinc two electrons per atom.

Stolica and Uhlig [647] measured the critical composition for passivity in copper-nickel-zinc solid solution alloys containing 32 to 45% nickel from potentiostatic measurements of critical and passive current densities in 1 N H_2SO_4. Alloys fall into two groups. The first behaves as pure copper. The second exhibits a passive current density which lies appreciably below the critical current density for anodic passivation and behaves more like passive nickel.

4.5.2.2. Nickel-Copper-Aluminum Alloys. Mansfeld and Uhlig [646] determined polarization characteristics of nickel-copper-aluminum ternary alloys. That of the 45% nickel-copper-aluminum alloy shows that aluminum contributes three electrons per atom [647]. Secondary passivity, which is observed for pure nickel in a potential region nobler than that for primary passivity, is similarly observed in these alloys, but again only when the d band of energy levels is unfilled, as confirmed by potential decay curves either in the primary or secondary region. Flade potentials are not greatly sensitive to aluminum content nor to the absolute number of electron vacancies. However, alloyed aluminum affects the passive current density for the 45% nickel ternary alloys, accounting for a minimum at 2.35% aluminum. The primary and secondary passive films are considered to have an adsorbed structure consisting mainly of oxygen in mono- or multilayers which increases the overvoltage for anodic dissolution of the alloy. The film corresponding to secondary passivity is less stable and presumably thicker than that corresponding to primary passivity. Both films are considered to be intermediates in the formation of stoichiometric metal oxides.

4.5.2.3. Chromium-Nickel-Iron Alloys. Feller and Uhlig [648] showed that a relationship exists between the electronic configuration and the passivity properties of chromium-nickel-iron alloys in 1.25 N H_2SO_4. Critical current densities for passivity were measured on single phase alloys as a function of chromium content for 20, 40, 50 and 60% nickel compositions. Major discontinuities in slope approximate the ratio 12/88 for Cr/Fe in the 20 and 40% nickel alloys series and 14/86 for Cr/Ni in the 50 and 60% nickel series. These ratios correspond to observed critical compositions for passivity in chromium-iron and chromium-nickel binary systems. Additional discontinuities occur at the critical binary chromium to iron ratio in the 50 and 60% nickel series. This behavior is interpreted in terms of separate electron interaction in the ternary system between chromium and iron distinct from the interaction between chromium and nickel. Flade and activation potentials become more active with increasing chromium content, corresponding to increased stability of passivity. In accordance with greater stability, average time for breakdown of passivity increased exponentially with chromium content. The pitting corrosion of 18.6% Cr-9.9% Ni-Fe alloys was investigated by Stolica [649].

4.5.3. Multicomponent Nickel Alloys

Greene [364] used the potentiostatic technique to study the behavior of nickel and of various nickel alloys in hydrochloric and sulfuric acids at various concentrations. The anodic curves indicated that some alloys, which exhibit the active-passive transition, are resistant to corrosion in oxidizing media. Greene and Leonard [650] measured potentiostatic anodic current/potential curves using 18 Cr-8 Ni stainless steel in 1 N H_2SO_4. Passivity is a continuous process that begins in the active region and becomes the controlling step at the primary passive potential [358].

Cartledge [651] determined potential/time curves of stainless steel electrodes (0.05 carbon, 9.75 nickel, 18.10 chromium, 0.71 manganese, 0.59 silicon, and 0.57 niobium), either preoxidized or previously saturated with hydrogen, in aerated 0.1 N H_2SO_4 at 85°C. The potential passes through characteristic inflections and exhibits a "critical recovery potential" above which spontaneous passivation occurs. This potential is approximately equal to the hydrogen potential for the same conditions.

Morris and Scaberry [652] applied an anodic rapid scan potential to measure the primary passive potentials and the anodic critical current densities for some nickel alloys in nonoxidizing test media. Bianchi, Mazza, Mussini, and Trasatti [653] studied the passive behavior of 18-8 stainless steel, and Bianchi, Barosi, and Trasatti [654] investigated stainless steel AISI-304 (iron, 70.88; chromium, 18.70; nickel, 8.36; manganese, 1.71; silicon, 0.30; and carbon, 0.05). Trabanelli, Zucchi, and Felloni [655] investigated the behavior of different nickel alloys. The results are summarized in Table 4.5.1.

The anodic behavior of nickel alloys N-2 (0.3% cobalt, 0.1% iron, 0.1% copper); N-K (0.13% silicon); 79-NM (16.8% iron, 4.1% molybdenum); and 50-N (48.5% iron, 0.15 to 0.30% silicon) was investigated by Parfenov and Kamyshev [656]. Potentiostatic anodic polarization curves in 1 N H_2SO_4 at 25, 35, and 50°C were reported. The passivation current and the critical passivation current were found to increase with temperature. The effects of added potassium iodide, urea, and urotropin on the critical passivation current were also studied. It was shown that alloying shifted the nickel passivation potential in the positive direction.

Hospadaruk and Petrocelli [657] investigated the pitting potentials of stainless steels 434 (0.18% nickel), 201 (4.6%), 301 (7.7%), and 314 (13.0%). The nucleation of pits on the passive electrode depends on electrode potential. This pitting potential indicates the tendency to pitting of different alloys which have different susceptibility toward it and are pH independent between $3 \leq pH \leq 8$. Pitting severity decreases in the order 434 > 201 > 301 > 314. Higher chromium content increases the pitting potential.

According to Leckie and Uhlig [658], for 18-8 stainless steel an increase of chloride ion concentration shifts the pitting potential to more active values. Perchlorate, sulfate, nitrate, and hydroxyl ions at sufficient concentrations act as pitting inhibitors. At 0°C, 15% chromium, 13% nickel stainless steel exhibits a pitting potential 0.5 V nobler than at 25°C, corresponding to greatly increased resistance to pitting [653]. This shift is less pronounced for stainless steels containing molybdenum; in fact, at or above 1.5% molybdenum, the critical potentials at 0°C are below those at 25°C. In 0.1 N NaBr, alloyed molybdenum shifts the potentials slightly in the active direction, contrary to a marked noble shift in 0.1 N NaCl. At 0°C the potentials are still more active than at 25°C. These trends correlate with observed

TABLE 4.5.1. Corrosion and Passivation Behavior of Several Alloys

	5 N HCl				5 N H_2SO_4							
	Nitrogen		Air		Nitrogen				Air			
	E_{corr} (V)	Weight loss [mg/(cm^2) (d)]	E_{corr} (V)	Weight loss [mg/(cm^2) (d)]	E_{corr} (V)	Weight loss [mg/(cm^2) (d)]	j_{crit} (mA/cm^2)	j_{pass} (μA/cm^2)	E_{corr} (V)	Weight loss [mg/(cm^2) (d)]	j_{crit} (mA/cm^2)	j_{pass} (μA/cm^2)
Nickel	-0.337	0.43	-0.298	2.82	-0.247	0.061	55	22	-0.055	1.33	140	50
Monel	-0.343	0.57	-0.295	3.25	-0.013	0.040	140	540	-0.001	1.21	180	3800
Inconel	-0.342	0.86	-0.324	2.70	-0.247	0.081	0.85	5	-0.092	1.40	1.2	6
Corronel	-0.080	0.014	-0.060	0.26	-0.052	0.076	300	-	+0.004	0.85	300	-
Ni-o-nel	-0.274	0.014	-0.250	1.18	+0.020	a	0.007	6	+0.153	a	0.008	7

[a]The change in the weight of the samples was negligible even after 72 hr in contact with the solution under either aerated or deaerated conditions.

pitting for 15% chromium-13% nickel stainless steel and the similar alloy containing 2.4% molybdenum in 10% $FeBr_3$ both at 0 and 25°C. Absence of pitting is observed in 10% $FeCl_3$ at 0°C for 15% chromium-13% nickel stainless steel, but not for a similar alloy containing 2.4% molybdenum, which presents pitting corrosion. The pitting potential is not affected appreciably by pH in the acid range but it moves markedly in the noble direction in the alkaline range, corresponding to observed resistance to pitting in alkaline chloride media. These results were interpreted in terms of competitive adsorption of chloride ions and other anions for sites on the alloy surface. Only at a sufficiently high surface concentration of chloride ion is oxygen, which makes up the passive film, displaced locally; the passivity is destroyed by this, resulting in a pit. A review on the subject has been published by Kolotyrkin [659].

Teeple [660], Dillon [661], and Bourrat [662] investigated from anodic polarization curves obtained by potentiostatic and potentiodynamic methods the corrosion of stainless steels, nickel, and nickel alloys in organic acids. The critical anodic current density, the primary passivation potential, and the passive current density are given in Tables 4.5.2 and 4.5.3. The anodic dissolution of the stainless steels Inconel and Ni-o-nel decreases along the series: formic > acetic > lactic > citric acid. For nickel and Monel, the intensity of attack decreases along the series: formic > lactic > citric > acetic acid. In chromium and nickel-chromium stainless steels a region of secondary passivation is observed at a nobler potential in the transpassivation zone, the region being clearly defined for AISI-430 and having an inflected shape for the other steels. This has diversely been explained by the competing adsorption of oxygen and the anions, by the formation of layers by the precipitated reaction products, and by the surface enrichment in carbon which retards the rate of diffusion of the reagents.

Hoar and Scully [663] studied the influence of metal yielding on the anodic dissolution of 18-8 stainless steel in 42 wt% $MgCl_2$ aqueous solutions at 14°C under controlled potential conditions. At -0.14 V (NHE) in solution flowing at 52 cm/sec to eliminate concentration polarization, yielding of the metal anode at ~ 100%/min increased the anode current density by a factor of 10^4 or higher. A preliminary discussion of the mechanism of the mechanochemical effect was given; probably the intense disarrangement of the deforming metal surface during the arrival of dislocations pile-ups, produced by restricted slip, results in a large transient increase in the density of anodically active surface sites. Paul and Wieland [664] showed that corrosion fatigue and the tensile test on Hastelloy B provokes a negative potential shift which is due to the slip planes which intercept the surface. On stopping the deformation, the potential returns quickly to almost the original value. The potential shift is dependent on the deformation rate.

4.6. ELECTROCHEMICAL BEHAVIOR OF NiSi, NiAs, NiSb, NiS, AND $NiTe_2$ AND THEIR CONSTITUENT ELEMENTS

Electrodes formed by these compounds were investigated by Huq and Rosenberg [665] and by Huq, Rosenberg, and Makrides [365]. The cathodic polarization behavior of NiSi, NiAs,

TABLE 4.5.2. Passivation of Some Alloys in Organic Acids

	5 N Formic acid ($K = 1.76 \times 10^{-4}$)			5 N Acetic acid ($K = 1.75 \times 10^{-5}$)			5 N Lactic acid ($K = 1.38 \times 10^{-4}$)			5 N Citric acid ($K = 8.5 \times 10^{-4}$)		
	E_c(SCE) (mV)	j_c ($\mu A/cm^2$)	j_p ($\mu A/cm^2$)	E_c(SCE) (mV)	j_c ($\mu A/cm^2$)	j_p ($\mu A/cm^2$)	E_c(SCE) (mV)	j_c ($\mu A/cm^2$)	j_p ($\mu A/cm^2$)	E_c(SCE) (mV)	j_c ($\mu A/cm^2$)	j_p ($\mu A/cm^2$)
Nickel	-175	5500	-	-110	240	7.5	-220	950	27	-230	420	7.5
Monel	-10	7800	-	-35	730	13	-70	1650	35	-48	730	9
Inconel	-60	75	4.5	-150	17	3	-175	8	3.6	-120	2.8	2.5
Ni-o-nel	-150	-	3	0	-	2	+70	-	2	-43	-	2

TABLE 4.5.3. Passivation of Some Nickel-Containing Alloys in Organic Acids

	5 N Formic acid				5 N Acetic acid				5 N Lactic acid				5 N Citric acid			
	E_c (SCE) (mV)	j_c ($\mu A/cm^2$)	E_{pp} (SCE) (mV)	j_p ($\mu A/cm^2$)	E_c (SCE) (mV)	j_c ($\mu A/cm^2$)	E_{pp} (SCE) (mV)	j_p ($\mu A/cm^2$)	E_c (SCE) (mV)	j_c ($\mu A/cm^2$)	E_{pp} (SCE) (mV)	j_p ($\mu A/cm^2$)	E_c (SCE) (mV)	j_c ($\mu A/cm^2$)	E_{pp} (SCE) (mV)	j_p ($\mu A/cm^2$)
AISI 430	-510	930	-400	6	-420	56	-360	3	-420	48	-360	3	-290	-	-	3
AISI 302	-435	410	-300	5.5	-373	41	-295	3	-365	17	-270	4	-140	-	-	3
AISI 347	-425	640	-270	8.5	-375	85	-255	4	-358	23	-280	4	-175	-	-	5
AISI 316	+55	-	-	3	-85	-	-	2.5	-230	-	-	2.5	+120	-	-	4

NiSb, $NiTe_2$, and NiS at room temperature has been determined in 1 M perchlorate solutions at pH 0.04 and 10.8 up to -1.0 V (NHE). Measurements were also made on nickel and on nickel exposed to As_2O_3 and H_2S. Ni + As_2O_3 exhibit kinetic peculiarities which are probably attributable to hydride formation. The hydrogen evolution reaction in alkaline solution exhibits a Tafel slope greater than 0.100 V/decade and an exchange current density between 10^{-5} and 10^{-8} A/cm^2. In acid solution, Ni, NiSi, NiSb, and Ni + As_2O_3 electrodes present slopes close to 0.120 V/decade while $NiTe_2$, Ni + H_2S, and NiS show slopes between 0.050 and 0.080 V/decade. The exchange current density in acid solution is about 10^{-5} A/cm^2 for nickel. For NiAs, $NiTe_2$, and NiSi, regions exist where the current densities at a given potential exceed those of either of the individual elements. The interpretation of results was given in terms of the role of atomic and crystal factors in the electrocatalytic activity of the electrode materials.

The anodic dissolution of nickel and of the compounds NiSi, NiSb, NiAs, and NiS was studied in 1 M perchlorate solutions at pH 0.04 and 10.8. The active dissolution at low anodic potentials is followed by a transition to a passive state where the anodic current density is nearly independent of potential. In acid solution, the critical potentials of NiSi, NiAs, and NiS are less positive than that of nickel and the critical currents are lower by two to three orders of magnitude. The critical potential of NiSb is about the same as for Sb, but the critical current density is substantially lower. The anodic current in the passive region for the compounds is greater than for passive nickel. NiSi, NiAs, and NiS show a transpassive region in which Sb, As, and S are probably oxidized to ions at a high valence state. In alkaline solutions, NiSi, NiSb, and NiS show an active-passive transition at the same potential as pure nickel. However, the rate of oxidation of these compounds in the passive state is again considerably greater than for passive nickel. It was suggested that in acid solution the passive film on nickel compounds is either a mixed oxide or an oxide having the basic structure of the second element. In alkaline solution, nickel oxide is first formed but it probably has a different structure than the oxide formed on passive nickel. The difference between acid and alkaline solutions arises from the difference in the stability of nickel oxide in the two electrolytes. In general, the oxide of the companion element or a mixed oxide is more stable than nickel oxide in acid solution at low potentials. This situation is reversed in alkaline solution, and in this case the initial passive film is nickel oxide. The electrocatalytic reduction of oxygen on nickel and nickel compounds occurs in the potential region where a passive film is stable. The order of catalytic activity for oxygen reduction is Ni $>$ NiAs $>$ NiSi in acid solution, and NiS $>$ Ni $>$ NiSi $\approx$ NiSb $>$ NiAs in alkaline solution.

5. APPLIED ELECTROCHEMISTRY

5.1. ELECTROWINNING AND ELECTROREFINING

Metal nickel is obtained from sulfide ores by means of different metallurgical processes. In the International Nickel Co. process the ores are converted into a final matte containing about 48% nickel, 27% copper, 2% iron, and 23% sulfur. This material, in the form of anodes, is refined electrolytically. In the Orford (top and bottom) process the resulting product is black nickel oxide with about 77.5% nickel, 0.1% copper, 0.25% iron, and 0.008% sulfur. The oxide is reduced in reverberatory furnaces to metallic nickel, approximately 99.4% nickel including less than 1% cobalt. Some of this metal is marketed in the form of pig or shot metal but most of it is further refined electrolytically. A summarized description of those processes and electrorefining is given in the literature [666, 667].

Nickel electrorefining involves the cathodic deposition of pure nickel from crude anodes. The nickel anodes are dissolved electrolytically in a nickel sulfate electrolyte in a two-compartment cell in which the anode and cathode compartments are separated by a porous diaphragm to avoid codeposition of nickel and the impurities. The impure anolyte is treated in a separated system for removal of impurities prior to return to the cell as catholyte. Starting sheets for cathodes are produced by plating and stripping thin sheets of nickel on stainless steel blanks. The product is 99.95% nickel including 0.3-0.5% cobalt. Sludge from the process contains platinum and other valuable metals which are recovered. Operating data on nickel refining are summarized in Table 5.1.1 as given by Mantell [666].

A different method of treating sulfide ores is the Falconbridge process. In this procedure the copper is not separated from the nickel in the smelting operation, but instead, the final matte contains both the copper and nickel that are separated by the Hybinette process, briefly, as follows: the high grade matte is roasted to remove most of the sulfur and is leached with dilute sulfuric acid to dissolve most of the copper and only a small amount of the nickel. The insoluble residue is melted and cast into anodes that contain about 65% nickel and from 3 to 8% sulfur. The anodes are dissolved electrolytically, copper is removed from the electrolyte by cementation, and refined nickel is deposited on the cathode.

Renzoni, McQuire, and Barker [668] described a process using directly the matte cast as sulfide anodes in electrorefining. This process (Fig. 5.1.1) eliminates the high temperature pyrometallurgy required for the oxidation of the matte and the reduction melting process to obtain metal anodes for electrorefining. The matte anodes contain 72% nickel, 23% sulfur, 3% copper, and 1% iron, in addition to other minor impurities. The anodic oxidation of sulfide anodes involves the oxidation of sulfide to elemental sulfur and the release of Ni(II) ions. The anode potential for the reaction $Ni_3S_2 = 3Ni^{2+} + 2S + 6e$ requires an increase in anode potential of 1.0 V as compared to the potential required for nickel metal anodes. The sludge containing 95% elemental sulfur formed during sulfide anode corrosion is facilitated by bagging each sulfide anode in open-woven, acid-resistant bags [666].

TABLE 5.1.1. Electrolytic Refining of Nickel

Electrolyte:	
Ni, g/l	50-60
Boric acid, g/l	20
Sodium sulfate, g/l	35
Sodium chloride, g/l	50
Temperature, °C	52-57
Circulation apparatus	Hard-rubber pumps
Current:	10 rotaries, each 1100 kva
A/tank	6500
V/tank	2.4
Current efficiency, %	98
kw-hr/lb metal	1.1
Anodes:	
Composition	95% Ni, 2.5% Cu, 0.75% Fe, 0.75% S, precious metal 3/4 oz/ton
Length, width, thickness, in.	36 x 27 x 2
Weight, lb	490
Mode of suspension	Cast lugs
Life, days	30-32
Scrap, %	35
Cathodes:	
Starting sheet blanks	Stainless steel
Starting sheet, length x width, in.	36 x 28
Current density, asf	15
Weight, lb	11-12
Mode of suspension	Nickel loops
Replaced after ? days	12-14
Weight, lb	135
Deposition vats:	
Length, width, depth	$16'9\frac{1}{2}'' \times 2'10\frac{1}{2}'' \times 5'2''$
Number of anodes, cathodes	31, 30
Tank material	Reinforced concrete, mastic lined
Anode mud:	
Composition:	
From primary anodes	Precious metals 12 oz/ton, Ni 15-20%
From secondary anodes	Precious metals 500 oz/ton

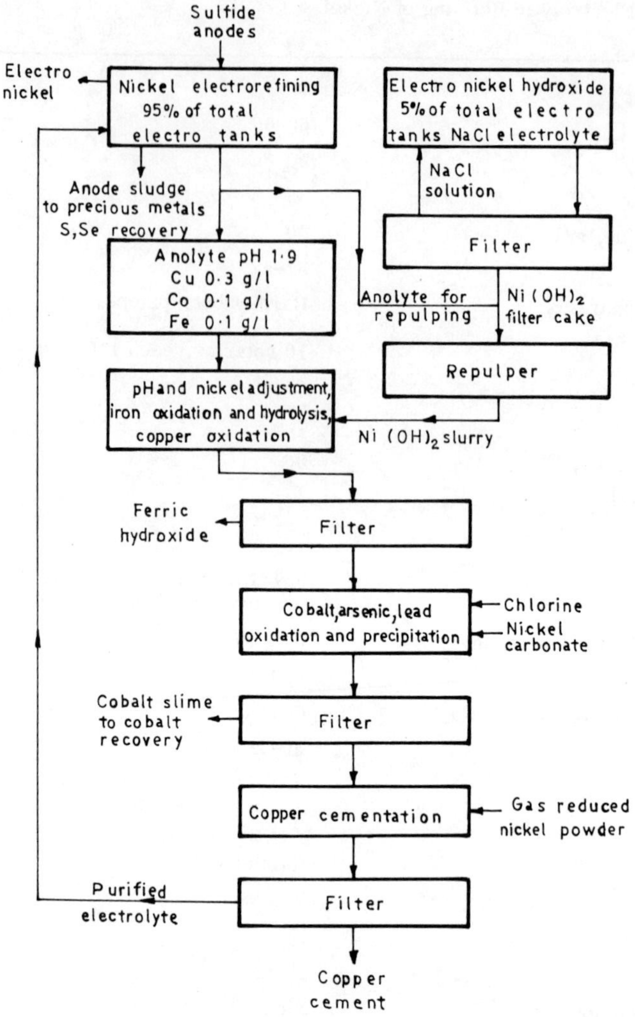

FIG. 5.1.1. Flowsheet for electrolytic nickel from sulfide anodes [666]. (By permission of McGraw-Hill.)

5.2. ELECTRODEPOSITION, ELECTROPLATING, AND ELECTROPOLISHING

5.2.1. Electrolytic Nickel Plating

The electrodeposition of nickel is used to produce, on the cathode object, coatings of nickel for decorative and protective purposes. It is applied to steel, brass, zinc, aluminum, and numerous alloys. An undercoat of copper is frequently employed to prepare a better surface for the subsequent application of the nickel coating. For many applications the nickel plating is covered with a thin coating of electrodeposited chromium for further improvement in appearance and resistance to tarnish. In modern practice, thicknesses of 0.001 or 0.002 in. are required when the nickel coat is the principal protection against atmospheric corrosion. Thinner coatings are used when the demands for protection against corrosion are not so severe. Flashes of copper are sometimes plated on copper and brass to insure cleanliness; heavy coatings of copper are applied to zinc prior to its plating in Watts-type baths or bright nickel baths; deposition of copper from a cyanide bath is preferred in plating steel.

Nickel anodes for plating operations are made of nearly pure nickel, sometimes with small additions of nickel oxide or carbon to depolarize or improve the dissolution behavior of the metal.

Several types of plating baths, each with several variations, are used for nickel plating. Bath formulas and preferred operating conditions are indicated in Table 5.2.1 [667, 669-679]. Deposits from all chloride baths are finer grained, harder, stronger, and less ductile; the throwing power is good and the conductivity is about twice that of the chloride-sulfate baths. For brightening operation, organic compounds are added and operating conditions properly adjusted [680-682]. Further details about black nickel baths are found in Refs. 683-685. Table 5.2.2 presents the mechanical characteristics of nickel electrodeposits obtained under different conditions, according to Raub [686].

5.2.2. Electroless Nickel Plating

Electroless nickel plating does not require any current, allowing nickel-plating on nonconductive materials, and yields excellent mechanical properties and uniform thicknesses of the deposits. Electroless nickel plating offers good protection from corrosion and abrasion which can be highly improved by heat treatment. This plating is, in fact, a binary alloy of nickel and phosphorus formed by many parallel thin lamellae, each thin lamella showing a structure of fibers perpendicular to the substrate metal. Electroless nickel plating was carried out under controlled conditions by Brenner et al. [687, 688]. The first baths were basic ones: $NiCl_2 . 6H_2O$ 30 g/l + $NaH_2PO_2 . H_2O$ 10 g/l + NH_4Cl 50 g/l + $Na_3C_6H_5O_7 . 5\frac{1}{2}H_2O$ 100 g/l, pH adjusted to 8 to 9 with NH_3. This led to a medium-bright deposit with a plating rate of ~6 μm/hr at 90°C. A bath with the composition: $NiCl_2 . 6H_2O$ 20 g/l + NaH_2PO_2 20 g/l + $CH_3CO_2H/CH_3CO_2NH_4$ buffer 0.5 M + $Pb(CH_3CO_2)_2$ 1 ppm, pH 4.5, and 95°C was used by Gabrielli and Raulin [689] to obtain current/potential curves at a nickel rotating disk electrode. Apparently the deposit occurs with not less than two heterogeneous steps, involving an adsorbed intermediate species, which could be $NiOH_{ad}$ [689, 690]. Technical information about electroless nickel plating is given in the literature [687].

TABLE 5.2.1. Conditions for Electrodeposition of Nickel

Solution	Substrate	Ingredients	Concentration oz/gal	Concentration g/l	pH (electrometric)	Temperature °F	Temperature °C	Normal cathode current density (A/sq ft)[a]
Electrotyping		Nickel sulfate	9	70	5.6–6.0	90	32	10–20
		Ammonium chloride	0.7	6				
Double salt	Steel, brass	Nickel sulfate	16	120	5.0–5.5	Room	Room	5–10
		Ammonium chloride	2	15				
		Boric acid	2	15				
	Steel, brass, aluminum, copper, zinc	Nickel sulfate	30–40		2.0–2.5 or 5.0–5.2	100–150		25–50
		Nickel chloride	6					
		Boric acid	4					
Sulfate-fluoride	Magnesium	Nickel sulfate	8		5.5	104		10–30
		Boric acid	4.6					
		Ammonium fluoride	4.6					
		Duponol ME	0.0067					
Fluorborate	Steel, copper	Nickel fluorborate		300	2.7–3.5	130		50
		Nickel		75				
		Boric acid		30				
Sulfamate	Cladding substrates	Nickel sulfamate		400–450	3.3–3.9	120–145		15–40 (cathodic)
		Boric acid		30–35				5–10 (anodic)
		Wetting agent		0.5–1.0				
Barrel plating		Nickel sulfate	20	150	5.0–5.5	75–90	24–32	-
		Ammonium chloride	4	30				
		Boric acid	4	30				

High sulfate	Nickel sulfate	13	100	5.3-5.8	70-90	21-32	10-35
	Ammonium chloride	4	30				
	Anhydrous sodium sulfate	13	100				
	Boric acid	2	15				
Black nickel	Nickel sulfate	10	75	5.6-5.9	120-130	49-54	5-20
	Nickel ammonium sulfate	6	45				
	Zinc sulfate	5	37				
	Sodium thiocyanate	2	15				
Hard nickel	Nickel sulfate	24	180	5.6-5.9	110-140	43-60	25-50
	Ammonium chloride	3.3	25				
	Boric acid	4	30				
Chloride	Nickel chloride	40	300	1.5-2.0	140	60	25-200
	Boric acid	4	30				
Chloride-sulfate	Nickel sulfate	26	200	1.5-2.0	115	46	25-100
	Nickel chloride	23	175				
	Boric acid	5.3	40				
Chloride-acetate	Nickel chloride	17.2	130	4.5	120	49	50
	Nickel acetate	18	135				
Pyrophosphate-chloride	Nickel chloride		118.9	9.5	-	60	6 A/dm^2
	Pyrophosphate (P_2O_7)		234.8				
	Ammonium citrate		33.3				
	Ni (0.5 M)		29.4				

[a]Cathode current density permissible in still baths or with agitation only sufficient to prevent stratification. Higher current densities are possible with increased rate of agitation.

TABLE 5.2.2. Mechanical Properties of Nickel Deposits Obtained Under Different Conditions

Bath no.	Type of bath	Tensile stress (kp/mm^2)
1	Watts bath, commercial chemicals, pH 3	22
2	As 1, after treatment with H_2O_2 and activated carbon, and preliminary electrolysis	10
3	As 2, pH 6.0	17
4	Hard nickel bath with NH_4^+, pH 5.8	11
5	As 4, pH 6.2	10
6	As 4, pH 6.8	23
7	Watts bath pH 4.2, 1 g/l H_2O_2 (30%)	Very high, cannot be measured
8	As 7, after standing for 1 hr at 54°C	43
9	As 7, after standing overnight at 54°C	10
10	Watts bath pH 4.2 + wetting agent + 0.02 g/l coumarin	25
11	As 10, + 2 g/l saccharin	3.1
12	As 11, + another 2 g/l saccharin	1.9
13	Nickel chloride bath, pH 5.0	22

5.2.3. Electropolishing of Nickel

Electrolytic polishing is usually applied, not in place of the entire mechanical process, but only in the final operations. As developed by Jacquet [691, 692], the roughly ground specimen is connected as an anode in an electrolytic bath under regulated conditions of composition, temperature, and current density. This causes a selective attack on the protruding irregularities and results in a highly polished surface. According to Higgins [693], the anodic dissolution in <7 N HCl solutions is diffusion controlled. At prepolarized anodes in >7 N HCl solutions the dissolution is also diffusion controlled, but at nonprepolarized anodes in these solutions the simple diffusion mechanism no longer holds. Table 5.2.3 lists some sets of conditions that have been used for electrolytic polishing of nickel [667, 692].

TABLE 5.2.3. Recommendations for the Electrolytic Polishing of Nickel [667]

Electrolyte	Current density	Temperature	Time
210 ml perchloric[a] acid (55° Be) 790 ml glacial acetic acid	50 A/dm^2		
Sulfuric acid (45 to 62° Be)	280 to 560 A/ft^2	85° to 140° F	10 to 25 sec
2 vols absolute methyl alcohol 1 vol concentrated nitric acid	5 to 10 A/in^2	20° to 30°C	5 to 20 sec
2 volt glacial acetic acid 1 vol perchloric[a] acid	High		1 to 5 min
Dilute sulfuric acid (73% by wt)	250 A/ft^2	30°C	0.5 to 2.5 min
15% sulfuric acid 63% orthophosphoric acid 22% water[b]	100 A/ft^2	40°C	20 min

[a]Employ caution when using. May be explosive under certain conditions.
[b]Also used for Monel, Nichrome, and Chromel.

5.3. ELECTROCHEMICAL PREPARATION OF NICKEL AND NICKEL COMPOUNDS

5.3.1. Nickel Powder

Electrolytic procedures have been applied to obtain nickel powder for powder metallurgy. Electrolytically precipitated nickel powder is formed as a nonadherent loose crystal structure material in cells. Ibl and Truempler [694] investigated the fundamental aspects for nickel powder production.

Nickel powder can be obtained from neutral or acid baths, as suggested by Mantell [666], employing an ammonium chloride circulating electrolyte, where the pH was adjusted by means of sodium hydroxide, using nickel anodes and nickel cathodes. The electrolyte composition is 300 g/l NH_4Cl + 2 g/l NaOH, operated at 70°C, and 20-30 A/ft^2 cathode current density. The final product is 98-99% Ni. Additives such as NH_4SCN, KNa-tartrate, urea, sugar and glycerol improve the characteristics of the process.

Acid baths are also described in the literature [35]. Sulfur-free nickel powder is obtained from a 0.6 $\underline{N}$ $NiCl_2$ + 1 $\underline{N}$ NH_4Cl solution at 50°C, pH 1, and current density of 10-20 A/dm^2.

Other acid baths contain $NiSO_4$, $(NH_4)_2SO_4$ and NH_4Cl at pH 2 to 4; they operate at 40 to 50° C and 50 A/dm^2.

Homogeneous powder of 0.62-0.88 specific gravity, suited for battery plates, is obtained from a $NiCl_2$ and NH_4Cl solution at pH 3 to 5, 5 A/dm^2, and 80°C.

Experimental studies on the preparation conditions of powdery or spongy nickel deposits from sulfamate baths were made [695] with a 0.1 to 0.125 M Ni(II), 0.6 to 0.7 M ammonium sulfamate, and 0.1 M NH_4Cl solution at pH 4.6 to 5.0, 25 to 35°C, and current density 5 to 20 A/dm^2.

5.3.2. Ultrahigh-Purity Ni-Single Crystals

The method proposed by Albano and Soden [696, 697] consists of the primary electrochemical winning of nickel from rigorously purified $NiCl_2$ solution. Further refinement of the plated nickel and consolidation into single crystal form was accomplished by zone melting of individual bars of the material after a surface oxidation treatment.

Ni-single crystals with resistance ratios ($R_{273°K}/R_{4.2°K}$) of about 4,000 have been prepared by electron beam float zone melting. These ratios were obtained by the selective removals of carbon by oxygen treatment during the zone melting of electrolytic nickel. Total carbon and iron contents of nickel, neglecting all other contributions to the residual resistance, must be less than 5 ppm to obtain the above-mentioned resistance ratio.

5.3.3. Electrochemical Preparation of Nickel Compounds

NiO can be obtained by electrolyzing nickel in molten NaOH at 650 to 850°C in alumina containers with a current density of 1 to 20 A/cm^2. The product contains metallic sodium as a $NaNiO_3$ solid solution. The electrolysis in molten KOH yields black NiO needles together with other by-products [698].

$Ni(OH)_2$ is electrolytically formed on a nickel anode using an alkaline solution. The electrolysis of a neutral KCl solution between a nickel anode and a platinum cathode initially gives Ni(II) which afterwards precipitates as $Ni(OH)_2$. The same product is obtained by electrolysis of aqueous solutions of NaOH, KOH, Na_2CO_3, and K_2CO_3, with addition of alcohol or aldehyde, between nickel electrodes. Without the addition agent the precipitate turns black. The $Ni(OH)_2$ current efficiency is acceptable with solution concentrations lower than 0.025 M [699]. The anodic current/potential curves of 0.1 and 5 N NaOH aqueous solutions, saturated with hydrogen at 80°C, present discontinuities at 0.07 to 1.41 V for the former and 0.18 to 1.31 V for the latter. Within these potential limits $Ni(OH)_2$ is formed. At higher potentials $Ni(OH)_2$ is oxidized [415]. $Ni(OH)_2$ is also formed at the cathode when 0.005 to 0.02 N $NiSO_4$ solutions are electrolyzed and hydrogen is evolved. At higher Ni(II) concentrations $Ni(OH)_2$ formation no longer occurs, as metallic nickel and nickel hydroxosulfate ion are formed [700]. The same process occurs with acetate buffered 0.925 N $NiCl_2$ on platinum or silver cathode and nickel anode at pH 4 to 6, 18 to 20°C, and 25 mA/cm^2 [701]. The conditions for flocculation of colloidal $Ni(OH)_2$ have also been investigated [702, 703].

Compounds such as $NiO_{1.33}.aq$, $Ni_3O_2(OH)_4$, and $Ni_3O_4.aq$ are formed anodically from acetate buffer $NiSO_4$ solutions at 40 to 60°C. Similar compounds are obtained from $Ni(NO_3)_2$ solutions [561, 562].

β-NiOOH is obtained by electrolysis of a 0.5 N $Ni(NO_3)_2$ + 1 N sodium acetate solution at 10 mA/cm^2. The preparation is improved with the addition of $LiNO_3$, $Ca(NO_3)_2$, and $Sr(NO_3)_2$ together with sodium acetate. Potassium acetate and KNO_3 can also be employed. The higher the sodium ion, or foreign ion concentrations, the larger the impurity content of the product [579, 704].

NiI_2 is formed by electrolysis of Cu_2I_2 dissolved in acetonitrile at room temperature between a platinum cathode and a nickel anode, under nitrogen. The efficiency is nearly 100% [705].

A compound approaching the stoichiometry Ni_3S_2 is deposited on a platinum cathode by electrolysis of a warm solution of thiosulfate-enriched nickel thiosulfate using a nickel anode. The concentration of the thiosulfate determines the sulfur content of the product. Ni_3S_2 is also obtained on a stainless steel cathode from a solution containing 100 g/l $Ni(NH_4)_2(SO_4)_2 \cdot 6H_2O$ + 10 g/l $Na_2S_2O_3 \cdot 7H_2O$ + 15 g/l Na citrate.$5H_2O$ (pH 6.4) at 30°C and 0.02 to 0.17 A/dm^2 with a lead anode. On increasing the current density the sulfur content diminishes [706, 707]. Cathodic discharge of nickel sulfide in a propylene carbonate-$LiClO_4$ electrolyte has also been reported [708].

Either commercial nickel or Raney-nickel anodes were proposed for the electrochemical oxidation of different organic compounds. The reader is referred to pertinent literature [709].

5.3.4. Electrodeposition of Binary and Ternary Nickel Alloys

Binary and ternary nickel alloys have been electrodeposited on different substrates. Various processes have been developed for the codeposition of nickel with cobalt, cadmium, zinc, iron, tungsten, copper, and other metals. These deposits are used to take advantage of the special properties they possess, particularly for colorful decorative purposes. Electrodeposits are influenced by composition of the solution, pH, current density, agitation, and temperature of the electrolyte [686, 710-713]. The electrodeposition of magnetic alloys has been studied by many authors. Details about plating bath and electrodeposition conditions are given in Refs. 712 and 714-720. Table 5.3.1 summarizes the conditions for electrodeposition of different alloys.

The conditions for electrodeposition of Ni-Al coatings [721], Ni-Re [722], Ni-Ru [723], and Ni-Cr (on Ti) [724], Ni-W and Ni-Mo have also been reported [725]. Ni-Cd [726], Cu-Zn-Ni [727], Cu-Ni-Fe [728], Ag-Ni [729, 730], and Au-Ni [731] alloys have been deposited from cyanide baths. Re-Ni alloys [732] were obtained from alkaline solutions [733].

Ternary thin-film alloys containing Ni, with Fe and Ni-Co as major components, and Mn, Ru, W, V, etc. as third component, are of interest because of their magnetic properties [723, 734-738].

TABLE 5.3.1. Conditions for Electrodeposition of Nickel-Containing Alloys

Alloy	Characteristics	Bath composition		Conditions	Refs.
Ni-Fe	80% Ni	$NiSO_4 \cdot 7H_2O$	200 g/l		[739]
		$FeSO_4 \cdot 7H_2O$	8 g/l		
		$NiCl_2 \cdot 6H_2O$	5 g/l		
		H_3BO_3	25 g/l		
		Saccharin	3 g/l		
Ni-Fe	40-87% Ni	$NiCl_2 \cdot 6H_2O$	71.4 g/l	60°C, 0.35-3.5 A/dm^2	[740]
		$FeCl_3 \cdot 6H_2O$	27.0 g/l		
		$P_2O_7^{4-}$	174.0 g/l		
		Nickel	17.6 g/l		
		Iron	5.6 g/l		
		pH	8.3		
Ni-Fe	74-84% Fe	Fe(II) sulfamate	186 g/l	0.5-5.0 A/dm^2, 0.7-1.65 V, Cu cathodes	[741]
		Ni(II) sulfamate	125 g/l		
		Iron	41.9 g/l		
		Nickel	29.3 g/l		
		Na-acetate.$3H_2O$	27.2 g/l		
		pH	3.5		
Ni-Cu	Mixed crystals Ni in Cu: $E_{elecdp} = -0.75$ V Cu in Ni: $E_{elecdp} = -0.95$ V	$CuSO_4$	0.1 M		[742]
		$NiSO_4$	0.1 M		
		H_3BO_3	30 g/l		

Ni–Cu	18–55% Ni	Nickel	0.3 M	60°C, 0.25–5.0 A/dm^2	[743]
		Copper	0.05 M		
		P_2O_7	145.4 g/l		
		Pyrophosphate/metal	7		
		pH	8.7		
Ni–Cu		Citrate	75 g/l	55°C, 2.16 A/dm^2	[744]
		Metal as sulfate	0.19 M		
		pH	~5		
Monel	-	$NiSO_4 \cdot 7H_2O$	100 g/l	0.4 A/dm^2	[745]
		$CuSO_4 \cdot 5H_2O$	6 g/l		
		$C_6H_8O_7 \cdot H_2O$	100 g/l		
		KCN	3 g/l		
		$Na_2CO_3 \cdot 10H_2O$	to pH 7.5–8.0		
Ni–Cu	Ni contents increase with NH_4Cl concentration	Cu(II)	0.5 N		[746]
		Ni(II)	0.5 N		
		$K_4P_2O_7$	0.75 M		
		NH_4Cl	0–0.5 N		
Ni–Co	~20% Ni	$CoCl_2$	50 g/l	67–72°C, 10.8 A/dm^2, superimposed ac	[747]
		$NiCl_2$	50 g/l		
		H_3BO_3	40 g/l		
		pH	4.9–5.1		

(continued)

TABLE 5.3.1 (continued)

Alloy	Characteristics	Bath composition		Conditions	Refs.
Ni-Co	69-74%	$NiCl_2$	71.4 g/l	60°C, 0.35-8.4 A/dm^2, 1.1-2.3 V, Pt-Cu cathode	[748]
		$CoCl_2$	23.8 g/l		
		Pyrophosphate	174 g/l		
		NH_4-citrate	20 g/l		
		Nickel	0.3 M		
		Cobalt	0.1 M		
		pH	9.3		
Ni-Co	41-64% Ni	Ni-sulfamate	425 g/l	40°C, 1-18 A/dm^2, 1.0-4.7 V	[749, 750]
		Co-sulfamate	43 g/l		
		Total sulfamate	358 g/l		
		NaF	4 g/l		
		H_3BO_3	30 g/l		
		Nickel	100 g/l		
		Cobalt	10 g/l		
		pH	5.3		
Ni-Cr	Thin films	$Cr_2(SO_4)_3$	3 N	50°C, 25-50 A/dm^2	[751]
		$NiSO_4 \cdot 7H_2O$	0.4 N		
		H_3BO_3	25 g/l		
		Trilon B	50 g/l		
		pH	1.3-1.4		

Ni-Mo	60% Mo, 35.5% Ni, 4.5% nonmetallic	Mo(IV)	0.5 N	60°C, 16.1 A/dm^2	[752]
		Ni(II)	0.05 N		
		Na-pyrophosphate	0.15 N		
		$N_2H_4 \cdot H_2SO_4$	3.0 g/l		
		pH	8.3		
Ni-Se	Se contents decreases as cd increases	$NiSO_4 \cdot 7H_2O$	0.67 M		[753]
		$Na_2SO_4 \cdot 10H_2O$	0.25 M		
		NaCl	0.34 M		
		H_3BO_3	0.32 M		
Ni-Sn		Ethylene glycol	200 ml		[754]
		H_2O	400 ml		
		HCl (concentrated)	500 ml		
		$SnCl_2 \cdot 2H_2O$	120 g		
		$NiCl_2 \cdot 6H_2O$	300 g		
Ni-Sn		Sn(II)-fluoride	25 g/l	60-95°C, 12-24 A/ft^2	[755]
		Ni-fluoride	saturated		
Ni-Sn	67-92% Sn	Sn(II)pyrophosphate	20.6 g/l	60°C, 2.1-3.0 V, 0.5-6.0 A/dm2	[756]
		$NiCl_2 \cdot 6H_2O$	47.6 g/l		
		Pyrophosphate	0.75 M		
		NH_4-citrate	10 g/l		
		Tin	11.9 g/l		
		Nickel	11.7 g/l		
		pH	8.7		

(continued)

TABLE 5.3.1 (continued)

Alloy	Characteristics	Bath composition		Conditions	Refs.
Ni-Sn		$NiCl_2$	250-300 g/l	55-65°C, 0.3-1.5 A/dm^2	[757]
		$SnCl_2$	40-50 g/l		
		NaF	25-30 g/l		
		NH_4F	35-38 g/l		
		pH	3-3.5		
Ni-Ti	Up to 8% Ti	$Ti(SO_4)_2$	0.0625 M	1 A/dm^2	[758]
		$NiCl_2$	0.125 M		
		Lactic acid	0.5 M		
		Aminoacetic acid	0.5 M		
		H_3BO_3	0.4 M		
		$NH_4F.HF$	0.1 M		
		pH	5		
Ni-W	47-94% Ni	Nickel	0.025-0.38 M	30-90°C, 30 A/dm^2, 1.9-5 V, Pt electrodes	[759]
		Tungsten	0.005-0.7 M		
		Pyrophosphate	0.0625-0.95 M		
		$(P_2O_7^{4-})/(Ni^{2+})$	2.5		
		NH_4-citrate	0.05 M		
		pH	7.7-10.3		
Ni-Zn	2-16% Zn	Ni-sulfate	240 g/l	50°C, 1-4 A/dm^2	[711]
		KCl	20 g/l		
		H_3BO_3	30 g/l		
		Zinc	0.08-0.48 g/l		
		p-Toluene-sulfonamide	1 g/l		

Ni-Zn	δ-Phase alloy	$NiSO_4 \cdot 6H_2O$	0.2 M	30°C	[760]
		$ZnSO_4 \cdot 7H_2O$	0.3 M		
		$(NH_4)_2SO_4$	0.1 M		
		pH	4.6		
Ni-Fe-As	6-20% Fe, 1-15% As, thin films, magnetic alloy	$NiSO_4 \cdot 6H_2O$	218 g/l		[712]
		$FeSO_4 \cdot 7H_2O$	6.0 g/l		
		$NaAsO_2$	0.4 g/l		
		H_3BO_3	25.0 g/l		
		NaCl	9.7 g/l		
		Saccharin	0.8 g/l		
		Na-lauryl sulfate	0.2 g/l		
		pH	2.2		
Ni-Fe-Cr		$CrK(SO_4)_2 \cdot 12H_2O$	400 g/l	25°C, Cu cathode	[761]
		$NiSO_4 \cdot 7H_2O$	56 g/l		
		$(NH_4)_2Fe(SO_4)_2 \cdot 6H_2O$	39 g/l		
		NaF	8 g/l		
		Na_3-citrate	71-140 g/l		
		pH	1.8		
Ni-Fe-W	22-24% W, 23-43% Fe, 29-55% Ni	Nickel	5.9 g/l	60°C, 1.0-6.0 A/dm^2	[762]
		Iron	5.6 g/l		
		Tungsten	36.8 g/l		
		Pyrophosphate	104.4 g/l		
		NH_4-citrate	11.3 g/l		
		pH	9.0		

5.4. CORROSION

5.4.1. Nickel

Nickel is not a chemically active element and requires the presence of an oxidizing agent for most of its corrosion reactions. In sulfurous atmospheres, nickel will corrode at a rate not greater than 0.0002 in./yr, and will become coated with colored tarnish films. Corrosion under relative humidity higher than 70% will form a film or blur on the surface. Light speeds up the formation of this blur [667].

Distilled water and ordinary tap water have little effect on nickel. Salt water is more corrosive than fresh waters, but the metal is resistant to a sufficient degree to be useful. Neutral and alkaline salts corrode nickel to a slight degree (<0.005 in./yr); nonoxidizing acid salts to a greater degree (~0.02 in./yr). Oxidizing acid salts and mixtures of these salts with mineral acids are severely corrosive, as are some oxidizing alkaline salts such as hypochlorites when the available chlorine concentration exceeds 3 g/l. The rate of corrosion in acid varies according to concentration, air contents, impurity contents, agitation, and temperature. Organic acids and acid compounds are moderately corrosive to nickel, but neutral and alkaline compounds are practically noncorrosive [667].

Most alkalies have little corrosive effect on nickel even at high temperatures and high concentrations, except in the presence of sulfur. Anhydrous ammonia is also noncorrosive, but ammonium hydroxide in concentrations over 1% corrodes nickel appreciably. The excellent resistance of nickel to alkalies results in many commercial applications.

Common dry gases do not corrode nickel, but many gases are corrosive when wet, including nitric acid, chlorine (above 1000°F), bromine, and sulfur dioxide. Sulfur gas attacks nickel above 600 to 700°F. The mixture of nitrogen, hydrogen, and ammonia during the synthesis of ammonia is destructive for nickel but the metal is useful in the nitriding process below 1075°F. Steam at temperatures below 450°C corrodes nickel only slightly, and most of the corrosive effect is due to carbon dioxide and other impurities in the steam. Nickel is not useful for handling molten metal, except for mercury below 371°C. Molten aluminum, tin, lead, solder, bismuth, antimony, brass, and zinc attack nickel rapidly [667].

Nickel becomes passive when it is immersed in strong oxidizing agents or when it is used as an anode in electrolytes containing salts of oxidizing acids. The passivity of nickel by simple immersion is not of practical importance because it may be destroyed by slight changes in the chemical environment or mechanically. Anodic passivity is affected by the presence of impurities in the metal or in the electrolyte. Chlorides in particular tend to destroy anodic passivity, and use is made of this phenomenon in maintaining activity of nickel anodes in electroplating (see Section 4).

Guillaume, Constant, Lemaire, Valensi, and Brisou [763] studied the role of certain bacteria in the corrosion of nickel in water and seawater. Certain bacteria are inhibitors; nonreducing bacteria give hydroxide and proteolytic reducing bacteria give a complex.

5.4.2. Corrosion of Alloys

5.4.2.1. Ferrous Alloys. In general, in those media where nickel is more resistant to corrosion than iron, the beneficial effects of nickel in iron-nickel alloys increase with increasing nickel content. Iron-nickel alloys are more resistant than carbon steel to atmospheric exposure but do not match stainless steels. Strauss recommended a 28% nickel steel for resistance both to marine and industrial atmospheres [667].

The resistance of iron-nickel alloys to corrosion in aqueous solutions of any kind has been widely studied. Strauss claimed almost unlimited serviceability in seawater for 28% nickel alloy. Studies of the behavior of iron-nickel alloys in dilute hydrochloric acid showed that nickel improves the resistance of iron, and the improvement is greatest for nickel contents above about 25%. The iron-nickel alloys are not as resistant as pure nickel to nitric acid. In general, the rate of corrosion of iron-nickel alloys in sulfuric acid decreases with increasing nickel content. The relation is not always regular, and minima in the corrosion rates at about 50% nickel have been reported [667].

5.4.2.2. Nonferrous Alloys. In general, the Monels are much more resistant to acids than commercial nickel, less resistant to alkalies and equally resistant to salts. Being solid-solution alloys, they are free from local galvanic corrosion [667]. The rate of atmospheric corrosion of Monel is less than that of nickel. Colored tarnish films are formed in industrial and humid atmospheres and slow blurring occurs indoors.

Monel is highly resistant to the corrosive action of both fresh and salt water, especially to cavitation and the effects of impingement in seawater under conditions of high velocity. In stagnant seawater, accumulations of marine organisms and other solids may result in subsequent pitting underneath.

Neutral and alkaline salts, including chlorides, carbonates, sulfates, nitrates, and acetates, rarely attack Monel at a rate greater than 0.005 in./yr. Nonoxidizing acid solutions, such as zinc chloride, ammonium sulfate, and aluminum sulfate, usually attack Monel at rates of less than 0.02 in./yr.

Oxidizing acid salts, such as ferric chloride, ferric sulfate and cupric chloride, are severely corrosive. However, dilute acids containing chromates are not highly corrosive. The oxidizing alkaline hypochlorites are highly corrosive when the available chlorine concentration exceeds 3 g/l.

Corrosion of Monel by mineral acids is largely dependent on the oxidizing characteristics of the acid, degree of concentration, amount of aeration, and temperature. Monel corrosion is practically nonexistent at room temperature in air-free sulfuric acid solutions with concentrations up to 80%. In air-saturated solutions the maximum rate of corrosion occurs at 5% concentration. Boiling sulfuric acid may be handled when in concentrations lower than 20%. In hydrochloric acid the rate of corrosion varies with concentration and air contents from 0.005 in./yr in 1% air-free cold acid up to 0.25 in./yr in air-saturated 30% cold acid. At high temperatures, Monel has been used with aerated hydrochloric acid in 2% concentrations at 50°C and 1% concentrations at 80°C. Other mineral acids which are not severely corrosive are hydrofluoric, throughout a wide range of concentrations and temperatures, and pure phosphoric acid in low concentrations. Nitric and other oxidizing acids are severely corrosive. Organic acids are moderately corrosive toward Monel and may become appreciably so when saturated with air [667].

All common dry gases have little effect on Monel at normal temperatures. Chlorine, bromine, sulfur dioxide, nitric oxide, and ammonia are detrimental when water is present. As in the case of nickel, the mixture of nitrogen, hydrogen, and ammonia formed in ammonia synthesis is destructive under the conditions of temperature and pressure used. Monel resists atmospheric oxidation at temperatures up to about 540°C [667].

Molten metals, such as aluminum, tin, lead, solder, bismuth, antimony, zinc, and brass, attack Monel quite rapidly. Amalgamation with mercury is resisted up to 375°C [667].

Galvanic corrosion and the methods for its prevention are essentially the same as for nickel. North and Pryor [764] studied the influence of the structure of corrosion products on the corrosion rate of copper-nickel alloys. The corrosion rate for nickel-chromium alloys is low in sulfuric acid at normal temperatures but may be severe in hot, strong solutions. Addition of small amounts of oxidizing salts to the acid decreases the rate of corrosion of these alloys. Their use with hydrochloric acid is restricted to dilute solutions and normal temperatures. Phosphoric acid may be handled at normal temperatures. Sulfurous acid is corrosive at any temperature. The alloys are highly resistant to nitric acid at normal temperatures, but when hot nitric acid should be used, iron-chromium-nickel alloys are preferred. Nickel-chromium alloys resist all concentrations of ammonia in solution. They resist corrosion by caustic alkalies and alkaline solutions in most concentrations although concentrated sodium or potassium hydroxide solutions are corrosive. They are completely resistant to common gases when dry and at normal temperatures but are attacked by moist chlorine.

The 80Ni-14Cr-6Fe alloy showed the best resistance to reducing and to alternately reducing and oxidizing atmospheres, followed in decreasing order by the 60-14-26 and 32-20-48 alloys [667].

Two nickel-chromium alloys that are particularly valued for corrosion resistance are Inconel (77% Ni, 15% Cr, 7% Fe) and Illium "G" (58% Ni, 22% Cr, 6% Cu, 6% Mo, 6% Fe, 1% W). Inconel resists progressive oxidation at temperatures below 2000° F. It can be used safely up to about 1500° F in oxidizing sulfur atmospheres and to about 1000° F in reducing sulfurous atmospheres. Inconel is resistant to aqueous hydrofluoric acid at normal temperatures [667].

Illium "G" is virtually immune to corrosion in the atmosphere, fresh water, salt water, and neutral and alkaline salts. It is highly resistant to severely oxidizing conditions. It resists sulfuric and nitric acids over a wide range of concentrations and conditions and has limited applications in hydrochloric acid and acid chlorides. It is highly resistant to phosphoric acid and is resistant to sulfurous acid, mixtures of sulfurous and sulfuric acids, and to hydrogen sulfide. It is highly resistant to organic acids and to neutral and alkaline organic compounds [667].

Nickel-copper alloys containing more than 50% of nickel are, in general, more resistant than nickel under reducing conditions and more resistant than copper under oxidizing conditions. They have excellent resistance to atmospheric exposure — except that sulfurous atmospheres produce superficial tarnishing — to natural waters of all kinds, to salt solutions, to organic acids and compounds, and to most alkaline solutions, including concentrated caustic soda solutions at temperatures below the boiling point. The alloys are resistant but not immune to attack by mineral acids including sulfurous acid and sulfites. Dry gases, including anhydrous ammonia, are not actively corrosive, and the alloys are resistant to chlorine up to about 850°F. However, in the presence of water the alloys are attacked by nitric oxide, chlorine and other halogens, sulfur dioxide, and ammonia. In reducing sulfur atmospheres the alloys are subject to intergranular attack at temperatures above 700°F [667].

The Hastelloy alloys include Hastelloy A (60% Ni, 20% Mo, 20% Fe); Hastelloy B (65% Ni, 30% Mo, 5% Fe); Hastelloy C (58% Ni, 17% Mo, 15% Cr, 5% W, 5% Fe); and Hastelloy D (85% Ni, 8 to 11% Si, 3% Cu).

Hastelloy A is, in general, resistant to corrosion in hydrochloric acid at any concentration and room temperature, and in unaerated solutions at temperatures below about 160°F, particularly in the absence of oxidizing salts. It is resistant to sulfuric acid in concentrations up to 50% at temperatures up to the boiling point and to higher concentrations at temperatures up to about 160°F; to organic acids and their compounds; to acid chloride, sulfate, and phosphate salt solutions; to atmospheric oxidation; and to oxidizing and reducing flue gases at temperatures below about 1450°F. However, it is not recommended for service with nitric or other strongly oxidizing acids [667].

Hastelloy B shows a high resistance to any concentration of hydrochloric acid at temperatures up to the boiling point. Hastelloy B is more resistant than A, C, or D to boiling sulfuric acid up to about 60%, but is less resistant than D to boiling acid of higher concentrations. Alloy B resembles Alloy A in its resistance to miscellaneous media and like Alloy A is not recommended for service with strongly oxidizing acids [667].

Hastelloy C is unusually resistant to oxidizing solutions, especially to those containing chlorides. It is unique among alloys in its resistance to hypochlorite solutions and moist chlorine below about 100°F. It resists all concentrations of hydrochloric acid at room temperature and is somewhat superior to Alloy A in resistance to sulfuric acid. Hastelloy C is resistant to the same gases as Alloy A at somewhat higher temperatures, up to about 1800°F [667].

The most important property of Hastelloy D is its resistance to concentrated solutions of sulfuric acid at elevated temperatures. It is resistant to many corrosive media but is only moderately resistant to hydrochloric acid and is not recommended for service with nitric or other strongly oxidizing acids [667].

5.5. USE IN BATTERIES AND CELLS

5.5.1. The Alkaline Batteries

The alkaline batteries consist of an anode of $Ni(OH)_2$ and a cathode of $Fe(OH)_2$ or $Cd(OH)_2$. The electrolyte is a 21% aqueous solution of potassium hydroxide to which a small amount of LiOH has been added. The following reactions take place at the cathode and at the anode, respectively [765, 766]:

$$Fe(OH)_2 + 2e \underset{\text{discharge}}{\overset{\text{charge}}{\rightleftharpoons}} Fe + 2OH^- \tag{5.5.1}$$

and

$$2Ni(OH)_2 + 2OH^- \underset{\text{discharge}}{\overset{\text{charge}}{\rightleftharpoons}} 2Ni(OH)_3 + 2e \tag{5.5.2}$$

or

$$2Ni(OH)_2 + 2OH^- \underset{\text{discharge}}{\overset{\text{charge}}{\rightleftharpoons}} 2NiOOH + 2H_2O + 2e \tag{5.5.3}$$

leading to the overall reaction:

$$2Ni(OH)_2 + Fe(OH)_2 \underset{\text{discharge}}{\overset{\text{charge}}{\rightleftharpoons}} Fe + 2Ni(OH)_3 \tag{5.5.4}$$

or

$$2Ni(OH)_2 + Fe(OH)_2 \underset{\text{discharge}}{\overset{\text{charge}}{\rightleftharpoons}} Fe + 2H_2O + 2NiOOH \tag{5.5.5}$$

For the corresponding nickel-cadmium system, cadmium replaces iron. Since the electrolyte KOH does not take part in the above reactions, its quantity remains unchanged during the charging and discharging process. However, it has recently been shown that the positive active nickel hydroxides undergo some change in hydration and that the positive ions, K^+ and Li^+, may take part in the electrode reaction at the positive electrode. Also on charge, some of the active material of the nickel electrode is presumably oxidized to Ni(IV) with additional complications in phase changes and subsequent behavior on standing in the charged state. Under discharge state the material is $Ni(OH)_2$ [30, 580, 591, 601] which during charging oxidizes to β-NiOOH [579, 580, 591]. The active material in the charged plate consists of a nonstoichiometric oxide with Ni(III), Ni(IV), OH^-, and O^{2-} in the lattice [578] (see Section 4). Oxygen is evolved from a completely charged nickel plate at open circuit [22-24, 767] due to the decomposition of Ni(IV) present in the β-NiOOH. The oxygen formed remains occluded in part in the active material [767]. The self-discharge process is slow and the self-discharge curve is similar to that of the cathodic discharge. The reactions which take place at the anode

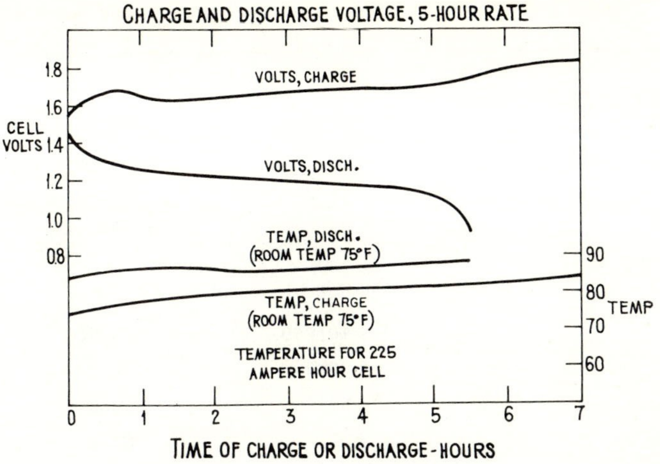

FIG. 5.5.1. Cell volts and temperature during charge and discharge [768]. (By permission of the Electrochemical Society.)

have been discussed from the kinetic point of view in Section 4.4. The thermodynamic reversible potentials of the cells are 1.35 V per cell for nickel-iron and 1.30 V per cell for nickel-cadmium. The maximum energy density in the case of the nickel-cadmium cell is equal to 0.22 kw-hr/kg. Numerous fundamental studies were published on different aspects of pocket, tubular, and sintered nickel positive plates (see references in Hoare's book [581]).

5.5.1.1. The Edison Cell. Although the efficiency of the Edison Cell is lower than that of the lead accumulator, it is much more durable and is therefore widely employed. Its steel construction and long life under severe operating conditions make it particularly suitable for heavy-duty industrial and railway applications. Vibration does not cause loss of active material from the plates nor are they affected by many of the accidental conditions of operation such as overcharging, overdischarging, short-circuiting, or reverse charging. Average life in heavy-duty services reaches up to 7 years.

The cell voltages measured during a typical charge and discharge at normal rate and room temperature are shown on the graph of Fig. 5.5.1. From these curves it will be noted that the average voltage over the discharge period is approximately 1.2 V, and the average charge voltage is about 1.69 V. Some technical data of commercially available accumulators are given in Table 5.5.1 [768].

The recommended procedure for storing batteries of this type over long periods is to discharge them completely, short-circuit, and maintain the solution level above the plate tops. There is no danger that they will be damaged by freezing.

TABLE 5.5.1. Typical Battery Assemblies; Cradle or Demountable Steel Box

Cell type	A8	C8	D8
Number of cells	30	30	30
Number of trays	5	5	5
Cells per tray	6	6	6
A-hr capacity	300	450	600
Watt-hour capacity			
per cell	360	540	720
per battery	10,800	16,200	21,600
Length: Cradle, in.	36.5	36.5	37.0
Box, in.	36.9	36.9	38.4
Width: Cradle, in.	31.1	31.1	31.6
Box, in.	31.5	31.5	31.9
Height: Cradle, in.	17.0	22.0	29.0
Box, in.	17.5	22.5	29.1
Weight in pounds			
Single cell	28.0	40. 9	54.8
Battery complete in cradle	1001	1140	1953
Battery complete in box	1114	1602	2130
Cubic feet			
Single cell	0.213	0.298	0.407
Battery in cradle	11.15	14.45	19.56
Battery in box	11.78	15.15	20.62
Pounds/ft^3			
Single cell	131.3	136.9	135.0
Battery in cradle	89.8	99.5	99.8
Battery in box	94.6	105.8	103.2
W-hr/ft^3			
Single cell	1688	1805	1772
Battery in cradle	968	1122	1105
Battery in box	917	1068	1047
W-hr/lb			
Single cell	12.85	13.20	13.14
Battery in cradle	10.78	14.20	11.05
Battery in box	9.69	10.12	10.13

5.5.1.2. The Nickel-Cadmium Batteries. In view of its characteristics, it is not surprising to find that a nickel-cadmium battery can be used as a replacement for practically any other type of battery, limited solely by cost and a slight difference in weight and volume. It has attained a wide usefulness in the industrial field for stand-by and emergency lighting, for diesel cranking, for telephone service, for railroad carlighting service, and, to a limited extent, for electric truck operations.

Cells of this type are the most highly developed sealed cells and are produced in large numbers and in many sizes and shapes. Flat button cells are available from about 50 mA-hr ($\frac{5}{8}$ in. diam) to 450 mA-hr ($1\frac{3}{4}$ in. diam). Cylindrical cells are produced in sizes ranging from the AA penlight size of 450 mA-hr up to the F-size of about 8 A-hr. A few special cylindrical cells have been made up to 50 A-hr. The prismatic or rectangular cells range in size from about 5 A-hr up to about 25 A-hr. Gasket-type seals are used for the button and cylindrical cells. However, ceramic insulating seals have been employed on some of the cylindrical and rectangular types. Some of the latter use safety relief vents for protection against excessive pressures which might develop due to some type of mishandling.

Sealed operation is made possible by using an excess of uncharged negative material (CdO) so that oxygen is evolved on the charged nickel oxide at the positive electrode. This diffuses, possibly through the separator and through the voids of the cell, to combine with the sponge cadmium metal of the negative electrode. Cadmium is a very desirable material since it neither reacts with the KOH electrolyte nor evolves gas on standing. Active materials are utilized as pressed powders sometimes enclosed in fine screens or are impregnated in sintered nickel plaques. An excellent cycle life of several thousand cycles is obtained at low depths of discharge such as 25%. However, the cycle life falls to a few hundred for deep discharges such as 75%. The addition of LiOH helps to maintain the voltage level in a continuous cycling performance. The cycle life is also adversely affected by extremes of temperature. On constant current charges, such as C/10, the voltage levels out at about 1.5 V. The cells discharge on moderate loads at about 1.2 V.

The internal pressures on change range up to about 40 psi. Protection against cell reversal is obtained by incorporating some cadmium oxide as an antipolar mass in the positive nickel electrode. Some positive active material is also sometimes incorporated in the negative plate. Lately, special high rate cylindrical cells have been produced by using spiral wound electrodes to obtain large surface areas. Some features of the sintered plate nickel-cadmium cell, such as the cell voltage vs discharge time and the relative watt-hour capacity vs discharge rate, are shown in Figs. 5.5.2 and 5.5.3, respectively. The temperature coefficients of nickel-cadmium cells are given in Table 5.5.2. At 25°C the emf of the cell is 1.27 V. The effect of LiOH in nickel-cadmium batteries has been considered by numerous authors. See Section 4.4.B and also references in Hoare's book [581].

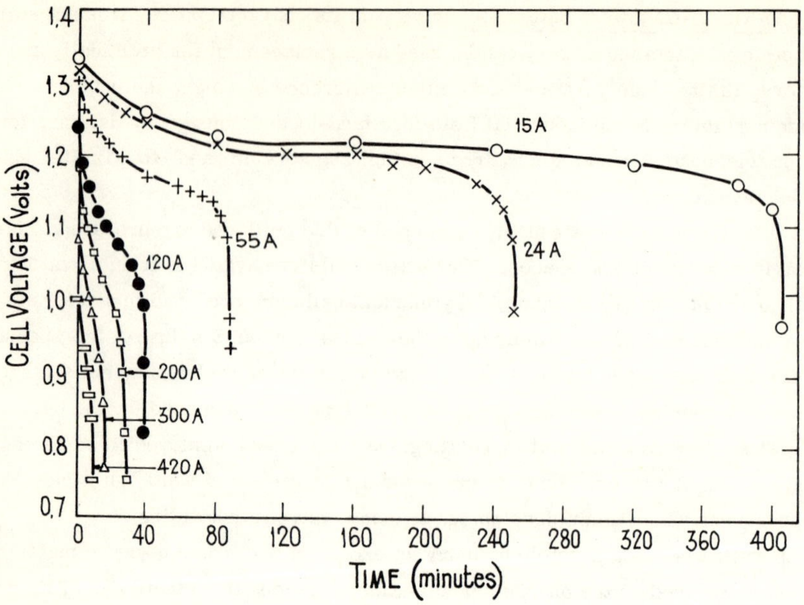

FIG. 5.5.2. Cell voltage vs discharge time, all discharge rates at 85 ± 5° F, for a nickel-cadmium sintered-plate cell. Rated capacity: 75.0 A-hr [768]. (By permission of the Electrochemical Society.)

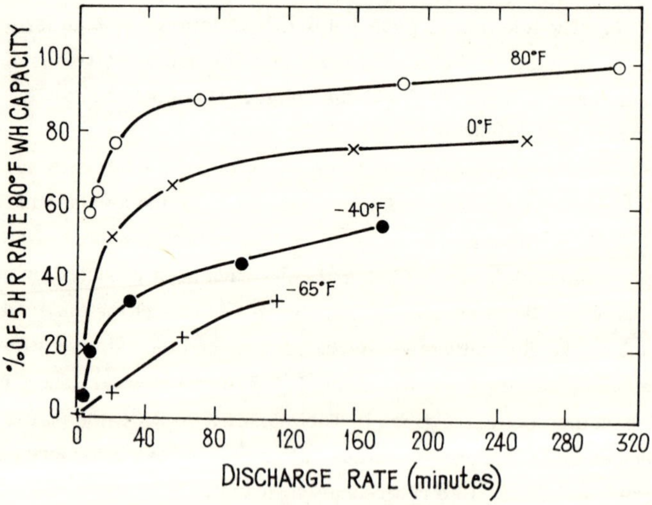

FIG. 5.5.3. Percent of 5-hr rate, 80° F W-hr capacity vs discharge rate (min) for a nickel-cadmium sintered-plate cell. Rated capacity: 75.0 A-hr [768]. (By permission of the Electrochemical Society.)

TABLE 5.5.2. Temperature Coefficients of Nickel-Cadmium Cells

Electrode	dE/dt (V/°C)	Refs.
Sintered nickel	-0.0004	[590]
Pocket nickel	-0.0003	[591]
Tubular nickel	-0.0004 to -0.0008	[769]

5.5.2. Nickel Catalyst Electrodes

A porous nickel electrode acting as an oxygen diffusion electrode was used by Bacon [770]. The electrode was obtained by sintering nickel at high temperature and pressure.

Various authors obtained fuel cell electrodes from nickel powder [771-774]. Appelt, Dominiczak, Nowacki, and Paszkiewicz [775] described the sintered structure, catalytic properties, overpotential for hydrogen evolution, and porosity conditions, at 500 and 700°C, of nickel powder electrodes. Freundlich [776] described the characteristics, inner area, and pore size distribution of a sintered nickel plate electrode.

The Raney-nickel hydrogen electrodes were made by pressing aluminum-nickel powder mixtures followed by simultaneous sintering of a nickel skeleton and alloying of the surface layer with aluminum present in the pores [777]. Nickel catalyst hydrogen electrodes have been described by Dousek for fuel cells [778]. A compressed Raney-nickel electrode was prepared by Kalberlah and Winsel [779]. The electrode was discharged in 6 N KOH solution and the effect of potential, electrical resistance, and apparent weight change in the electrolyte was determined. From the potential/charge, resistance/charge, and power/charge curves, four zones can be distinguished: 1) the anodic oxidation of chemisorbed hydrogen and residual aluminum (capacity 0.3 electrons per Ni atom); 2) the oxidation of nickel atoms at the surface to $Ni(OH)_2$; 3) the oxidation of nickel in the internal phase; and 4) the evolution of oxygen and formation of higher valency nickel oxides.

Wiesener [780] reported that porous carbon electrodes impregnated with catalyst such as nickel or nickel boride have excellent properties for the oxidation of hydrazine in alkali media.

REFERENCES

1. L. Colombier, C. R. Acad. Sci., Paris, 199, 273 (1934).
2. K. Murata, Bull. Chem. Soc. Japan, 3, 57 (1928).
3. M. M. Haring and E. G. van den Bosche, J. Phys. Chem., 33, 161 (1929).
4. D. S. Carr and C. F. Bonilla, J. Electrochem. Soc., 99, 475 (1952).

5. M. A. Lopez-Lopez, C. R. Acad. Sci., Paris, 256, 2594 (1963); 255, 3170 (1962).

6. W. M. Latimer, Oxidation Potentials, 2nd ed., Prentice-Hall, Englewood Cliffs, New Jersey, 1952, p. 200.

7. J. W. Larson, P. Cerutti, H. K. Garber, and L. O. Kepler, J. Phys. Chem., 72, 2902 (1968).

8. A. J. de Béthune and N. A. S. Loud, Standard Aqueous Electrode Potentials and Temperature Coefficients, Hampel, Skokie, Illinois, 1964.

9. M. Pourbaix, Atlas of Electrochemical Equilibria in Aqueous Solutions, Pergamon-CEBELCOR, Brussels, 1966.

10. T. Markovic and M. Ahmedbasic, Werkst. Korros., 16, 212 (1965).

11. T. Markovic, ibid., 16, 570 (1965).

12. M. Pourbaix, N. de Zoubov, and E. Deltombe, Proceedings of the Meeting of the International Committee of Electrochemical Thermodynamics and Kinetics, 7th, Lindau, 1955, Butterworths, London, 1956, p. 193.

13. B. E. Conway and E. Gileadi, Can. J. Chem., 40, 1933 (1962).

14. W. M. Vogel, Electrochim. Acta, 13, 1815 (1968).

15. A. Prokopcikas, Liet. TSR Mokslu. Akad. Dar. Ser., B-1962(2), 31; C.A. 58, 1915 (1963).

16. J. Einerhand, Z. Elektrochem., 54, 302 (1950).

17. G. Milazzo, Electrochemistry, Elsevier, New York, 1963, p. 167.

18. P. L. Bourgault and B. E. Conway, Can. J. Chem., 38, 1557 (1960).

19. J. Zedner, Z. Elektrochem., 11, 809 (1905).

20. J. Zedner, ibid., 12, 463 (1906).

21. J. Zedner, ibid., 13, 752 (1907).

22. F. Foerster, ibid., 13, 414 (1907).

23. F. Foerster, ibid., 14, 17 (1908).

24. F. Foerster and F. Krueger, ibid., 33, 406 (1927).

25. A. Hickling and J. E. Spice, Trans. Faraday Soc., 43, 762 (1947).

26. K. Georgi, Z. Elektrochem., 38, 681 (1932); 38, 714 (1932).

27. S. E. S. El Wakkad and S. E. Emara, J. Chem. Soc., 1953, 3504.

28. M. deKay Thompson and H. K. Richardson, Trans. Amer. Electrochem. Soc., 7, 95 (1905).

29. W. Pfanhauser, Z. Elektrochem., 7, 698 (1901).

30. F. Kornfeil, Proc. Ann. Battery Res. Develop. Conf., 12th, Fort Monmouth, 18 (1958).

31. A. I. Zurin, Tr. Leningr. Politekhn. Inst., 6, 38 (1953).

32. A. A. Bulakh and O. A. Khan, Zh. Prikl. Khim., 27, 166 (1954).

33. B. Z. Ustinskii and D. M. Chizhikov, ibid., 22, 1249 (1949).

34. N. S. Fortunatov and V. I. Mikhailovskaya, Ukr. Khim. Zh., 16, 667 (1951).

35. Gmelins Handbuch der Anorganische Chemie, 8th ed., Nickel, Part B-2, GMBH chem. ed., Weinheim, 1966.

36. K. H. Gayer and A. B. Garret, J. Amer. Chem. Soc., 71, 2973 (1949).

37. H. J. Wijs, Rec. Trav. Chim. Pays-Bas, 44, 663 (1925).

38. R. Näsänen, Ann. Acad. Fennicae, A-59, 3 (1943).

39. J. Bjerrum, G. Schwarzenbach, and L. G. Sillén, Stability Constants, II, Inorganic Ligands, The Chemical Society, London, 1958, p. 13.

40. H. G. Denham, J. Chem. Soc., 93, 41 (1908).

41. S. Chaberek, R. C. Courtney, and A. E. Martell, J. Amer. Chem. Soc., 74, 5557 (1952).

42. Z. Ksandr and M. Hejtmanek, Sb. Celostatni Pracovni Konf. Anal. Chemiku, 1st, Prague, 1952, p. 42; (1952); Chem. Abstr., 50, 3150g (1956).

43. F. Cuta, Z. Ksandr, and M. Hejtmanek, Chem. Listy, 50, 1064 (1956).

44. K. H. Gayer and N. Woonter, J. Amer. Chem. Soc., 74, 1437 (1952).

45. C. Kullgren, Z. Phys. Chem., 85, 466 (1913).

46. J. A. Bolzan, E. A. Jáuregui, and A. J. Arvía, Electrochim. Acta, 8, 841 (1963).

47. D. H. Hume and I. M. Kolthoff, J. Amer. Chem. Soc., 72, 4423 (1950).

48. F. Ageno and E. Valla, Atti Accad. Naz. Lincei, Rend., Cl. Sci. Fis, Mat. Nat., [5] 20, 706 (1911).

49. A. Thiel and H. Gessner, Z. Anorg. Allg. Chem., 86, 49 (1914).

50. J. Goret and B. Trémillon, Electrochim. Acta, 12, 1065 (1967).

51. A. Delgado, D. Posadas, and A. J. Arvía, ibid., In Press.

52. W. J. Hamer, M. S. Malmberg, and B. Rubin, J. Electrochem. Soc., 112, 750 (1965).

53. W. J. Hamer, M. S. Malmberg, and B. Rubin, ibid., 103, 8 (1956).

54. W. J. Hamer, ibid., 10, 140 (1965).

55. M. D. Ingram and G. J. Janz, Electrochim. Acta, 10, 786 (1965).

56. K. Grjotheim, Z. Phys. Chem., (NF), 11, 150 (1957).

57. H. W. Jenkins, G. Mamantov, and D. L. Manning, J. Electroanal. Chem., 19, 385 (1968).

58. H. W. Jenkins, G. Mamantov, and D. L. Manning, ibid., 117, 183 (1970).

59. H. A. Laitinen and C. H. Liu, J. Amer. Chem. Soc., 80, 1015 (1958).

60. H. C. Gaur and H. L. Jindal, Electrochim. Acta, 13, 835 (1968).

61. H. C. Gaur and W. K. Behl, ibid., 8, 107 (1963).

62. H. C. Gaur and W. K. Behl, Proceeding of the First Australian Conference on Electrochemistry (A. Friend and F. Gutman, eds.), Pergamon, London, 1954, p. 543.

63. H. C. Gaur and H. L. Jindal, Electrochim. Acta, 15, 1113 (1970).

64. H. C. Gaur and H. L. Jindal, ibid., 15, 1127 (1970).

65. Yu. K. Delimarskii and A. A. Kolotii, Zh. Fiz. Khim., 23, 97 (1949).

66. Yu. K. Delimarskii and A. A. Kolotii, ibid., 23, 437 (1949).

67. D. C. Hamby and A. B. Scott, J. Electrochem. Soc., 115, 704 (1968).

68. H. A. Laitinen and B. B. Bathia, ibid., 107, 705 (1960).

69. D. G. Hill, B. Porter, and A. S. Gillespie, ibid., 105, 408 (1958).

70. S. M. Selis, G. R. B. Elliot, and L. P. McGinnis, ibid., 106, 134 (1959).

71. S. M. Selis and L. P. McGinnis, ibid., 106, 900 (1959).

72. R. E. Panzer, Electrochem. Technol., 2, 10 (1964).

73. Ch. Ilschner-Gensch, J. Electrochem. Soc., 105, 635 (1958).

74. W. J. Treadwell, Z. Elektrochem., 22, 414 (1916).

75. A. F. Kapustinski and N. J. Nowesselezew, Zh. Fiz. Khim., 11, 61 (1938).
76. F. D. Richardson and J. E. Jeffes, J. Iron Steel Inst., 160, 261 (1948).
77. D. P. Bogatski, J. Gen. Chem. USSR, 21, 1 (1951).
78. N. Andreeva, Ukr. Khim. Zh., 21, 569 (1955).
79. A. A. Kolotii, ibid., 28, 188 (1962).
80. S. M. Selis and L. P. McGinnis, J. Electrochem. Soc., 108, 191 (1961).
81. O. P. Penyagina, T. I. Manukhina, I. N. Ozeryanaya, and M. V. Smirnov, Electrochem. Molten Solid Electrolytes, 8, 73 (1970).
82. V. E. Dmitrenko, M. L. Ezerkii, G. L. Rezhikov, and A. V. Zyrin, Novoe v Proizvod. Khim. Istochnikov. Toka, 8, 51 (1968).
83. S. I. Rempel and I. N. Ozeryanaya, Zh. Fiz. Khim., 25, 1181 (1951).
84. L. L. Chermak, Tsvet. Metal, 31, 37 (1958); Chem. Abstr., 53, 4963 (1959).
85. L. T. Sryvalin and O. A. Esin, Zh. Fiz. Khim., 26, 371 (1952).
86. N. I. Kornilov, N. G. Ilyushchenko, B. G. Rossokhin, and G. I. Belyaeva, Electrochem. Molten Solid Electrolytes, 6, 74 (1968).
87. N. G. Ilyushchenko, N. I. Kornilov, and B. G. Rossokhin, ibid., 7, 60 (1969).
88. A. Kisza, Bull. Acad. Pol. Sci., Ser. Sci., Chim., 13, 415 (1965).
89. A. Rahmel, Electrochim. Acta, 13, 495 (1968).
90. R. Littlewood, J. Electrochem. Soc., 109, 525 (1962).
91. S. L. Marchiano and A. J. Arvía, Electrochim. Acta, 17, 861 (1972).
92. S. L. Marchiano and A. J. Arvía, An. Soc. Cienti. Arg., 192, 263 (1971).
93. K. Kiukkola and C. Wagner, J. Electrochem. Soc., 104, 379 (1957).
94. S. Pizzini and R. Morlotti, ibid., 114, 1179 (1967).
95. T. L. Markin, R. J. Bones, and V. J. Wheeler, Proc. Brit. Ceram. Soc., 8, 51 (1967).
96. C. B. Alcock and S. Zador, Electrochim. Acta, 12, 673 (1967).
97. G. G. Charette and S. N. Flengas, J. Electrochem. Soc., 115, 796 (1968).
98. S. Pizzini, R. Morlotti, and V. Wagner, ibid., 116, 915 (1969).
99. J. P. Coughlin, U.S. Bur. Mines Bull. 542 (1954).
100. G. Horsley, U. K. At. Energy Authority Rep. AERE-R 3427 (1956).
101. D. J. Cameron and A. E. Unger, Metall. Trans., 1, 2615 (1970).
102. K. Goto and Y. Matsushita, J. Electrochem. Soc. Japan, 35, 1 (1967).
103. H. H. Moebius, Z. Phys. Chem., 230, 396 (1965).
104. H. Rickert and H. Wagner, Electrochim. Acta, 11, 83 (1966).
105. J. W. Patterson, E. C. Bogren, and R. Rapp, J. Electrochem. Soc., 114, 752 (1967).
106. D. Mehandjiev, Dokl. Bolgar. Akad. Nauk, 22, 1253 (1969).
107. A. M. Lacy and J. A. Pask, J. Amer. Ceram. Soc., 53, 559 (1970).
108. A. M. Lacy and J. A. Pask, ibid., 53, 676 (1970).
109. A. M. Lacy and J. A. Pask, ibid., 54, 236 (1971).
110. I. M. Kolthoff and J. J. Lingane, Polarography, Interscience, New York, 1952.
111. L. Meites, Handbook of Analytical Chemistry, McGraw-Hill, New York, 1962.
112. G. W. C. Milner, The Principles and Applications of Polarography, Longmans, London, 1962.

113. P. W. West and J. F. Dean, Ind. Eng. Chem. Anal. Ed., 17, 686 (1945).

114. P. W. West, J. F. Dean, and E. J. Breda, Collect. Czech. Chem. Commun., 13, 1 (1948).

115. E. Wahlin, Acta Chem. Scand., 7, 956 (1953).

116. P. Kivalo, K. B. Oldham, and H. A. Laitinen, J. Amer. Chem. Soc., 75, 4148 (1953).

117. T. Takahashi and H. Shirai, J. Electroanal. Chem., 3, 313 (1962).

118. K. Zutshi, ibid., 6, 198 (1963).

119. D. S. Turnham, ibid., 10, 19 (1965).

120. I. A. Korsunov and M. K. Scennikova, Z. Anal. Chem., 4, 5 (1949).

121. P. Delahay and C. C. Mattax, J. Amer. Chem. Soc., 76, 874 (1954).

122. R. Tanaka, R. Tamamushi, and M. Kodama, Bull. Chem. Soc. Japan, 33, 14 (1960).

123. J. Lingane and H. Kerlinger, Ind. Eng. Chem. Anal. Ed., 17, 686 (1945).

124. R. Tanaka and R. Tamamushi, Sb. Mezinarod. Polarogr. Sjezdu, Praze, 1st Congr., 1951, p. 486.

125. D. N. Hume and I. M. Kolthoff, J. Amer. Chem. Soc., 72, 4423 (1950).

126. J. J. Lingane, Ind. Eng. Chem. Anal. Ed., 18, 430 (1946).

127. P. N. Kovalenko and L. S. Nadezina, Zh. Obshch. Khim., 22, 740 (1952).

128. A. Calusaru, J. Electroanal. Chem., 23, 157 (1969).

129. G. Jessop, Nature, 158, 59 (1946).

130. O. N. Srivastava, J. K. Gupta, and C. M. Gupta, Electrochim Acta, 16, 585 (1971).

131. C. Nishihara and H. Matsuda, J. Electroanal. Chem., 28, 17 (1970).

132. G. B. Jones, Anal. Chim. Acta, 7, 578 (1952).

133. P. N. Kovalenko and L. S. Nadezina, Dokl. Akad. Nauk, SSSR, 78, 1165 (1951).

134. D. C. Olson, Anal. Chem., 39, 1785 (1967).

135. E. Jacobsen and K. Schroeder, J. Phys. Chem., 66, 134 (1962).

136. J. Dolezal, V. Petrus, and J. Zyka, J. Electroanal. Chem., 3, 169 (1962).

137. C. V. Banks and J. P. Laplante, Anal. Chim. Acta, 27, 101 (1962).

138. D. G. Davis and E. A. Boudreaux, J. Electroanal. Chem., 8, 434 (1964).

139. W. Kémula, L. J. Jeftic, and Z. Galus, ibid., 10, 387 (1965).

140. R. Peticha, Sb. Mezinarod. Polarogr. Sjezdu, Praze, 1st Congr., 1951, p. 539.

141. J. Tirouflet, E. Laviron, R. Dabard, and J. Komeda, Bull. Soc. Chim. Fr., 62, 857 (1963).

142, N. E. Vanderborgh and D. E. Sellers, J. Amer. Chem. Soc., 86, 2790 (1964).

143. P. D. T. Coulder and R. T. Iwamoto, J. Electroanal. Chem., 13, 21 (1967).

144. H. E. Ulery, J. Electrochem. Soc., 113, 479 (1966).

145. E. J. Kuta, Anal. Chem., 32, 1065 (1960).

146. J. C. Goudeau, G. Berthon, M. Camps, and M. L. Bernard, Electrochim. Acta, 13, 309 (1968).

147. P. K. Migal and V. G. Agasleva, Zh. Obshch. Khim., 29, 8 (1959).

148. D. L. McMasters, R. B. Dunlap, J. R. Kuempel, L. W. Kreider, and T. R. Shearer, Anal. Chem., 39, 103 (1967).

149. G. H. Brown and R. Al-Urfali, J. Amer. Chem. Soc., 80, 2113 (1958).

150. J. B. Headridge, M. Ashraf, and H. L. H. Dodds, J. Electroanal. Chem., 16, 114 (1968).

151. A. Ciana and C. Furlani, Electrochim. Acta, 10, 1149 (1965).

152. G. P. Kumar and D. A. Pantony, J. Polarogr. Soc., 14, 84 (1968).

153. R. Schmid and V. Gutmann, Chem. Zvesti, 23, 746 (1969).

154. R. C. Larson and R. T. Iwamoto, J. Electroanal. Chem., 6, 234 (1963).

155. J. F. Coetzee and D. K. MacGuire, J. Phys. Chem., 67, 1814 (1963).

156. I. M. Kolthoff and J. F. Coetzee, J. Amer. Chem. Soc., 79, 1852 (1957).

157. A. I. Popov and D. H. Geske, ibid., 79, 2074 (1957).

158. R. C. Larson and R. T. Iwamoto, ibid., 82, 3526 (1960).

159. J. F. Coetzee and J. L. Hedrick, J. Phys. Chem., 67, 221 (1963).

160. R. C. Larson and R. T. Iwamoto, J. Amer. Chem. Soc., 82, 3239 (1960).

161. I. M. Kolthoff and T. B. Reddy, J. Electrochem. Soc., 108, 980 (1961).

162. J. B. Headridge, D. Pletcher, and M. Callingan, J. Chem. Soc., A, 684 (1967).

163. I. M. Kolthoff and J. F. Coetzee, J. Amer. Chem. Soc., 82, 575 (1960).

164. S. Ikeda and E. Itabashi, Bull. Chem. Soc. Japan, 41, 1844 (1968).

165. G. H. Brown and H. S. Hsiung, J. Electrochem. Soc,, 107, 57 (1960).

166. G. B. Bachman and N. J. Astle, J. Amer. Chem. Soc., 64, 1303 (1942).

167. W. Hubicki and A. Stasiewicz, Ann. Univ. M. Curie-Sklodowska, Lublin, AA-16, 53 (1961).

168. G. Schoeber and V. Gutmann, Monatsh. Chem., 92, 292 (1961).

169. V. Gutmann and E. Nedbalek, ibid., 88, 320 (1957).

170. T. A. Pinfold and F. Sebba, J. Amer. Chem. Soc., 78, 5193 (1956).

171. A. Davison, J. A. McCleverty, E. T. Schawl, and E. J. Wharton, ibid., 89, 830 (1967).

172. M. J. Barker-Hawkes, E. Billig, and H. B. Gray, ibid., 88, 4870 (1966).

173. D. C. Olson, V. P. Mayweg, and G. N. Schrauzer, ibid., 88, 4876 (1966).

174. A. Davison, M. Edelstein, R. H. Holm, and A. H. Maki, Inorg. Chem., 2, 1227 (1963).

175. A. L. Balch, F. Roehrscheid, and R. H. Holm, J. Amer. Chem. Soc., 87, 2301 (1965).

176. A. L. Balsh and R. H. Holm, ibid., 88, 5201 (1966).

177. J. B. Headridge, Electrochemical Techniques for Inorganic Chemists, Academic, London, 1969.

178. R. M. de Fremont et al. in Polarography 1964 (G. J. Hills, ed.), Macmillan, London, 1966.

179. V. K. Schwabe and E. Ross, Z. Anorg. Allg. Chem., 325, 181 (1963).

180. E. E. Colichman, Anal. Chem., 27, 1559 (1955).

181. N. Steinberg and N. H. Nachtrieb, J. Amer. Chem. Soc., 72, 3558 (1950).

182. Yu. K. Delimarskii and I. D. Pachenko, Ukr. Khim. Zh., 19, 47 (1953).

183. J. E. B. Randles and W. White, Z. Elektrochem., 59, 666 (1955).

184. G. C. Barker and R. Faircloth, AERE C/R 2032 (1956).

185. M. Franchini and S. Martini, EURATOM Rep. EUR-1808-E (1964).

186. D. Inman, D. G. Lovering, and R. Narayan, Trans. Faraday Soc., 63, 3017 (1967).

187. J. Dandoy and L. Gierst, J. Electroanal. Chem., 2, 116 (1961).

188. S. I. Woodburn and R. J. Magee, Aust. J. Chem., 20, 439 (1967).

189. N. S. Anikina and V. S. Kuzub, Zh. Anal. Khim., 18, 1502 (1963).

190. R. C. Rooney, Tech. Bull. Southern Analytical Ltd., England, 1962.

191. Z. Galus and L. J. Jeftic, J. Electroanal. Chem., 14, 415 (1967).

192. G. K. Budnikov, V. A. Mikhailov, N. A. Gainutdinova, V. A. Belavin, and V. F. Toropova, Elektrokhimiya, 6, 887 (1970).

193. J. L. Jones and H. A. Fritsche, J. Electroanal. Chem., 12, 334 (1966).

194. A. Davison, N. Edelstein, R. H. Holm, and A. H. Maki, Inorg. Chem., 3, 814 (1964).

195. R. Williams, E. Billig, J. H. Waters, and H. B. Gray, J. Amer. Chem. Soc., 88, 43 (1966).

196. R. H. Holm, A. R. Balch, A. Davison, A. H. Maki, and T. E. Berry, ibid., 89, 2866 (1967).

197. E. I. Stiefel, J. H. Waters, E. Billig, and H. B. Gray, ibid., 87, 3016 (1965).

198. Yu. K. Delimarskii and B. F. Markov, Electrochemistry of Fused Salts, Sigma Press, Washington, D.C., 1961.

199. G. J. Janz, Molten Salts Handbook, Academic, New York, 1967.

200. Yu. K. Delimarskii and K. M. Kabalina, Ukr. Khim. Zh., 24, 435 (1958).

201. Yu. K. Delimarskii, K. M. Bojko, and G. W. Schilina, Electrochim. Acta, 6, 215 (1962).

202. E. D. Black and T. DeVries, Anal. Chem., 27, 906 (1955).

203. Yu. K. Delimarskii and V. V. Kuzimovich, Zh. Neorg. Khim., 4, 1213 (1959).

204. Yu. K. Delimarskii and V. V. Kuzimovich, Teoriya i Prakt. Polarogr. Anal. Nauk Moldausk. SSSR, Materialy Pergobo Vses. Soveshch., 1962, 77.

205. D. L. Manning, J. Electroanal. Chem., 7, 302 (1964).

206. V. Bartocci, R. Marassi, and F. Pucciarelli, Chim. Ind. (Milan), 52, 1201 (1970).

207. K. E. Johnson and H. A. Laitinen, J. Electrochem. Soc., 110, 314 (1963).

208. E. Schmidt, Electrochim. Acta, 8, 23 (1963).

209. Yu. K. Delimarskii, reported in H. D. Graves and G. H. Hills, Advances in Electrochemistry and Electrochemical Engineering, Vol. 4 (P. Delahay and W. C. Tobias, eds.), Wiley (Interscience) New York, 1965.

210. D. L. Maricle and D. N. Hume, Anal. Chem., 33, 1188 (1961).

211. H. C. Gaur and W. K. Behl, J. Electroanal. Chem., 5, 261 (1963).

212. J. Christie and R. A. Osteryoung, J. Amer. Chem. Soc., 82, 1841 (1960).

213. R. Narayan and D. Inman, J. Polarogr. Soc., 11, 27 (1965).

214. N. G. Chovnik and A. N. Fomichev, Ukr. Khim. Zh., 36, 60 (1970).

215. G. Mamantov, P. Papov, and P. Delahay, J. Amer. Chem. Soc., 79, 4034 (1957).

216. M. M. Nicholson, ibid., 79, 7 (1957).

217. A. I. Zebreva and M. T. Kozlovskii, Collect. Czech. Chem. Commun., 25, 3188 (1960).

218. W. Kemula, Z. Galus, and Z. Kublik, Bull. Acad. Pol. Sci., Ser.Sci. Chim., 6, 661 (1958).

219. W. Kemula, Z. Galus, and Z. Kublik, Nature, 182, 1228 (1958).

220. W. Kemula and Z. Galus, Bull. Acad. Pol. Sci., Ser. Sci. Chim., 7, 553 (1959).
221. W. Kemula and Z. Galus, ibid., 7, 729 (1959).
222. Z. Kublik, Acta Chim. Hung. 27, 79 (1961).
223. V. F. Ivanov and Z. A. Iofa, Dokl. Akad. Nauk, SSSR, 137, 1149 (1961).
224. V. F. Ivanov and Z. A. Iofa, ibid., 140, 1369 (1962).
225. J. T. Porter and W. D. Cooke, J. Amer. Chem. Soc., 77, 1481 (1955).
226. I. E. Krasnova and A. I. Zebreva, Elektrokhimiya, 2, 247 (1966).
227. A. Sevcik, Collect. Czech. Chem. Commun., 13, 349 (1948).
228. J. Mindowicz, Electrochim. Acta, 5, 202 (1961).
229. G. Jangg and H. Jedlicka, ibid., 13, 679 (1968).
230. M. M. Nicholson, Anal. Chem., 32, 1058 (1960).
231. Y. Okinaka and I. M. Kolthoff, J. Amer. Chem. Soc., 82, 324 (1960).
232. H. A. Laitinen and W. J. Subcasky, ibid., 80, 2623 (1958).
233. K. Morinaga, Bull. Chem. Soc. Japan, 29, 793 (1956).
234. H. Matsuda and Y. Ayabe, ibid., 29, 134 (1956).
235. K. Morinaga, J. Chem. Soc. Japan, 76, 133 (1955).
236. I. T. Oiwa, N. Tanaka, and R. Tamamushi, Electrochim. Acta, 9, 981 (1964).
237. H. Imai, J. Sci. Hiroshima Univ., A-22, 291 (1958).
238. S. K. Ja and S. N. Srivastava, Bull. Chem. Soc. Japan, 40, 810 (1967).
239. R. Parsons, Handbook of Electrochemical Constants, Butterworths, London, 1959.
240. M. Hollnager and R. Landsberg, Z. Phys. Chem., 212, 94 (1959).
241. B. E. Conway, Electrochemical Data, Elsevier, Amsterdam, 1952.
242. F. W. Salt, Discussions Faraday Soc., 1, 107 (1947).
243. H. A. Laitinen, R. Tischer, and R. K. Roe, J. Electrochem. Soc., 107, 546 (1960).
244. N. Tanaka and R. Tamamushi, Electrochim. Acta, 9, 963 (1964).
245. P. Lukovtsev, S. Levina, and A. N. Frumkin, Acta Physicochim. USSR, 11, 21 (1939).
246. A. Legran and S. Levina, ibid., 12, 243 (1940).
247. V. L. Kheifets and B. S. Krasikov, Zh. Fiz. Khim., 31, 1992 (1957).
248. B. Jakuszewski and Z. Koslowski, Rocz. Chem., 36, 1873 (1962).
249. B. Jakuszewski and Z. Koslowski, ibid., 38, 93 (1964).
250. B. Jakuszewski and Z. Koslowski, Acta Chim. Soc. Sci. Lodz, 10, 5 (1965).
251. J. O'M. Bockris, S. D. Argade, and E. Gileadi, Electrochim. Acta, 14, 1267 (1969).
252. L. Campanella, J. Electroanal. Chem., 28, 228 (1970).
253. A. N. Frumkin, Colloid Symp. USSR, 11, 815 (1939).
254. Yu. K. Delimarskii and V. S. Kikhno, Ukr. Khim. Zh., 30, 1156 (1964).
255. Yu. K. Delimarskii and V. S. Kikhno, ibid., 31, 116 (1965).
256. L. V. Volkov, A. N. Ponomarev, and B. P. Yuriev, Tr. Leningrad Politekh. Inst., Nr. 304, 94 (1970).
257. Yu. K. Delimarskii and V. S. Kikhno, Ukr. Khim. Zh., 35, 468 (1969).
258. P. N. Andersen, J. L. Anderson, and H. Eyring, J. Phys. Chem., 73, 3562 (1969).

259. A. Rakov, T. Borisova, and B. V. Ershler, Zh. Fiz. Khim., 22, 1390 (1948).

260. J. O'M. Bockris and E. C. Potter, J. Chem. Phys., 20, 614 (1952).

261. M. L. Kronenberg, J. C. Banter, E. Yeager, and F. Hovorka, J. Electrochem. Soc., 110, 1007 (1963).

262. R. R. Sayano and K. Nobe, Water Resources Center Contrib. 104; Rep. No. 65-38, (1965).

263. D. L. Piron and K. Nobe, Water Resources Center Contrib. 115, Rep. No. 66-37, (1966).

264. R. J. Brodd and N. Hackerman, J. Electrochem. Soc., 104, 704 (1957).

265. V. V. Batrakov, A. Gamilikhana, E. I. Mikhailova, and Z. A. Iofa, Elektrokhimiya, 4, 601 (1968).

266. T. Yoshida, S. Nomoto and Y. Okada, Denki Kagaku, 37, 633 (1969).

267. D. Berndt, Electrochim. Acta, 10, 1067 (1965).

268. E. I. Mikhailova and Z. A. Iofa, Elektrokhimiya, 1, 107 (1965).

269. K. Schwabe and W. Schwenke, Electrochim. Acta, 9, 231 (1964).

270. D. C. Grahame, J. Electrochem. Soc., 98, 343 (1951).

271. L. K. Partridge, A. C. Tansley, and A. S. Porter, Electrochim. Acta, 12, 1573 (1967).

272. R. G. Barradas and B. E. Conway, J. Electroanal. Chem., 6, 314 (1963).

273. J. O'M. Bockris and D. A. J. Swinkels, J. Electrochem. Soc., 111, 736 (1964).

274. J. O'M. Bockris, M. Green, and D. A. J. Swinkels, ibid., 111, 743 (1964).

275. J. O'M. Bockris, M. A. V. Devanathan, and K. Mueller, Proc. Roy. Soc., A, 274, 55 (1963).

276. J. P. Strassner and P. Delahay, J. Amer. Chem. Soc., 78, 6232 (1952).

277. P. Delahay, Double Layer and Electrode Kinetics, Wiley (Interscience), New York, 1966, p. 127.

278. L. Gierst, Transactions of the Symposium on Electrode Processes (E. Yeager, ed.), Wiley, New York, 1961, p. 109.

279. J. M. Hale and R. Parsons, Collect. Czech. Chem. Commun., 27, 2444 (1962).

280. J. Koutecky, Chem. Listy, 47, 9, 323 (1953).

281. N. S. Hush and J. W. Scarrot, J. Electroanal. Chem., 7, 26 (1964).

282. T. J. Swift and R. E. Connick, J. Chem. Phys., 37, 307 (1962).

283. E. Verdier and R. Pernicelli, J. Chim. Phys., 67, 965 (1970).

284. J. Heyrovsky and J. Kuta, Principles of Polarography, Academic, New York, 1966.

285. N. V. Emelianova, Rec. Trav. Chim. Pays-Bas, 44, 528 (1925).

286. V. Cagloti, G. Sartori, and P. Silvestroni, Atti. Accad. Naz. Lincei, Cl. Sci. Fiz. Mat. Nat., 3, 448 (1947).

287. A. A. Vlcek, Collect. Czech. Chem. Commun., 22, 948 (1957).

288. A. A. Vlcek, ibid., 22, 1736 (1957).

289. G. Torsi and P. Papoff, J. Electroanal. Chem., 20, 231 (1969).

290. Ya. I. Turiyan and D. N. Malyavinskaya, Elektrokhimiya, 5, 106 (1969).

291. Ya. I. Turiyan, Zh. Fiz. Khim., 31, 2423 (1957).

292. Ya. I. Turiyan and G. F. Serova, ibid., 34, 1009 (1960).

293. Ya. I. Turiyan and G. F. Serova, ibid., 31, 1976 (1957).

294. H. B. Mark and C. N. Reilley, J. Electroanal. Chem., 4, 189 (1963).

295. H. B. Mark and C. N. Reilley, Anal. Chem., 35, 195 (1963).

296. Ya. I. Turiyan and O. N. Malyaviskaya, J. Electroanal. Chem., 23, 69 (1969).

297. S. G. Mairanovskii, V. P. Gulimyai and N. K. Lisitsina, Elektrokhimiya, 5, 752 (1969).

298. A. A. Vlcek, Nature, 177, 1043 (1956).

299. H. B. Mark and H. G. Schwartz, J. Electroanal. Chem., 6, 493 (1963).

300. M. Brezina, Advan. Polarogr., 3, 933 (1959).

301. M. Brezina, Biochem. Z., 272, 104 (1934).

302. H. B. Mark, J. Electroanal. Chem., 7, 276 (1964).

303. P. D. DeMars, J. Electrochem. Soc., 108, 779 (1961).

304. Ya. I. Turiyan and O. E. Ruvinskii, Elektrokhimiya, 4, 1446 (1968).

305. Ya. I. Turiyan and O. E. Ruvinskii, J. Electroanal. Chem., 23, 61 (1969).

306. E. Itabashi and S. Ikeda, ibid., 26, 103 (1970).

307. F. Foerster and F. Krueger, Z. Elektrochem., 33, 418 (1927).

308. N. Sato and G. Okamoto, J. Electrochem. Soc., 111, 897 (1964).

309. K. Schwabe and C. H. Voigt, ibid., 113, 886 (1966).

310. J. Matulis and R. Slizys, Electrochim. Acta, 9, 1177 (1964).

311. R. Slizys, Galvanotechnik, 61, 653 (1970).

312. K. E. Heusler and L. Gaiser, Electrochim. Acta, 13, 59 (1968).

313. D. D. Condit, Aerospace Research Laboratory, Wright Patterson A.F. Base, Ohio, 69-0018-Project 7021.

314. E. J. Kelly, J. Electrochem. Soc., 112, 124 (1965).

315. E. J. Kelly, ibid., 115, 1111 (1968).

316. R. C. V. Piatti, A. J. Arvía, and J. J. Podestá, Electrochim. Acta, 14, 541 (1969).

317. R. C. V. Piatti, J. J. Podestá, and A. J. Arvía, An. Asoc. Quím. Argentina, 57, 71 (1969).

318. W. A. Wesley, J. Electrochem. Soc., 103, 296 (1956).

319. G. A. DiBari and J. V. Petrocelli, ibid., 112, 99 (1965).

320. A. T. Vagramyan and L. A. Uvarov, Izv. Akad. Nauk SSSR, 1962 (9), 1520.

321. A. T. Vagramyan and L. A. Uvarov, Dokl. Akad. Nauk, SSSR, 146, 635 (1962).

322. A. T. Vagramyan, M. A. Zhamagortsyan, and L. A. Uvarov, Izv. Akad. Nauk SSSR, 1964 (2), 301.

323. A. T. Vagramyan, M. A. Zhamagortsyan, L. A. Uvarov, and A. A. Yavich, Zashch. Metal., 5, 74 (1969).

324. A. T. Vagramyan, M. A. Zhamagortsyan, L. A. Uvarov, and A. A. Yavich, Elektrokhimiya, 6, 733 (1970).

325. A. T. Vagramyan, M. A. Zhamagortsyan, L. A. Uvarov, and A. A. Yavich, ibid., 6, 755 (1970).

326. F. Ovari and A. L. Rotinyan, Zh. Prikl. Khim., 42, 227 (1969); Elektrokhimiya, 6, 755 (1970).

327. F. Ovari and A. L. Rotinyan, Elektrokhimiya, 6, 516 (1970).

328. N. P. Fedotiev and Z. I. Dmitreshova, Zh. Prikl. Khim., 30, 221 (1957).

329. A. I. Oshe and V. A. Lovachev, Elektrokhimiya, 6, 1416 (1970).

330. J. O'M. Bockris, A. K. N. Reddy, and B. Rao, J. Electrochem. Soc., 113, 1133 (1966).

331. V. A. Lovachev, A. I. Oshe, and B. N. Kabanov, Elektrokhimiya, 5, 1206 (1969).

332. G. G. Penov, Z. Ya. Kosakovskaya, A. P. Botneva, L. A. Andreeva, and N. P. Zhuk, Zashch. Metal., 6, 544 (1970).

333. I. Epelboin, 21st Meeting of the International Committee on Thermodynamics and Electrochemical Kinetics, Prague, 1970.

334. N. M. Dorofeeva and I. I. Kalinichenko, Izv. Vyssh. Uchev. Zh. Fiz., 13, 756 (1970).

335. A. R. Yamuna and N. Subramanyan, Werkst. Korros., 21, 607 (1970).

336. A. S. Valeev and A. M. Ushanova, Elektrokhimiya, 5, 809 (1969).

337. M. Le Blanc and M. G. Levi, Z. Elektrochem. (Boltzmann Festchrift), 183 (1904).

338. A. Schweitzer, Z. Elektrochem., 15, 612 (1909).

339. K. Georgi, ibid., 39, 736 (1933).

340. E. Mueller and K. Schwabe, ibid., 40, 868 (1934).

341. K. Hauffe and I. Pfeiffer, Z. Metallkunde, 45, 554 (1954).

342. W. J. Mueller, H. K. Cameron, and W. Machu, Z. Elektrochem., 33, 401 (1927).

343. W. J. Mueller, H. K. Cameron, and W. Machu, Monatsh. Chem., 52, 475 (1929).

344. M. J. Mueller, H. K. Cameron, and W. Machu, ibid., 59, 73 (1932).

345. W. Machu and A. Raegheb, Werkst. Korros., 4, 429 (1953).

346. W. Machu and A. Raegheb, ibid., 5, 217 (1954).

347. W. Machu and A. Raegheb, Z. Metallkunde, 47, 176 (1956).

348. R. Landsberg and M. Hollnager, Z. Elektrochem., 58, 680 (1954).

349. R. Landsberg and M. Hollnager, ibid., 60, 1098 (1956).

350. G. Truempler and H. Meyer, Helv. Chim. Acta, 35, 1034 (1952).

351. G. Truempler and W. Saxer, ibid., 36, 1630 (1953).

352. R. Piontelli and G. Serravalle, Z. Elektrochem., 62, 759 (1958).

353. J. Osterwald and H. G. Feller, J. Electrochem. Soc., 107, 473 (1960).

354. D. MacGillavry, J. J. Singer, and J. H. Rosenbaum, J. Amer. Chem. Soc., 73, 1388 (1951).

355. D. MacGillavry, J. H. Rosenbaum, and R. W. Swenson, J. Electrochem. Soc., 99, 22 (1956).

356. D. R. Turner, ibid., 98, 434 (1951).

357. G. Okamoto and N. Sato, J. Electrochem. Soc. Japan, 27, 125 (1959).

358. Ya. M. Kolotyrkin, Z. Elektrochem., 62, 664 (1958).

359. N. Sato and G. Okamoto, J. Electrochem. Soc., 110, 605 (1963).

360. K. Arnold and K. J. Vetter, Z. Elektrochem., 64, 407 (1960).

361. H. Pfisterer, A. Politycky, and E. Fuchs, ibid., 63, 257 (1959).

362. G. Okamoto, T. Takaishi, and N. Sato, J. Electrochem. Soc. Japan, 26, 615 (1958).

363. E. Raub and A. Disam, Metalloberflaeche, 15, 193 (1961).

364. N. D. Greene, First International Congress on Corrosion, Butterworths, London, 1963, p. 113.

365. A. K. Huq, A. Rosenberg, and A. C. Makrides, J. Electrochem. Soc., 111, 278 (1964).

366. J. L. Weininger and M. W. Breiter, ibid., 110, 484 (1963).

367. J. L. Weininger and M. W. Breiter, ibid., 111, 707 (1964).

368. D. E. Davies and W. Barker, Corrosion, 20, 47t (1964).

369. T. S. De Gromoboy and L. L. Shreir, Electrochim. Acta, 11, 895 (1966).

370. W. J. Mueller, Z. Elektrochem., 33, 401 (1927).

371. G. Okamoto, H. Kobayashi, M. Nagayama, and N. Sato, ibid., 62, 775 (1958).

372. K. Schwabe and G. Dietz, ibid., 62, 751 (1958).

373. L. F. Trueb, G. Truempler, and N. Ibl, Helv. Chim. Acta, 44, 1433 (1961).

374. M. Keddam and I. Epelboin, C. R. Acad. Sci., Paris, 259, 137 (1964).

375. K. Schwabe, Electrochim. Acta, 3, 186 (1960).

376. L. Kandler, D. Romer and E. Heusler, ibid., 8, 233 (1963).

377. K. J. Vetter, J. Electrochem. Soc., 110, 598 (1963).

378. K. J. Vetter and K. Arnold, Z. Elektrochem., 64, 244 (1960).

379. H. H. Uhlig and H. G. Feller, J. Electrochem. Soc., 107, 865 (1960).

380. T. Ishikawa and G. Okamoto, Electrochim. Acta, 9, 1259 (1964).

381. K. G. Weil, Z. Elektrochem., 62, 638 (1958).

382. K. J. Vetter, ibid., 62, 642 (1958).

383. P. Guenther, ibid., 62, 619 (1958).

384. G. Trusov, E. P. Gochalieva, and U. M. Novakovskii, Elektrokhimiya, 4, 366 (1968).

385. G. W. D. Briggs and W. F. K. Wynne-Jones, Electrochim. Acta, 7, 241 (1962).

386. K. Schwabe, Werkst. Korros., 18, 961 (1967).

387. K. Schwabe, Angew. Chem., 78, 253 (1966).

388. U. Ebersbach, K. Schwabe, and K. Ritter, Electrochim. Acta, 12, 927 (1967).

389. U. Ebersbach, K. Schwabe, and P. Koenig, ibid., 14, 773 (1969).

390. M. Kesten and H. G. Feller, ibid., 16, 761 (1971).

391. G. Gilli, P. Borea, F. Zucchi, and G. Trabanelli, Corros. Sci., 9, 673 (1969).

392. F. Zucchi and G. Trabanelli, Electrochim. Metal., 4, 313 (1969).

393. G. Truempler and R. Keller, Helv. Chim. Acta, 44, 1691 (1961).

394. U. R. Evans, An Introduction to Metallic Corrosion, Arnold, London, 1963, p. 40.

395. D. L. Piron and K. Nobe, Department of Engineering, Univ. California (Los Angeles), Rep. 66-37 (1966).

396. J. Tousek, Collect. Czech. Chem. Commun., 31, 3083 (1966).

397. I. A. Ammar and S. Darwish, Electrochim. Acta, 12, 225 (1967).

398. I. A. Ammar and S. Darwish, ibid., 11, 1541 (1966).

399. I. A. Ammar and S. Darwish, ibid., 13, 781 (1968).

400. J. Postlethwaite, ibid., 12, 333 (1967).

401. J. Postlethwaite and D. R. Hurp, Corros. Sci., 7, 435 (1967).

402. J. Postlethwaite and L. B. Freese, Corrosion, 23, 109 (1967).

403. I. A. Ammar, S. Darwish, and S. Riad, Electrochim. Acta, 13, 1875 (1969).

404. C. J. Chatfield and L. L. Shreir, ibid., 14, 1015 (1969).

405. J. Osterwald and H. H. Uhlig, J. Electrochem. Soc., 108, 505 (1961).

406. G. Okamoto and N. Sato, Trans. Japan Inst. Metal., 1, 16 (1960).

407. A. Desestret, I. Epelboin, M. Froment, M. Keddam, and P. H. Morel, Electrochim. Acta, 8, 433 (1963).

408. A. B. Ijzermans, Corros. Sci., 10, 113 (1970).

409. M. C. Petit, Electrochim. Acta, 13, 557 (1968).

410. A. Jouanneau and M. C. Petit, ibid., 15, 1325 (1970).

411. I. Garz, H. Worch, and W. Schatt, Corros. Sci., 9, 71 (1969).

412. T. P. Hoar, ibid., 7, 341 (1967).

413. E. Kunze and K. Schwabe, ibid., 4, 109 (1964).

414. R. Gressmann, ibid., 8, 325 (1968).

415. M. Volchkova, K. G. Antonova, and A. I. Krasilchikov, Zh. Fiz. Khim., 23, 714 (1949).

416. K. Schwabe and R. Radeglia, Werkst. Korros., 5, 281 (1969).

417. W. Schatt and H. Worch, Corros. Sci., 9, 869 (1969).

418. T. Tokuda and M. B. Ives, ibid., 11, 297 (1971).

419. N. Watanabe and B. Chang, International Meeting on Fluorine Chemistry, Moscow, 1969.

420. N. Hackerman, E. S. Snavely, and L. C. Fiel, Electrochim. Acta, 12, 535 (1967).

421. J. A. Donohue and A. Zletz, J. Electrochem. Soc., 115, 1039 (1968).

422. N. Hackerman, E. S. Snavely, and L. C. Fiel, Corros. Sci., 7, 39 (1967).

423. A. K. Vijh, Electrochim. Acta, 16, 441 (1971).

424. R. Piontelli, U. Bertocci, and G. Sternheim, Z. Elektrochem., 62, 772 (1958).

425. E. Heitz, Electrochim. Acta, 10, 49 (1965).

426. K. Schwabe and W. Schmidt, Corros. Sci., 10, 143 (1970).

427. R. Littlewood and E. J. Argent, Electrochim. Acta, 4, 155 (1961).

428. J. Amosse, J. Bouteillon, and N. J. Barbier, C.R. Acad. Sci., Paris, 267, 22 (1968).

429. S. Pizzini, R. Morlotti, and E. Roemer, J. Electrochem. Soc., 113, 1305 (1966).

430. G. J. Janz and A. Conte, Electrochim. Acta, 9, 1269 (1964); 9, 1279 (1964).

431. P. Drossbach and D. Sauermann, ibid., 9, 1373 (1963).

432. E. I. Gurovich, Zh. Prikl. Khim., 29, 1358 (1956).

433. T. Notoya and R. Midorikawa, Denki Kagaku, 37, 291 (1969).

434. H. S. Swofford and A. H. Laitinen, J. Electrochem. Soc., 110, 814 (1963).

435. A. J. Arvía, J. J. Podestá, and R. C. V. Piatti, Electrochim. Acta, 17, 901 (1972).

436. A. J. Arvía, J. J. Podestá, and R. C. V. Piatti, ibid., 17, 889 (1972).

437. A. J. Arvía and H. A. Videla, ibid., 11, 537 (1966).

438. H. J. Krueger, A. Rahmel, and W. Schwenk, ibid., 13, 625 (1968).

439. R. Piontelli, G. Poli, and G. Serravalla, Transaction of the Symposium on Electrode Processes (E. Yeager, Ed.), Wiley, New York, 1961, 67.

440. R. Piontelli, M. Lazzari, and B. Rivolta, Ist. Lomb. (Rend. Sci.), A-99, 277 (1965).

441. R. Piontelli, Electrochim. Metall., 1, 1 (1966).

442. J. Mieluch, Bull. Acad. Pol. Sci. Ser. Sci. Chim., 17, 43 (1969).

443. R. Piontelli, L. Peraldo Bicelli, and A. LaVecchia, Rend. Acc. Naz. Lincei VIII, 27, 312 (1959).

444. R. Piontelli, L. Peraldo Bicelli, and C. Romagnani, ibid., 34, 233 (1963).

445. B. E. Conway and J. O'M. Bockris, Proc. Roy. Soc., 248, 394 (1958).

446. N. A. Pangarov, Electrochim. Acta, 9, 721 (1964).

447. I. Garz and U. Haefke, Werks. Korros., 22, 293 (1971).

448. L. N. Yagopoliskaya, Zashch. Metal., 6, 674 (1970).

449. A. Windfeldt, Electrochim. Acta, 9, 1139 (1964).

450. R. Glocker and E. Kaupp, Z. Phys., 24, 121 (1924).

451. G. I. Finch and C. H. Sun, Trans. Faraday Soc., 32, 852 (1936).

452. S. R. Makarieva, Izv. Akad. Nauk, Ser. Khim. SSSR, 5/6, 1211 (1938).

453. B. C. Banerjee and A. Goswami, J. Electrochem. Soc., 106, 590 (1959).

454. B. C. Banerjee and A. Goswami, ibid., 106, 23 (1959).

455. H. Schloetterer, Z. Kristallogr., 119, 321 (1964).

456. G. I. Finch, N. Wilman, and L. Yang, Discussions Faraday Soc., 1, 144 (1947).

457. H. Wilman, Trans. Inst. Metal Finish., 32, 281 (1955).

458. H. Wilman, J. Imp. Coll. Chem. Eng. Soc., 9, 96 (1955).

459. A. K. N. Reddy and H. Wilman, Trans. Inst. Metal Finish., 36, 97 (1959).

460. A. K. N. Reddy and S. R. Rajagopalan, J. Electroanal. Chem., 6, 153 (1963).

461. A. K. N. Reddy and S. R. Rajagopalan, ibid., 6, 159 (1963).

462. B. C. Banerjee and P. L. Walker, J. Electrochem. Soc., 108, 449 (1961).

463. B. C. Banerjee and P. L. Walker, ibid., 109, 436 (1962).

464. B. C. Banerjee, ibid., 107, 80 (1960).

465. N. A. Pangarov, S. D. Vitkova, and I. Uzunova, Electrochim. Acta, 11, 1747 (1966).

466. N. A. Pangarov, L. A. Uvarov, and A. T. Vagramyan, ibid., 13, 1905 (1968).

467. S. Rashkov and N. A. Pangarov, ibid., 14, 17 (1969).

468. E. R. Thompson and K. R. Lawless, ibid., 14, 269 (1969).

469. R. M. Bozorth, Phys. Rev., 26, 390 (1925).

470. N. A. Pangarov, J. Electroanal. Chem., 9, 70 (1965).

471. D. J. Evans, Trans. Faraday Soc., 54, 1086 (1958).

472. A. G. Ives, J. W. Edington, and G. P. Rothwell, Electrochim. Acta, 15, 1797 (1970).

473. M. H. Jones and M. G. Kenez, Plating, 53, 995 (1966).

474. I. Epelboin and M. Froment, 19th Meeting of the International Committee on Electrochemical Kinetics and Thermodynamics, Detroit, 1968.

475. M. Froment, G. Maurin, and J. Thévenin, C.R. Acad. Sci., Paris, C, 264, 1520 (1967).

476. M. Froment and G. Maurin, ibid., 266, 1017 (1968).

477. M. Froment and G. Maurin, ibid., 266, 1125 (1968).

478. M. Froment, G. Maurin, and J. Thévenin, ibid., 269, 1367 (1969).

479. A. K. N. Reddy, J. Electroanal. Chem., 6, 141 (1963).

480. S. Tajima and M. Ogata, Electrochim. Acta, 15, 61 (1970).

481. J. W. Diggle, ibid., 15, 1559 (1970).

482. S. Tajima and M. Ogata, ibid., 13, 1845 (1968).

483. G. Gardam, J. Electrodep. Tech. Soc., 22, 115 (1947).

484. H. Leidheiser, Z. Elektrochem., 59, 756 (1955).

485. D. C. Foulke and O. Kardos, Proc. Amer. Electroplat. Soc., 43, 172 (1956).

486. O. Kardos, ibid., 43, 181 (1956).

487. J. D. Thomas, ibid., 43, 60 (1956).

488. S. A. Watson and J. Edwards, Trans. Inst. Metal Finish, 34, 167 (1957).

489. E. Raub, Metalloberflaeche, 12, 353 (1958).

490. S. E. Beacom, Plating, 46, 814 (1959).

491. S. E. Beacom and B. J. Riley, J. Electrochem. Soc., 106, 309 (1959); 107, 785 (1960); 108, 758 (1961).

492. S. A. Watson, Trans. Inst. Metal Finish., 37, 144 (1960).

493. J. Edwards, ibid., 38, 33 (1962).

494. A. A. Voronko and V. A. Kaikaris, Zh. Prikl. Khim., 34, 2582 (1961).

495. A. A. Voronko and V. A. Kaikaris, ibid., 35, 2863 (1962).

496. N. T. Kudriavtsev, S. S. Kruglikov, and R. P. Sobolev, Tr. Moskovsk. Khim. Tekhnol. Inst. im. Mendeleeva, 32, 259 (1961).

497. N. T. Kudriavtsev, S. S. Kruglikov, G. F. Vorobiova, and M. M. Zubov, Zh. Prikl. Khim., 35, 777 (1962).

498. N. T. Kudriavtsev, S. S. Kruglikov, G. F. Vorobiova, and V. M. Levovsky, ibid., 35, 781 (1962); Dokl. Akad. Nauk, SSSR, 140, 877 (1961).

499. S. S. Kruglikov, N. T. Kudriavtsev, G. F. Vorobiova, and A. Ya. Antonov, Dokl. Akad. Nauk, SSSR, 149, 911 (1963).

500. E. B. Leffler and H. Leidheiser, Plating, 44, 388 (1957).

501. A. T. Vagramyan and Z. A. Solovyova, Itogi Nauki Elektrokhim., 1, 166 (1964).

502. O. Kardos and D. G. Foulkes, in Advances in Electrochemistry and Electrochemical Engineering, Vol. 2 (P. Delahay and C. W. Tobias, eds.), Wiley (Interscience), New York, 1962.

503. H. Fischer, Electrochim. Acta, 2, 72 (1960).

504. O. Volk and H. Fischer, ibid., 4, 251 (1961).

505. M. Froment and R. Wiart, ibid., 8, 481 (1963).

506. G. T. Rogers and K. J. Taylor, ibid., 8, 887 (1963).

507. G. T. Rogers and K. J. Taylor, ibid., 11, 1685 (1966).

508. G. T. Rogers and K. J. Taylor, ibid., 13, 109 (1968).

509. G. T. Rogers and K. J. Taylor, ibid., 13, 2189 (1968).

510. L. K. Partridge, A. C. Tansley, and A. S. Porter, ibid., 11, 517 (1966).

511. S. S. Kruglikov, N. T. Kudriavtsev, and R. P. Sobolev, ibid., 12, 1263 (1967).

512. E. Raub, A. Knoedler, A. Dissam, and H. Kawase, Metalloberflaeche, 23, 293 (1969).

513. J. Vegys, A. Skuaene, and A. Boduevas, Issl. Obl. Elektroosazhdeniya Metal. Mater. Respub. Konf. Elektrokhim. Litov. SSR, 10th., 13 (1968).

514. R. Wiart, Oberflaeche-Surface, 9, 241 (1968).

515. J. Edwards and M. J. Levett, Trans. Inst. Metal Finish., 44, 27 (1969).

516. Z. Alaune and A. Lazauskiene, Liet. TSR Mokslu Akad. Darb., Ser. B, 1970, 59.

517. Z. Talaikyte and Z. Alaune, ibid., Ser. B., 1970, 51.

518. S. S. Kruglikov and V. A. Volkov, Elektrokhimiya, 6, 1765 (1970).

519. A. Knoedler, Oberflaeche, 10, 390 (1969).

520. J. Matulis, Liet. TSR Mokslu Akad. Darb., Ser. B, 1969, 95.

521. A. Jugarav, S. Guruviah, and K. Parameswara Iyer, Metal Finish., 64, 82 (1966).

522. J. J. Hockstra and D. Trivich, J. Electrochem. Soc., 111, 162 (1964).

523. O. Zeimyte, A. Boduevas, and J. Matulis, Liet. TSR Mokslu Akad. Darb., Ser. B, 1969, 93.

524. V. Reklyte, J. Bubelis, and J. Matulis, ibid., Ser. B, 1970, 113.

525. V. Bukaveckas and J. Matulis, ibid., Ser. B, 1969, 29.

526. O. Galdikiene, V. Dagyte-Ukeliene, and J. Matulis, ibid., Ser. B, 1969, 65.

527. V. Bukaveckas and J. Matulis, ibid., Ser. B, 1970, 19.

528. S. S. Kruglikov and V. A. Volkov, Elektrokhimiya, 6, 1033 (1970).

529. S. S. Kruglikov, Electrochemistry, 1965, Israel Program Scientific Translations, Jerusalem, 1970, p. 95.

530. S. S. Kruglikov and L. M. Antipova, Tr. Mosk. Khim. Tekhnol. Inst., 62, 198 (1969).

531. J. A. Crossley, P. A. Brook, and J. W. Cuthbertson, Electrochim. Acta, 11, 1153 (1966).

532. P. A. Brook and J. A. Crossley, ibid., 11, 1189 (1966).

533. R. Weil and W. N. Jacobus, Plating, 53, 102 (1966).

534. I. Dubsky and P. Kozak, Metalloberflaeche, 24, 423 (1970).

535. Z. Alaune and A. Lazauskiene, Galvanotechnik, 61, 656 (1970).

536. J. L. Dye and D. J. Klingenmaier, J. Electrochem. Soc., 104, 275 (1957).

537. W. A. Wood, Trans. Faraday Soc., 31, 1248 (1935).

538. W. Hume-Rothery and M. R. J. Wyllie, Proc. Roy. Soc., A, 181, 331 (1943).

539. C. L. Clark and S. H. Simonson, J. Electrochem. Soc., 98, 110 (1951).

540. W. Smith, J. H. Kaeler, and H. J. Read, Plating, 36, 355 (1959).

541. K. M. Gorbunova and A. A. Sutyagina, Zh. Fiz. Khim., 32, 785 (1958).

542. R. Weil and R. Paquin, J. Electrochem. Soc., 107, 87 (1960).

543. R. Weil and H. C. Cook, ibid., 109, 295 (1962).

544. D. R. Cliffe and J. P. C. Farr, ibid., 111, 299 (1964).

545. G. P. Rothwell and T. P. Hoar, Electrochim. Acta, 10, 403 (1965).

546. S. S. Kruglikov, N. T. Kudriavtsev, G. F. Vorobiova, and A. Ya. Antonov, ibid., 10, 253 (1965).

547. W. Feitknecht, W. H. Studer, and H. Meyer, Kolloid-Z., 139, 131 (1954).

548. W. Bédert, ibid., 139, 133 (1954).
549. J. Besson, C.R. Acad. Sci., Paris, C, 219, 130 (1944).
550. J. Besson, ibid., 220, 320 (1945).
551. J. Besson, ibid., 222, 390 (1946).
552. J. Besson, ibid., 223, 28, 288 (1946).
553. J. Besson, Ann. Chim. (Paris), 2, 527 (1947).
554. I. Belluci and E. Clavari, Gazz. Chim. Ital., 36, 58 (1906); 37, 409 (1907).
555. O. R. Howell, J. Chem. Soc., 123, 669, 1772 (1923).
556. M. LeBlanc and R. Mueller, Z. Elektrochem., 39, 204 (1933).
557. W. M. Vogel, Electrochim. Acta, 13, 1815 (1968); 13, 1821 (1968).
558. G. F. Huettig and A. Peter, Z. Anorg. Allg. Chem., 189, 183 (1930); 189, 190 (1930).
559. D. K. Goralevich, Zh. Obshch. Khim., 1, 973 (1931).
560. R. W. Cairns and E. Ott, J. Amer. Chem. Soc., 55, 534 (1933).
561. O. Glemser and J. Einerhand, Z. Anorg. Allg. Chem., 261, 26, 43 (1950).
562. W. Feitknecht, H. R. Christen, and H. Studer, ibid., 283, 88 (1956).
563. H. Bode, Angew. Chem., 73, 553 (1961).
564. H. Bartl, H. Bode, G. Sterr, and J. Witte, Electrochim. Acta, 16, 615 (1971).
565. M. Aia, J. Electrochem. Soc., 113, 1045 (1966); 114, 418 (1967).
566. H. Bode, K. Dehmelt, and J. Witte, Electrochim. Acta, 11, 1079 (1966).
567. W. Dennstedt and W. Loeser, ibid., 16, 429 (1971).
568. R. Brownsword and J. P. G. Farr, ibid., 16, 845 (1971).
569. T. O. Rouse and J. L. Weininger, J. Electrochem. Soc., 113, 184 (1966).
570. L. M. Elina, T. I. Borisova, and T. I. Zalkind, Zh. Fiz. Khim., 28, 785 (1954).
571. Yu. N. Chernykh and A. A. Yakuvleva, Elektrokhimiya, 6, 1671 (1970).
572. G. Grube and A. Vogt, Z. Elektrochem., 44, 353 (1938).
573. E. M. Kuchinskii and B. V. Ershler, Zh. Fiz. Khim., 20, 539 (1949).
574. G. W. D. Briggs, E. Jones, and W. F. K. Wynne-Jones, Trans. Faraday Soc., 51, 1433 (1955).
575. P. D. Lukovtsev and G. J. Slaidin, Electrochim. Acta, 6, 17 (1962).
576. H. Ewe and A. Kalberlah, ibid., 15, 1885 (1970).
577. L. N. Sagoyan, Izv. Akad. Nauk Arm. (Khim. Nauki), 17, 3 (1964); Chem. Abstr., 61, 3914 (1964).
578. E. Jones and W. F. K. Wynne-Jones, Trans. Faraday Soc., 52, 1260 (1956).
579. G. W. D. Briggs and W. F. K. Wynne-Jones, ibid., 52, 1272 (1956).
580. O. Glemser and J. Einerhand, Z. Elektrochem., 54, 302 (1950).
581. J. P. Hoare, The Electrochemistry of Oxygen, Wiley (Interscience), New York, 1968.
582. S. Okada, T. Shiraishi, and K. Watanabe, J. Chem. Soc. Japan, 51, 129 (1948).
583. S. Okada and T. Shiraishi, ibid., 52, 37 (1949).
584. S. Okada, T. Shiraishi, and T. Yasuhara, ibid., 53, 5 (1950).
585. S. Okada, T. Shiraishi, and T. Yasuhara, ibid., 53, 578 (1950).
586. S. Okada. T. Shiraishi, and T. Yasuhara, ibid., 54, 16 (1951).

587. G. W. D. Briggs and M. Fleischmann, Trans. Faraday Soc., 62, 3217 (1966).

588. G. W. D. Briggs and M. Fleischmann, ibid., 67, 2397 (1971).

589. D. M. McArthur, J. Electrochem. Soc., 117, 422, 729 (1970).

590. A. J. Salkind and P. F. Bruins, ibid., 109, 356 (1962).

591. S. U. Falk, ibid., 107, 661 (1960).

592. R. S. McEwen, J. Phys. Chem., 75, 1782 (1971).

593. G. W. D. Briggs, G. W. Stott, and W. F. K. Wynne-Jones, Electrochim. Acta, 7, 244 (1962).

594. B. V. Ershler, G. S. Tyurikov, and A. D. Smirnova, Zh. Fiz. Khim., 14, 985 (1940).

595. A. P. Rollet, Ann. Chim. (Paris), 13, 202 (1930).

596. F. P. Kober, J. Electrochem. Soc., 112, 1064 (1965).

597. F. P. Kober, ibid., 114, 215 (1967).

598. D. Tuomi, ibid., 112, 1 (1965).

599. K. Appelt, Electrochim. Acta, 13, 1727 (1968).

600. P. D. Lukovtsev, Proceedings of the 4th Conference on Electrochemistry of the Academy of Sciences, Moscow, 1956 (Published 1959).

601. G. W. D. Briggs and W. F. K. Wynne-Jones, Electrochim. Acta, 7, 249 (1962).

602. S. A. Aleshkevich, E. I. Golovchenko, V. P. Morozov, and L. N. Sagoyan, Elektrokhimiya, 4, 1237 (1968).

603. G. Feuillade and R. Jacoud, Electrochim. Acta, 14, 1297 (1969).

604. B. E. Conway and P. L. Bourgault, Can. J. Chem., 37, 292 (1959).

605. B. E. Conway and P. L. Bourgault, ibid., 40, 1690 (1962).

606. B. E. Conway, M. A. Sattar, and D. Gilroy, Electrochim. Acta, 14, 677 (1969).

607. B. E. Conway and M. A. Sattar, J. Electroanal. Chem., 19, 351 (1968).

608. E. Jost and F. Rufenacht, J. Electrochem. Soc., 113, 97 (1966).

609. L. A. Kukoz and M. F. Skalozubov, Tr. Novocherk. Politekh. Inst., 134, 19 (1962); Chem. Abstr., 58, 12163 (1963).

610. I. Maxim and T. Braun, J. Phys. Chem. Solids, 24, 537 (1963).

611. A. Riga, R. Greef, and E. Yeager, U.S. Clearinghouse Fed. Sci. Tech. Inform. AD646456.

612. K. Nii, Corros. Sci., 10, 571 (1970).

613. H. Yoneyama and H. Tamura, Denki Kagaku, 38, 644 (1970).

614. S. H. Toshima, ibid., 34, 564 (1966).

615. F. J. Morin, Bell System Tech. J., 37, 1047 (1958).

616. E. J. W. Verwey, P. W. Haayman, and F. C. Romeyin, Chem. Weekbl., 44, 705 (1948).

617. R. R. Heikes and W. D. Johnston, J. Chem. Phys., 26, 582 (1957).

618. A. M. Adams, F. T. Bacon, and R. G. H. Watson, Fuel Cells (W. Mitchell, Jr., ed.), Academic, London, 1963, p. 129.

619. D. Yohe, A. Riga, R. Greef, and E. Yeager, Electrochim. Acta, 13, 1351 (1968).

620. A. C. C. Tseung, B. S. Hobbs, and A. D. S. Tantran, ibid., 15, 473 (1970).

621. D. Tuomi and G. J. B. Crawford, J. Electrochem. Soc., 115, 459 (1968).

622. F. P. Kober and P. Lublin, ibid., 113, 396 (1966).

623. G. J. Slaidin and P. D. Lukovtsev, Dokl. Akad. Nauk, SSSR, 142, 1130 (1962).

624. D. Bruce and P. Hancock, J. Inst. Metals, 97, 140, 148 (1969).

625. P. Hancock, Werkst. Korros., 21, 1002 (1970).

626. H. H. Baumbach and C. Wagner, Z. Phys. Chem., B, 24, 59 (1934).

627. S. P. Mitoff, J. Chem. Phys., 35, 882 (1961).

628. H. G. Sockel and H. Schmalzried, Ber. Bunsenges. Phys. Chem., 72, 745 (1968).

629. C. Wagner, J. Chem. Phys., 21, 1819 (1953).

630. J. G. Aiken and A. G. Jordan, J. Phys. Chem. Solids, 29, 2153 (1968).

631. A. D. Neuimin, S. F. Paliguev, V. N. Strekalovskii, and G. V. Burov, Electrochem. Molten and Solid Electrolytes, 2, 66 (1964).

632. I. A. Makhnovetskaya and R. T. Tochilkina, Zh. Prikl. Khim. Lening., 43, 1608 (1970).

633. H. H. Uhlig, Electrochim. Acta, 16, 1939 (1971).

634. B. Rubin, Ph.D. Thesis, University of Pennsylvania, 1969; J. O'M. Bockris and P. K. Subramayan, Corros. Sci., 10, 435 (1970).

635. P. A. Bond and H. H. Uhlig, J. Electrochem. Soc., 107, 488 (1960).

636. J. Horvath and H. H. Uhlig, ibid., 115, 791 (1968).

637. M. H. Tikkanen and O. Hyavarinen, Corrosion, 26, 169 (1970).

638. H. Dahms and I. M. Croll, J. Electrochem. Soc., 112, 771 (1965).

639. A. T. Vagramyan, T. A. Fedoseeva, D. V. Fedoseev, and L. A. Uvarov, Elektrokhimiya, 6, 1773 (1970).

640. T. A. Fedoseeva and A. T. Vagramyan, Dokl. Akad. Nauk, SSSR, 196, 396 (1971).

641. G. Economy, R. Speiser, F. H. Beck, and M. G. Fontana, J. Electrochem. Soc., 108, 337 (1961).

642. H. H. Uhlig, P. Bond, and H. G. Feller, ibid., 110, 650 (1963).

643. P. Borggraefe and H. G. Feller, Corros. Sci., 7, 265 (1967).

644. Tin Nickel Alloy Plating, Tin Research Institute, Greenford, Great Britain, 1954.

645. M. Clarke, Corros. Sci., 6, 1 (1966).

646. M. Mansfeld and H. H. Uhlig, J. Electrochem. Soc., 115, 900 (1968).

647. D. N. Stolica and H. H. Uhlig, ibid., 110, 1215 (1963).

648. H. G. Feller and H. H. Uhlig, ibid., 110, 650 (1963).

649. D. N. Stolica, Corros. Sci., 9, 455 (1969).

650. N. D. Greene and B. R. Leonard, Electrochim. Acta, 9, 45 (1964).

651. G. H. Cartledge, J. Electrochem. Soc., 104, 420 (1957).

652. P. E. Morris and R. C. Scaberry, Corrosion, 26, 169 (1970).

653. G. Bianchi, F. Mazza, T. Mussini, and S. Trasatti, 2nd International Congress on Metallurgical Corrosion, New York, 1963.

654. G. Bianchi, A. Barosi, and S. Trasatti, Electrochim, Acta, 10, 557 (1965).

655. G. Trabanelli, F. Zucchi, and L. Felloni, Corros. Sci., 5, 221 (1965).

656. G. S. Parfenov and A. K. Kamyshev, Uch. Zap. Omsk. Gos. Pedagog. Inst., 26, 103 (1967).

657. V. Hosparaduk and J. V. Petrocelli, J. Electrochem. Soc., 113, 878 (1966).

658. H. P. Leckie and H. H. Uhlig, ibid., 113, 1262 (1966).

659. Ya. Kolotyrkin, Corrosion, 19, 261t (1963).

660. H. O. Teeple, ibid., 8, 14 (1952).

661. C. P. Dillon, ibid., 13, 124, 138 (1957).

662. J. Bourrat, Corros. Anticorros., 11, 140 (1963).

663. T. P. Hoar and J. C. Scully, J. Electrochem. Soc., 111, 348 (1964).

664. M. Paul and H. Wieland, Electrochim. Acta, 14, 1025 (1969).

665. A. K. M. Shamsul Huq and A. J. Rosenberg, J. Electrochem. Soc., 111, 270 (1964).

666. C. L. Mantell, Electrochemical Engineering, 4th ed., McGraw-Hill, New York, 1960.

667. NBS Circular 485, Nickel and Its Alloys, U.S. Department of Commerce, National Bureau of Standards, Washington, D.C., 1953.

668. L. S. Renzoni, R. C. McQuire, and M. V. Barker, J. Metals, 10, 414 (1958).

669. International Nickel Co. Inc., Practical Nickel Plating, 2nd ed., 1959.

670. International Nickel Co. Mond, Ltd., Nickel Plating. A Short Account of Technique and Applications, Publication 2563, 1963.

671. K. E. Volk, Metalloberflache, 16, 6 (1962).

672. E. Raub and O. Loebich, ibid., 16, 1 (1962).

673. G. Fitzgerald-Lee, Electropl. Metal Finish., 15, 90 (1962).

674. W. L. Pinner, K. G. Soderberg, and W. A. Wesley, in Modern Electroplating (A. J. Gray, ed.), Wiley, New York, 1953.

675. S. K. Panikkar and T. L. Rama-Char, J. Electrochem. Soc., 106, 494 (1959).

676. E. J. Roehl and W. A. Wesley, Plating, 37, 1 (1950).

677. L. Cambi and R. Piontelli, Rend. Ist. Lomb. Sci., 72, 128 (1939).

678. R. C. Barrett, Plating, 41, 1027 (1954).

679. M. B. Diggin, Trans. Inst. Metal Finish., 31, 243 (1954).

680. E. E. Halls, Metal Treatment, 7, 11 (1941).

681. W. Blum, ASM Metals Handbook, 1948, p. 716.

682. M. B. Diggen, Rev. Amer. Electroplaters Soc., 33, 500 (1946).

683. J. G. Poor, Metal Finish., 41, 694 (1943).

684. B. A. Shenoi and K. S. Indira, ibid., 61, 65 (1963).

685. K. S. Indira, S. R. Rajagopalan, M. I. A. Siddigi, and K. S. G. Doss, Electrochim. Acta, 9, 1301 (1964).

686. E. Raub and K. Mueller, Fundamentals of Metal Deposition, Elsevier, Amsterdam, 1967.

687. A. Brenner, C. Cohen, G. Gutzeit, A. Krieg, W. H. Metzger, W. H. Safranek, E. B. Saubestre, C. F. Waite, and W. E. Moline, Symposium on Electroless Nickel Plating, Special Technical Publ. No. 265, ASTM, 1959.

688. A. Brenner and G. E. Riddell, J. Res. Nat. Bur. Stand., 39, 385 (1947).

689. G. Gabrielli and F. Raulin, J. Appl. Electrochem., 1, 167 (1971).

690. F. M. Dunahue and C. U. Yu, Electrochim. Acta, 15, 237 (1970).

691. P. A. Jacquet, Metall. Rev., 1, 157 (1956).

692. W. J. Tegart, The Electrolytic and Chemical Polishing of Metals in Research and Industry, Pergamon, London, 1956.

693. J. K. Higgins, J. Electrochem. Soc., 106, 989 (1959).

694. N. Ibl and G. Truempler, Helv. Chim. Acta, 35, 363 (1952).

695. B. Mazza, P. Pedeferri, R. Piontelli, and F. Siniscalco, Proceedings of the Symposium on Sulphamic Acid, Electrochim. Metallorum, 1, 297 (1966).

696. V. J. Albano and R. R. Soden, J. Electrochem. Soc., 113, 511 (1966).

697. R. R. Soden and V. J. Albano, ibid., 115, 766 (1968).

698. M. Dodero and C. Déportes, C.R. Acad. Sci., Paris, C, 242, 2939 (1956).

699. S. Iki, Ind. Eng. Chem., 20, 472 (1928).

700. A. Nicol, Ann. Chim., 2, 670 (1947); C.R. Acad. Sci., Paris, C, 222, 1034 (1946).

701. K. M. Oesterle, Z. Elektrochem., 35, 505 (1929).

702. D. J. Macnaughtan, G. E. Gardam, and R. A. F. Hammond, Trans. Faraday Soc., 29, 729 (1933).

703. A. L. Rotinyan and V. A. Zelides, Zh. Prikl. Khim., 24, 604 (1951).

704. G. W. D. Briggs, J. Chem. Soc., 1957, 1846.

705. H. Schmidt, Z. Anorg. Allg. Chem., 271, 305 (1953).

706. R. Brill and F. Halle, Angew. Chem., 48, 785 (1935).

707. H. Kersten and W. T. Young, J. Appl. Phys., 8, 133 (1937).

708. R. Jasinski and B. Burrows, Private Communication.

709. N. L. Weimberg and H. R. Weimberg, Chem. Rev., 68, 449 (1968).

710. E. Raub and F. Sautter, Metalloberflaeche, 13, 129 (1959).

711. E. Raub, Tech. Rundschau, 45, 9 (1957).

712. N. E. Khomutov, ed., Electrochemistry, Electrodeposition of Metals and Alloys, Issue 1, Israel Program Scientific Translations, Jerusalem, 1969.

713. M. M. Melinikova, ed., Electrochemistry, 1966, Israel Program Scientific Translations, Jerusalem, 1970.

714. G. Fedde, Proc. AFIPS Conf., Washington, D.C., 22, 213 (1962).

715. A. Brenner, Electrodeposition of Alloys, Principles and Practice, Academic, New York, 1963.

716. A. J. Hardwich, J. Appl. Phys., 34, 818 (1963).

717. M. H. Francombe and A. J. Noreika, ibid., 32, 5 (1961).

718. I. W. Wolf, J. Electrochem. Soc., 108, 959 (1961); J. Appl. Phys., 33, 1152 (1962).

719. W. O. Freitag, J. S. Mathias, and G. DiGiulio, J. Electrochem. Soc., 111, 35 (1964).

720. W. O. Freitag and J. S. Mathias, ibid., 112, 64 (1965).

721. D. E. Couch, H. Shapiro, and A. Brenner, ibid., 105, 485 (1958).

722. A. M. Pisuquin and M. Ya. Popereka, Elektrolit. Osazh. Splav., 2, 15 (1968).

723. L. I. Kadaner, R. B. Avakyan, and Z. N. Kiryukhina, ibid., 2, 43 (1968).

724. K. P. Batashev and I. A. Belozerova, ibid., 1, 41 (1968).

725. W. H. Safranek and L. E. Vaaler, Plating, 46, 133 (1959).

726. E. Raub, Metalloberflaeche, 7, 17 (1953).

727. C. L. Faust and G. W. Montillon, Trans. Electrochem. Soc., 67, 281 (1935); 65, 361 (1934).

728. L. E. Stout and C. L. Faust, ibid., 61, 341 (1931); 60, 271 (1931).

729. C. H. Proctor, Metal Ind., 18, 13 (1920).

730. K. Masaki, J. Chem. Soc. Japan, 7, 158 (1932).

731. E. Raub and K. Bihlmaier, Mitt. Forsch. Inst. Prob. Edelmetalle Staatl. Hoch. Fachschule Schwaeb. Gmund, 11, 59 (1937).

732. L. E. Netherton and M. L. Holt, Trans. Electrochem. Soc., 66, 453 (1934).

733. C. G. Fink and F. L. Jones, ibid., 56, 461 (1931).

734. R. S. Smith, L. E. Godycki, and J. C. Lloyd, J. Electrochem. Soc., 108, 996 (1961).

735. I. W. Wolf, ibid., 108, 959 (1961).

736. R. D. Fisher, ibid., 109, 479 (1962).

737. K. L. Vasanth, T. L. Rama-Char, and L. Tirumale, J. Electrochem. Soc. India, 18, 186 (1969).

738. H. J. Feller, Ch. Kamp, and M. Kesten, Corros. Sci., 10, 687 (1970).

739. G. H. Cockett and E. S. Spencer-Timms, J. Electrochem. Soc., 108, 906 (1961).

740. V. Sree and T. L. Rama-Char, Metal Finish., 59, 49 (1961).

741. S. S. Misra and T. L. Rama-Char, 6th International Metal Finishing Conference, London, 1964.

742. J. M. Polukarov, K. M. Gorbunova and W. W. Bondar, Electrochim. Acta, 1, 362 (1959).

743. S. K. Panikkar and T. L. Rama-Char, J. Sci. Ind. Res. India, 19-A, 510 (1960).

744. B. H. Priscott, Trans. Inst. Metal Finish., 36, 93 (1959).

745. A. P. Van Peteghen and H. F. Demeyre, Galvano, 29, 715 (1960); Nickel Bull., 34, 90 (1961).

746. V. S. Galinker, P. V. Savenko, O. K. Kudra, and V. A. Loboda, Ukr. Khim. Zh. 36, 735 (1970).

747. V. Zentner, Proc. Amer. Electroplaters Soc., 47, 166 (1960).

748. V. Sree and T. L. Rama-Char, J. Electrochem. Soc., 108, 64 (1961).

749. S. Venkatachalan and T. L. Rama-Char, Electroplating Metal Finish., 15, 233 (1962).

750. D. W. Endicott and J. R. Knapp, Plating, 53, 43 (1966).

751. L. I. Soboleva, T. E. Tsupak, and N. T. Kudryavtsev, Tr. Mosk. Khim. Tekhnol. Inst., 1969, 191.

752. N. W. Jovey, A. Krohn, and G. M. Hanneken, J. Electrochem. Soc., 110, 362 (1963).

753. S. V. Krupin and A. L. Rotinyan, Elektrokhim. Protsessy Elektroosazhdenii Anndnom Rastvorenii Metal, 1969, 106.

754. T. T. Campbell and R. Abel, U.S. Bur. Mines Rep. Invest. 5482 (1959); See Nickel Bull., 33, 6 (1960).

755. M. Clarke, Trans. Inst. Metal Finish., 38, 186 (1961).

756. T. L. Rama-Char and J. Vaid, Electroplating Metal Finish., 14, 367 (1961).

757. E. S. Bruk and R. G. Kovaleva, Elektrolyt Osazh. Splav., 1, 32 (1968).

758. G. Kadziene, J. Bubelis, and J. Matulis, Liet. TSR Mokslu Akad. Derb., Ser. B, 1970, 67.

759. K. I. Vasee and T. L. Rama-Char, J. Sci. Ind. Res. India, 19-B, 413 (1960).

760. K. Nagaraja-Rao, M. I. A. Siddigi, and C. V. Surynarayana, Electrochim. Acta, 10, 557 (1965).

761. K. N. Pimenova and P. F. Kalyuzhnaya, Zh. Prikl. Khim. Lening. 43, 2105 (1970).

762. T. L. Rama-Char, Electroplating Metal Finish., 15, 40 (1962).

763. I. Guillaume, H. Constant, J. C. Lemaire, G. Valensi, and J. Brisou, Electrochim. Acta, 15, 1445 (1970).

764. R. F. North and M. J. Pryor, Corros. Sci., 10, 297 (1970).

765. G. Kortum, Treatise on Electrochemistry, Elsevier, Amsterdam, 1965.

766. C. A. Hampel, ed., The Encyclopedia of Electrochemistry, Reinhold, New York, 1964.

767. H. Winkler, 8th Meeting of the International Committee on Thermodynamics and Electrochemical Kinetics, Butterworths, London, 1956, p. 383.

768. F. C. Andersen, J. Electrochem. Soc., 99, 244C (1952).

769. T. Osono and K. Watanabe, J. Electrochem. Soc. Japan, 19, 14 (1951).

770. F. T. Bacon, Electrochim. Acta, 14, 569 (1969).

771. C. D. Mantell, J. Electrochem. Soc., 106, 70 (1959).

772. W. Tabor, Chem. Techn. 9, 645 (1957).

773. A. I. Levin, Zh. Prikl. Khim., 19, 779 (1946).

774. H. Schadner and H. Tannenberger, Rev. Energ. Primaire, 2, 49 (1966).

775. K. Appelt, Z. Dominiczak, A. Nowacki, and M. Paszkiewicz, Electrochim. Acta, 10, 617 (1965).

776. J. Freundlich, ibid., 6, 47 (1962).

777. A. R. Despic, D. M. Drazic, C. M. Petrovic, and V. L. Vujcic, J. Electrochem. Soc., 111, 1109 (1964).

778. F. P. Dousek, Rev. Energ. Primaire, 1, 101 (1965).

779. A. Kalberlah and A. Winsel, Electrochim. Acta, 13, 1689 (1968).

780. W. Wiesener, ibid., 15, 1065 (1970).

SUBJECT INDEX

K

L

M

N

O

P

Q

R

S